Bioinorganic Chemistry: Concepts and Models

Bioinorganic Chemistry: Concepts and Models

Editor: Isaac Walker

Callisto Reference
www.callistoreference.com

Callisto Reference,
118-35 Queens Blvd., Suite 400,
Forest Hills, NY 11375, USA

Visit us on the World Wide Web at:
www.callistoreference.com

ISBN: 978-1-64116-540-2 (Hardback)

Cataloging-in-Publication Data

Bioinorganic chemistry : concepts and models / edited by Isaac Walker.
p. cm.
Includes bibliographical references and index.
ISBN 978-1-64116-540-2
1. Bioinorganic chemistry. 2. Biochemistry. 3. Chemistry, Inorganic. I. Walker, Isaac.
QP531 .B56 2021
572.51--dc23

Table of Contents

Preface

Bioinorganic chemistry is the field concerned with the examination of the role of metals in biology. It is an integrated study of inorganic chemistry and biochemistry that uses inorganic models for representing the behavior of metalloproteins. It facilitates the understanding of electron-transfer proteins, atom and group transfer chemistry, substrate bindings and activation, and metal properties in biological chemistry. Inorganic elements, most importantly sodium, calcium, potassium, magnesium, phosphate, bicarbonate, etc. act as ionic electrolytes. Transition metals are present as trace elements in organisms. These metals are used as cofactors in proteins that enable the activity of enzymes. This book elucidates the concepts and innovative models around prospective developments with respect to bioinorganic chemistry. It explores all the important aspects of this field in the present day scenario. A number of latest researches have been included to keep the readers up-to-date with the global concepts in this area of study.

This book has been the outcome of endless efforts put in by authors and researchers on various issues and topics within the field. The book is a comprehensive collection of significant researches that are addressed in a variety of chapters. It will surely enhance the knowledge of the field among readers across the globe.

It gives us an immense pleasure to thank our researchers and authors for their efforts to submit their piece of writing before the deadlines. Finally in the end, I would like to thank my family and colleagues who have been a great source of inspiration and support.

Editor

1

The Advancing of Zinc Oxide Nanoparticles for Biomedical Applications

Jinhuan Jiang,[1] Jiang Pi,[1] and Jiye Cai[1,2]

[1]*State Key Laboratory of Quality Research in Chinese Medicines, Macau University of Science and Technology, Macau, China*
[2]*Department of Chemistry, Jinan University, Guangzhou, China*

Correspondence should be addressed to Jiye Cai; tjycai@jnu.edu.cn

Academic Editor: Francesco Paolo Fanizzi

Zinc oxide nanoparticles (ZnO NPs) are used in an increasing number of industrial products such as rubber, paint, coating, and cosmetics. In the past two decades, ZnO NPs have become one of the most popular metal oxide nanoparticles in biological applications due to their excellent biocompatibility, economic, and low toxicity. ZnO NPs have emerged a promising potential in biomedicine, especially in the fields of anticancer and antibacterial fields, which are involved with their potent ability to trigger excess reactive oxygen species (ROS) production, release zinc ions, and induce cell apoptosis. In addition, zinc is well known to keep the structural integrity of insulin. So, ZnO NPs also have been effectively developed for antidiabetic treatment. Moreover, ZnO NPs show excellent luminescent properties and have turned them into one of the main candidates for bioimaging. Here, we summarize the synthesis and recent advances of ZnO NPs in the biomedical fields, which will be helpful for facilitating their future research progress and focusing on biomedical fields.

1. Introduction

Recently, biomedical nanomaterials have received more concerns because of their prominent biological characteristics and biomedical applications. With the development of nanomaterials, metal oxide nanoparticles show promising and far-ranging prospect for biomedical field, especially for antibacteria, anticancer drug/gene delivery, cell imaging, biosensing, and so on [1].

Zinc oxide nanoparticles (ZnO NPs), as one of the most important metal oxide nanoparticles, are popularly employed in various fields due to their peculiar physical and chemical properties [2, 3]. ZnO NPs are firstly applied in the rubber industry as they can provide wearproof of the rubber composite, improve performance of high polymer in their toughness and intensity and antiaging, and other functions [4, 5]. Because of the strong UV absorption properties of ZnO, they are increasingly used in personal care products, such as cosmetics and sunscreen [6]. In addition, ZnO NPs have superior antibacterial, antimicrobial, and excellent UV-blocking properties. Therefore, in the textile industry, the finished fabrics by adding ZnO NPs exhibited the attractive functions of ultraviolet and visible light resistance, antibacteria, and deodorant [7]. Apart from the applications mentioned above, zinc oxide can also be used in other branches of industry, including concrete production, photocatalysis, electronics, electrotechnology industries, and so on [4, 8].

It is generally known that zinc as an essential trace element extensively exists in all body tissues, including the brain, muscle, bone, skin, and so on. As the main component of various enzyme systems, zinc takes part in body's metabolism and plays crucial roles in proteins and nucleic acid synthesis, hematopoiesis, and neurogenesis [2–5]. Nano-ZnO, with small particle size, makes zinc more easily to be absorbed by the body. Thus, nano-ZnO is commonly used as a food additive. Moreover, ZnO is graded as a "GRAS" (generally recognized as safe) substance by the US Food and Drug Administration (FDA) [9]. With these properties, ZnO NPs have received more attention in biomedical applications. Compared with other metal oxide NPs, ZnO NPs with the comparatively inexpensive and relatively less toxic property exhibit excellent biomedical applications, such as anticancer, drug delivery, antibacterial, and diabetes treatment; anti-inflammation; wound healing; and bioimaging [1, 10–12].

Herein, in this review, we will summarize the methods of synthesis and recent exciting progress on the use of ZnO NPs in the biomedical fields.

2. Synthesis of ZnO NPs

The biological activity of nanoparticles depends on factors including surface chemistry, size distribution, particle morphology, and particle reactivity in solution. Therefore, the development of nanoparticles with controlled structures that are uniform in size, morphology, and functionality is essential for various biomedical applications.

The ZnO NPs occurring in a very rich variety of size and shape will provide a wide range of properties. The methods for stable ZnO NPs preparation have been widely developed in recent years, which mainly include the chemical precipitation method, sol-gel method, solid-state pyrolytic method, solution-free mechanochemical method, and biosynthesis method.

2.1. Chemical Precipitation. The most popular method for ZnO NPs preparation is chemical precipitation, which usually involves two reaction reagents: a highly purified zinc forerunner such as zinc acetate ($Zn(CH_3COO)_2{\cdot}2H_2O$), zinc nitrate ($Zn(NO_3)_2$), or zinc sulfate ($ZnSO_4$) and a solution of precipitator such as sodium hydroxide (NaOH) or ammonium hydroxide ($NH_3{\cdot}H_2O$) [13]. Typically, the precipitator is added dropwise to the dissolved zinc precursor until the pH level reached about 10. Then, completely mix these solutions to get a white intermediate of zinc hydroxide. Ultimately, the sample of zinc hydroxide ($Zn(OH)_2$) was converted to ZnO after sintering at high temperature.

Controlled parameters in this method mainly include the concentration of zinc forerunner and precipitator, the molar ratio of two reagents, and the reaction and calcination temperature.

Bisht et al. synthesized ZnO NPs using the chemical precipitation method using $Zn(CH_3COO)_2{\cdot}2H_2O$ and NaOH in a molar ratio of 1 : 5. Intermediate products were calcined at 200°C for 2 h in a muffle furnace to obtain white fine powder ZnO with the size of 18.67 ± 2.2 nm [14].

Bettini et al. introduced a simplified precipitation approach in ZnO NPs using $ZnSO_4$ and NaOH solution with a molar ratio of 1 : 2, which was carried out under vigorous stirring for 12 h at room temperature. The obtained white precipitate was washed several times and separated by centrifugation [15–17]. Finally, the precipitate (ZnO) was dried in an oven at 100°C for 6 h. The prepared ZnO NPs with a flake-like structure showed a size distribution about 100 nm.

The obtaining of ZnO NPs by the chemical precipitation method is not only simple and easily controlled but also easy industrialized. However, due to the surface effect of nanoparticles, the precursor of nanooxide prepared by the chemical precipitation method could form agglomerates easily.

2.2. Sol-Gel Method. A novel sol-gel synthesis of ZnO NPs was firstly presented by Spanhel and Anderson [18], which mainly involved three major steps:

(1) Preparation of zinc precursor

A sample of $Zn(CH_3COO)_2{\cdot}2H_2O$ was dissolved in ethanol, placed into a distillation apparatus, and then refluxed for few hours at atmospheric pressure. The solution was boiled nearly at 80°C and stirred to obtain the condensate and hygroscopic reaction mixture.

(2) Preparation of ZnO clusters

The hygroscopic mixture was diluted to ethanolic solution with the addition of $LiOH{\cdot}H_2O$ powder. The suspension will become transparent with the help of an ultrasonic bath. This procedure could accelerate the release of OH ions, and the reaction at low temperature under air conditions could prevent rapid particle growth and get ZnO sols.

(3) Crystal growth

Crystal growth is a self-induced procedure occurring at room temperature. But the amount of LiOH which could strongly affect the crystals growth rate, shape, and size should be well controlled. LiOH-induced ZnO growth can be briefly summarized as follows:

$$
\begin{aligned}
Zn(CH_3COO)_2 \cdot 2H_2O + 2LiOH &\longrightarrow Zn(OH)_2 \\
&\quad + 2CH_3COOLi + 2H_2O \\
Zn(OH)_2 + 2H_2O &\longrightarrow Zn(OH)_4{}^{2-} + 2H^+ \\
Zn(OH)_4{}^{2-} &\longleftrightarrow ZnO + H_2O + 2OH^-
\end{aligned}
\tag{1}
$$

Some other alkalis can also be used for ZnO growth; for instance, Rani et al. synthesized ZnO NPs using NaOH instead of LiOH and successfully obtained ZnO NPs with the largest crystallite size of 14 nm at 9 pH value [19].

The sol-gel method is the subject of much interest, in view of the simplicity, low cost, and relatively mild conditions of synthesis, which could offer a simple route to quantum size ZnO particles.

2.3. Solid-State Pyrolytic Method. The solid-state pyrolytic method was first developed by Wang et al. with advantages of low cost and easy operation for the growth of high-quality ZnO nanoparticles [20].

The typical synthesis procedure is as follows: $Zn(CH_3COO)_2{\cdot}2H_2O$ and $NaHCO_3$ are mixed at room temperature. The mixture is pyrolysed at the reaction temperature. $Zn(CH_3COO)_2{\cdot}2H_2O$ is converted into ZnO, while $NaHCO_3$ is transformed into CH_3COONa and can be eventually cleaned out with deionized water. Subsequently, white ZnO NPs can be obtained via the thermal decomposition process. The particle sizes can be regulated by selecting different pyrolytic temperatures. Using this method, Wang et al. obtained ZnO NPs of different sizes in the range of 8–35 nm.

2.4. Solution-Free Mechanochemical Method. Solution-free mechanochemical preparation of ZnO NPs is a two-step synthesis method. The first step is to grind $Zn(CH_3COO)_2$ and $H_2C_2O_4{\cdot}2H_2O$ powder mixtures for a certain time to form $ZnC_2O_4{\cdot}2H_2O$ nanoparticles [21].

$Zn(CH_3COO)_2$ (solid, large particles) + $H_2C_2O_4{\cdot}2H_2O$ (solid, large particles) = $ZnC_2O_4{\cdot}2H_2O$ (solid particles) + $2CH_3COOH$ (liquid and gas) + $H_2C_2O_4{\cdot}2H_2O$ (solid particles).

The second step is the thermal decomposition of $ZnC_2O_4{\cdot}2H_2O$ nanoparticles at a very high temperature to get ZnO NPs:

$$\begin{aligned} ZnC_2O_4{\cdot}2H_2O &= ZnC_2O_4 + 2H_2O\uparrow \\ ZnC_2O_4{\cdot}2H_2O &= ZnO + CO_2\uparrow + CO\uparrow + 2H_2O\uparrow \\ CO + O_2 &= CO_2\uparrow \\ 2H_2C_2O_4{\cdot}2H_2O + O_2 &= 4CO_2\uparrow + 4H_2O\uparrow \\ CH_3COOH\ (\text{liquid}) &= CH_3COOH\ (\text{gas})\uparrow \end{aligned} \tag{2}$$

The advantages of this method are the low production costs and high homogeneity of the crystalline structure and morphology. But the morphology of the ZnO NPs strongly depends on the milling time of the reactant mixture, a longer time of milling led to a smaller particle size. The obtained ZnO NPs show an average size ranging from 24 to 40 nm.

Pardeshi and Patil synthesized ZnO NPs with different morphologies and crystallite sizes using this method by varying the calcination temperature from 400°C to 900°C. It was found that zinc oxide calcined from 400°C to 550°C exhibited the same crystallite growth rate (38–50 nm) [22].

2.5. Biological Methods. Physical and chemical methods for ZnO NPs preparations have widely developed. Nowadays, the development of green chemistry has attracted more and more attention because it is mostly environmentally friendly [23]. A broad variety of plant extract are used for the biosynthesis of ZnO NPs such as the leaf of *Azadirachta indica* (L.) [23], *Cochlospermum religiosum* (L.) [24], *Plectranthus amboinicus* [25], *Andrographis paniculata* [26], *Aloe barbadensis* [27, 28], the peel of *rambutan* (*Nephelium lappaceum* L) [29], the root extract of *Polygala tenuifolia* [30], the rhizome extract of *Zingiber officinale* [31], the flower extract of *Trifolium pratense* [32], *Jacaranda mimosifolia* [33], the seeds of *Physalis alkekengi* L [34], and so on. Biosynthetic and environment friendly technology for the synthesis of ZnO NPs are believed to be more ecofriendly, economical (low priced), nontoxic, and biocompatible than chemical and physical methods. ZnO NPs prepared by this method exhibited strong potential for biomedical applications such as its excellent anticancer and antibacterial activity.

3. Biomedical Applications of ZnO Nanoparticles

ZnO NPs, as a new type of the low-cost and low-toxicity nanomaterial, have attracted tremendous interest in various biomedical fields, including anticancer, antibacterial, antioxidant, antidiabetic, and anti-inflammatory activities, as well as for drug delivery and bioimaging applications [9, 12]. Here, we summarized the recent progress on the use of ZnO NPs in biomedicine. ZnO NPs less than 100 nm are considered to be relatively biocompatible, which support their biomedical applications and represent a powerful property in promoting the biomedical research.

3.1. Anticancer Activity. Cancer, a condition of uncontrolled malignant cell proliferation, is typically treated by chemotherapy, radiotherapy, and surgery in the past several decades. Although all these therapies seem to be very effective for killing cancer cells in theory, these nonselective therapy methods also introduce a lot of serious side effects [35]. Recently, nanomaterial-based nanomedicine, with high biocompatibility, easily surface functionalization, cancer targeting, and drug delivery capacity, has demonstrated the potential to overcome these side effects. Zn^{2+} is an essential nutrient for adults, and ZnO nanomaterials are considered to be safe in vivo. Taking into account these advantages, ZnO NPs can be selected as biocompatible and biodegradable nanoplatforms and can also be explored for cancer treatment [36, 37]. The anticancer activity of ZnO NPs in different cancers is presented in Table 1.

3.1.1. Anticancer Activity by Inducing Cancer Cell Apoptosis. The mitochondrial electron transport chain is known to be associated with intracellular ROS generation, and anticancer agents entering into cancer cells could destroy the electron transport chain and release huge amounts of ROS [58, 59]. However, excessive ROS will lead to mitochondrial damage and result in the loss of protein activity balance that finally causes cell apoptosis [60]. ZnO NPs present certain cytotoxicity in cancer cells mainly by themselves based on a higher intracellular release of dissolved zinc ions, followed by increased ROS induction and induced cancer cell death via the apoptosis signaling pathway.

Sharma et al. explored the effects of ZnO NPs on human liver cancer HepG2 cells and its possible pharmacological mechanism [42]. ZnO NPs-exposed HepG2 cells presented higher cytotoxicity and genotoxicity, which were associated with cell apoptosis mediated by the ROS triggered mitochondrial pathway. The loss of the mitochondrial membrane potential could open outer membrane pores which would result in the release of some related apoptotic proteins including cytochrome c into the cytosol and activate the caspase. Mechanistic studies had proved that the loss of mitochondrial membrane potential-mediated HepG2 cell apoptosis was mainly due to the decrease in mitochondrial membrane potential and Bcl-2/Bax ratios as well as accompanying with the activation of caspase-9. Besides, ZnO NPs could noticeably activate p38 and JNK and induce and attract $p53^{ser15}$ phosphorylation but was not dependent on

Table 1: The anticancer effects of ZnO NPs in different human cancer cell lines.

Cancer type	Effect and mechanism
Colon cancer	ZnO NPs suppressed cell viability in Caco-2 cell line via increased ROS and induced IL-8 release [38] ZnO NPs and fatty acids could induce lysosomal destabilization in Caco-2 cells [39] ZnO NPs induced Caco-2 cells cytotoxicity associated with increased intracellular Zn ions [40] ZnO NPs conjugated with peptides had a higher antiproliferation in HT-29 colon cancer cells than other Au NPs and Fe_3O_4 NPs [41]
Hepatocarcinoma	ZnO NPs caused ROS generation and oxidative DNA damage and lead to mitochondrial-mediated apoptosis in HepG2 cells [42] ZnO NPs selectively induce apoptosis in HepG2 cells, which was also mediated by ROS via the p53 pathway [43] Dox-ZnO nanocomplex can act as a drug delivery system to increase the internalization of the anticancer drug Dox in SMMC-7721 cells [44]
Breast cancer	Ecofriendly formulated ZnO NPs arrest the cell cycle in the G2/M phase and upregulated proapoptotic genes p53, p21, Bax, and JNK and downregulated antiapoptotic genes Bcl-2, AKT1, and ERK1/2 in a dose-dependent manner in MCF-7 cells [45] A doxorubicin delivery system based on zinc oxide nanomaterials can bypass the P-gp increase in the drug accumulation in resistant MCF-7R and MCF-7S cells [46] RGD (Arg-Gly-Asp)-targeted ZnO NPs can target integrin $\alpha v\beta 3$ receptors to increase the toxicity of the ZnO NPs to MDA-MB-231 cells at lower doses [47] ZnO-Fe_3O_4 magnetic composite nanoparticles site-specific have no significant toxicity towards noncancerous NIH 3T3 cells but show obvious toxicity at similar concentration to MDA-MB-231 cells [14] FA-functionalized PTX-ZnO NPs released ~75% of the paclitaxel payload within six hours in acidic pH, improved chemotherapy tolerance, and increased antitumor efficacy [48]
Lung cancer	ZnO NPs incorporated in liposomes not only rendered pH responsivity to the delivery carrier but also exhibited synergetic chemo-photodynamic anticancer action [49] Human lung adenocarcinoma cells with an EGFR mutation are sensitive to ZnO NP20 and Al-ZnO NP20, which resulted in nonautophagic cell death [50]
Ovarian cancer	ZnO NPs are able to induce significant cytotoxicity, apoptosis, and autophagy in SKOV3 cells through reactive oxygen species generation and oxidative stress [51]
Cervical cancer	DOX-ZnO/PEG nanocomposites exhibited better dose-dependent toxicity towards HeLa cell lines [52] ZnO nanoparticles showed a dynamic cytotoxic effect in cervical carcinoma cells which induced the apoptosis through the increased intracellular ROS level and upregulated apoptotic gene p53 and caspase-3 expression [53]

Table 1: Continued.

Cancer type	Effect and mechanism
Gastric cancer	PMMA-AA/ZnO NPs and PMMA-PEG/ZnO were able to carry a large amount of the hydrophobic drug (curcumin) showing highly anti-gastric cancer activity [54, 55]
Human epidermal cancer	ZnO NPs induce cell death at high concentrations, and at lower concentrations, they induce cell cycle arrest in the S and G 2/M phase by intracellular ROS generation in A431 cells [56]
Acute promyelocytic leukemia	HA/ZnO nanocomposite caused G2/M cell cycle arrest and stimulated apoptosis-related increase in caspase-3 and −7 activities of the HL-60 cells [57]

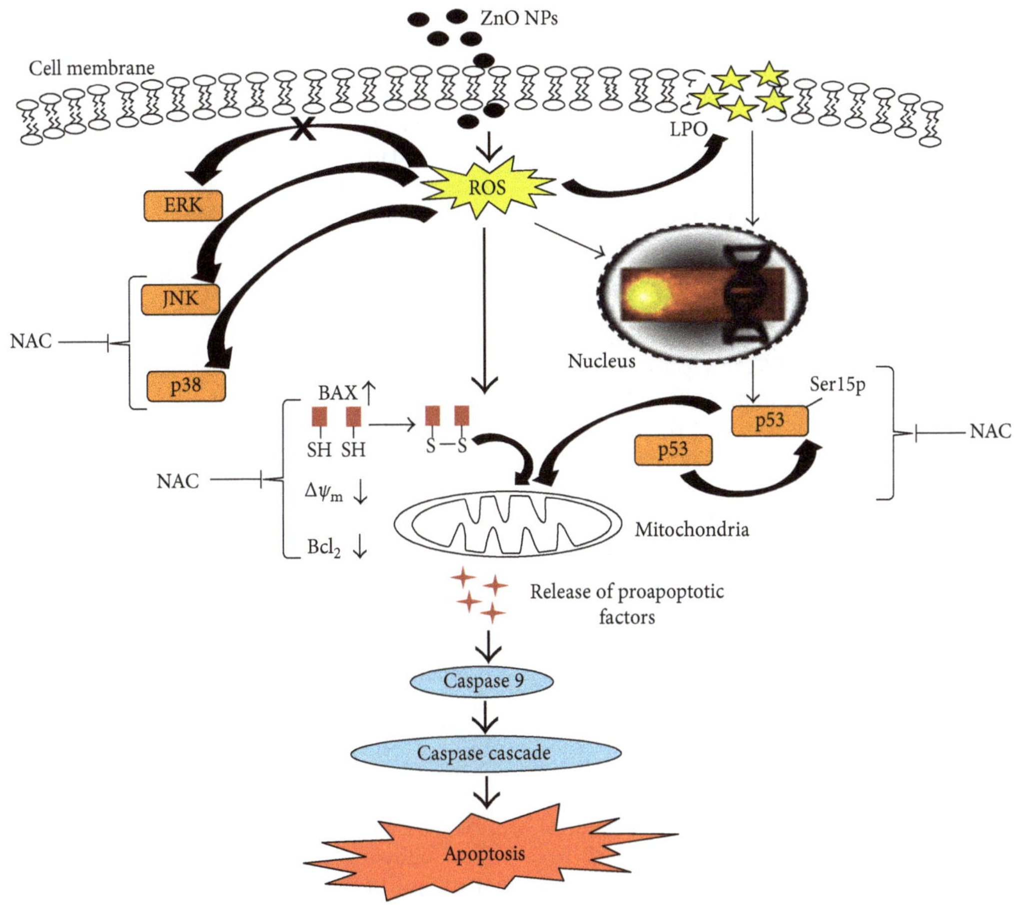

Figure 1: The mechanism of ZnO NPs-induced toxicity in human liver cells [42].

JNK and p38 pathways (Figure 1). These results afforded valuable insights into the mechanism of ZnO NPs-induced apoptosis in human liver HepG2 cells.

Moghaddam et al. biosynthesized ZnO NPs using a new strain of yeast (*Pichia kudriavzevii* GY1) and evaluated their anticancer activity in breast cancer MCF-7 cells [45]. ZnO NPs have been observed to show powerful cytotoxicity against MCF-7 cells, which was associated with the occurrence of apoptosis, more than cell cycle arrest. The ZnO NPs-induced apoptosis was mainly through both extrinsic and intrinsic apoptotic pathways, and some antiapoptotic genes of Bcl-2, AKT1, and JERK/2 were downregulated, while proapoptotic genes of p21, p53, JNK, and Bax were upregulated.

ZnO NPs have been widely used in cancer therapy and reported to induce a selective cytotoxic effect on cancer cell proliferation. Chandrasekaran and Pandurangan investigated the cytotoxicity of ZnO nanoparticles against cocultured C2C12 myoblastoma cancer cells and 3T3-L1 adipocytes, which showed that ZnO NPs could be more cytotoxic to C2C12 myoblastoma cancer cells than 3T3-L1 cells. Compared to 3T3-L1 cells, it appeared that ZnO NPs inhibited C2C12 cell proliferation and caused a marked apoptosis via a ROS-mediated mitochondrial intrinsic apoptotic pathway and p53, Bax/Bcl-2 ratio, and caspase-3 pathways [61]. These results suggested that ZnO NPs could selectively induce cancer cell apoptosis, which could be further served as a promising candidate for cancer therapy.

3.1.2. Anticancer by Autophagy. Autophagy is a highly regulated catabolic process that activated in response to different kinds of stresses like damaged organelles, ROS, anticancer agents, and protein aggregation. Excessive cellular damage may lead to cell death by the extension of autophagy and cellular self-consumption and result in cancer cell apoptosis [62, 63]. Hence, autophagy not only promotes cell survival but also activates lethal mechanisms in cancer cells, thus be considered as an important event in nanoparticle-induced cytotoxicity.

Bai et al. found that ZnO NPs with a crystal size of 20 nm resulted in a concentration-dependent loss of ovarian cancer SKOV3 cell viability [51]. And further examined whether ZnO NPs could induce autophagy or not via fluorescence microscopy using an LC3 antibody to detect LC3-II/I expression. Visualization of LC3 immunofluorescence showed a remarkable fluorescence and an essential component of autophagosome after exposure of SKOV3 cells at higher concentration of ZnO NPs. In addition, ZnO NPs-treated SKOV3 cells resulted in an upregulation of LC3-I/II and p53 expression, which further induced autophagic cell death.

Arakha et al. fabricated ZnO NPs using the chemical precipitation method and further evaluated their anticancer activity [64], which found that ZnO NPs with different sizes could obviously inhibit the proliferation of fibrosarcoma HT1080 cells. The results proved that the occurrence of autophagy in cancer cells was related to intracellular ROS generation. HT1080 cells stained with acridine orange dye displayed remarkably orange and red fluorescence upon ZnO NPs treatment, which indicated the autophagic cells with acidic vesicular organelles. Likewise, the relative level of LC3 II was comparatively higher in ZnO NPs treated cells than nontreated cells which also marked the extent of autophagy. Interaction ZnO NPs with HT1080 cell has relatively higher ROS generation. Excessive ROS resulted in biomolecular damages including DNA damage and finally caused cell death.

Previous studies have indicated that ROS and autophagy are involved in the cytotoxicity of ZnO NPs, but the regulatory mechanisms between autophagy and ROS remain to be elucidated. Zhang et al. investigated the regulatory mechanism of autophagy and the link between autophagy and ROS in ZnO NPs-treated lung epithelial cells [65]. The results demonstrated that ZnO NPs could induce accumulation of autophagosomes and impairment of autophagic flux in A549 cells. This autophagy induction was positively correlated with the dissolution of ZnO NPs in lysosomes to release zinc ions, and zinc ions released from ZnO NPs were able to damage lysosomes, leading to impaired autophagic flux and mitochondria. Impaired autophagic flux resulted in the accumulation of damaged mitochondria, which could generate excessive ROS to cause cell death. This research provided a novel insight into the regulation mechanisms of autophagy-lysosomes-mitochondria-ROS axis, which would contribute to a better understanding of the toxicity of nanomaterials.

3.1.3. Anticancer Drug Delivery. Using nanoparticles in targeted drug delivery provides exciting opportunities for much more safety and effective cancer treatment. By targeting the specific sites of cancer cells, nanoparticle-based drug delivery could reduce the overall amount of drugs used and thus minimize undesirable side effects [9, 66]. Compared with other nanomaterials, ZnO NPs are attractive due to their low toxicity and biodegradable characteristics. ZnO NPs have acquired tremendous interest in cancer drug delivery. Different types of drugs such as doxorubicin, paclitaxel, curcumin, and baicalin or DNA fragments could be loaded onto the ZnO NPs to show better solubility, higher toxicity compared with individual agents, and effective delivery into cancer cells [48, 67–69].

Hariharan et al. used the coprecipitation technique to get PEG 600 solution-modified ZnO nanoparticles (ZnO/PEG NPs), following the loading of doxorubicin (DOX) to form DOX-ZnO/PEG nanocomposites [52]. DOX-ZnO/PEG nanocomposites not only enhanced the intracellular accumulation of DOX but also presented a concentration-dependent inhibition on cervical cancer HeLa cell proliferation. Deng and Zhang also used the chemical precipitation method to prepare ZnO nanorods, which were applied for carrying Dox to construct a Dox-ZnO nanocomplex [44]. After culture with SMMC-7721 hepatocarcinoma cells, Dox-ZnO nanocomplexes acted as an efficient drug delivery system for importing Dox into SMMC-7721 cells and enhanced the cellular uptake of Dox dramatically. Furthermore, coupled with ultraviolet (UV) illumination, Dox-ZnO nanocomplexes caused more cell death through photocatalytic properties and synergistically triggered caspase-dependent apoptosis.

Puvvada et al. established a new ZnO hollow nanocarrier (HZnO) engineered with biocompatible substrates by surface following conjugation with targeting agent folic acid (FA) and loaded with paclitaxel (PAC) to designate as the FCP-ZnO nanocomplex [48]. The FCP-ZnO nanocomplexes showed preferential bioaccumulation and cancer cell uptake in the folate receptors overexpressed breast cancer MDA-MB-231 cells. Because of FA-mediated endocytosis and intracellular release within the acidic endolysosome, the FCP-ZnO nanocomplexes not only exhibited significantly higher cytotoxicity in vitro MDA-MB-231 cells but also reduced MDA-MB-231 xenograft tumours in nude mice.

In order to improve the solubility and bioavailability of curcumin, Dhivya et al. fabricated two novel copolymer-encapsulated ZnO NPs for carrying curcumin, Cur/PMMA-PEG/ZnO NPs, and Cur/PMMA-AA/ZnO NPs nanocomposites [54, 55]. By means of the experimental study, PMMA-PEG/ZnO nanocomposites with the average size less than 80 nm could release curcumin more quickly in the acidic conditions at pH ~5.8. Compared to constituent nanomaterials (nanocurcumin, PMMA-PEG, ZnO NPs, and PMMA-PEG/ZnO), the Cur/PMMA-PEG/ZnO nanocomposite performed largest observable inhibition on human gastric cancer AGS cell viability (IC_{50} ~0.01 $\mu g/mL^{-1}$) and induced cell cycle arrest at the S phase. For another nanocomposite, PMMA-AA/ZnO NPs with a size of 42 ± 5 nm could carry a large amount of curcumin and also had obvious antiproliferation on AGS cancer cells.

3.1.4. Targeting Functionalization. Targeted nanoparticles (NPs) also provide more therapeutic benefits besides specificity and specific localization like high payload, multidrug conjugation, easy tuning of release kinetics, selective localization, and bypass of multidrug resistance mechanism [70]. In order to increase the targeting effects and selectivity against cancer cells, plenty of functionalization techniques have been reported for nanoparticle modification. Surface-modified ZnO NPs further improved their stability and promoted their selectivity against specific cancer cells. The central attention is on the functionalization of the ZnO NPs surface with different kinds of biological molecules comprising different types of proteins, peptides, nucleic acids, folic acid, hyaluronan, and so on [47, 57, 71–73]. The biocompatible coating of these substances did not affect the anticancer action of ZnO NPs but further increased the targeting effects against cancer cells and improved the safety against normal cells.

For example, Chakraborti et al. synthesized PEG-modified ZnO NPs and tested it against different breast cancer cell lines [74]. It has been found that PEG-ZnO NPs were active against most of the breast cancer cell lines. The main mechanism by which PEG-ZnO kills a cancer cell is by generating ROS and triggering p53-dependent apoptosis leading to cell death.

Namvar et al. produced hyaluronan/ZnO nanocomposite (HA/ZnO) through green synthesis for the first time for cancer treatment [57]. The HA/ZnO nanocomposites caused morphological changes and inhibited proliferation of cancer cells (pancreatic adenocarcinoma PANC-1 cell, ovarian adenocarcinoma CaOV-3 cell, colonic adenocarcinoma COLO205 cell, and acute promyelocytic leukemia HL-60 cell) in dose- and time-dependent manner. Encouraging, HA/ZnO nanocomposite treatment for 72 hours did not cause toxicity to the normal human lung fibroblast (MRC-5) cell line. Compared to bare ZnO NPs, RGD peptide modification also increased the targeting effects of ZnO NPs on integrin $\alpha v\beta 3$ receptors overexpressed MDA-MB-231 cells [47]. It appeared to increase the toxicity of the ZnO NPs to breast cancer MCF-7 and MDA-MB-231 cells at lower doses.

In general, the anticancer activity of nanoscaled ZnO materials with prominent functionality may provide a new opportunity for exploiting ZnO NPs in treating cancer diseases. The theory analysis and experimental research proved that ZnO NPs with less side effect present greater selectivity among normal and cancerous cells. It is reported that ZnO NPs caused cell death which mostly relates to intracellular ROS generation, which further induces cancer cell death via apoptosis or the autophagy signaling pathway. But up to now, the advanced anticancer mechanism study of ZnO NPs is still lacked of, especially in cellular and molecular mechanism strengthening. Therefore, in subsequent research work, we should attach more importance to their molecular mechanism in vitro and vivo and overcome its limitations in cancer therapy.

3.2. Antibacterial Activity. ZnO NPs can be selected as an antibacterial material because of its superior properties, such as high specific surface area and high activity to block a wide scope of pathogenic agents. But recently, the antibacterial activity of ZnO NPs is still scarcely known. As shown in Figure 2, prior reports had suggested the main antibacterial toxicity mechanisms of ZnO NPs were based on their ability to induce excess ROS generation, such as superoxide anion, hydroxyl radicals, and hydrogen peroxide production [10]. The antibacterial activity may involve the accumulation of ZnO NPs in the outer membrane or cytoplasm of bacterial cells and trigger Zn^{2+} release, which would cause bacterial cell membrane disintegration, membrane protein damage, and genomic instability, resulting in the death of bacterial cells [75–77].

Presently, Gram-negative *Escherichia coli* (*E. coli*) and Gram-positive *Staphylococcus aureus* (*S. aureus*) are mainly chosen as model bacteria to evaluate the antibacterial activity of ZnO NPs [77, 78]. Some other Gram-negative bacteria such as *Pseudomonas aeruginosa* (*P. aeruginosa*) [24, 79], *Proteus vulgaris* (*P. vulgaris*) [80], *Vibrio cholerae* (*V. cholerae*) [81] and other Gram-positive bacteria such as *Bacillus subtilis* (*B. subtilis*) [82] and *Enterococcus faecalis* (*E. faecalis*) [83] are also investigated. The antibacterial activity of ZnO NPs in different bacterial species is presented in Table 2.

Jiang et al. reported the potential antibacterial mechanisms of ZnO NPs against *E. coli* [76]. It showed that ZnO NPs with an average size about 30 nm caused cell death by directly contacting with the phospholipid bilayer of the membrane, destroying the membrane integrity. The addition of radical scavengers such as mannitol, vitamin E, and glutathione could block the bactericidal action of ZnO NPs, potentially revealing that ROS production played a necessary function in the antibacterial properties of ZnO NPs. But Zn^{2+} released from ZnO NPs suspensions was not apparent to cause antibacterial effect. Reddy prepared ZnO NPs with sizes of ~13 nm and examined their antibacterial (*E. coli* and *S. aureus*) activities [78]. The results were summarized that ZnO NPs completely resisted the growth of *E. coli* at concentrations of about 3.4 mM but inhibited growth of *S. aureus* at much lower concentrations (≥1 mM). Moreover, Ohira and Yamamoto also found the antibacterial (*E. coli* and *S. aureus*) activity of ZnO NPs with small crystallite sizes was stronger than those with large crystallite sizes [97]. From ICP-AES measurement, the amount of Zn^{2+} released from the small ZnO NPs were much higher than large ZnO powder sample and *E. coli* was more sensitive to Zn^{2+} than

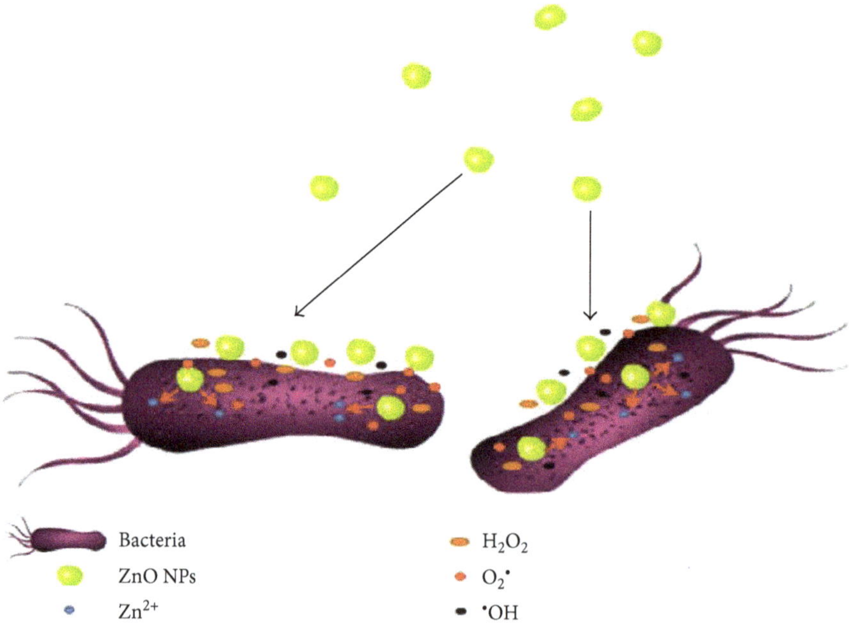

FIGURE 2: Schematic illustration of antibacterial activity of ZnO NPs.

S. aureus. So we can believe that eluted Zn^{2+} from ZnO NPs also take a key role in antibacterial action.

Iswarya et al. extracted crustacean immune molecule β-1,3-glucan binding protein (Phβ-GBP) from the heamolymph of *Paratelphusa hydrodromus* and then successfully fabricated the Phβ-GBP-coated ZnO NPs. The Phβ-GBP-ZnO NPs were spherical in shape with a particle size of 20–50 nm and restrained the growth of *S. aureus* and *P. vulgaris*. It should be noted that *S. aureus* was more susceptible to Phβ-GBP-ZnO NPs than *P. vulgaris*. In addition, Phβ-GBP-ZnO NPs could alter cell membrane permeability and trigger high level of ROS formation both in *S. aureus* and *P. vulgaris* [87]. Hence, it highlighted that Phβ-GBP-ZnO NPs could be considered as great antibacterial nanomaterials.

Epidemic disease *cholera*, a serious diarrheal disease caused by the intestinal infection of Gram-negative bacterium *V. cholera*, mainly affects populations in the developing countries [81, 94]. Aiming at the development of nanomedicine against *cholera*, Sarwar et al. carried out a detailed study about ZnO NPs against *Vibrio cholerae* (two biotypes of cholera bacteria (classical and El Tor)). ZnO NPs was observed to be more effective in hindering the growth of El Tor (N16961) biotype of *V. cholera*, which was closely associated with ROS production. These results would damage bacterial membrane, increase permeabilization, and substantially modify their morphology [85]. They also detected the antibacterial activity of the ZnO NPs in cholera toxin (CT) mouse models. It was found that ZnO NPs could induce the CT secondary structure collapsed gradually and interact with CT by interrupting CT binding with the GM1 anglioside receptor [98].

Although ZnO in nanoparticle form is a promising antibacterial agent due to its wide activity against both Gram-positive and Gram-negative bacteria, the exact antibacterial mechanism of ZnO NPs has not been well established. Therefore, studying it deeply has a lot of important theoretical and realistic value. In the future, we believe ZnO NPs can be explored as antibacterial agents, such as ointments, lotions, and mouthwashes. In addition, it can be coated on various substrates to prevent bacteria from adhering, spreading, and breeding in medical devices.

3.3. ZnO NPs for Diabetes Treatment. Diabetes mellitus is a serious public health problem, and the WHO has estimated that, in 2014, there were more than 400 million adults with diabetes all over the world [99]. Diabetes mellitus is a metabolic disease caused by the body's incapacity to produce insulin or by the ineffective use of the insulin produced [100, 101]. Zinc is a trace element and abundantly found mineral in all human tissues and tissue fluids. Zinc is well known to keep the structural integrity of insulin and has an active role in the secretion of insulin from pancreatic cells. It also participates in insulin synthesis, storage, and secretion [102]. Therefore, ZnO NPs as a novel agent in order for zinc delivery have been developed and evaluated for their antidiabetic potential.

Kitture et al. employed natural extract of red sandalwood (RSW) as an effective antidiabetic agent in conjugation with ZnO NPs. The antidiabetic activity was assessed with the help of α-amylase and α-glucosidase inhibition assay with murine pancreatic and small intestinal extracts [103]. Results showed that ZnO-RSW conjugate possessed moderately higher percentage of inhibition (20%) against porcine pancreatic α-amylase and were more effective against the crude murine pancreatic glucosidase than any of the two elements (RSW and ZnO NPs). The conjugated ZnO-RSW displayed 61.93% of inhibition in glucosidase while the bare ZnO NPs and RSW showed 21.48% and 5.90%, respectively.

TABLE 2: The antibacterial effects of ZnO NPs in different bacterial species.

Material	Size (nm)	Bacterial species	Antibacterial mechanism
ZnO NPs	30	*E. coli*	Destroy the membrane integrity and ROS production [76]
Ag-ZnO composite	64	*S. aureus* and GFP *E. coli*	ROS and the release of Ag^+ and Zn^{2+} [84]
ZnO NPs	~80	*V. cholera*	Depolarization of the membrane structure, increased permeabilization and damage of DNA, and generation of ROS [85]
ZnO NPs	~20	*E. coli* 11634	Hydrogen peroxide (H_2O_2) [86]
ZnO NPs		*E. coli* and *S. aureus*	Release of Zn^{2+} [78]
Phβ-GBP-coated ZnO NPs (Phβ-GBP-ZnO NPs)	20~50	*S. aureus* and *P. vulgaris*	Alter the bacterial cell membrane permeability and high level of ROS [87]
ZnO nanocatalyst	~18	*B. subtilis*, *E. coli*, *K. pneumonia*, and *S. typhimurium*	H_2O_2, OH^-, and other ROS [88]
CdO-ZnO nanocomposite	27	*E. coli*, *P. aeruginosa*, *Klebsiella pneumonia*, *S. aureus*, *P. vulgaris*, and *Bacillus* spp	ROS (OH^-, H_2O_2, and O_2^{2-}) and the release of Zn^{2+} and Cd^{2+} [89]
ZnO quantum dots	4	*E. coli* MG1655, *Cupriavidus metallidurans* CH34	The toxicity is mainly from Zn^{2+} [90]
ZnO/kaoline nanocomposites		*S. aureus*, *E. coli*, *E. faecalis*, and *P. aeruginosa*	Zn^{2+} and consequent diffusion of these ions into the cytoplasm [91]
ZnO nanostructures (ZnO-NSs)	70~80	*S. aureus*, *S. typhimurium*, *P. vulgaris*, and *K. pneumoniae*	ROS damage to cell membranes [92]
ZnO NPs	40	*Streptococcus mutans* (MTCC497), *S. pyogenes* (MTCC1926), *Vibrio cholerae* (MTCC3906), *Shigella flexneri* (MTCC1457), and *Salmonella typhi* (MTCC1252)	ROS and the release of Zn^{2+} [93]
ZnO NPs	90~100	*V. cholera* and enterotoxic *E. coli* (ETEC)	Inhibit adenylyl cyclase activity, and cAMP levels are decreased [94]
Ge-ZnO NPs	20	*P. aeruginosa* and *E. faecalis*	Penetrated the cell and caused bacterial cell death [83]
SA/ZnO composites		*E. coli* and *S. aureus*	ROS production [95]
ZnO@GA NPs	11.5 ± 4.4	*E. coli* and *S. aureus*	Attributed to the high affinity of GA for the bacterial cell membrane and the increased lipophilicity upon the addition of GA [96]

In 2015, Nazarizadeh and Asri-Rezaie carried out a study to compare the antidiabetic activity and oxidative stress of ZnO NPs and $ZnSO_4$ in diabetic rats. It found that ZnO NPs with small dimensions at higher doses (3 and 10 mg/kg) had a much greater antidiabetic effect compared to $ZnSO_4$ (30 mg/kg). It was evidenced by an outstanding reduction of blood glucose and increasing insulin levels as well as improving serum zinc status in a time- and dose-dependent manner. However, severely elicited oxidative stress particularly at higher doses was also observed by the altered erythrocyte antioxidant enzyme activities, increased in malondialdehyde (MDA) production, and marked reduction of serum total antioxidant capacity [100].

The hyperglycemia can directly enhance an inflammatory state by regulating C-reactive protein (CRP) and cytokines, such as interleukins, which is involved in the development of cardiovascular diseases. Hussein et al. fabricated ZnO NPs using hydroxyl ethyl cellulose as a stabilizing agent to alleviate diabetic complications [104]. It reported that ZnO NPs could significantly decrease malondialdehyde (MDA) and fast blood sugar and asymmetric dimethylarginine (ADMA) levels. The inflammatory markers, interleukin-1 (IL-1α) and CRP, were also notably decreased after ZnO NPs treatment, concomitant with an increase in nitric oxide (NO) and serum antioxidant enzyme (PON-1) levels in diabetic rats.

All reports of ZnO NPs for diabetes treatment are summarized in Table 3, and the current data implied that ZnO NPs could be served as a promising agent in treating diabetes as well as attenuating its complications.

3.4. Anti-Inflammatory Activity. Inflammation is part of the complex biological response of body tissues to harmful stimuli, such as pathogens, damaged cells, or irritants [111]. Since the advent of nanoparticles and considering these biological activities of zinc ions, the anti-inflammatory effects of ZnO NPs have also attracted much attention.

Atopic dermatitis (AD) is a chronic inflammatory skin disease characterized by the impairment of the skin-barrier

Table 3: ZnO NPs for diabetes treatment.

Type of NPs	Size	Drug loaded/synergy	Effects
ZnO NPs	60–95 nm spherical	—	Mitigated the diabetic complications [104]
ZnO NPs	~20 nm spherical	—	A significant decrease in fasting blood glucose and increase in high-density lipoprotein levels [105]
ZnO NPs	10–30 nm	Thiamine	ZnO NPs in combination with thiamine-improved diabetes therapy [106]
ZnO NPs	—	—	ZnO NPs effectively reversed diabetes-induced pancreatic injury [107]
ZnO-RSW NPs	~20 nm	Conjugated red sandalwood (RSW)	ZnO-RSW NPs showed excellent activity against the crude murine pancreatic glucosidase as compared to the individual ZnO NPs and the RSW extract [103]
ZnO NPs	—	—	ZnO NPs acted as a potent antidiabetic agent evidenced by improved glucose disposal, insulin levels, and zinc status in diabetic rats [100]
ZnO NPs	~10 nm	—	ZnO NPs presented pleiotropic antidiabetic effects via improved insulin signaling, enhanced glucose uptake, decreased hepatic glucose output, decreased lipolysis, and enhanced pancreatic beta cell mass [108]
ZnO NPs, CeO_2 NPs, Ag NPs	ZnO NPs: 55 nm, CeO_2 NPs: 54 nm, Ag NPs: 22.5 nm	—	ZnO NPs and Ag NPs had more potent antihyperglycemic activity than CeO_2 NPs [109]
ZnO NPs	Spherical: 96–115 nm; hexagonal: 57 ± 0.3 nm	—	ZnO NPs displayed better antidiabetic potential (IC_{50}: 66.78 μg/mL) than $ZnNO_3$ (IC_{50}: 91.33 μg/mL) in terms of the α-amylase inhibition activity [26]
ZnO NPs	—	Vildagliptin	ZnO NPs and vildagliptin have synergistic effects on the therapy of type-2 diabetes [110]
ZnO NPs	10–15 nm spherical	—	ZnO NPs could improve glucose tolerance and higher serum insulin and reduce blood glucose, nonesterified fatty acids, and triglycerides [101]

functions, which was involved with complex interaction between genetic and environmental factors [112, 113]. Textiles have the longest and most intense contact with the human skin. Wiegand explored the role of ZnO-functionalized textile fibers in the control of oxidative stress in AD in vitro and in vivo [114]. The study found that it is an obvious improvement of AD pruritus and subjective sleep quality when AD patients wore the ZnO textiles overnight on 3 consecutive days. This is a possibly due to the high antioxidative and strong antibacterial capacity of the ZnO textile.

Ilves et al. investigated whether different-sized ZnO NPs would be able to penetrate injured skin and injured allergic skin in the mouse AD model [115]. Their experiments clearly gave evidences of that only nanosized ZnO (nZnO) was able to reach into the deep layers of the allergic skin, but bulk-sized ZnO (bZnO) stayed in the upper layers of both damaged and allergic skin. Compared with bZnO, nZnO exerted higher anti-inflammatory properties by decreasing drastically on proinflammatory cytokines (IL-10, IL-13, IFN-γ, and Th2 cytokines) in the mouse model of AD. These results demonstrated that ZnO NPs with a small size had great effects on reducing skin inflammation in AD models.

The anti-inflammatory activity of ZnO NPs is not confined to atopic dermatitis treatment but has also shown to be very effective for other inflammatory diseases. Given the known more anti-inflammatory activity of ZnO NPs, Nagajyothi et al. described a straightforward, inexpensive, and ecofriendly ZnO NPs using the root extract of *P. tenuifolia* and the anti-inflammatory activities were investigated in LPS-stimulated RAW 264.7 macrophages [30]. ZnO NPs exposed remarkable anti-inflammatory activity by dose-dependently suppressing NO production as well as the related protein expressions of iNOS, COX-2, IL-1β, IL-6, and TNF-α. Thatoi et al. prepared the ZnO NPs under photocondition using the aqueous extracts of two mangrove plants, *Heritiera fomes* and *Sonneratia apetala*, and found

Table 4: ZnO NPs with anti-inflammatory activity.

Type of NPs	Size	Effects
ZnO-functionalized textile fibers	—	A rapid improvement of AD severity, pruritus, and subjective sleep quality when AD patients wore the ZnO textiles overnight on 3 consecutive days [114]
ZnO NPs	—	ZnO NPs exerted higher anti-inflammatory properties by decreasing drastically on proinflammatory cytokines in the mouse model of AD [115]
ZnO NPs	Spherical: 33.03–73.48 nm	ZnO NPs relieved inflammation and displayed a dose-dependent effect in the suppression of related protein expressions [30]
ZnO NPs	Spherical: 96–115 nm; hexagonal: 57 ± 0.3 nm	ZnO NPs exhibited excellent anti-inflammatory activity with 66.78 μg/mL IC_{50} value [26]
ZnO incorporated into TiO_2 nanotubes (TNTs/ZnO)	—	TNTs/ZnO had a significant inhibitory effect on the proliferation and adhesion of macrophages [117]
ZnO NPs and Ag NPs	—	ZnO NPs had a higher potential for anti-inflammatory (79%) in comparison with Ag NPs (69.1%) [116]
ZnO NPs	69.4 ± 13.0 nm	The anti-inflammatory abilities of ZnO NPs to suppress proinflammatory cytokines IL-1β and TNF-α and myeloperoxidase (MPO) in the colitic mice and activate Nrf2 signaling [118]
ZnO NPs	—	Dietary supplementation with ZnO NPs was effective in inhibiting mRNA expression of inflammatory cytokines (IFN-γ, IL-1β, TNF-α, and NF-κB) in the ileum in weaned piglets [119]

that ZnO NPs had a higher potential for anti-inflammatory (79%) in comparison with silver nanoparticles (69.1%) [116].

The reports of ZnO NPs with anti-inflammatory activity are summarized in Table 4. Hence, ZnO NPs also have the potential to be utilized for anti-inflammatory treatment.

3.5. ZnO NPs for Bioimaging. ZnO NPs exhibit efficient blue emissions and near-UV emissions, which have green or yellow luminescence related to oxygen vacancies, therefore further extending its application into bioimaging field [12, 36, 120].

Using a simple sol-gel method, Xiong et al. prepared stable aqueous ZnO@polymer core-shell nanoparticles (ZnO@poly(MAA-co-PEGMEMA))for the first time. The ZnO@polymer core-shell nanoparticles exhibited high quantum yield and very stable broad photoluminescence in aqueous solutions. As shown in Figure 3, within human hepatoma cells, ZnO-1 (derived from LiOH) with an average size of 3 nm showed green fluorescence, while ZnO-2 (derived from NaOH) with an average size of 4 nm appeared yellow. It was worth to note that these nanoparticles did not show any remarkable toxicity for human hepatoma cells when their concentrations were less than 0.2 mg/mL. Furthermore, the luminescence was very stable during cell culturing and the cells were alive at 45 min of exposure. So, as a type of safe and cheap luminescent labels, the ZnO@polymer core-shell nanoparticles can be used as fluorescent probes for cell imaging in vitro [121].

Jiang et al. constructed ZnO nanosheets for the imaging of cultured cells. They treated drug sensitive leukemia line K562 cells with ZnO nanosheets, and the yellow-orange light emission was clearly observed around or inside the cells under UV irradiation (365 nm) at room temperature [122]. ZnO nanostructures were successfully attached onto or penetrated into the cells, which suggested that ZnO nanosheets with visible yellow-orange emission could act as a feasible label for the bioimaging.

Tang et al. prepared ZnO NPs by using the chemical precipitation method. It exhibited emission colors of blue, green, yellow, and orange [123]. The emission color can be changed via adjusting the pH of the precipitation solutions. To stabilize ZnO NPs in water, they encapsulated the ZnO NPs with silica to form ZnO@silica core-shell nanostructures. The obtained ZnO@silica core-shell nanoparticles exhibited excellent water stability, and the visible emissions of ZnO were retained. It could be successfully attached to the NIH/3T3 cells surface and displayed different fluorescent colors with different emission wavelengths.

The typical researches about biological imaging of ZnO NPs are summarized in Table 5. Based on its advanced intrinsic fluorescence, ZnO nanomaterial can also be used as a promising candidate for cell imaging and pathological studies.

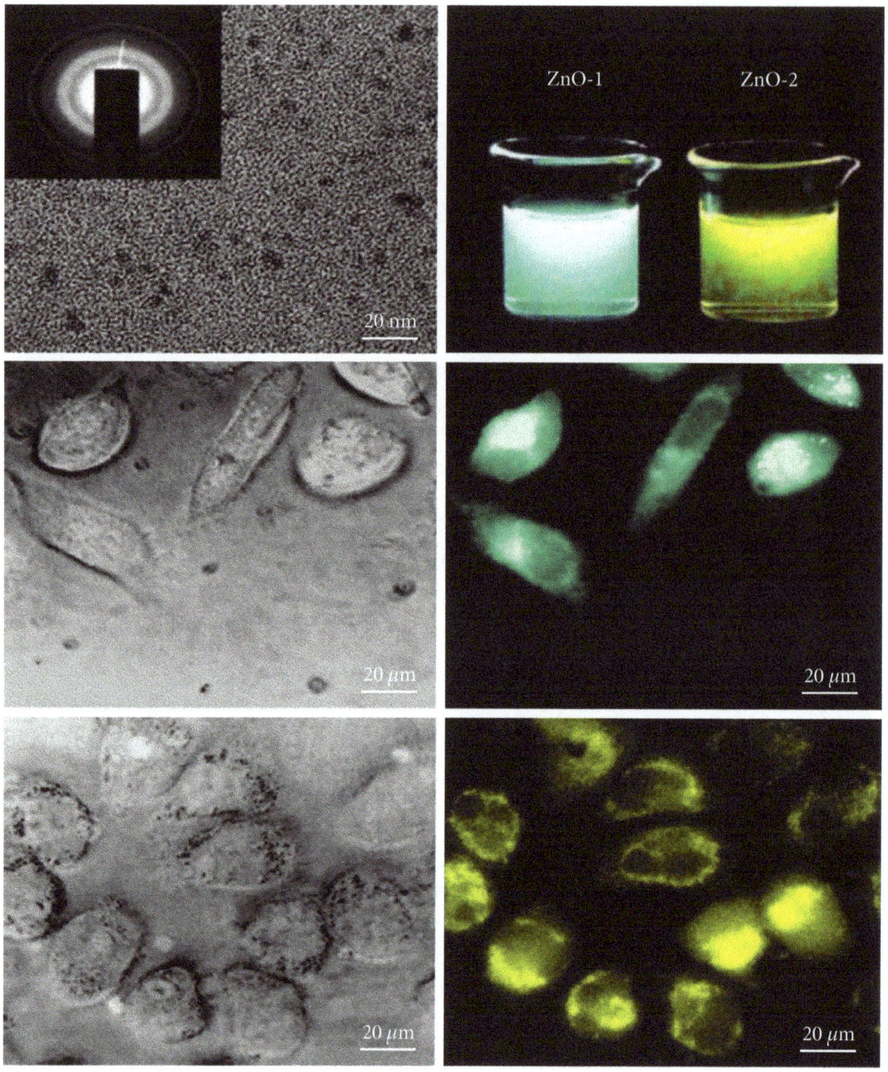

FIGURE 3: The upper part is the high-resolution transmission electron microscopy (HRTEM) image of the ZnO@polymer core-shell nanoparticles and the aqueous solutions of ZnO-1 and ZnO-2 under a UV light; the middle part is the DIC picture and the fluorescent image of the human hepatoma cells labeled by ZnO-1; and the lower part is the DIC picture and the fluorescent image of the hepatoma cells labeled by ZnO-2 [121].

4. Conclusions and Future Perspectives

ZnO NPs have exhibited promising biomedical applications based on its anticancer, antibacterial, antidiabetic, anti-inflammatory, drug delivery, as well as bioimaging activity. Due to inherent toxicity of ZnO NPs, they possess strong inhibition effects against cancerous cell and bacteria, by inducing intracellular ROS generation and activating apoptotic signaling pathway, which makes ZnO NPs a potential candidate as anticancer and antibacterial agents. In addition, ZnO NPs have also been well known to promote the bioavailability of therapeutic drugs or biomolecules when functioning as drug carriers to achieve enhanced therapy efficiency. Moreover, with the ability to decrease blood glucose and increase in insulin levels, ZnO NPs have shown the promising potential in treating diabetes and attenuating its complications, which can be further evaluated.

ZnO NPs are listed as a kind of safe substance by the FDA. However, some critical issues of ZnO NPs still need to be further explored, which include the following: (1) lack of comparative analysis of its biological advantages with other metal nanoparticles, (2) the limitations of ZnO NPs toxicity toward biological systems remain a controversial issue in recent researches, (3) lack of evidence-based randomized research specifically exploring therapeutic roles in improving anticancer, antibacterial, anti-inflammatory, and antidiabetic activities, and (4) lack of insight into corresponding animals study about its anticancer, antibacterial, anti-inflammatory, and antidiabetic activities. Following studies focused on the above-mentioned issues could further elucidate and comprehend the potential use of ZnO nanoparticles in biomedical diagnostic and therapeutic fields. We believe that nanomaterials would dramatically promote the development of

Table 5: Typical researches about biological imaging of ZnO NPs.

Materials	Size and models	Biological imaging
ZnO@MAA-co-PEGMEMA	Human hepatoma cells	With tunable photoluminescence emission and high quantum yield, under UV light, ZnO-1 showed green fluorescence, while ZnO-2 appeared yellow [121]
ZnO@PMAA-co-PDMAEMA	Spherical: 4 nm/COS-7 cells	The PDMAEMA-modified ZnO QDs emitted strong yellow luminescence under UV light [124]
ZnO-Au@PEG NPs	<100 nm/B16F10 cells	ZnO-Au@PEG NPs can penetrate into the living cells and exhibit bright yellow fluorescence [125]
ZnO nanosheets	46.6 ± 8.5 nm/K562 cells	Yellow-orange light emission was observed around or inside the cells under UV light [122]
ZnO@silica NPs	2.7–4.4 nm/HeLa cells	Monodispersed ZnO@silica NPs with blue, green, and yellow emission through using VTES, TEOS, and APS as modification materials [126]
Fe_3O_4-ZnO NPs	15.7 nm/dendritic cells	Fe_3O_4-ZnO NPs emit green fluorescence under UV light irradiation [127].
ZnO@silica NPs	NIH/3T3 cells	ZnO@silica NPs exhibited emission colors of blue, green, yellow, and orange under 365 nm excitation via the adjustment of the pH of the precipitation solutions [123]
ZnO@silica NPs	55 nm/nerve cells, Caco-2 cells	ZnO@silica QD colloidal solution exhibited a significant blue emission [128]
ZnO nanowires	20–50 nm/U87MG cells, MCF-7 cells	ZnO nanowires exhibited green fluorescent. RGD peptide-conjugated green fluorescent ZnO NWs can be specifically targeted to cell surface receptors in vitro [129]

medicine, and ZnO nanoparticles are expected to make more exciting contributions in these fields.

Acknowledgments

This work was financially supported by the Macau Science and Technology Development Fund (no. 028/2014/A1) and China Postdoctoral Science Foundation (2018M631026).

References

[1] P. K. Mishra, H. Mishra, A. Ekielski, S. Talegaonkar, and B. Vaidya, "Zinc oxide nanoparticles: a promising nanomaterial for biomedical applications," *Drug Discovery Today*, vol. 22, no. 12, pp. 1825–1834, 2017.

[2] T. G. Smijs and S. Pavel, "Titanium dioxide and zinc oxide nanoparticles in sunscreens: focus on their safety and effectiveness," *Nanotechnology, Science and Applications*, vol. 4, pp. 95–112, 2011.

[3] J. A. Ruszkiewicz, A. Pinkas, B. Ferrer, T. V. Peres, A. Tsatsakis, and M. Aschner, "Neurotoxic effect of active ingredients in sunscreen products, a contemporary review," *Toxicology Reports*, vol. 4, pp. 245–259, 2017.

[4] A. Kolodziejczak-Radzimska and T. Jesionowski, "Zinc oxide–from synthesis to application: a review," *Materials*, vol. 7, no. 4, pp. 2833–2881, 2014.

[5] S. Sahoo, M. Maiti, A. Ganguly, J. J. George, and A. K. Bhowmick, "Effect of zinc oxide nanoparticles as cure activator on the properties of natural rubber and nitrile rubber," *Journal of Applied Polymer Science*, vol. 105, no. 4, pp. 2407–2415, 2007.

[6] M. D. Newman, M. Stotland, and J. I. Ellis, "The safety of nanosized particles in titanium dioxide- and zinc oxide-based sunscreens," *Journal of the American Academy of Dermatology*, vol. 61, no. 4, pp. 685–692, 2009.

[7] A. Hatamie, A. Khan, M. Golabi et al., "Zinc oxide nanostructure-modified textile and its application to biosensing, photocatalysis, and as antibacterial material," *Langmuir*, vol. 31, no. 39, pp. 10913–10921, 2015.

[8] F. X. Xiao, S. F. Hung, H. B. Tao, J. Miao, H. B. Yang, and B. Liu, "Spatially branched hierarchical ZnO nanorod-TiO_2 nanotube array heterostructures for versatile photocatalytic and photoelectrocatalytic applications: towards intimate integration of 1D-1D hybrid nanostructures," *Nanoscale*, vol. 6, no. 24, pp. 14950–14961, 2014.

[9] J. W. Rasmussen, E. Martinez, P. Louka, and D. G. Wingett, "Zinc oxide nanoparticles for selective destruction of tumor cells and potential for drug delivery applications," *Expert Opinion on Drug Delivery*, vol. 7, no. 9, pp. 1063–1077, 2010.

[10] Z. Y. Zhang and H. M. Xiong, "Photoluminescent ZnO nanoparticles and their biological applications," *Materials*, vol. 8, no. 6, pp. 3101–3127, 2015.

[11] S. Kim, S. Y. Lee, and H. J. Cho, "Doxorubicin-wrapped zinc oxide nanoclusters for the therapy of colorectal adenocarcinoma," *Nanomaterials*, vol. 7, no. 11, p. 354, 2017.

[12] H. M. Xiong, "ZnO nanoparticles applied to bioimaging and drug delivery," *Advanced Materials*, vol. 25, no. 37, pp. 5329–5335, 2013.

[13] M. A. Majeed Khan, M. Wasi Khan, M. Alhoshan, M. S. AlSalhi, and A. S. Aldwayyan, "Influences of Co doping on the structural and optical properties of ZnO nanostructured," *Applied Physics A*, vol. 100, no. 1, pp. 45–51, 2010.

[14] G. Bisht, S. Rayamajhi, B. Kc, S. N. Paudel, D. Karna, and B. G. Shrestha, "Synthesis, characterization, and study of in vitro cytotoxicity of ZnO-Fe_3O_4 magnetic composite nanoparticles in human breast cancer cell line (MDA-MB-231) and mouse fibroblast (NIH 3T3)," *Nanoscale Research Letters*, vol. 11, no. 1, p. 537, 2016.

[15] S. Bettini, R. Pagano, V. Bonfrate et al., "Promising piezoelectric properties of new ZnO@octadecylamine adduct," *Journal of Physical Chemistry C*, vol. 119, no. 34, pp. 20143–20149, 2015.

[16] R. Pagano, A. Quarta, S. Pal, A. Licciulli, L. Valli, and S. Bettini, "Enhanced solar-driven applications of ZnO@Ag patchy nanoparticles," *Journal of Physical Chemistry C*, vol. 121, no. 48, pp. 27199–27206, 2017.

[17] S. Bettini, R. Pagano, L. Valli, and G. Giancane, "Enhancement of open circuit voltage of a ZnO-based dye-sensitized solar cell by means of piezotronic effect," *Chemistry-An Asian Journal*, vol. 11, no. 8, pp. 1240–1245, 2016.

[18] L. Spanhel and M. A. Anderson, "Semiconductor clusters in the sol-gel process-quantized aggregation, gelation, and crystal-growth in concentrated ZnO colloids," *Journal of the American Chemical Society*, vol. 113, no. 8, pp. 2826–2833, 1991.

[19] S. Rani, P. Suri, P. Shishodia, and R. Mehra, "Synthesis of nanocrystalline ZnO powder via sol–gel route for dye-sensitized solar cells," *Solar Energy Materials and Solar Cells*, vol. 92, no. 12, pp. 1639–1645, 2008.

[20] Z. J. Wang, H. M. Zhang, L. G. Zhang, J. S. Yuan, S. G. Yan, and C. Y. Wang, "Low-temperature synthesis of ZnO nanoparticles by solid-state pyrolytic reaction," *Nanotechnology*, vol. 14, no. 1, pp. 11–15, 2003.

[21] L. Shen, N. Bao, K. Yanagisawa, K. Domen, A. Gupta, and C. A. Grimes, "Direct synthesis of ZnO nanoparticles by a solution-free mechanochemical reaction," *Nanotechnology*, vol. 17, no. 20, pp. 5117–5123, 2006.

[22] S. K. Pardeshi and A. B. Patil, "Effect of morphology and crystallite size on solar photocatalytic activity of zinc oxide synthesized by solution free mechanochemical method," *Journal of Molecular Catalysis A: Chemical*, vol. 308, no. 1-2, pp. 32–40, 2009.

[23] K. Elumalai and S. Velmurugan, "Green synthesis, characterization and antimicrobial activities of zinc oxide nanoparticles from the leaf extract of *Azadirachta indica* (L.)," *Applied Surface Science*, vol. 345, pp. 329–336, 2015.

[24] C. Mahendra, M. Murali, G. Manasa et al., "Antibacterial and antimitotic potential of bio-fabricated zinc oxide nanoparticles of *Cochlospermum religiosum* (L.)," *Microbial Pathogenesis*, vol. 110, pp. 620–629, 2017.

[25] L. Fu and Z. X. Fu, "*Plectranthus amboinicus* leaf extract-assisted biosynthesis of ZnO nanoparticles and their photocatalytic activity," *Ceramics International*, vol. 41, no. 2, pp. 2492–2496, 2015.

[26] G. Rajakumar, M. Thiruvengadam, G. Mydhili, T. Gomathi, and I. M. Chung, "Green approach for synthesis of zinc oxide nanoparticles from *Andrographis paniculata* leaf extract and evaluation of their antioxidant, anti-diabetic, and anti-inflammatory activities," *Bioprocess and Biosystems Engineering*, vol. 41, no. 1, pp. 21–30, 2018.

[27] Y. G. Qian, J. Yao, M. Russel, K. Chen, and X. Y. Wang, "Characterization of green synthesized nano-formulation (ZnO-A. vera) and their antibacterial activity against pathogens," *Environmental Toxicology and Pharmacology*, vol. 39, no. 2, pp. 736–746, 2015.

[28] K. Ali, S. Dwivedi, A. Azam et al., "Aloe vera extract functionalized zinc oxide nanoparticles as nanoantibiotics against multi-drug resistant clinical bacterial isolates," *Journal of Colloid and Interface Science*, vol. 472, pp. 145–156, 2016.

[29] R. Yuvakkumar, J. Suresh, A. J. Nathanael, M. Sundrarajan, and S. I. Hong, "Novel green synthetic strategy to prepare ZnO nanocrystals using rambutan (*Nephelium lappaceum* L.) peel extract and its antibacterial applications," *Materials Science and Engineering: C*, vol. 41, pp. 17–27, 2014.

[30] P. C. Nagajyothi, S. J. Cha, I. J. Yang, T. V. Sreekanth, K. J. Kim, and H. M. Shin, "Antioxidant and anti-inflammatory activities of zinc oxide nanoparticles synthesized using *Polygala tenuifolia* root extract," *Journal of Photochemistry and Photobiology B, Biology*, vol. 146, pp. 10–17, 2015.

[31] A. C. Janaki, E. Sailatha, and S. Gunasekaran, "Synthesis, characteristics and antimicrobial activity of ZnO nanoparticles," *Spectrochimica Acta Part A: Molecular and Biomolecular Spectroscopy*, vol. 144, pp. 17–22, 2015.

[32] R. Dobrucka and J. Dugaszewska, "Biosynthesis and antibacterial activity of ZnO nanoparticles using *Trifolium pratense* flower extract," *Saudi Journal of Biological Sciences*, vol. 23, no. 4, pp. 517–523, 2016.

[33] D. Sharma, M. I. Sabela, S. Kanchi et al., "Biosynthesis of ZnO nanoparticles using *Jacaranda mimosifolia* flowers extract: synergistic antibacterial activity and molecular simulated facet specific adsorption studies," *Journal of Photochemistry and Photobiology B: Biology*, vol. 162, pp. 199–207, 2016.

[34] J. Qu, X. Yuan, X. H. Wang, and P. Shao, "Zinc accumulation and synthesis of ZnO nanoparticles using *Physalis alkekengi* L," *Environmental Pollution*, vol. 159, no. 7, pp. 1783–1788, 2011.

[35] H. Sharma, K. Kumar, C. Choudhary, P. K. Mishra, and B. Vaidya, "Development and characterization of metal oxide nanoparticles for the delivery of anticancer drug," *Artificial Cells, Nanomedicine, and Biotechnology*, vol. 44, no. 2, pp. 672–679, 2016.

[36] Y. Zhang, T. R. Nayak, H. Hong, and W. Cai, "Biomedical applications of zinc oxide nanomaterials," *Current molecular medicine*, vol. 13, no. 10, pp. 1633–1645, 2013.

[37] M. Martinez-Carmona, Y. Gun'ko, and M. Vallet-Regi, "ZnO nanostructures for drug delivery and theranostic applications," *Nanomaterials*, vol. 8, no. 4, p. 268, 2018.

[38] I. De Angelis, F. Barone, A. Zijno et al., "Comparative study of ZnO and TiO_2 nanoparticles: physicochemical characterisation and toxicological effects on human colon carcinoma cells," *Nanotoxicology*, vol. 7, no. 8, pp. 1361–1372, 2013.

[39] Y. Cao, M. Roursgaard, A. Kermanizadeh, S. Loft, and P. Moller, "Synergistic effects of zinc oxide nanoparticles and fatty acids on toxicity to Caco-2 cells," *International Journal of Toxicology*, vol. 34, no. 1, pp. 67–76, 2015.

[40] X. Fang, L. Jiang, Y. Gong, J. Li, L. Liu, and Y. Cao, "The presence of oleate stabilized ZnO nanoparticles (NPs) and reduced the toxicity of aged NPs to Caco-2 and HepG2 cells," *Chemico-Biological Interactions*, vol. 278, pp. 40–47, 2017.

[41] J. Bai Aswathanarayan and R. Rai Vittal, "Muddegowda U: anticancer activity of metal nanoparticles and their peptide conjugates against human colon adenorectal carcinoma cells," *Artificial Cells, Nanomedicine, and Biotechnology* pp. 1–8, 2017, In press.

[42] V. Sharma, D. Anderson, and A. Dhawan, "Zinc oxide nanoparticles induce oxidative DNA damage and ROS-triggered mitochondria mediated apoptosis in human liver cells (HepG2)," *Apoptosis*, vol. 17, no. 8, pp. 852–870, 2012.

[43] M. J. Akhtar, M. Ahamed, S. Kumar, M. M. Khan, J. Ahmad, and S. A. Alrokayan, "Zinc oxide nanoparticles selectively induce apoptosis in human cancer cells through reactive oxygen species," *International Journal of Nanomedicine*, vol. 7, pp. 845–857, 2012.

[44] Y. Deng and H. Zhang, "The synergistic effect and mechanism of doxorubicin-ZnO nanocomplexes as a multimodal agent integrating diverse anticancer therapeutics," *International Journal of Nanomedicine*, vol. 8, pp. 1835–1841, 2013.

[45] A. B. Moghaddam, M. Moniri, S. Azizi et al., "Eco-friendly formulated zinc oxide nanoparticles: induction of cell cycle arrest and apoptosis in the MCF-7 cancer cell line," *Genes*, vol. 8, no. 10, p. 281, 2017.

[46] J. Liu, X. Ma, S. Jin et al., "Zinc oxide nanoparticles as adjuvant to facilitate doxorubicin intracellular accumulation and visualize pH-responsive release for overcoming drug resistance," *Molecular Pharmaceutics*, vol. 13, no. 5, pp. 1723–1730, 2016.

[47] B. A. Othman, C. Greenwood, A. F. Abuelela et al., "Correlative light-electron microscopy shows RGD-targeted ZnO nanoparticles dissolve in the intracellular environment of triple negative breast cancer cells and cause apoptosis with intratumor heterogeneity," *Advanced Healthcare Materials*, vol. 5, no. 11, pp. 1310–1325, 2016.

[48] N. Puvvada, S. Rajput, B. N. Kumar et al., "Novel ZnO hollow-nanocarriers containing paclitaxel targeting folate-receptors in a malignant pH-microenvironment for effective monitoring and promoting breast tumor regression," *Scientific Reports*, vol. 5, no. 1, article 11760, 2015.

[49] N. Tripathy, R. Ahmad, H. A. Ko, G. Khang, and Y. B. Hahn, "Enhanced anticancer potency using an acid-responsive ZnO-incorporated liposomal drug-delivery system," *Nanoscale*, vol. 7, no. 9, pp. 4088–4096, 2015.

[50] K. J. Bai, K. J. Chuang, C. M. Ma, T. Y. Chang, and H. C. Chuang, "Human lung adenocarcinoma cells with an EGFR mutation are sensitive to non-autophagic cell death induced by zinc oxide and aluminium-doped zinc oxide nanoparticles," *Journal of Toxicological Sciences*, vol. 42, no. 4, pp. 437–444, 2017.

[51] D. P. Bai, X. F. Zhang, G. L. Zhang, Y. F. Huang, and S. Gurunathan, "Zinc oxide nanoparticles induce apoptosis and autophagy in human ovarian cancer cells," *International Journal of Nanomedicine*, vol. 12, pp. 6521–6535, 2017.

[52] R. Hariharan, S. Senthilkumar, A. Suganthi, and M. Rajarajan, "Synthesis and characterization of doxorubicin modified ZnO/PEG nanomaterials and its photodynamic action," *Journal of Photochemistry and Photobiology B: Biology*, vol. 116, pp. 56–65, 2012.

[53] M. Pandurangan, G. Enkhtaivan, and D. H. Kim, "Anticancer studies of synthesized ZnO nanoparticles against human cervical carcinoma cells," *Journal of Photochemistry and Photobiology B: Biology*, vol. 158, pp. 206–211, 2016.

[54] R. Dhivya, J. Ranjani, J. Rajendhran, J. Mayandi, and J. Annaraj, "Enhancing the anti-gastric cancer activity of curcumin with biocompatible and pH sensitive PMMA-AA/ZnO nanoparticles," *Materials Science and Engineering: C*, vol. 82, pp. 182–189, 2018.

[55] R. Dhivya, J. Ranjani, P. K. Bowen, J. Rajendhran, J. Mayandi, and J. Annaraj, "Biocompatible curcumin loaded PMMA-PEG/ZnO nanocomposite induce apoptosis and cytotoxicity in human gastric cancer cells," *Materials Science and Engineering: C*, vol. 80, pp. 59–68, 2017.

[56] P. Patel, K. Kansara, V. A. Senapati, R. Shanker, A. Dhawan, and A. Kumar, "Cell cycle dependent cellular uptake of zinc oxide nanoparticles in human epidermal cells," *Mutagenesis*, vol. 31, no. 4, pp. 481–490, 2016.

[57] F. Namvar, S. Azizi, H. S. Rahman et al., "Green synthesis, characterization, and anticancer activity of hyaluronan/zinc oxide nanocomposite," *OncoTargets and therapy*, vol. 9, pp. 4549–4559, 2016.

[58] D. F. Stowe and A. K. S. Camara, "Mitochondrial reactive oxygen species production in excitable cells: modulators of mitochondrial and cell function," *Antioxidants and Redox Signaling*, vol. 11, no. 6, pp. 1373–1414, 2009.

[59] E. Moghimipour, M. Rezaei, Z. Ramezani et al., "Transferrin targeted liposomal 5-fluorouracil induced apoptosis via mitochondria signaling pathway in cancer cells," *Life Sciences*, vol. 194, pp. 104–110, 2017.

[60] C. Y. Guo, L. Sun, X. P. Chen, and D. S. Zhang, "Oxidative stress, mitochondrial damage and neurodegenerative diseases," *Neural Regeneration Research*, vol. 8, no. 21, pp. 2003–2014, 2013.

[61] M. Chandrasekaran and M. Pandurangan, "In vitro selective anti-proliferative effect of zinc oxide nanoparticles against co-cultured C2C12 myoblastoma cancer and 3T3-L1 normal cells," *Biological Trace Element Research*, vol. 172, no. 1, pp. 148–154, 2016.

[62] K. N. Yu, T. J. Yoon, A. Minai-Tehrani et al., "Zinc oxide nanoparticle induced autophagic cell death and mitochondrial damage via reactive oxygen species generation," *Toxicology in Vitro*, vol. 27, no. 4, pp. 1187–1195, 2013.

[63] S. Hackenberg, A. Scherzed, A. Gohla et al., "Nanoparticle-induced photocatalytic head and neck squamous cell carcinoma cell death is associated with autophagy," *Nanomedicine*, vol. 9, no. 1, pp. 21–33, 2014.

[64] M. Arakha, J. Roy, P. S. Nayak, B. Mallick, and S. Jha, "Zinc oxide nanoparticle energy band gap reduction triggers the oxidative stress resulting into autophagy-mediated apoptotic cell death," *Free Radical Biology and Medicine*, vol. 110, pp. 42–53, 2017.

[65] J. Zhang, X. Qin, B. Wang et al., "Zinc oxide nanoparticles harness autophagy to induce cell death in lung epithelial cells," *Cell Death and Disease*, vol. 8, no. 7, article e2954, 2017.

[66] N. Erathodiyil and J. Y. Ying, "Functionalization of inorganic nanoparticles for bioimaging applications," *Accounts of Chemical Research*, vol. 44, no. 10, pp. 925–935, 2011.

[67] J. Wang, J. S. Lee, D. Kim, and L. Zhu, "Exploration of zinc oxide nanoparticles as a multitarget and multifunctional anticancer nanomedicine," *ACS Applied Materials and Interfaces*, vol. 9, no. 46, pp. 39971–39984, 2017.

[68] S. B. Ghaffari, M. H. Sarrafzadeh, Z. Fakhroueian, S. Shahriari, and M. R. Khorramizadeh, "Functionalization of ZnO nanoparticles by 3-mercaptopropionic acid for aqueous curcumin delivery: synthesis, characterization, and anticancer assessment," *Materials Science and Engineering: C*, vol. 79, pp. 465–472, 2017.

[69] Y. Li, C. Zhang, L. Liu, Y. Gong, Y. Xie, and Y. Cao, "The effects of baicalein or baicalin on the colloidal stability of ZnO nanoparticles (NPs) and toxicity of NPs to Caco-2 cells," *Toxicology Mechanisms and Methods*, vol. 28, no. 3, pp. 167–176, 2017.

[70] N. Kamaly, Z. Xiao, P. M. Valencia, A. F. Radovic-Moreno, and O. C. Farokhzad, "Targeted polymeric therapeutic nanoparticles: design, development and clinical translation," *Chemical Society Reviews*, vol. 41, no. 7, pp. 2971–3010, 2012.
[71] Z. Han, X. Wang, C. Heng et al., "Synergistically enhanced photocatalytic and chemotherapeutic effects of aptamer-functionalized ZnO nanoparticles towards cancer cells," *Physical Chemistry Chemical Physics*, vol. 17, no. 33, pp. 21576–21582, 2015.
[72] K. C. Biplab, S. N. Paudel, S. Rayamajhi et al., "Enhanced preferential cytotoxicity through surface modification: synthesis, characterization and comparative in vitro evaluation of TritonX-100 modified and unmodified zinc oxide nanoparticles in human breast cancer cell (MDA-MB-231)," *Chemistry Central Journal*, vol. 10, no. 1, 2016.
[73] Y. Y. Ma, H. Ding, and H. M. Xiong, "Folic acid functionalized ZnO quantum dots for targeted cancer cell imaging," *Nanotechnology*, vol. 26, no. 30, article 305702, 2015.
[74] S. Chakraborti, S. Chakraborty, S. Saha et al., "PEG-functionalized zinc oxide nanoparticles induce apoptosis in breast cancer cells through reactive oxygen species-dependent impairment of DNA damage repair enzyme NEIL2," *Free Radical Biology and Medicine*, vol. 103, pp. 35–47, 2017.
[75] L. E. Shi, Z. H. Li, W. Zheng, Y. F. Zhao, Y. F. Jin, and Z. X. Tang, "Synthesis, antibacterial activity, antibacterial mechanism and food applications of ZnO nanoparticles: a review," *Food Additives and Contaminants: Part A*, vol. 31, no. 2, pp. 173–186, 2014.
[76] Y. Jiang, L. Zhang, D. Wen, and Y. Ding, "Role of physical and chemical interactions in the antibacterial behavior of ZnO nanoparticles against *E. coli*," *Materials Science and Engineering: C*, vol. 69, pp. 1361–1366, 2016.
[77] R. K. Dutta, B. P. Nenavathu, M. K. Gangishetty, and A. V. Reddy, "Antibacterial effect of chronic exposure of low concentration ZnO nanoparticles on *E. coli*," *Journal of Environmental Science and Health: Part A*, vol. 48, no. 8, pp. 871–878, 2013.
[78] K. M. Reddy, K. Feris, J. Bell, D. G. Wingett, C. Hanley, and A. Punnoose, "Selective toxicity of zinc oxide nanoparticles to prokaryotic and eukaryotic systems," *Applied Physics Letters*, vol. 90, no. 21, article 213902, 2007.
[79] B. N. Singh, A. K. Rawat, W. Khan, A. H. Naqvi, and B. R. Singh, "Biosynthesis of stable antioxidant ZnO nanoparticles by *Pseudomonas aeruginosa* rhamnolipids," *PLoS One*, vol. 9, no. 9, Article ID e106937, 2014.
[80] R. Ishwarya, B. Vaseeharan, S. Kalyani et al., "Facile green synthesis of zinc oxide nanoparticles using *Ulva lactuca* seaweed extract and evaluation of their photocatalytic, antibiofilm and insecticidal activity," *Journal of Photochemistry and Photobiology B: Biology*, vol. 178, pp. 249–258, 2018.
[81] T. Chatterjee, S. Chakraborti, P. Joshi, S. P. Singh, V. Gupta, and P. Chakrabarti, "The effect of zinc oxide nanoparticles on the structure of the periplasmic domain of the *Vibrio cholerae* ToxR protein," *FEBS Journal*, vol. 277, no. 20, pp. 4184–4194, 2010.
[82] Y. H. Hsueh, W. J. Ke, C. T. Hsieh, K. S. Lin, D. Y. Tzou, and C. L. Chiang, "ZnO nanoparticles affect *Bacillus subtilis* cell growth and biofilm formation," *PLoS One*, vol. 10, no. 6, Article ID e0128457, 2015.
[83] M. Divya, B. Vaseeharan, M. Abinaya et al., "Biopolymer gelatin-coated zinc oxide nanoparticles showed high antibacterial, antibiofilm and anti-angiogenic activity," *Journal of Photochemistry and Photobiology B: Biology*, vol. 178, pp. 211–218, 2018.
[84] I. Matai, A. Sachdev, P. Dubey, S. U. Kumar, B. Bhushan, and P. Gopinath, "Antibacterial activity and mechanism of Ag-ZnO nanocomposite on *S. aureus* and GFP-expressing antibiotic resistant *E. coli*," *Colloids and Surfaces B: Biointerfaces*, vol. 115, pp. 359–367, 2014.
[85] S. Sarwar, S. Chakraborti, S. Bera, I. A. Sheikh, K. M. Hoque, and P. Chakrabarti, "The antimicrobial activity of ZnO nanoparticles against *Vibrio cholerae*: variation in response depends on biotype," *Nanomedicine: Nanotechnology, Biology and Medicine*, vol. 12, no. 6, pp. 1499–1509, 2016.
[86] K. Ghule, A. V. Ghule, B. J. Chen, and Y. C. Ling, "Preparation and characterization of ZnO nanoparticles coated paper and its antibacterial activity study," *Green Chemistry*, vol. 8, no. 12, pp. 1034–1041, 2006.
[87] A. Iswarya, B. Vaseeharan, M. Anjugam et al., "Multipurpose efficacy of ZnO nanoparticles coated by the crustacean immune molecule beta-1,3-glucan binding protein: toxicity on HepG2 liver cancer cells and bacterial pathogens," *Colloids and Surfaces B: Biointerfaces*, vol. 158, pp. 257–269, 2017.
[88] M. Shaban, F. Mohamed, and S. Abdallah, "Production and characterization of superhydrophobic and antibacterial coated fabrics utilizing ZnO nanocatalyst," *Scientific Reports*, vol. 8, no. 1, p. 3925, 2018.
[89] K. Karthik, S. Dhanuskodi, C. Gobinath, and S. Sivaramakrishnan, "Microwave-assisted synthesis of CdO-ZnO nanocomposite and its antibacterial activity against human pathogens," *Spectrochimica Acta Part A: Molecular and Biomolecular Spectroscopy*, vol. 139, pp. 7–12, 2015.
[90] X. Bellanger, P. Billard, R. Schneider, L. Balan, and C. Merlin, "Stability and toxicity of ZnO quantum dots: Interplay between nanoparticles and bacteria," *Journal of Hazardous Materials*, vol. 283, pp. 110–116, 2015.
[91] K. Dedkova, B. Janikova, K. Matejova et al., "Preparation, characterization and antibacterial properties of ZnO/kaoline nanocomposites," *Journal of Photochemistry and Photobiology B: Biology*, vol. 148, pp. 113–117, 2015.
[92] M. Ramani, S. Ponnusamy, C. Muthamizhchelvan, J. Cullen, S. Krishnamurthy, and E. Marsili, "Morphology-directed synthesis of ZnO nanostructures and their antibacterial activity," *Colloids and Surfaces B: Biointerfaces*, vol. 105, pp. 24–30, 2013.
[93] S. Soren, S. Kumar, S. Mishra, P. K. Jena, S. K. Verma, and P. Parhi, "Evaluation of antibacterial and antioxidant potential of the zinc oxide nanoparticles synthesized by aqueous and polyol method," *Microbial Pathogenesis*, vol. 119, pp. 145–151, 2018.
[94] W. Salem, D. R. Leitner, F. G. Zingl et al., "Antibacterial activity of silver and zinc nanoparticles against *Vibrio cholerae* and enterotoxic *Escherichia coli*," *International Journal of Medical Microbiology*, vol. 305, no. 1, pp. 85–95, 2015.
[95] W. Wu, T. Liu, H. He et al., "Rheological and antibacterial performance of sodium alginate/zinc oxide composite coating for cellulosic paper," *Colloids and surfaces B: Biointerfaces*, vol. 167, pp. 538–543, 2018.
[96] J. Lee, K. H. Choi, J. Min, H. J. Kim, J. P. Jee, and B. J. Park, "Functionalized ZnO nanoparticles with gallic acid for antioxidant and antibacterial activity against methicillin-resistant *S. aureus*," *Nanomaterials*, vol. 7, no. 11, p. 365, 2017.

[97] T. Ohira and O. Yamamoto, "Correlation between antibacterial activity and crystallite size on ceramics," *Chemical Engineering Science*, vol. 68, no. 1, pp. 355–361, 2012.

[98] S. Sarwar, A. Ali, M. Pal, and P. Chakrabarti, "Zinc oxide nanoparticles provide anti-cholera activity by disrupting the interaction of cholera toxin with the human GM1 receptor," *Journal of Biological Chemistry*, vol. 292, no. 44, pp. 18303–18311, 2017.

[99] S. N. Seclen, M. E. Rosas, A. J. Arias, and C. A. Medina, "Elevated incidence rates of diabetes in Peru: report from PERUDIAB, a national urban population-based longitudinal study," *BMJ Open Diabetes Research and Care*, vol. 5, no. 1, article e000401, 2017.

[100] A. Nazarizadeh and S. Asri-Rezaie, "Comparative study of antidiabetic activity and oxidative stress induced by zinc oxide nanoparticles and zinc sulfate in diabetic rats," *AAPS PharmSciTech*, vol. 17, no. 4, pp. 834–843, 2016.

[101] R. D. Umrani and K. M. Paknikar, "Zinc oxide nanoparticles show antidiabetic activity in streptozotocin-induced Type 1 and 2 diabetic rats," *Nanomedicine*, vol. 9, no. 1, pp. 89–104, 2014.

[102] R. Malizia, A. Scorsone, P. D'Angelo, C. Lo Pinto, L. Pitrolo, and C. Giordano, "Zinc deficiency and cell-mediated and humoral autoimmunity of insulin-dependent diabetes in thalassemic subjects," *Journal of Pediatric Endocrinology and Metabolism: JPEM*, vol. 11, no. 3, pp. 981–984, 1998.

[103] R. Kitture, K. Chordiya, S. Gaware et al., "ZnO nanoparticles-red sandalwood conjugate: a promising anti-diabetic agent," *Journal of Nanoscience and Nanotechnology*, vol. 15, no. 6, pp. 4046–4051, 2015.

[104] J. Hussein, M. El-Banna, T. A. Razik, and M. E. El-Naggar, "Biocompatible zinc oxide nanocrystals stabilized via hydroxyethyl cellulose for mitigation of diabetic complications," *International Journal of Biological Macromolecules*, vol. 107, pp. 748–754, 2018.

[105] A. Bayrami, S. Parvinroo, A. Habibi-Yangjeh, and S. Rahim Pouran, "Bio-extract-mediated ZnO nanoparticles: microwave-assisted synthesis, characterization and antidiabetic activity evaluation," *Artificial Cells, Nanomedicine, and Biotechnology*, vol. 46, no. 4, pp. 730–739, 2018.

[106] A. Amiri, R. A. F. Dehkordi, M. S. Heidarnejad, and M. J. Dehkordi, "Effect of the zinc oxide nanoparticles and thiamine for the management of diabetes in alloxan-induced mice: a stereological and biochemical study," *Biological Trace Element Research*, vol. 181, no. 2, pp. 258–264, 2018.

[107] N. S. Wahba, S. F. Shaban, A. A. Kattaia, and S. A. Kandeel, "Efficacy of zinc oxide nanoparticles in attenuating pancreatic damage in a rat model of streptozotocin-induced diabetes," *Ultrastructural Pathology*, vol. 40, no. 6, pp. 358–373, 2016.

[108] S. C. Asani, R. D. Umrani, and K. M. Paknikar, "In vitro studies on the pleotropic antidiabetic effects of zinc oxide nanoparticles," *Nanomedicine*, vol. 11, no. 13, pp. 1671–1687, 2016.

[109] K. Shanker, J. Naradala, G. K. Mohan, G. S. Kumar, and P. L. Pravallika, "A sub-acute oral toxicity analysis and comparative in vivo anti-diabetic activity of zinc oxide, cerium oxide, silver nanoparticles, and *Momordica charantia* in streptozotocin-induced diabetic Wistar rats," *RSC Advances*, vol. 7, no. 59, pp. 37158–37167, 2017.

[110] R. M. El-Gharbawy, A. M. Emara, and S. E. Abu-Risha, "Zinc oxide nanoparticles and a standard antidiabetic drug restore the function and structure of beta cells in type-2 diabetes," *Biomedicine and Pharmacotherapy*, vol. 84, pp. 810–820, 2016.

[111] L. Ferrero-Miliani, O. H. Nielsen, P. S. Andersen, and S. E. Girardin, "Chronic inflammation: importance of NOD2 and NALP3 in interleukin-1 beta generation," *Clinical and Experimental Immunology*, vol. 147, no. 2, pp. 227–235, 2006.

[112] M. Boguniewicz and D. Y. Leung, "Atopic dermatitis: a disease of altered skin barrier and immune dysregulation," *Immunological Reviews*, vol. 242, no. 1, pp. 233–246, 2011.

[113] R. Jurakic Toncic and B. Marinovic, "The role of impaired epidermal barrier function in atopic dermatitis," *Acta Dermatovenerologica Croatica: ADC*, vol. 24, no. 2, pp. 95–109, 2016.

[114] C. Wiegand, U. C. Hipler, S. Boldt, J. Strehle, and U. Wollina, "Skin-protective effects of a zinc oxide-functionalized textile and its relevance for atopic dermatitis," *Clinical, Cosmetic and Investigational Dermatology*, vol. 2013, pp. 115–121, 2013.

[115] M. Ilves, J. Palomaki, M. Vippola et al., "Topically applied ZnO nanoparticles suppress allergen induced skin inflammation but induce vigorous IgE production in the atopic dermatitis mouse model," *Particle and Fibre Toxicology*, vol. 11, no. 1, p. 38, 2014.

[116] P. Thatoi, R. G. Kerry, S. Gouda et al., "Photo-mediated green synthesis of silver and zinc oxide nanoparticles using aqueous extracts of two mangrove plant species, *Heritiera fomes* and *Sonneratia apetala* and investigation of their biomedical applications," *Journal of Photochemistry and Photobiology B: Biology*, vol. 163, pp. 311–318, 2016.

[117] S. Yao, X. Feng, J. Lu et al., "Antibacterial activity and inflammation inhibition of ZnO nanoparticles embedded TiO_2 nanotubes," *Nanotechnology*, vol. 29, no. 24, article 244003, 2018.

[118] J. Q. Li, H. Q. Chen, B. Wang et al., "ZnO nanoparticles act as supportive therapy in DSS-induced ulcerative colitis in mice by maintaining gut homeostasis and activating Nrf2 signaling," *Scientific Reports*, vol. 7, article 43126, 2017.

[119] T. Xia, W. Lai, M. Han, M. Han, X. Ma, and L. Zhang, "Dietary ZnO nanoparticles alters intestinal microbiota and inflammation response in weaned piglets," *Oncotarget*, vol. 8, no. 39, pp. 64878–64891, 2017.

[120] P. Zhu, Z. Weng, X. Li et al., "Biomedical applications of functionalized ZnO nanomaterials: from biosensors to bioimaging," *Advanced Materials Interfaces*, vol. 3, no. 1, article 1500494, 2016.

[121] H. M. Xiong, Y. Xu, Q. G. Ren, and Y. Y. Xia, "Stable aqueous ZnO@polymer core-shell nanoparticles with tunable photoluminescence and their application in cell imaging," *Journal of the American Chemical Society*, vol. 130, no. 24, pp. 7522-7523, 2008.

[122] H. Jiang, H. Wang, and X. Wang, "Facile and mild preparation of fluorescent ZnO nanosheets and their bioimaging applications," *Applied Surface Science*, vol. 257, no. 15, pp. 6991–6995, 2011.

[123] X. Tang, E. S. G. Choo, L. Li, J. Ding, and J. Xue, "Synthesis of ZnO nanoparticles with tunable emission colors and their cell labeling applications," *Chemistry of Materials*, vol. 22, no. 11, pp. 3383–3388, 2010.

[124] P. Zhang and W. Liu, "ZnO QD@PMAA-co-PDMAEMA nonviral vector for plasmid DNA delivery and bioimaging," *Biomaterials*, vol. 31, no. 11, pp. 3087–3094, 2010.

[125] W. Wu, J. Shen, P. Banerjee, and S. Zhou, "A multifuntional nanoplatform based on responsive fluorescent plasmonic

ZnO-Au@PEG hybrid nanogels," *Advanced Functional Materials*, vol. 21, no. 15, pp. 2830–2839, 2011.

[126] H. J. Zhang, H. M. Xiong, Q. G. Ren, Y. Y. Xia, and J. L. Kong, "ZnO@silica core-shell nanoparticles with remarkable luminescence and stability in cell imaging," *Journal of Materials Chemistry*, vol. 22, no. 26, pp. 13159–13165, 2012.

[127] N. H. Cho, T. C. Cheong, J. H. Min et al., "A multifunctional core–shell nanoparticle for dendritic cell-based cancer immunotherapy," *Nature Nanotechnology*, vol. 6, no. 10, pp. 675–682, 2011.

[128] K. Matsuyama, N. Ihsan, K. Irie, K. Mishima, T. Okuyama, and H. Muto, "Bioimaging application of highly luminescent silica-coated ZnO-nanoparticle quantum dots with biotin," *Journal of Colloid and Interface Science*, vol. 399, pp. 19–25, 2013.

[129] H. Hong, J. Shi, Y. Yang et al., "Cancer-targeted optical imaging with fluorescent zinc oxide nanowires," *Nano Letters*, vol. 11, no. 9, pp. 3744–3750, 2011.

2

Structural Properties and Reactive Site Selectivity of Some Transition Metal Complexes of 2,2′(1E,1′E)-(ethane-1,2-diylbis (azan-1-yl-1-ylidene))bis (phenylmethan-1-yl-1-ylidene) dibenzoic Acid: DFT, Conceptual DFT, QTAIM, and MEP Studies

Fritzgerald Kogge Bine, Nyiang Kennet Nkungli, Tasheh Stanley Numbonui, and Julius Numbonui Ghogomu

Department of Chemistry, Faculty of Science, Research Unit of Noxious Chemistry and Environmental Engineering, University of Dschang, P.O. Box 67, Dschang, Cameroon

Correspondence should be addressed to Julius Numbonui Ghogomu; ghogsjuju@hotmail.com

Academic Editor: Konstantinos Tsipis

Herein is presented a density functional theory (DFT) study of reactivity and structural properties of transition metal complexes of the Schiff base ligand 2,2′(1E,1′E)-(ethane-1,2-diylbis(azan-1-yl-1-ylidene))bis(phenylmethan-1-yl-1-ylidene)dibenzoic acid (hereafter denoted EDA2BB) with Cu(II), Mn(II), Ni(II), and Co(II). The quantum theory of atoms-in-molecules (QTAIM), conceptual DFT, natural population analysis (NPA), and molecular electrostatic potential (MEP) methods have been used. Results have revealed a distorted octahedral geometry around the central metal ion in each gas phase complex. In the DMSO solvent, a general axial elongation of metal-oxygen bonds involving ancillary water ligands has been observed, suggestive of loosely bound water molecules to the central metal ion that may be acting as solvent molecules. Weak, medium, and strong intramolecular hydrogen bonds along with hydrogen-hydrogen and van der Waals interactions have been elucidated in the complexes investigated via geometric and QTAIM analyses. From the chemical hardness values, the complex $[Co(EDA2BB)(OH_2)_2]$ is the hardest, while $[Cu(EDA2BB)(OH_2)_2]$ is the softest. Based on the global electrophilicity index, the complexes $[Ni(EDA2BB)(OH_2)_2]$ and $[Cu(EDA2BB)(OH_2)_2]$ are the strongest and weakest electrophiles, respectively, among the complexes studied. In conclusion, the reactivity of the complexes is improved vis-à-vis the ligand, and stable geometries of the complexes are identified, alongside their prominent electrophilic and nucleophilic sites.

1. Introduction

The development of resistance to currently used pharmaceutical drugs by pathogens is a direct consequence of drug abuse, thereby necessitating the search for new and more potent chemical agents with increased therapeutic properties against disease-causing microorganisms [1]. This also explains the appearance of superbugs which are drug-resistant strains of microorganisms that only respond to the most powerful antibiotics, or in some cases, to no antibiotics. In the search for new drugs, transition metals have been identified to possess unique biological activities on pathogenic organisms and thus provide a pathway for the synthesis of new drugs [2].

Transition metal complexes derived from salen-type ligands are of particular interest in medicinal chemistry for the development of antimicrobial agents. Indeed, Schiff base compounds and their metal complexes are outstanding in the domain of metal-based drugs. They have been extensively investigated in this perspective firstly because of their wide range of applications in medicine [3]. Secondly, much interest has been dedicated to their study because of their synthetic flexibility, selectivity, and sensitivity towards the central metal atom/ion. Interestingly, the structural similarities of Schiff base reagents (Schiff base ligands and their metal complexes) to biomolecules facilitate elucidation of the biological mechanisms of transformation and racemization reactions [4].

Recently, Al-Shemary and Zaidan [5] reported the successful synthesis and characterization of a new tetradented Schiff base ligand known as 2,2′(1E,1′E)-(ethane-1,2-diylbis(azan-1-yl-1-ylidene))bis(phenylmethan-1-yl-1-ylidene)dibenzoic acid (EDA2BB) and its Cu^{II}, Ni^{II}, Mn^{II}, Co^{II}, and Hg^{II} complexes. Although EDA2BB and its metal complexes showed good antibacterial activities on some selected targets, a comprehensive analysis of their geometric and structural properties, as well as their chemical reactivity and stability, has not been carried out till date, to the best of our knowledge. To address these issues, we embarked on a quantum chemical study of the structural properties and reactive site selectivity of EDA2BB and its metal complexes. The success of the molecular docking of a drug into its receptor site, which usually involves an interplay of precise molecular interactions, depends greatly on the geometry, conformation, and electronic properties of the two molecules [6]. Therefore, our current theoretical study on the transition metal complexes of EDA2BB is a particularly important prerequisite for any future docking investigations using these molecules as "Ligands".

The main objective of the present study was to perform a geometric and structural analysis on some transition metal complexes of the ligand EDA2BB, comprising $[Cu(EDA2BB)(OH_2)_2]$ (**A**), $[Ni(EDA2BB)(OH_2)_2]$ (**B**), $[Mn(EDA2BB)(OH_2)_2]$ (**C**), and $[Co(EDA2BB)(OH_2)_2]$ (**D**), using the density functional theory (DFT) method and the quantum theory of atoms-in-molecules (QTAIM) analysis. The general molecular structure of these complexes is depicted in Figure 1. Thereafter, the chemical reactivity and reactive site selectivity of these complexes were elucidated via conceptual DFT, molecular electrostatic potential (MEP), and natural population analysis (NPA) studies. The choice of the DFT over the Hatree–Fock and semiempirical methods was motivated by its good compromise between cost (with respect to computational time and hardware requirements) and accuracy [7]. Today, parallel with the impressive development of computational hardware, the quantum chemical software required to perform DFT calculations has progressed to the state where calculations can be performed with high efficiency and in a user-friendly manner [8].

2. Computational Details

All theoretical calculations were carried out with the ORCA 3.0.3 computational package [9]. The input files were prepared using Avogadro 1.1.1 [10]. Geometry optimization and frequency calculations were performed using the DFT method in conjunction with the def2-TZVP(-f) Ahlrichs basis set [11]. The GGA functional BP86 [12, 13] was chosen for these calculations because it provides excellent geometries and vibrational frequencies [8]. To speed-up the geometry optimization and frequency calculations, but with marginal loss in accuracy, the resolution-of-the-identity (RI-J) approximation [14] was used in conjunction with appropriate auxiliary basis sets. Solvent effects were simulated using the conductor-like screening model (COSMO) [15] that is well-integrated into the ORCA program system. Dimethylsulfoxide (DMSO) was used as a solvent in this work in order to employ the same solvent as that used by Al-Shemary and Zaiden [5] in their experimental studies. Furthermore, long-range dispersion interactions were incorporated via Grimme's atom-pairwise dispersion correction using the Becke–Johnson damping scheme (D3BJ) [16, 17], since the pure exchange-correlation functionals fail to properly account for such interactions.

The geometries of the Cu(II), Co(II), and Mn(II) complexes of EDA2BB were optimized using the Unrestricted Kohn–Sham (UKS) formalism, while the Ni(II) complex was optimized using the Restricted Kohn–Sham (RKS) formalism. No constraints on symmetry, bond lengths, bond angles, or dihedral angles were applied in the geometry optimization calculations. In order to be sure that we obtained reliable and stable structures (minima on the PES), vibrational frequencies were calculated for all optimized structures, and their analyses showed no imaginary frequencies.

In the implementation of conceptual DFT for the evaluation of ionization energies and electron affinities, the hybrid *Meta*-GGA functional, M06 [18] was used for single point energy calculations based on the BP86-optimized geometries. This functional was chosen because it incorporates an inbuilt dispersion correction and performs well for energy calculations [18]. The M06 functional was employed along with the basis set ma-def2-TZVP for the metal ions and ma-def2-SVP for every other element. To speed-up these calculations, the RI-J approximation was combined with the chain-of-spheres (COSX) approximation, giving rise to the RIJCOSX approximation [19]. Throughout our ORCA calculations, numerical quadrature grids of at least 4 were used. NPA calculations were performed on the geometries optimized at BP86/def2-TZVP(-f) level, with the aid of the JANPA computational package [20].

3. Results and Discussion

3.1. Structural and Geometric Analysis. The input geometries to ORCA consisted of octahedral complexes, each containing one EDA2BB ligand and two water molecules serving as ancillary ligands to a transition metal(II) ion, as previously suggested by Al-Shemary and Zaiden [5]. Selected geometric parameters (bond lengths, bond angles, and dihedral or torsional angles) of the complexes investigated herein are listed in Table 1. The optimized geometries of the complexes are shown in Figure 2, as visualized using the Chemcraft 1.8 visualization software [21].

It can be seen from Table 1 that nearly all metal-ligand (M-L) bond lengths calculated in both gas phase and DMSO fall within the range of typical M-L bond distances, 1.849–2.087 au [22]. The relatively large values of the M-O_{55} and M-O_{58} bond lengths in the complexes currently investigated, 2.888–3.531 au suggest that the water ligands are loosely bound to the central metal ions. A general axial elongation of the M-O bonds involving the ancillary water ligands is observed in DMSO, suggesting that the water ligands are much more loosely bound to the central metal ions in the solvent phase and may instead be acting as

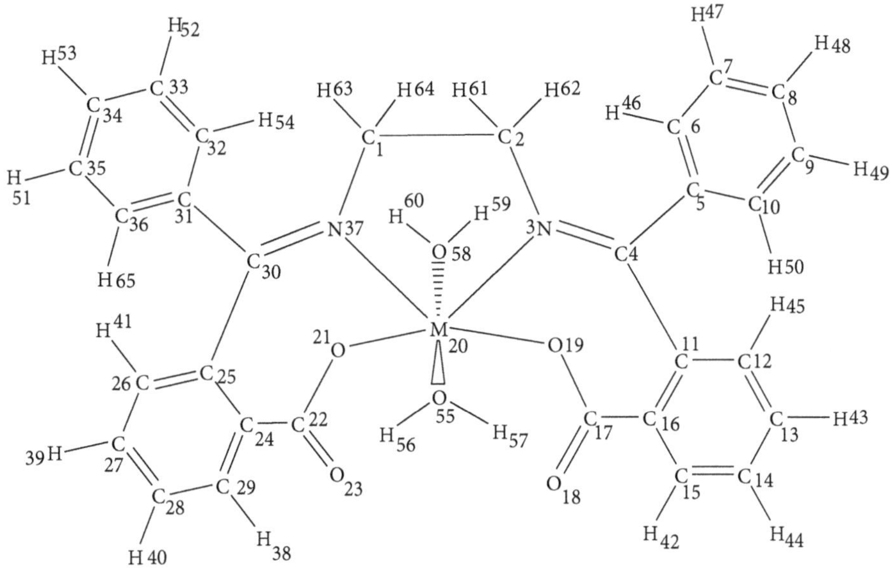

FIGURE 1: Input geometries with formula $[\mathbf{M}(\mathbf{N_2O_4C_{30}H_{22}})(\mathbf{OH_2})_2]$, where $\mathbf{M}_{20}$ = **Cu(II)**, **Ni(II)**, **Mn(II)** and **Co(II)**.

TABLE 1: Selected optimized geometric parameters of the complexes investigated in gas phase and DMSO.

Geometric parameters	Gas phase				DMSO			
	A	B	C	D	A	B	C	D
Bond lengths (Å)								
$\mathbf{M_{20}}$-$\mathbf{N_{37}}$	1.973	1.849	1.912	1.864	1.972	1.845	1.911	1.879
$\mathbf{M_{20}}$-$\mathbf{N_3}$	2.087	1.913	1.997	1.935	2.066	1.895	1.981	1.900
$\mathbf{O_{21}}$-$\mathbf{M_{20}}$	1.965	1.857	1.943	1.918	2.022	1.877	2.019	1.923
$\mathbf{O_{19}}$-$\mathbf{M_{20}}$	2.015	1.890	2.021	1.981	1.975	1.872	2.045	1.917
$\mathbf{O_{55}}$-$\mathbf{M_{20}}$	2.452	3.268	2.203	2.346	2.288	3.336	2.103	2.203
$\mathbf{O_{58}}$-$\mathbf{M_{20}}$	2.846	2.888	2.138	2.360	3.243	3.531	2.168	3.463
Bond angles (°)								
$\mathbf{O_{21}}$-$\mathbf{M_{20}}$-$\mathbf{N_{37}}$	89.4	91.0	90.5	89.9	87.3	90.9	89.2	91.6
$\mathbf{N_3}$-$\mathbf{M_{20}}$-$\mathbf{O_{19}}$	98.2	99.3	95.2	99.9	98.1	99.8	97.8	93.9
$\mathbf{N_3}$-$\mathbf{M_{20}}$-$\mathbf{N_{37}}$	81.3	82.7	84.2	84.5	81.5	83.4	84.7	83.3
$\mathbf{O_{58}}$-$\mathbf{M_{20}}$-$\mathbf{O_{19}}$	77.1	78.8	87.2	82.5	65.5	58.1	83.3	55.0
$\mathbf{O_{55}}$-$\mathbf{M_{20}}$-$\mathbf{O_{19}}$	74.8	63.4	77.3	75.4	95.0	61.0	87.5	99.8
Dihedral angles (°)								
$\mathbf{O_{21}}$-$\mathbf{M_{20}}$-$\mathbf{N_{37}}$-$\mathbf{C_{30}}$	45.5	66.0	40.6	49.3	52.8	67.8	42.7	61.3
$\mathbf{O_{19}}$-$\mathbf{M_{20}}$-$\mathbf{N_3}$-$\mathbf{C_4}$	−41.3	−43.9	−26.6	−33.0	−50.3	−39.9	−32.9	−54.7
$\mathbf{O_{21}}$-$\mathbf{M_{20}}$-$\mathbf{N_{37}}$-$\mathbf{C_1}$	−118.4	−119.7	−135.4	−127.7	−117.2	−119.7	−135.0	−119.3
$\mathbf{O_{19}}$-$\mathbf{M_{20}}$-$\mathbf{N_3}$-$\mathbf{C_2}$	147.3	144.0	164.1	153.7	140.2	147.4	159.3	129.2
$\mathbf{M_{20}}$-$\mathbf{O_{21}}$-$\mathbf{C_{22}}$-$\mathbf{C_{24}}$	−4.9	−11.7	−5.6	−9.8	−9.1	−12.6	−10.3	−25.2
$\mathbf{M_{20}}$-$\mathbf{O_{19}}$-$\mathbf{C_{17}}$-$\mathbf{C_{16}}$	90.1	90.1	103.5	95.1	78.8	78.0	97.1	27.0
$\mathbf{C_1}$-$\mathbf{N_{37}}$-$\mathbf{C_{30}}$-$\mathbf{C_{31}}$	11.2	−1.4	23.9	18.8	11.1	−1.3	23.1	−6.5
$\mathbf{C_2}$-$\mathbf{N_3}$-$\mathbf{C_4}$-$\mathbf{C_5}$	2.5	8.5	6.1	3.3	3.0	6.0	5.3	−0.7
Hydrogen bond lengths (Å)								
$\mathbf{H_{56}}\cdots\mathbf{O_{19}}$	1.917	1.996	1.978	1.879	—	2.928	3.103	—
$\mathbf{H_{59}}\cdots\mathbf{O_{18}}$	1.920	1.998	1.673	1.833	2.045	2.052	1.743	—
$\mathbf{H_{60}}\cdots\mathbf{O_{21}}$	2.105	2.338	2.004	2.010	2.024	2.061	2.158	1.960
$\mathbf{O_{19}}\cdots\mathbf{H_{59}}$	—	—	—	—	—	—	—	1.960
$\mathbf{N_3}\cdots\mathbf{H_{57}}$	3.317	—	2.655	2.962	—	—	—	—
Hydrongen bond angles (°)								
$\mathbf{O_{55}}$-$\mathbf{H_{56}}$-$\mathbf{O_{19}}$	138.1	164.1	122.1	133.5	—	99.8	—	—
$\mathbf{O_{18}}$-$\mathbf{H_{59}}$-$\mathbf{O_{58}}$	160.9	162.6	157.8	155.5	173.6	175.6	156.6	
$\mathbf{O_{58}}$-$\mathbf{H_{60}}$-$\mathbf{O_{21}}$	138.0	135.8	114.1	123.0	163.2	166.2	110.2	146.4
$\mathbf{N_3}$-$\mathbf{H_{57}}$-$\mathbf{O_{55}}$	90.1	—	97.8	91.2	—	—	—	—
$\mathbf{O_{19}}$-$\mathbf{H_{59}}$-$\mathbf{O_{58}}$	—	—	—	—	—	—	—	146.8

M = Cu(II), Ni(II), Mn(II), and Co(II). For atom numbering, refer to Figure 1.

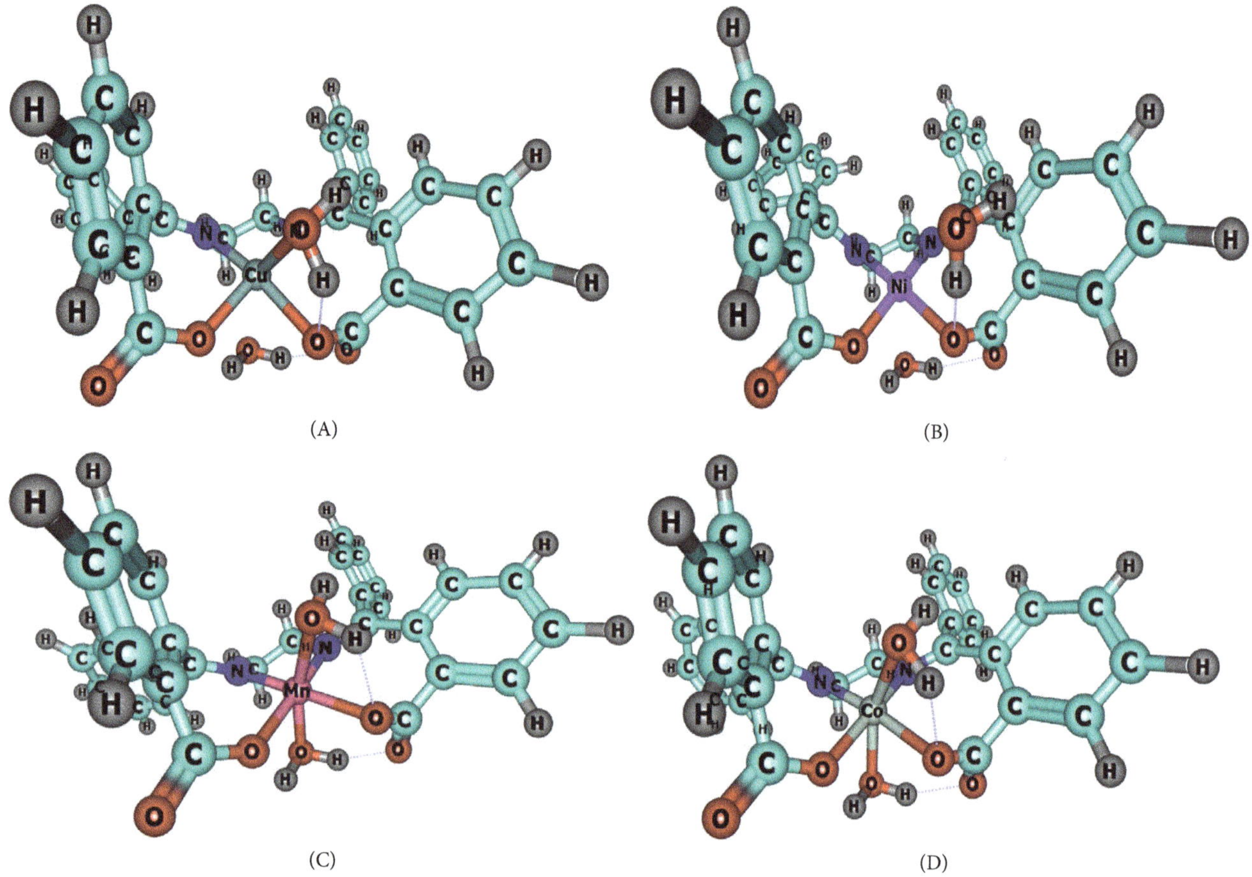

FIGURE 2: The optimized geometries of $[\mathbf{Cu(EDA2BB)(OH_2)_2}]$ (**A**), $[\mathbf{Ni(EDA2BB)(OH_2)_2}]$ (**B**), $[\mathbf{Mn(EDA2BB)(OH_2)_2}]$ (**C**), and $[\mathbf{Co(EDA2BB)(OH_2)_2}]$ (**D**) at the RI-BP86-D3(BJ)/def2-TZVP(-f) level of theory in gas phase.

solvent molecules. This is probably caused by some intermolecular interactions between the water molecules and the DMSO solvent molecules. Moreover, ligand substitution reactions in which the water ligands are replaced by DMSO molecules may occur.

The M-L bond angles indicate distorted octahedral geometries around the central metal ions in the complexes studied. These distortions are probably imposed by the chelate rings. If the complexes under investigation were to be perfectly octahedral, all of the dihedral angles listed in Table 1 will be 0° each, except the angles O_{21}-M-N_{37}-C_1 and O_{19}-M-N_3-C_2 which will be 180° instead. Upon inspection of Table 1, however, the dihedral angles O_{21}-M-N_{37}-C_1 and O_{19}-M-N_3-C_2 are found to have values in the range -117.2° to -135.4° and 129.2° to 164.1° respectively, which deviate significantly from 180°. Moreover, the values of the other torsional angles show a maximum deviation of 103.5° from 0°. These results confirm that the optimized structures of the complexes investigated are distorted octahedrons.

The intramolecular hydrogen bond (HB) parameters in the complexes studied, which were determined based on the categorization of HBs by Jeffrey [23], are also listed in Table 1. Geometric cutoff limits for HB interactions (herein denoted X–H···A, where X–H is the donor and A is the acceptor) were employed in identifying potential hydrogen bonding cases. The cut-off limits used in this paper are H···A distances up to 3.2 Å and X–H···A angles $\geq 90^\circ$. From Table 1, it can be seen that the HB lengths in the complexes optimized in the gas phase are shorter than those in the complexes optimized in the solvent phase, implying that the HBs in the gas phase complexes could be stronger than those in the solvent phase counterparts. It is worth noting that the HB donors of all the intramolecular HBs described in Table 1 are the water ligands. Hence, the elongation of the HBs in the solvent phase can be attributed to some intermolecular HB interactions between the water Ligands and the polar solvent (DMSO) molecules, which probably act as the HB acceptors. Based on the aforementioned geometric cut-off limits for HB interactions, a complete disappearance of the HB interactions ongoing from the gas phase to the solvent phase is observed in some complexes, notably the Co(II), Mn(II), and Cu(II) complexes.

3.2. Infrared Spectroscopy. In the realm of theoretical chemistry, frequency calculations not only characterize stationary points (as minima, transition states, etc.), but are also used to predict the IR spectra of molecules. In this regard, the IR spectra of the molecules currently investigated have been calculated, and some pertinent vibrational frequencies of these molecules are listed in Table 2. Also listed in Table 2 are the experimentally determined vibrational

TABLE 2: Calculated IR frequencies (scaled) at RI-BP86/def2-TZVP level, along with the corresponding experimental IR frequencies for EDA2BB and its metal(II) complexes.

Vibrational assignment	A		B		C		D		EDA2BB	
	Calc.	Exp.	Calc.	Exp.	Calc.	Exp.	Calc.	Exp.	Calc.	Exp.
ν(OH)	3386	3367	3451.7	3423	3392	3321	3363	3482	3113	2520–3337
$\nu(\text{C-H})_{\text{aroma.}}$	3080	3042	3081	3082	3075	3026	3079	3018	3081	3020
$\nu(\text{C-H})_{\text{aliph.}}$	2925	2910	2960	2904	2966	2993	2972	2964	2962	2993
$\nu(\text{HC=N})_{\text{imine}}$	1601	1609	1598	1614	1582	1604	1570	1612	1611	1627
$\nu(\text{C=O})_{\text{asym}}$	1611	1533	1628	1519	1644	1529	1636	1529	1649	—
ν(M-O)	368	432	387	412	401	418	413	421	—	—
ν(M-N)	512	520	506	493	493	499	477	489	—	—

Calc. stands for the calculated and scaled IR wavenumbers. Exp. represents the experimentally determined counterparts.

frequencies obtained from Al-Shemary and Zaiden [5]. The theoretical and experimental IR frequencies are compared in this work as a means of determining the credibility of the level of theory used in geometry optimization. The computed IR spectra have also been used to provide some insights into geometric structures of the molecules studied.

Usually, calculated IR frequencies are larger than the corresponding experimental values. To obtain a good agreement between the experimental and the calculated wavenumbers, the latter are usually corrected with appropriate scale or correction factors [24]. In the present study, the scale factor 0.9953, which is suitable for correcting wavenumber computed at the BP86/def2-TZVP level of theory [25], has been used.

To determine the validity of the calculated frequencies with respect to the experimental values, the correlation Equation (1) was established:

$$\nu_{\text{calc.}} = 0.980\nu_{\text{exp.}} + 27.20, \qquad (1)$$

where $\nu_{\text{calc.}}$ and $\nu_{\text{exp.}}$ represent the calculated (scaled) and the experimental IR frequencies, respectively. The correlation coefficient ($R^2 = 0.999$) obtained between the theoretical and experimental FT-IR frequencies shows a good linear relationship between these values, thus affirming the suitability of the level of theory used for geometry optimization. In particular, a good agreement is found to exist between the O-H vibrational frequencies of the water ligands and the experimental values. Good agreements between theoretical and experimental frequencies are also found to exist for the aliphatic C-H, aromatic C-H, the azomethine C=N, and M-N vibrations.

The IR spectrum of the ligand is found to be more or less modified by metal coordination. Upon complexation, the emergence of O-H bands due to the water ligands is observed in the IR spectrum of EDA2BB in the region 3300–3600 cm^{-1}. Furthermore, IR bands due to the metal-ligand (Cu-O and Cu-N) vibrations also appeared in the spectrum of EDA2BB in the range 350–700 cm^{-1}. Moreover, the strengthening of the carbonyl and the imine IR bands of EDA2BB was observed after complexation, which confirms that the ligand coordinates to the central metal ions via the carbonyl oxygen and the azomethine nitrogen. The disappearance of the IR bands corresponding to the O-H vibrations of EDA2BB indicates the loss of its carboxylic protons upon complexation with the metal ions.

3.3. QTAIM Analyses on the Complexes Investigated. To ascertain the existence of HBs alongside other interactions such as van der Waals interactions in the complexes studied, Bader's quantum theory of atoms-in-molecules (QTAIM) [26] approach was employed. QTAIM analyses on the complexes were carried out using the Multiwfn 3.3.9 software [27]. It is worthy of note that the nature of the chemical bonds in transition metal complexes remains an area of ongoing research [28]. Critical points were searched via topology analysis of the electron density. In topology analysis language, the points at which the gradient of the electron density goes to zero (except at infinity) are known as critical points (CPs). In this paper, bond CPs (BCP) were of most interest. The Pointcaré–Hopf relationship was satisfied each time a critical point search was performed, indicating that all CPs may have been found. Figure 3 shows the molecular graphs of the EDA2BB-metal complexes studied, based on their gas phase optimized geometries at RI-BP86-D3(BJ)/def2-TZVP(-f) level of theory.

Bader's QTAIM analysis [26] has become the first choice tool used by quantum chemists to analyze the nature and strengths of bonding interactions. In the present study, topological analyses of the electron density and its Laplacian have been performed at bond critical points. Large $\rho(r)$ values and $\nabla^2\rho(r) < 0$ indicate polar and nonpolar covalent bonding interactions, whereas small $\rho(r)$ values and $\nabla^2\rho(r) > 0$ indicate closed-shell interactions [29]. Generally, $\rho(r)$ is greater than 0.20 au for covalent bonding interactions and less than 0.10 au for closed-shell interactions [29]. The values of $\nabla^2\rho(r)$ and $-G(r)/V(r)$ were also used to characterize the bonding interactions in the complexes of EDA2BB investigated. Generally, $G(r)$ is the kinetic energy density at the BCP (always positive), and $V(r)$ is the potential energy density at the BCP (always negative) [30]. When $\nabla^2\rho(r) > 0$ and $-G(r)/V\ (r) > 1$, the interactions are noncovalent, whereas when $\nabla^2\rho(r) > 0$ and $0.5 < -G(r)/V(r) < 1$, the interactions are partially or partly covalent [29].

In the context of QTAIM analysis, Popelier [31] developed some useful criteria for characterizing HBs, which have been exploited in this work. According to Popelier, the formation an HB depends on the electron density and its Laplacian at the BCP, which should lie in the range of 0.002–0.040 au and 0.024–0.139 au, respectively. Espinosa [32] formulated an equation of the form $E_{\text{int}} = 0.5V(r)$, which is useful for estimating weak interatomic interaction

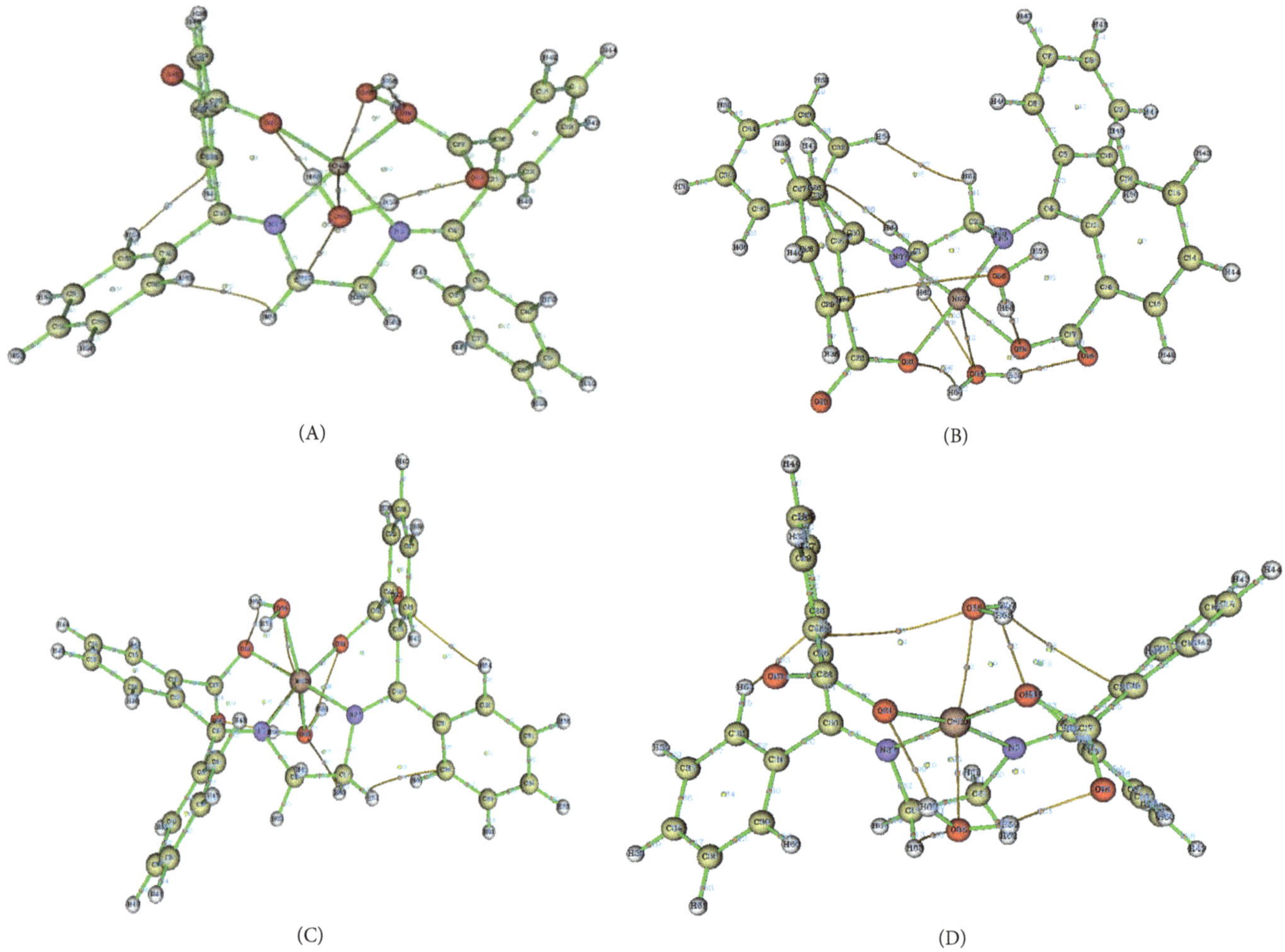

FIGURE 3: Molecular graphs of EDA2BB complexes computed based on their gas phase optimized geometries at the BP86-D3(BJ)/def2-TZVP(-f) level of theory. The BCPs are the small red spheres around the centre of the bond paths. The bond paths are the brown or green colored lines.

energies, particularly HB energies. In this equation, $V(r)$ is the potential energy density.

The computed parameters at some BCPs in the complexes studied are listed in Table 3. It is clear from this table that all $\nabla^2\rho(r)$ values are positive, except that for the interaction $H_{64}\cdots C_{36}$ in C (−0.584 au) which is negative. Furthermore, the $\rho(r)$ value for this interaction is greater than 0.20 au, but less than 0.10 au for all the other interactions in this complex. These results show that the $H_{64}\cdots C_{36}$ in C is the only covalent interaction among the interactions list in Table 3. This conclusion is further supported by the large interaction energy (−80.949 kcal/mol) for the $H_{64}\cdots C_{36}$ interaction, as compared to the interaction energies of the other interactions, which are in the range −0.314 to −67.143 kcal/mol.

Based on the interaction energies listed in Table 3 and on Jeffrey's classification of HBs [23], the conventional HBs in the complexes under investigation have been classified as weak (0–4 kcal/mol), medium (4–14 kcal/mol), or strong (14–40 kcal/mol). For the complex C, the HBs $H_{63}\cdots O_{58}$, $H_{56}\cdots O_{19}$, and $H_{59}\cdots O_{18}$ are weak, medium, and strong, respectively. For the complexes A and B, the HBs $H_{63}\cdots O_{58}$ and $O_{21}\cdots H_{60}$ are weak, whereas $H_{59}\cdots O_{18}$ and $H_{56}\cdots O_{19}$ are medium. For D, the HB $H_{63}\cdots O_{58}$ is weak, while $H_{59}\cdots O_{18}$, $H_{56}\cdots O_{19}$, and $O_{21}\cdots H_{60}$ are medium. Additionally, hydrogen-hydrogen (H–H) interactions have been found to exist in B ($H_{54}\cdots H_{61}$) and D ($H_{54}\cdots H_{41}$). Furthermore, some interactions of the form C$\cdots$H in the complexes studied have been identified as intramolecular van der Waal's interactions, since their interaction energies (−0.914 to −2.510 kcal/mol) are lower than the interaction energies of virtually all of the hydrogen bonds identified.

3.4. Electronic Properties

3.4.1. Natural Population Analysis (NPA). To determine the charge redistribution and the extent of metal-ligand charge transfer in the complexes investigated, NPA charges (also known as natural charges) were computed at BP86-D3 (BJ)/def2-TZVP(-f) level of theory in gas phase. The natural charges on selected atoms of the investigated compounds are listed in Table 4.

It is clear from Table 4 that the formal charge (+2) on each central metal ion decreases upon complexation with EDA2BB. The formal charge on Ni^{2+}, Cu^{2+}, Mn^{2+}, and Co^{2+}

TABLE 3: Bonding interactions and values of the electron density [$\rho(r)$] and its Laplacian [$\nabla^2\rho(r)$], kinetic energy density [$G(r)$], potential energy density [$V(r)$], and interatomic interaction energy (E_{int}) for the noncovalent and covalent bonding interactions in the complexes studied.

Complex	Bonding interaction	$\rho(r)$	$\nabla^2\rho(r)$	$G(r)$	$V(r)$	$-G(r)/V(r)$	E_{int} (kcal/mol)
A	$H_{54}\cdots C_{26}$	0.011	0.036	0.008	−0.006	1.239	−1.883
	$H_{64}\cdots C_{36}$	0.012	0.043	0.009	−0.008	1.213	−2.510
	$H_{63}\cdots O_{58}$	0.018	0.061	0.013	−0.011	1.19	−3.451
	$H_{56}\cdots O_{19}$	0.030	0.102	0.025	−0.025	1.009	−7.844
	$H_{59}\cdots O_{18}$	0.028	0.087	0.021	−0.021	1.022	−6.589
	$O_{21}\cdots H_{60}$	0.017	0.071	0.015	−0.012	1.23	−3.765
	$Cu_{20}\cdots O_{55}$	0.030	0.119	0.030	−0.031	0.983	−9.726
	$Cu_{20}\cdots O_{58}$	0.015	0.044	0.011	−0.011	0.997	−3.451
	$Cu_{20}\cdots N_{3}$	0.072	0.267	0.084	−0.101	0.831	−31.689
	$Cu_{20}\cdots N_{37}$	0.094	0.346	0.116	−0.144	0.8	−45.181
	$Cu_{20}\cdots O_{19}$	0.074	0.324	0.098	−0.116	0.849	−36.395
	$Cu_{20}\cdots O_{21}$	0.086	0.364	0.114	−0.136	0.834	−42.671
B	$H_{54}\cdots C_{26}$	0.011	0.036	0.008	−0.006	1.239	−1.883
	$H_{64}\cdots C_{36}$	0.012	0.043	0.009	−0.008	1.213	−2.510
	$H_{54}\cdots H_{61}$	0.006	0.001	0.004	−0.003	1.295	−0.941
	$H_{64}\cdots C_{31}$	0.014	0.046	0.010	−0.008	1.181	−2.510
	$H_{63}\cdots O_{58}$	0.013	0.042	0.009	−0.007	1.262	−2.196
	$H_{56}\cdots O_{19}$	0.024	0.078	0.018	−0.016	1.092	−5.020
	$H_{59}\cdots O_{18}$	0.023	0.076	0.017	−0.015	1.111	−4.706
	$O_{21}\cdots H_{60}$	0.011	0.044	0.009	−0.007	1.327	−2.196
	$C_{24}\cdots O_{55}$	0.003	0.011	0.002	−0.001	1.611	−0.314
	$Ni_{20}\cdots O_{58}$	0.015	0.043	0.011	−0.012	0.962	−3.765
	$Ni_{20}\cdots N_{3}$	0.106	0.450	0.143	−0.174	0.823	−54.593
	$Ni_{20}\cdots N_{37}$	0.125	0.520	0.172	−0.214	0.803	−67.143
	$Ni_{20}\cdots O_{19}$	0.097	0.494	0.147	−0.171	0.862	−53.652
C	$Ni_{20}\cdots O_{21}$	0.111	0.519	0.160	−0.190	0.842	−59.613
	$H_{54}\cdots C_{26}$	0.013	0.041	0.009	−0.007	1.200	−2.196
	$H_{64}\cdots C_{36}$	0.246	−0.584	0.056	−0.258	0.217	−80.949
	$H_{63}\cdots O_{58}$	0.012	0.050	0.010	−0.008	1.253	−2.510
	$H_{56}\cdots O_{19}$	0.027	0.108	0.025	−0.023	1.084	−7.216
	$H_{59}\cdots O_{18}$	0.051	0.113	0.039	−0.050	0.779	−15.688
	$O_{21}\cdots O_{58}$	0.026	0.121	0.027	−0.024	1.129	−7.530
	$Mn_{20}\cdots O_{55}$	0.048	0.216	0.059	−0.064	0.921	−20.080
	$Mn_{20}\cdots O_{58}$	0.056	0.276	0.075	−0.081	0.929	−25.414
	$Mn_{20}\cdots N_{3}$	0.091	0.391	0.115	−0.132	0.871	−41.416
	$Mn_{20}\cdots N_{37}$	0.115	0.465	0.144	−0.172	0.838	−53.966
	$Mn_{20}\cdots O_{19}$	0.073	0.379	0.104	−0.112	0.922	−35.140
	$Mn_{20}\cdots O_{21}$	0.092	0.468	0.130	−0.144	0.907	−45.181
D	$H_{54}\cdots H_{41}$	0.012	0.040	0.009	−0.007	1.225	−2.196
	$H_{57}\cdots C_{11}$	0.006	0.020	0.004	−0.003	1.336	−0.914
	$H_{63}\cdots O_{58}$	0.016	0.059	0.013	−0.010	1.22	−3.138
	$H_{56}\cdots O_{19}$	0.033	0.114	0.029	−0.029	0.985	−9.099
	$H_{59}\cdots O_{18}$	0.034	0.100	0.027	−0.029	0.927	−9.099
	$O_{21}\cdots H_{60}$	0.024	0.102	0.023	−0.020	1.141	−6.275
	$C_{24}\cdots O_{55}$	0.007	0.028	0.005	−0.004	1.386	−1.255
	$Co_{20}\cdots O_{58}$	0.037	0.143	0.039	−0.043	0.916	−13.491
	$Co_{20}\cdots N_{3}$	0.100	0.468	0.142	−0.166	0.852	−52.083
	$Co_{20}\cdots N_{37}$	0.121	0.545	0.172	−0.207	0.829	−64.947
	$Co_{20}\cdots O_{19}$	0.077	0.406	0.116	−0.130	0.889	−40.788
	$Co_{20}\cdots O_{21}$	0.094	0.471	0.138	−0.158	0.872	−49.573
	$Co_{20}\cdots O_{55}$	0.037	0.153	0.041	−0.045	0.93	−14.119

decreased to +0.9308, +1.2323, +0.9438, and +1.0003, respectively. From these results, it is obvious that electron transfer from the ligand to each central metal ion occurred during complex formation. Clearly, these metal-ligand charge transfers are substantial in the Ni(II) and Mn(II) complexes, and less significant in the Cu(II) and Co(II) complexes. Note that in metal complexes, the more the reduction in polarity of the central metal, the more delocalized are π-electrons over the chelate ring, and this enhances the lipophilicity of the complexes [33]. This increased lipophilicity in turn enhances the penetration of the complexes into lipid membranes and the blocking of the metal binding sites in some enzymes found in microorganisms [33]. From this viewpoint, it can be affirmed from Table 4 that the Ni(II) and Mn(II) complexes of EDA2BB are potentially highly lipophilic toward the penetration of the cell membranes of microorganisms, while the Cu(II) and Co(II) complexes are least penetrating.

TABLE 4: NPA charges on the ligand donor atoms and the central metal ions in the complexes investigated, computed at BP86-D3 (BJ)/def2-TZVP(-f) level of theory in gas phase.

Atom/metal ion	A	B	C	D
M_{20}	1.2323	0.9308	0.9438	1.0003
N_3	−0.4590	−0.4023	−0.3932	−0.3837
N_{37}	−0.4502	−0.3540	−0.3452	−0.3457
O_{19}	−0.7521	−0.7049	−0.6761	−0.7007
O_{21}	−0.7058	−0.6312	−0.6514	−0.6663
O_{55}	−0.9494	−0.9434	−0.8375	−0.9144
O_{58}	−0.9840	−0.9715	−0.8639	−0.9531

M = Cu(II), Ni(II), Mn(II), and Co(II).

3.4.2. Global Reactivity Descriptors (GRDs). The stability and chemical reactivity of the complexes investigated were studied via conceptual density functional theory (CDFT), using the global reactivity descriptors. The energies necessary for the computation of these global reactivity descriptors were obtained from single-point calculations at M06/ma-def2-TZVP level of theory for the metal ions and at M06/ma-def2-SVP level of theory for every other element, based on the optimized geometries of the complexes. The Minnesota functional, M06 has been chosen here because it has been parameterized to provide accurate energies of transition metal complexes [18].

Within the framework of CDFT, global reactivity descriptors such as Chemical potential (μ), Electronegativity index (χ), Chemical hardness (η), and Chemical softness (S) have been defined. A detailed description of these reactivity descriptors can be found in our previous publication [34]. Note that all of these reactivity parameters are related to the response of chemical systems to changes in either their number of electrons or changes in the external potential, or both, which provide information about their reactivity [35]. To better describe the molecular reactivity of the complexes studied, our discussions have been based on the molecular electron density theory (MEDT) put forth by Domingo and coworkers [35]. This theory states that "while the electron density distribution at the ground-state is responsible for physical and chemical molecular properties, the capability for changes in electron density, and not molecular orbital interactions, is responsible for molecular reactivity".

The electronic chemical potential, μ, is associated with the feasibility of a system to exchange electron density with the environment at the ground state. The electronegativity, χ, is a measure of the resistance to electron density loss. The chemical hardness, η, is thought of as the resistance of a molecule to electron density exchange with the environment. On the other hand, the chemical softness, S, is the inverse of the chemical hardness, η. The values of the global reactivity descriptors computed in this work are reported in Table 5.

TABLE 5: Global reactivity descriptors (GRD) of the complexes investigated in eV and eV^{-1} for global softness, calculated at M06/ma-def2-TZVP level of theory for the metal ions and M06/ma-def2-SVP level of theory for every other element, in the gas phase.

GRD	A	B	C	D
IP_v	−0.13	0.29	0.24	0.28
EA_v	0.08	0.06	0.02	0.02
μ	0.023	−0.17	−0.13	−0.15
χ	−0.023	0.17	0.13	0.15
η	−0.10	0.12	0.11	0.13
S	−9.63	8.61	8.82	7.63
ω	-3.31×10^{-5}	1.73×10^{-3}	9.42×10^{-4}	1.41×10^{-3}

A careful inspection of Table 5 revealed that μ varies in the order **B** < **D** < **C** < **A**, which indicates that the Cu(II) complex is the most reactive with respect to electron-donation, while the Ni(II) complex is the most reactive in terms of electron acquisition. On the other hand, χ varies thus **A** < **C** < **D** < **B**, which confirms the Ni(II) complex as the most susceptible to electron gain, while the Cu(II) complex is the most resistant to electron density gain. It is clear from the ranking of η values **A** < **C** < **B** < **D**, that of all the complexes studied, the Co(II) complex is the hardest (chemically) and the Cu(II) complex is the most chemically soft. The global electrophilicity index, ω, a measure of the energy stabilization of a molecule when it acquires an additional amount of electron density, is one of the most important reactivity indices. Generally, electrophiles are characterized by high ω values [35]. On this basis, the strongest electrophile among the molecules studied is the Ni(II) complex, while the Cu(II) complex is the weakest electrophile.

3.5. Molecular Electrostatic Potential (MEP) Surfaces. Molecular Electrostatic Potential, $\Phi(r)$, has been widely used for predicting the sites that are susceptible to nucleophilic, electrophilic, and free radical attacks on molecular species, as well as molecular docking modes for quite a long time. $\Phi(r)$ is calculated using

$$\Phi(r) = \int \frac{\rho_{\text{total}}(r)}{|r - r'|} dr, \qquad (2)$$

where ρ_{total} represents both the nuclear and the electronic charge density. The integration is over the molecular volume, and r' represents the atomic positions relative to the same origin. The integration includes the atoms of only one molecule and therefore does not directly include the effects of charge distributions of the neighboring molecules [36].

To elucidate the sites for molecular reactivity in the investigated complexes in the gas phase, their MEP maps were computed at RIJCOSX-M06/def2-TZVP level of theory in the gas phase, and visualized (as shown in Figure 4) via the Molekel 4.3 graphical user interface [37]. In this figure, the color code red to yellow indicates regions of low electrostatic potential and is related to electrophilic reactivity, while blue

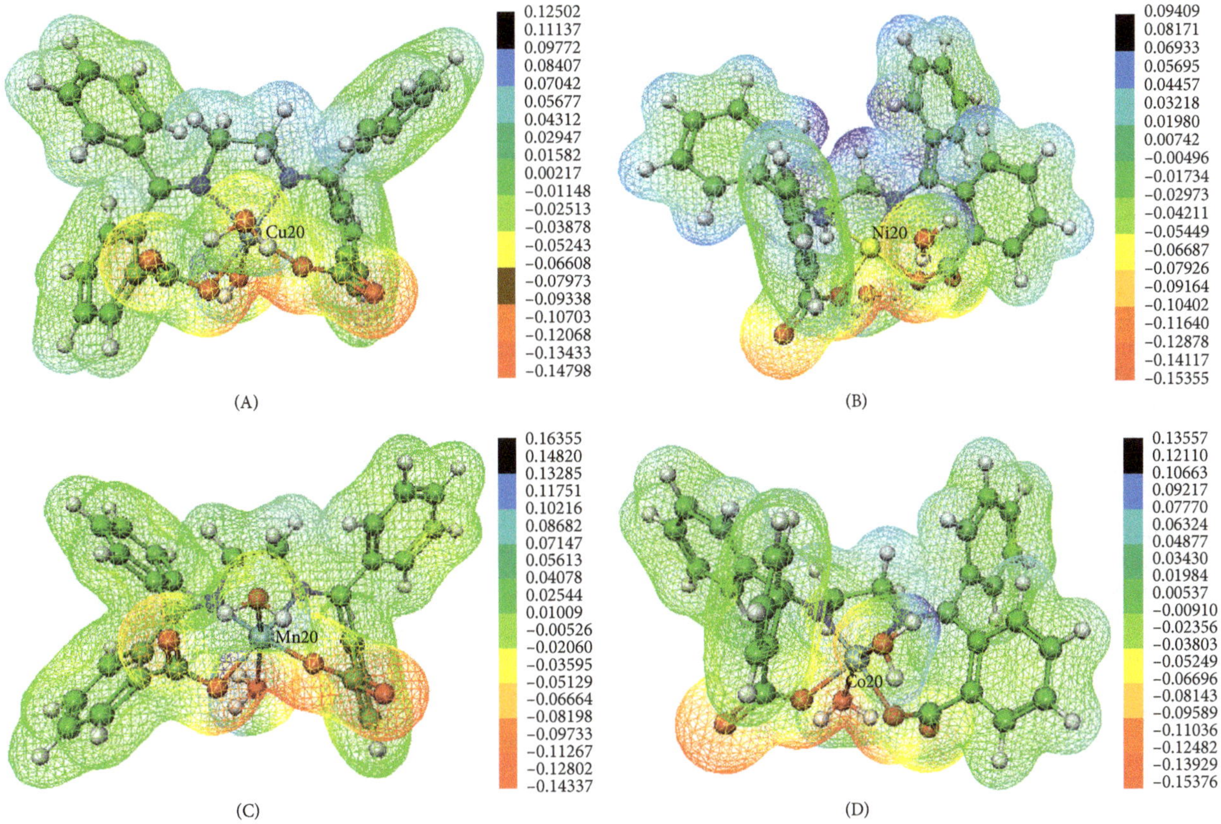

FIGURE 4: Molecular electrostatic potential (MEP) surfaces of EDA2BB complexes, mapped onto electron density isosurfaces of value 0.01 au, using Molekel 4.3.

to green indicates regions of high electrostatic potential related to nucleophilic reactivity sites. It should be noted that the values reported in the legend are in au.

The most negative regions of the MEP maps are associated with the oxygen atoms of both the water ligands and the Schiff base ligand. In particular, the oxygen atoms O_{23} and O_{18} in each complex (with electrostatic potentials ranging from −0.1537 au in D to −0.14337 au in B) are found to be the most suitable sites for electrophilic attack. The $C_1 \cdots H_{63}$, $C_1 \cdots H_{64}$, $C_2 \cdots H_{61}$, and $C_2 \cdots H_{62}$ bonds, as well as the hydrogen atoms (H_{57}, H_{60}, H_{59}, and H_{56}) of the water ligands in all of the complexes are seemingly the most probable sites for nucleophilic attack. This is shown by the relatively high electrostatic potentials in the neighborhood of these atoms, which range from 0.16355 au in C to 0.09409 au in B. It is worthy to note that these hydrogen atoms are responsible for both intramolecular and intermolecular HB formation by the complexes studied. Such intermolecular HBs could be formed with electronegative atoms in protein molecules present in the cell walls of microorganisms, resulting in interference with the normal cell processes [5].

4. Conclusion

Literature holds that some synthetic first-row transition metal complexes of the tetradentate Schiff base ligand 2,2′(1E,1′E)-(ethane-1,2-diylbis(azan-1-yl-1-ylidene))bis(phenylmethan-1-yl-1-ylidene)dibenzoic acid (herein denoted EDA2BB) and two water molecules acting as ancillary ligands possess interesting antimicrobial effects [5]. This work addresses a detailed analysis of their structural properties and chemical reactivity via theoretical studies on some transition metal complexes of EDA2BB: *[Cu(EDA2BB)(OH$_2$)$_2$]*, *[Ni(EDA2BB)(OH$_2$)$_2$]*, *[Mn(EDA2BB)(OH$_2$)$_2$]*, and *[Co(EDA2BB)(OH$_2$)$_2$]* via the DFT method. The DFT studies have been performed using the exchange-correlation functionals BP86 and M06. These functionals were employed together with the basis sets ma-def2-SVP, ma-def2-TZVP, and def2-TZVP(-f). Calculations were performed in gas phase, as well as in DMSO in order to simulate solvent effect. The quantum theory of atoms-in-molecules (QTAIM), conceptual DFT, natural population analysis (NPA), and molecular electrostatic potential (MEP) methods were also employed. The results obtained have revealed a distorted octahedral geometry around the central metal ion in each complex studied in gas phase. In DMSO solvent, a general axial elongation of the metal-oxygen bonds involving the water ligands has been observed, suggesting that the water ligands are loosely bound to the central metal ions and might be acting as solvent molecules. Weak, medium, and strong intramolecular hydrogen bonds, along with hydrogen-hydrogen and van der Waals interactions, have been elucidated in the complexes, via geometric and

QTAIM analyses. From the NPA charges on the central metal ions, it was found that electron transfer from the ligand EDA2BB to the central metal ions occurred during complex formation. Clearly, these metal-ligand charge transfers are substantial in the Ni(II) and Mn(II) complexes and less significant in the Cu(II) and Co(II) complexes. From the chemical hardness values, the complex *[Co(EDA2BB)(OH$_2$)$_2$]* is the hardest, while *[Cu(EDA2BB)(OH$_2$)$_2$]* is the softest. Based on the global electrophilicity index, the strongest electrophile among the complexes studied is *[Ni(EDA2BB)(OH$_2$)$_2$]*, whereas *[Cu(EDA2BB)(OH$_2$)$_2$]* is the weakest electrophile. The most negative regions of the MEP maps are associated with the oxygen atoms of both the water ligands and the Schiff base ligand. In particular, the oxygen atoms O_{23} and O_{18} in each complex are found to be the most suitable sites for electrophilic attack. The $C_1 \cdots H_{63}$, $C_1 \cdots H_{64}$, $C_2 \cdots H_{61}$, and $C_2 \cdots H_{62}$ bonds, as well as the hydrogen atoms (H_{57}, H_{60}, H_{59}, and H_{56}) of the water ligands in all of the complexes studied are seemingly the most probable sites for nucleophilic attack. In a nutshell, therefore, we have shown that the activities of the complexes are improved vis-à-vis the ligands, and stable geometries of the complexes have been identified, alongside their prominent electrophilic and nucleophilic sites.

Acknowledgments

This work benefitted from the research modernization allowance offered by the Ministry of Higher Education of Cameroon to faculty members, for which the authors are grateful.

Supplementary Materials

The Cartesian coordinates of atoms in the optimized geometries of the ligand (EDA2BB) in gas phase, along with those of its complexes in both gas phase and DMSO, all of which calculated at BP86-D3(BJ)/def2-TZVP(-f) level of theory, are presented in Tables S1–S9. The Harmonic vibrational frequencies of EDA2BB and its complexes in gas phase calculated at the same level of theory are presented in Tables S10–S14. The IR spectra based on unscaled vibrational frequencies of EDA2BB and its complexes in gas phase are shown in Figures S1–S5. (*Supplementary Materials*)

References

[1] G. Sugimoto, *Antibiotics Overuse and Rise of Superbugs*, KitsapSun, Bremerton, WA, USA, 2015, http://www.ghc.org/html/piblic/health-wellness/sugimoto/antibiotics.

[2] P. C. A. Bruijnincx and P. J. Sadler, "New trends for metal complexes with anticancer activity," *Current Opinion in Chemical Biology*, vol. 12, no. 2, pp. 197–206, 2008.

[3] W. Zoubi, "Biological activities of schiff bases and their complexes: a review of recent works," *International Journal of Organic Chemistry*, vol. 3, no. 3, pp. 73–95, 2013.

[4] H. Sharghi and M. A. Nasseri, "Schiff-base metal(II) complexes as new catalysts in the efficient, mild and regioselective conversion of 1,2-epoxyethanes to 2-hydroxyethyl thiocyanates with ammonium thiocyanate," *Bulletin of the Chemical Society of Japan*, vol. 7, no. 1, pp. 137–142, 2003.

[5] R. K. Al-Shemary and B. A. Zaidan, "Preparation and Characterization of some transition metal complexes of new tetradentate Schiff ligand type N_2O_2," *Scholars Academic Journal of Biosciences*, vol. 4, no. 1, pp. 18–26, 2016.

[6] T. Nogrady and D. F. Weaver, *Medical Chemistry: A molecular and Biochemical Approach*, Oxford University Press, Oxford, UK, 3rd edition, 2005.

[7] M. Orio, D. A. Pantazis, and F. Neese, "Density functional theory," *Photosynthesis Research*, vol. 102, no. 2-3, pp. 443–453, 2009.

[8] F. Neese, "Prediction of molecular properties and molecular spectroscopy with density functional theory: from fundamental theory to exchange-coupling," *Coordination Chemistry Reviews*, vol. 253, no. 5-6, pp. 526–563, 2009.

[9] F. Neese, "The ORCA program system," *Wiley Interdisciplinary Review: Computational Molecular Science*, vol. 2, no. 1, pp. 73–78, 2012.

[10] M. D. Hanwell, D. E. Curtis, D. C. Lonie, T. Vandermeersch, E. Zurek, and G. R. Hutchison, "Avogadro: an advanced semantic chemical editor, visualization, and analysis platform," *Journal of Cheminformatics*, vol. 4, no. 1, p. 17, 2012.

[11] F. Weigend and R. Ahlrichs, "Balanced basis sets of split-valence, triple zeta valence and quadruple zeta valence quality for H to Rn: design and assessment of accuracy," *Physical Chemistry Chemical Physics*, vol. 7, no. 18, pp. 3297–3305, 2005.

[12] A. D. Becke, "Density-functional exchange-energy approximation with correct asymptotic behavior," *Physical Review A*, vol. 38, no. 6, pp. 3098–3100, 1988.

[13] J. P. Perdew, "Density-functional approximation for the correlation energy of the inhomogeneous electron gas," *Physical Review B*, vol. 34, no. 12, pp. 8822–8824, 1986.

[14] F. Neese, "An improvement of the resolution of the identity approximation for the calculation of the coulomb matrix," *Journal of Computational Chemistry*, vol. 24, no. 14, pp. 1740–1747, 2003.

[15] A. Klamt and G. Schüürmann, "COSMO: A new approach to dielectric screening in solvents with explicit expressions for the screening energy and its gradient," *Journal of the Chemical Society, Perkin Transactions*, vol. 2, no. 5, pp. 799–805, 1993.

[16] S. Grimme, J. Antony, S. Ehrlich, and H. Krieg, "A consistent and accurate ab initio parametrization of density functional dispersion correction (DFT-D) for the 94 elements H–Pu," *Journal of Chemical Physics*, vol. 132, no. 15, article 154104, 2010.

[17] S. Grimme, S. Ehrlich, and L. Goerigk, "Effect of the damping function in dispersion corrected density functional theory," *Journal of Computational Chemistry*, vol. 32, no. 7, pp. 1456–1465, 2011.

[18] Z. Yan and D. G. Truhlar, "The M06 suite of density functionals for main group thermochemistry, thermochemical kinetics, noncovalent interactions, excited states, and transition elements: two new functionals and systematic testing of

four M06-class functionals and 12 other functional," *Theoretical Chemistry Account*, vol. 120, no. 1–3, pp. 215–241, 2008.

[19] F. Neese, F. Wennmohs, A. Hansen, and U. Becker, "Efficient, approximate and parallel Hartree–Fock and hybrid DFT calculations. A 'chain-of-spheres' algorithm for the Hartree–Fock exchange," *Chemical Physics*, vol. 356, no. 1–3, pp. 98–109, 2009.

[20] T. Y. Nikolaienko, L. A. Bulavin, and D. M. Hovorun, "JANPA: an open source cross-platform implementation of the natural population analysis on the java platform," *Computional Theoretical Chemistry*, vol. 1050, pp. 15–22, 2014.

[21] G. A. Zhurko, *Chemcraft-Graphical Program for Visualization of Quantum Chemistry Computations*, https://chemcraftprog.com.

[22] O. E. Sherif and N. S. Abdel-Kader, "DFT calculations, spectroscopic studies, thermal analysis and biological activity of supramolecular Schiff base complexes," *Arabian Journal of Chemistry*, vol. 11, no. 5, pp. 700–713, 2015.

[23] G. A. Jeffrey, *An Introduction to Hydrogen Bonding*, Oxford University Press, Oxford, UK, 1997.

[24] G. E. Lewars, *Computational Chemistry: Introduction to the Theory and Applications of Molecular and Quantum Mechanics*, Klumer academy publishers, New York, NY, USA, 2003.

[25] K. K. Manoj, B. Brina, and M. L. M. Jan, "Frequency and zero-point vibrational energy scale factors for double hybrid density functional (and selected other methods): can anharmonic force fields be avoided?," *Journal of Physical Chemistry A*, vol. 119, no. 9, pp. 1701–1714, 2014.

[26] R. W. F. Bader, *Atoms in Molecules: A Quantum Theory*, Oxford University Press, Oxford, UK, 1994.

[27] L. Tian and C. Feiwu, "Multiwfn: a multifunctional wavefunction analyzer," *Journal of Computational Chemistry*, vol. 33, no. 5, pp. 580–592, 2012.

[28] A. Sirohiwal, V. R. Hathwar, D. Dey, and D. Chopra, "Investigation of chemical bonding in *in situ* cryocrystallized organometallic liquids," *Chemical Physical Chemistry*, vol. 18, no. 20, pp. 2859–2863, 2017.

[29] M. Bayat, F. Yaghoobi, S. Salehzadeh, and S. Hokmi, "A theoretical study on the interaction of $[Al(H2O)_6]^{3+}$ and $[Mg(H2O)_6]^{2+}$ cations with fullerene (C60), coronene and benzene π-systems," *Polyhedron*, vol. 30, no. 17, pp. 2809–2814, 2011.

[30] D. Cremer and E. Kraka, "Theoretical determination of molecular structure and conformation. 15. Three-membered rings: bent bonds, ring strain, and surface delocalization," *Journal of the American Chemical Society*, vol. 107, no. 13, pp. 3800–3810, 1985.

[31] P. L. A. Popelier, "Characterization of a dihydrogen bond on the basis of the electron density," *Journal of Physical Chemistry A*, vol. 102, no. 10, pp. 1873–1880, 1998.

[32] E. Espinosa, E. Molins, and C. Lecomte, "Hydrogen bond strength revealed by topological analyses of experimentally observed electron densities," *Chemical Physics Letters*, vol. 285, no. 3-4, pp. 170–173, 1998.

[33] M. Sivakami, B. Natarajan, and M. Vijayachandrasekar, "Synthesis, characterisation and biological activity of a new Mannich base and it's metal complexes," *Chemical Science Transactions*, vol. 3, no. 3, pp. 1110–1114, 2014.

[34] N. K. Nkungli and J. N. Ghogomu, "Theoretical analysis of the binding of iron (III) protoporphyrin IX to 4 methoxyacetophenone thiosemicarbazone via DFT-D3, MEP, QTAIM, NCI, ELF, and LOL studies," *Journal of Molecular Modeling*, vol. 23, no. 17, 2017.

[35] L. R. Domingo, M. Ríos-Gutiérrez, and P. Pérez, "Applications of the conceptual density functional theory indices to organic chemistry reactivity," *Molecules*, vol. 21, no. 6, p. 748, 2016.

[36] M. Drissi, N. Benhalima, Y. Megrouss, R. Rachida, A. Chouaih, and F. Hamzaoui, "Theoretical and experimental electrostatic potential around them-nitrophenol molecule," *Molecules*, vol. 20, pp. 4042–4054, 2015.

[37] S. Portmann and H. P. Lüthi, "MOLEKEL: an interactive molecular graphics tool," *Chimia*, vol. 54, no. 12, pp. 766–770, 2000.

3

Coordination Behavior of Ni^{2+}, Cu^{2+}, and Zn^{2+} in Tetrahedral 1-Methylimidazole Complexes: A DFT/CSD Study

Samuel Tetteh

Department of Chemistry, School of Physical Sciences, College of Agriculture and Natural Sciences, University of Cape Coast, Cape Coast, Ghana

Correspondence should be addressed to Samuel Tetteh; stoshgh2001@yahoo.com

Academic Editor: Spyros P. Perlepes

The interaction between nickel (Ni^{2+}), copper (Cu^{2+}), and zinc (Zn^{2+}) ions and 1-methylimidazole has been studied by exploring the geometries of eleven crystal structures in the Cambridge Structural Database (CSD). The coordination behavior of the respective ions was further investigated by means of density functional theory (DFT) methods. The gas-phase complexes were fully optimized using B3LYP/GENECP functionals with 6-31G* and LANL2DZ basis sets. The Ni^{2+} and Cu^{2+} complexes show distorted tetrahedral geometries around the central ions, with Zn^{2+} being a perfect tetrahedron. Natural bond orbital (NBO) analysis and natural population analysis (NPA) show substantial reduction in the formal charge on the respective ions. The interaction between metal d-orbitals (donor) and ligand orbitals (acceptor) was also explored using second-order perturbation of the Fock matrix. These interactions followed the order $Ni^{2+} > Cu^{2+} > Zn^{2+}$ with Zn^{2+} having the least interaction with the ligand orbitals. Examination of the frontier orbitals shows the stability of the complexes in the order $Ni^{2+} > Cu^{2+} < Zn^{2+}$ which is consistent with the Irving–Williams series.

1. Introduction

Metalloproteins play important roles in the structure and physiology of cells. They account for nearly half of all proteins in nature [1]. Some of their cellular functions include water oxidation, photosynthesis, respiration, molecular oxygen reduction, and zinc fingers [1, 2]. The binding stability of divalent transition metals in metalloproteins has been largely studied in terms of semiempirical and qualitative theories such as the hard and soft acid and base principles of Parr and Pearson and the Irving–Williams series of stability constants [3]. Yet the fundamental role of metal ions in the structure and function of metalloproteins is still a matter of ongoing research [4, 5]. The amino acid histidine is a common ligand in metalloproteins [6]. The imidazole ring of histidine is an important five-membered heterocycle that is widely present in natural products and most synthetic molecules. Because of its unique electronic structure, imidazole-based compounds are widely used as anticancer, antifungal, antibacterial, anti-inflammatory, antihistaminic, and other medicinal agents [7–10]. It readily binds with a variety of enzymes and receptors in biological systems through hydrogen bonds, coordinate covalent bonds, ion-dipole interactions, π-π^* interactions, and van der Waals forces, thereby exhibiting broad bioactivities [7].

Imidazole is largely considered an *N*-donor in most of its coordination complexes [11]. This interaction largely affects the formal charge of the coordinating electrophile. According to Hasegawa et al., the total charge of 4-methylimidazole increased upon binding to Zn^{2+}, indicating that some negative charge is transferred to the Zn^{2+} in the complex. This charge delocalization has a tendency of reducing the formal charge on Zn^{2+} to 1.70. Similar atomic charges were observed in the nitrogen atoms of the 4-methylimidazole ligand [12]. This charge delocalization could greatly affect the reactivity and catalytic activity of imidazole-based transition metal complexes.

A study by Rulíšekr and Vondrášek on the coordination geometries of selected metal ions in metalloproteins with data from the Protein Data Bank suggests that Ni^{2+}, Cu^{2+}, and Zn^{2+} tend to bind in tetrahedral coordination modes although there are other geometries such as octahedral and square planar [3]. A search in the Cambridge Structural Database version 5.38 (November 2016) plus one update revealed one Ni complex bearing 1-methylimidazole ligand [13]. In this complex, the nickel ion is octahedrally coordinated with the trien ligand bonded through its four nitrogen atoms with the remaining two coordination sites occupied by 1-methylimidazole ligands. Ten copper complexes bearing 1-methylimidazole ligands were also found. These include molecules with the following reference codes: BEJGUS [14], CAHJAW [15], and CUSHON [16] with square-planar coordination around the copper ion through two carboxylic oxygen atoms and two 1-methylimidazole nitrogen atoms; MACCUA10 [17] of square-pyramidal geometry with the four nitrogen atoms of the macrocycle forming the basal plane and 1-methylimidazole nitrogen at the apex; GALLAG [18] of square-planar coordination through four 1-methylimidazole ligands; GALLOU [18], a tetragonally distorted octahedral complex with two water molecules *trans* to each other and four 1-methylimidazole ligands in the square plane; KAYPEF [19] (chloro(glycinato)(1-methylimidazole)copper (II) complex) with a square-pyramidal structure having four close ligating atoms (N2OCl) and an axial chlorine ligand; CEZLOI [20], a Cu^{+} complex with S_4 site symmetry in which the Cu ion is tetrahedrally coordinated with four 1-methylimidazole ligands; and finally, the complexes GALLEK and GALLIO with tetrahedral geometries in which the copper ions are linked to a central oxygen atom with μ_2-Cl atoms above each edge of the tetrahedron and terminal 1-methylimidazole ligand. However, there was no zinc complex bearing 1-methylimidazole ligand at the time of preparing this manuscript. Considering the importance of nickel, copper, and zinc ions in the structure and function of metalloproteins, this article investigates the effect of bonding on the energies of the d-orbitals of Ni^{2+}, Cu^{2+}, and Zn^{2+} in tetrahedral ligand fields. It also reports the effect on the formal charges of the ions upon coordination to the 1-methylimidazole (1-MeIm) ligand.

2. Methodology

2.1. CSD Analysis. The ligand and transition metal complexes were analyzed with version 5.38 of the CSD (November 2016) plus one update. The CSD program ConQuest Version 1.19 was used to perform substructure searches of Ni, Cu, and Zn complexes bearing 1-methylimidazole ligands. The accepted entries met the following criteria: 3D coordinates determined; crystallographic R factor ≤ 0.05; no disorder in the crystal structure; no errors in the structure; no polymeric bonding; no ions; and no powder structures and only organometallic structures (according to standard CSD definitions) [21]. The search revealed eleven crystal structures: one nickel and ten copper complexes. No zinc complex was found in the database. The geometry of *N*-methylimidazole in a novel 4- and 5-coordinated silicon complex (reference code: GAGXER) [22] was used as the reference. The 3D structures of all the complexes were visualized using version 3.9 of the CSD program *Mercury* [23].

2.2. Computational Studies. All input files were prepared using the GaussView 5.0.8 molecular structure viewer [24]. The Gaussian 09 program was used to perform all the computations. All structures were optimized with the density functional theory (DFT) using B3LYP three-parameter hybrid functionals with no constraints in the respective geometries. Effective core potentials (ECPs) were used to represent the valence electrons of Ni^{2+}, Cu^{2+}, and Zn^{2+} and the basis set of double-ζ quality associated with pseudopotentials known as LANL2DZ [25]. Because of the size of the complexes and the accompanying computational cost, the 6-31G* basis set was used for all other atoms. Similar basis sets have been used to study the geometries and molecular orbitals of transition metal complexes [26]. Frequency analyses were employed to confirm that the optimized geometries were at stationary points corresponding to local minima with no imaginary frequencies [21]. The optimized geometries with the accompanying numbering scheme and the Cartesian atomic coordinates are shown in Figure 1.

3. Results and Discussion

3.1. Geometry Optimization. Figure 1 shows the gas-phase optimized geometries and the selected atomic numbering scheme of 1-methylimidazole and the respective complexes. It also indicates the Cartesian atomic coordinates. Selected bond lengths of the optimized structures are shown in Table 1. A search in the version 5.38 of the CSD (November 2016) showed that there was no crystal structure of 1-methylimidazole in the database. However, the geometry of *N*-methylimidazole in a novel 4- and 5-coordinated silicon complex (reference code: GAGXER) [22] was used as the reference. The bond lengths did not differ significantly from that of the optimized structure. The slight differences could be attributed to the fact that the optimization was carried out on a single molecule in the gas phase with no intermolecular interactions, whereas in the crystal structure, there are lattice interactions which affect the bond parameters [21].

Generally, the intramolecular bond lengths in 1-MeIm were not affected by the coordination. Notable exceptions include the C1-N4, C2-N8, and C3-N8 bonds which recorded slight decreases in bond lengths as a result of coordination to the metal ion. The metal-ligand bond lengths (M-N) ranged between 2.018 and 2.036 Å in the Ni-1MeIm complex and between 2.017 and 2.038 Å in the Cu-1MeIm analogue but remained the same (2.085 Å) in the zinc complex. The differences in the Ni-N and Cu-N bond lengths could be attributed to Jahn–Teller distortions for the $(e)^4(t_2)^4$ and $(e)^4(t_2)^5$ electronic configurations in the respective Ni-1MeIm and Cu-1MeIm complexes [27]. Zn^{2+} has fully occupied d-orbitals and interacts weakly with ligand orbitals and maintains a spherical isotropy. The C3-H7 bond length did not show much variation as a result of the bonding interaction of the ligand with the metal ions. The observed differences in bond lengths were further investigated by a series of density functional theory (DFT) calculations.

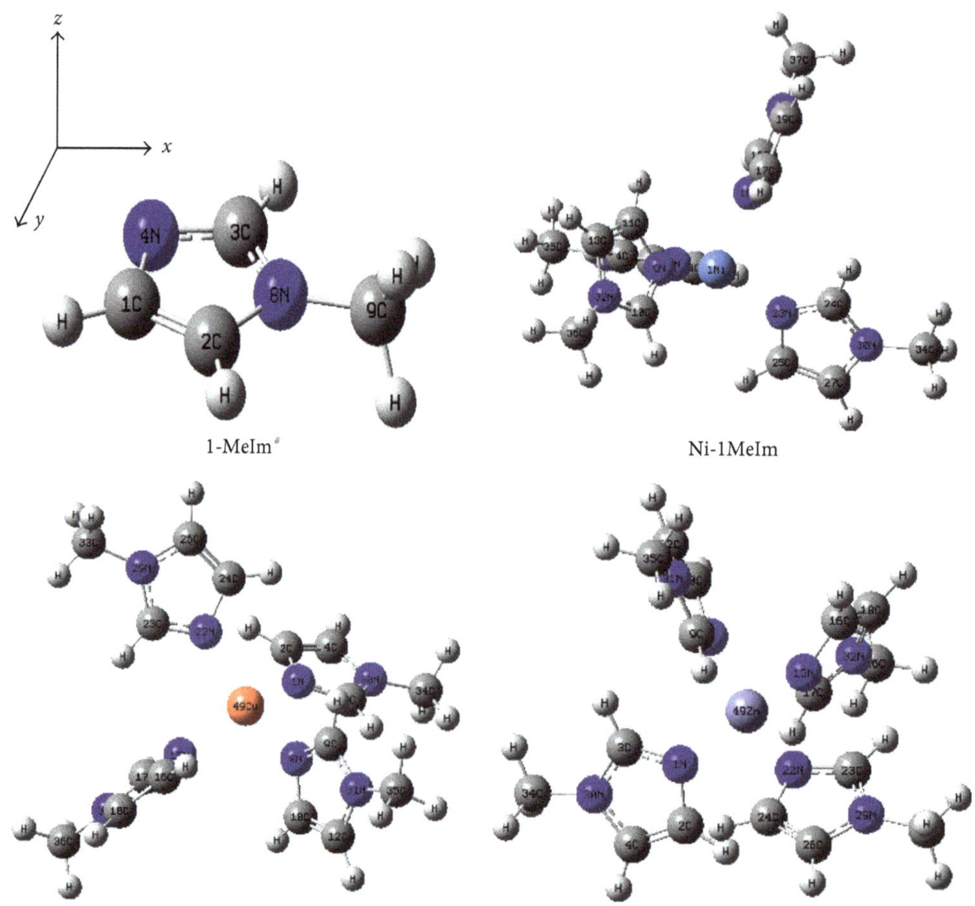

FIGURE 1: Optimized geometries of 1-methylimidazole and the respective complexes.

3.2. Natural Bond Orbital (NBO) Analysis and Charge Distribution. The effect of coordination and electron distribution on the ligands and metal d-orbitals was assessed by the NBO and second-order perturbation of the Fock matrix. This provides details about the type of hybridization, nature of bonding, and strength of interaction between the metal ion and ligand [28]. Table 2 shows the occupancy and electron density (ED) of molecular orbitals in 1-MeIm. The C1-N4 σ-bond in the uncoordinated ligand has an electron density of 1.980 and an occupancy of 41.4% C1 and 58.6% N4. Upon coordination, the electron density reduced to 0.990 in both Ni-1MeIm and Cu-1MeIm with the occupancies also changing to 38.1% and 61.9% for C1 and N4, respectively.

The reduced electron density in the C1-N4 σ-bond could have been delocalized onto the metal ions [29] to reduce their formal charges from +2 to −0.2985 and +0.2132, respectively, as shown in Table 3. Despite the fact that occupancy changed to 38.8% C1 and 61.7% N4 in Zn-1MeIm, the ED was not affected. Similar effects were observed in the C3-N4 σ-bond. The C3-N4 π-bond observed in the free ligand was, however, nonexistent in the Ni^{2+} and Cu^{2+} complexes. These π-electrons could have contributed to reducing the strength of the formal charges on the respective metal ions as observed in Table 3. Generally, although there were substantial changes in the occupancy of the bonds in the 1-MeIm upon coordination to the Zn^{2+} ion, the ED on the ligand remained unchanged as compared to the free ligand.

From Table 3, the calculated formal charges on the central metal ions have been reduced to −0.2985, +0.2132, and +1.3850, respectively. This reduction in the ionic charge is higher in Ni-1MeIm and Cu-1MeIm than in Zn-1MeIm. As shown in Table 2, there is higher delocalization of ligand electrons onto Ni^{2+} and Cu^{2+} than in the case of Zn^{2+}. The natural charge on the N4 is −0.4842 in the free ligand but −0.03501 and −0.3524 in the Ni^{2+} and Cu^{2+} complexes, respectively. It however increased to −0.6791 in the Zn^{2+} complex.

Similarly, natural charges on C1, C2, C3, N8, and C9 which decreased in the Ni-1MeIm and Cu-1MeIm complexes, however, increased in the Zn^{2+} analogue. These support the fact that although there is substantial delocalization of ligand-bonding electrons onto Ni^{2+} and Cu^{2+} d-orbitals [30], these electrons are, however, localized on the ligand ring of the Zn-1MeIm complex, making it more nucleophilic.

The 3d-orbital occupancy and energies of the complexes are assessed in Table 4. It is shown that the t_2 orbitals (d_{xy}, d_{xz}, and d_{xy}) of Ni-1MeIm have higher energies (−0.611 eV),

Table 1: Selected bond lengths of the optimized structures.

Bond	Length (Å)				
	1-MeIm		Ni-1MeIm	Cu-1MeIm	Zn-1MeIm
	Experimental	Calculated			
C1-C2	1.312	1.380	1.367	1.367	1.367
C1-N4	1.327	1.396	1.385	1.386	1.387
C1-H5	0.946	1.077	1.079	1.079	1.080
C2-H6	0.957	1.078	1.080	1.080	1.080
C2-N8	1.349	1.393	1.382	1.381	1.381
C3-N4	1.301	1.333	1.335	1.335	1.335
C3-H7	1.031	1.078	1.080	1.079	1.081
C3-N8	1.348	1.381	1.347	1.346	1.367
C9-N8	1.474	1.461	1.467	1.467	1.467
C9-H10	0.919	1.095	1.092	1.092	1.092
C9-H11	0.995	1.095	1.092	1.092	1.092
C9-H12	1.219	1.093	1.091	1.091	1.091
M-N1	—	—	2.018	2.038	2.085
M-N2	—	—	2.036	2.018	2.085
M-N3	—	—	2.032	2.017	2.085
M-N4	—	—	2.027	2.038	2.085

M-N represents the metal-nitrogen bond.

Table 2: NBO population analysis for the respective compounds.

Bond	Type	1-Methylimidazole		Ni-1MeIm		Cu-1MeIm		Zn-1MeIm	
		ED	Occupancy	ED	Occupancy	ED	Occupancy	ED	Occupancy
C1-C2	σ	1.986	48.8% C1-51.2% C2	0.992	49.6% C1-50.4% C2	0.992	49.6% C1-50.4% C2	1.982	49.6% C1-50.4% C2
C1-C2	π	1.861	48.5% C1-51.5% C2	0.921	50.1% C1-49.9% C2	0.921	50.0% C1-50.0% C2	1.843	50.3% C1-49.7% C2
C1-N4	σ	1.980	41.4% C1-58.6% N4	0.990	38.1% C1-61.9% N4	0.990	38.1% C1-61.9% N4	1.981	38.3% C1-61.7% N4
C1-H5	σ	1.986	62.7% C1-37.3% H5	0.992	62.9% C1-37.1% H5	0.992	63.0% C1-37.0% H5	1.985	62.7% C1-37.3% H5
C2-H6	σ	1.986	62.6% C2-37.4% H6	0.993	63.5% C2-36.5% H6	0.992	63.6% C2-36.4% H6	1.985	63.5% C2-36.5% H6
C2-N8	σ	1.984	36.0% C2-64.0% N8	0.991	36.1% C2-63.9% N8	0.991	36.0% C2-64.0% N8	1.982	36.1% C2-63.9% N8
C3-N4	σ	1.987	41.9% C3-58.1% N4	0.991	38.9% C3-61.1% N4	0.992	38.8% C3-61.2% N4	1.985	39.1% C3-60.9% N4
C3-N4	π	1.871	43.5% C3-56.5% N4	—	—	—	—	1.891	34.7% C3-65.3% N4
C3-H7	σ	1.985	62.2% C3-37.8% H7	0.992	62.6% C3-37.4% H7	0.992	62.5% C3-36.8% H7	1.984	62.5% C3-37.5% H7
C3-N8	σ	1.989	35.6% C3-64.4% N8	0.993	36.4% C3-63.6% N8	0.993	36.4% C3-63.6% N8	1.985	36.4% C3-63.6% N8
C9-N8	σ	1.993	36.4% C9-63.6% N8	0.995	34.8% C9-65.2% N8	0.995	34.8% C9-65.2% N8	1.990	34.8% C9-65.2% N8
C9-H10	σ	1.989	62.8% C9-37.2% H10	0.995	63.2% C9-36.8% H12	0.995	63.2% C9-36.8% H12	1.990	62.6% C9-37.4% H10
C9-H11	σ	1.989	62.8% C9-37.2% H11	0.993	63.2% C9-36.8% H11	0.993	63.2% C9-36.8% H11	1.987	63.2% C9-36.8% H11
C9-H12	σ	1.990	62.7% C9-37.3% H12	0.993	62.6% C9-37.4% H10	0.993	62.6% C9-37.4% H10	1.987	63.2% C9-36.8% H12

Table 3: NPA atomic charge distributions of some selected atoms in the complexes and free ligand.

Atom	Charge			
	1-MeIm	Ni-1MeIm	Cu-1MeIm	Zn-1MeIm
C1	−0.1277	−0.0356	−0.0338	−0.0741
C2	−0.1175	−0.0300	−0.0289	−0.0559
C3	+0.1396	+0.1241	+0.1264	+0.2410
N4	−0.4842	−0.3501	−0.3524	−0.6791
H5	+0.2505	+0.1271	+0.1282	+0.2523
H6	+0.2488	+0.1347	+0.1350	+0.2697
H7	+0.2395	+0.1236	+0.1226	+0.2457
N8	−0.4123	−0.1783	−0.1771	−0.3545
C9	−0.4957	−0.1789	−0.2357	−0.4711
H10	+0.2520	+0.1318	+0.1316	+0.2630
H11	+0.2520	+0.1312	+0.1318	+0.2626
H12	+0.2552	+0.1270	+0.2132	+0.2542
Ni^{2+}	—	−0.2985	—	—
Cu^{2+}	—	—	+0.2132	—
Zn^{2+}	—	—	—	+1.3850

TABLE 4: 3d-orbital occupancy and energy of the complexes.

Orbital	Ni-1MeIm		Cu-1MeIm		Zn-1MeIm	
	Occupancy	Energy (eV)	Occupancy	Energy (eV)	Occupancy	Energy (eV)
d_{xz}	0.9945	−0.609	0.9963	−0.612	1.9937	−0.817
d_{xy}	0.9950	−0.611	0.9958	−0.660	1.9962	−0.818
d_{yz}	0.9964	−0.611	0.9963	−0.612	1.9937	−0.817
$d_{x^2-y^2}$	0.9939	−0.626	0.9968	−0.628	1.9936	−0.818
d_{z^2}	0.9942	−0.623	0.9923	−0.605	1.9956	−0.818

TABLE 5: Second-order perturbation theory of the Fock matrix in the NBO basis for Ni-1MeIm.

Donor	Acceptor							
Ni1	N2	E(2)	N9	E(2)	N16	E(2)	N23	E(2)
LP(1)	RY * (1)	0.24	RY * (1)	0.03	RY * (2)	0.16	RY * (1)	0.24
LP(2)	RY * (1)	0.27	RY * (2)	0.03	RY * (1)	0.45	RY * (2)	0.08
LP(3)	RY * (2)	0.13	RY * (3)	0.05	RY * (1)	0.10	RY * (1)	0.13
LP(4)	RY * (1)	0.07	RY * (1)	0.07	RY * (1)	0.07	RY * (1)	0.68
LP(5)	RY * (2)	0.04	RY * (1)	0.32	—	—	RY * (3)	0.06

Notable interactions in Ni-1MeIm include LP(2)Ni1 → RY * (1)N2 and LP(2)Ni1 → RY * (1)N16 of energies 0.27 and 0.45 kcal/mol, respectively, and also LP(4)Ni1 → RY * (1)N23 of energy 0.68 kcal/mol. This delocalization of d-electrons onto non-Lewis orbitals could contribute to the decrease in occupancy of the Ni^{2+} d-orbitals [31]. In the case of Cu-1MeIm, similar interaction energies were observed: LP(1)Cu49 → RY*(1)N1, LP(1)Cu49 → RY * (1)N22, LP(4) Cu49 → RY * (1)N1, and LP(4)Cu49 → RY * (1)N22 of energies 0.32, 0.32, 0.29, and 0.29 kcal/mol, respectively.

TABLE 6: Second-order perturbation theory of the Fock matrix in the NBO basis for Cu-1MeIm.

Donor	Acceptor							
Cu49	N1	E(2)	N8	E(2)	N15	E(2)	N22	E(2)
LP(1)	RY * (1)	0.32	RY * (1)	0.19	RY * (1)	0.19	RY * (1)	0.32
LP(2)	—	—	RY * (1)	0.11	RY * (1)	0.09	—	—
LP(2)	—	—	RY * (2)	0.04	RY * (2)	0.03	—	—
LP(3)	—	—	RY * (1)	0.07	—	—	—	—
LP(4)	RY * (1)	0.29	RY * (1)	0.23	RY * (1)	0.22	RY * (1)	0.29
LP(4)	RY * (5)	0.07	RY * (5)	0.05	RY * (5)	0.05	RY * (5)	0.06
LP(5)	RY * (2)	0.07	—	—	—	—	RY * (2)	0.07

TABLE 7: Second-order perturbation theory of the Fock matrix in the NBO basis for Zn-1MeIm.

Donor	Acceptor							
Zn49	N1	E(2)	N8	E(2)	N15	E(2)	N22	E(2)
LP(4)	RY * (2)	0.07	RY * (2)	0.07	RY * (2)	0.07	RY * (2)	0.07
LP(4)	RY * (3)	0.09	RY * (3)	0.09	RY * (3)	0.09	RY * (3)	0.09
LP(5)	RY * (2)	0.11	RY * (2)	0.10	RY * (2)	0.10	RY * (2)	0.10

with the *e* orbitals ($d_{x^2-y^2}$ and d_{z^2}) having lower energy (−0.626 eV). This is typical of metal d-orbitals in a tetrahedral field [27].

There is, however, a reduced occupancy (≈0.9950) in all the orbitals. These electrons could be delocalized onto antibonding or non-Lewis type (Rydberg) molecular orbitals [31]. Similar occupancies were observed in the Cu^{2+} ion. With regard to the energy levels, the d_{xz} and d_{yz} orbitals were degenerated (−0.612 eV), but the d_{xy} orbital sunk low in energy (−0.660 eV) below the *e* energy levels. The $d_{x^2-y^2}$ orbital was also lower in energy than the d_{z^2} orbital. These differences in the energy support the Jahn–Teller distortions observed in Table 1 as a result of unequal occupancy of the t_2 orbitals in a tetrahedral field. Similarly, the Cu-1MeIm bond lengths also showed variations as observed in Table 1. In the case of Zn-1MeIm, the d-orbital occupancy was the same (≈2), showing that the d-electrons were localized within the Zn^{2+} d-orbitals [12]. Furthermore, the d-orbitals were degenerated with no d-d splitting. This explains the uniformity in the Zn-1MeIm bond lengths observed in Table 1.

To further probe the low occupancy of the Ni^{2+} and Cu^{2+} d-orbitals, second-order perturbation analysis of the Fock matrix was performed to investigate the interaction between metal d-orbitals (donor) and non-Lewis orbitals (acceptor) on the coordinating atom of the ligands. Tables 5–7 summarize the interaction between the d-electrons on the metal ions and the acceptor sites on the respective ligands. The highest energies, E(2), were recorded for all the complexes.

On the contrary, weaker interactions were observed in the case of Zn-1MeIm. Notable interactions include donation of

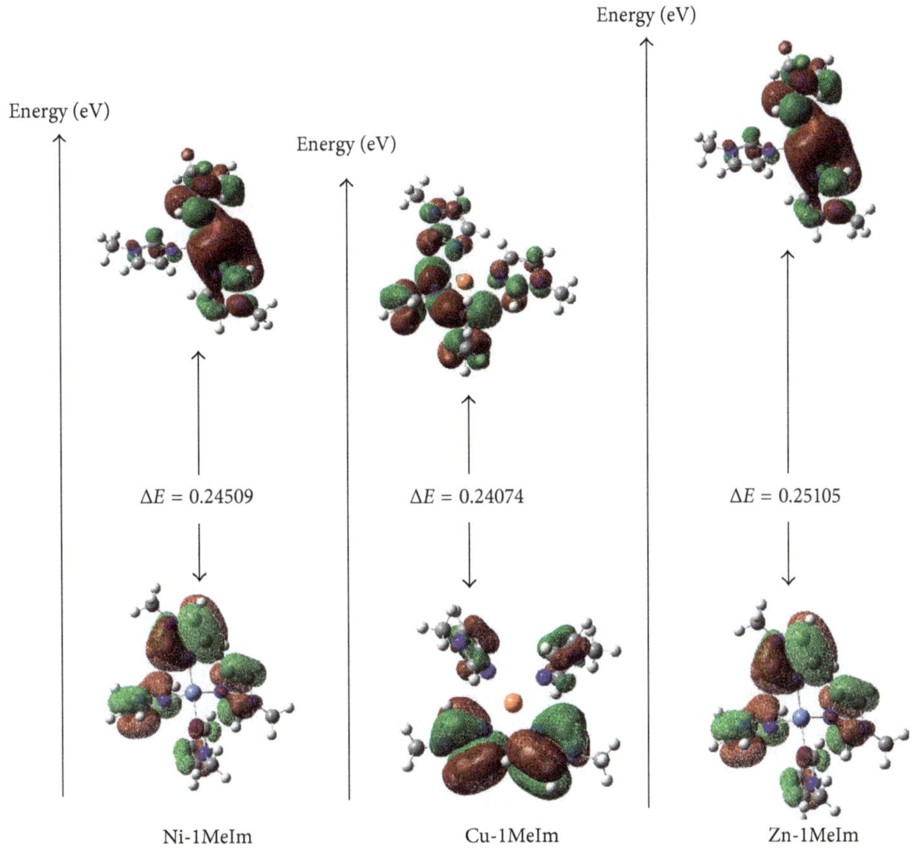

FIGURE 2: Frontier orbital energies of the complexes.

LP(5) on Zn49 to RY ∗ (2)N1, RY ∗ (2)N8, RY ∗ (2)N15, and RY ∗ (2)N22 of energies 0.11, 0.10, 0.10, and 0.10 kcal/mol, respectively.

These interaction energies are, however, lower than those observed in the Ni-1MeIm and Cu-1MeIm. This explains the higher occupancies of the Zn^{2+} d-orbitals compared with those of the Ni^{2+} and Cu^{2+} analogues as shown in Table 4.

The energies of the frontier orbitals are important in describing the chemical properties of molecules [32]. The energy of the highest occupied molecular orbital (HOMO) gives an indication of the electron-donating ability of the complex [33].

The higher the energy of the HOMO, the easier it is to donate electrons into the lowest unoccupied molecular orbital (LUMO). Also, the larger the energy difference between these frontier orbitals, the more stable the complex. Figure 2 shows the energy gap between the frontier orbitals in the respective complexes. The $E_{\text{HOMO}} - E_{\text{LUMO}}$ gap is in the order Ni-1MeIm > Cu-1MeIm < Zn-1MeIm. This order is in agreement with the Irving–Williams series [27].

4. Conclusion

This article assessed the coordination behavior of Ni^{2+}, Cu^{2+}, and Zn^{2+} ions in tetrahedral 1-methylimidazole ligand fields. Measurement of the bond lengths of the optimized structures revealed Jahn–Teller distortions in the Ni-1MeIm and Cu-1MeIm complexes. These distortions were, however, absent in Zn-1MeIm. A natural bond orbital (NBO) analysis of the molecular orbitals shows substantial decrease in the formal charges on the metal ions upon coordination to the ligands. This was in the order $Ni^{2+} < Cu^{2+} < Zn^{2+}$ with Zn^{2+} being the most electropositive one. Second-order perturbation of the Fock matrix shows higher interaction energies between Ni^{2+} 3d-orbitals (donor) and ligand orbitals (acceptor) than in the case of Zn^{2+} 3d-orbitals (donor). Finally, the frontier orbitals were assessed as a measure of the stability of the complexes. The energy of the $E_{\text{HOMO}} - E_{\text{LUMO}}$ gap was in the order Ni-1MeIm > Cu-1MeIm < Zn-1MeIm which is consistent with the Irving–Williams series.

Acknowledgments

The author is grateful to the Cambridge Crystallographic Data Centre (CCDC) for the opportunity to use the Cambridge Structural Database (CSD) for the substructure molecular searches and visualizations.

References

[1] K. J. Waldron, J. C. Rutherford, D. Ford, and N. J. Robinson, "Metalloproteins and metal sensing," *Nature*, vol. 460, no. 7257, p. 823, 2009.

[2] Y. Lu, N. Yeung, N. Sieracki, and N. M. Marshall, "Design of functional metalloproteins," *Nature*, vol. 460, no. 7257, p. 855, 2009.

[3] L. Rulíšekr and J. Vondrášek, "Coordination geometries of selected transition metal ions (Co^{2+}, Ni^{2+}, Cu^{2+}, Zn^{2+}, Cd^{2+}, and Hg^{2+}) in metalloproteins," *Journal of Inorganic Biochemistry*, vol. 71, no. 3-4, pp. 115–127, 1998.

[4] E. Nieboer, G. G. Fletcher, and Y. Thomassen, "Relevance of reactivity determinants to exposure assessment and biological monitoring of the elements," *Journal of Environmental Monitoring*, vol. 1, no. 1, pp. 1–14, 1999.

[5] P. Deschamps, P. Kulkarni, M. Gautam-Basak, and B. Sarkar, "The saga of copper (II)-l-histidine," *Coordination Chemistry Reviews*, vol. 249, no. 9-10, pp. 895–909, 2005.

[6] S. M. Sarkar, Y. Uozumi, and Y. Yamada, "A highly active and reusable self-assembled poly (imidazole/palladium) catalyst: allylic arylation/alkenylation," *Angewandte Chemie International Edition*, vol. 50, no. 40, pp. 9437–944s1, 2011.

[7] L. Zhang, X. M. Peng, G. L. Damu, R. X. Geng, and C. H. Zhou, "Comprehensive review in current developments of imidazole-based medicinal chemistry," *Medicinal Research Reviews*, vol. 34, no. 2, pp. 340–437, 2014.

[8] B. F. Abdel-Wahab, G. E. Awad, and F. A. Badria, "Synthesis, antimicrobial, antioxidant, anti-hemolytic and cytotoxic evaluation of new imidazole-based heterocycles," *European Journal of Medicinal Chemistry*, vol. 46, no. 5, pp. 1505–1511, 2011.

[9] E. Reisner, V. B. Arion, M. F. C. Guedes da Silva et al., "Tuning of redox potentials for the design of ruthenium anticancer drugs–an electrochemical study of [trans-RuCl4L (DM SO)]$^-$ and [trans-RuCl4L2]$^-$complexes, where L = imidazole, 1,2,4-triazole, indazole," *Inorganic Chemistry*, vol. 43, no. 22, pp. 7083–7093, 2004.

[10] A. Levina, A. Mitra, and P. A. Lay, "Recent developments in ruthenium anticancer drugs," *Metallomics*, vol. 1, no. 6, pp. 458–470, 2009.

[11] G. Sini, O. Eisenstein, and R. H. Crabtree, "Preferential C-binding versus N-binding in imidazole depends on the metal fragment involved," *Inorganic Chemistry*, vol. 41, no. 3, pp. 602–604, 2002.

[12] K. Hasegawa, T. Ono, and T. Noguchi, "Ab initio density functional theory calculations and vibrational analysis of zinc-bound 4-methylimidazole as a model of a histidine ligand in metalloenzymes," *Journal of Physical Chemistry A*, vol. 106, no. 14, pp. 3377–3390, 2002.

[13] M. R. Burke, M. F. Richardson, and P. J. McCarthy, "Crystal structure and single-crystal polarized spectrum of 1,4,7,10-tetraazadecanebis(N-methylimidazole)-nickel(II) perchlorate," *Journal of Crystallographic and Spectroscopic Research*, vol. 14, no. 2, pp. 143–155, 1984.

[14] L. P. Battaglia, A. B. Corradi, L. Menabue, G. C. Pellacani, P. Prampolini, and M. Saladini, "Ternary complexes of copper (II) with N-protected amino-acids and N-methylimidazole. Crystal and molecular structures of bis(N-acetyl-α-alaninato) bis(N-methylimidazole)copper(II) dihydrate," *Journal of the Chemical Society, Dalton Transactions*, no. 4, pp. 781–785, 1982.

[15] L. Battaglia, A. B. Corradi, G. Marcotrigiano, L. Menabue, and G. Pellacani, "Role of the tosyl group in the coordination ability of N-protected amino acids. 3. Ternary complexes of copper (II) with N-tosylglycine and aromatic amines. Crystal and molecular structure of bis(N-tosylglycinato)bis(N-methylimidazole)copper(II)," *Inorganic Chemistry*, vol. 22, no. 13, pp. 1902–1906, 1983.

[16] L. Antolini, L. Menabue, M. Saladini, M. Sola, L. P. Battaglia, and A. B. Corradi, "Imidazole-containing ternary complexes of N-benzyloxycarbonyl-aminoacids. Crystal and molecular structure of bis(N-benzyloxycarbonyl-alaninato)bis-(N-methylimidazole)copper(II) ethanol solvate," *Inorganica Chimica Acta*, vol. 93, no. 2, pp. 61–66, 1984.

[17] A. E. Elia, B. Santarsiero, E. Lingafelter, and V. Schomaker, "(N-Methylimidazole)(2,3,9,10-tetramethyl-1,4,8,11-tetraazacyclotetradeca-1,3,8,10-tetraene)copper(II) hexafluorophosphate," *Acta Crystallographica Section B: Structural Crystallography and Crystal Chemistry*, vol. 38, no. 12, pp. 3020–3023, 1982.

[18] W. Clegg, J. Nicholson, D. Collison, and C. Garner, "Structures of four complexes of copper with N-methylimidazole and chloro ligands," *Acta Crystallographica Section C: Crystal Structure Communications*, vol. 44, no. 3, pp. 453–461, 1988.

[19] R. E. Norman, N. J. Rose, and R. Stenkamp, "Mono-amino-acid–copper complexes: syntheses and structures of chloro (glycinato)(methanol)copper(II) and chloro(glycinato)(1-methylimidazole)copper(II)," *Acta Crystallographica Section C: Crystal Structure Communications*, vol. 46, no. 1, pp. 1–6, 1990.

[20] W. Clegg, S. R. Acott, and C. D. Garner, "Structure of tetrakis (N-methylimidazole-N')copper(I) perchlorate,[Cu(C4H6N2) 4][ClO4]," *Acta Crystallographica Section C: Crystal Structure Communications*, vol. 40, no. 5, pp. 768-769, 1984.

[21] S. Tetteh and R. Zugle, "Theoretical study of terminal vanadium(V) chalcogenido complexes bearing chlorido and methoxido ligands," *Journal of Chemistry*, vol. 2017, Article ID 6796321, 8 pages, 2017.

[22] K. Hensen, T. Zengerly, T. Müller, and P. Pickel, "Ionische strukturen von 4-bzw. 5fach koordiniertem Silicium Neue ionische Festkörperstrukturen von 4-bzw. 5fach koordiniertem Silicium:[Me3Si (NMI)]+ Cl–,[Me2HSi (NMI) 2]+ Cl–,[Me2Si (NMI) 3] 2+ 2 Cl–· NMI," *Zeitschrift für Anorganische und Allgemeine Chemie*, vol. 558, no. 1, pp. 21–27, 1988.

[23] S. L. Childs, P. A. Wood, N. Rodrĩguez-Hornedo, L. S. Reddy, and K. I. Hardcastle, "Analysis of 50 crystal structures containing carbamazepine using the materials module of mercury CSD," *Crystal Growth and Design*, vol. 9, no. 4, pp. 1869–1888, 2009.

[24] S. Fatma, A. Bishnoi, V. Singh et al., "Spectroscopic and electronic structure calculation of a potential antibacterial agent incorporating pyrido-dipyrimidine-dione moiety using first principles," *Journal of Molecular Structure*, vol. 1110, pp. 128–137, 2016.

[25] A. A. A. Aziz, F. M. Elantabli, H. Moustafa, and S. M. El-Medani, "Spectroscopic, DNA binding ability, biological activity, DFT calculations and non linear optical properties (NLO) of novel Co (II), Cu (II), Zn (II), Cd (II) and Hg (II) complexes with ONS Schiff base," *Journal of Molecular Structure*, vol. 1141, pp. 563–576, 2017.

[26] A. V. Mitin, J. Baker, and P. Pulay, "An improved 6-31G∗ basis set for first-row transition metals," *Journal of Chemical Physics*, vol. 118, no. 17, pp. 7775–7782, 2003.

[27] S. I. Gorelsky, L. Basumallick, J. Vura-Weis et al., "Spectroscopic and DFT investigation of [M{HB(3,5-iPr2pz)3}

(SC6F5)](M = Mn, Fe, Co, Ni, Cu, and Zn) model complexes: periodic trends in metal–thiolate bonding," *Inorganic Chemistry*, vol. 44, no. 14, pp. 4947–4960, 2005.
[28] A. D. Kulkarni and D. G. Truhlar, "Performance of density functional theory and Møller–Plesset second-order perturbation theory for structural parameters in complexes of Ru," *Journal of Chemical Theory and Computation*, vol. 7, no. 7, pp. 2325–2332, 2011.
[29] A. M. Mansour, "Coordination behavior of sulfamethazine drug towards Ru(III) and Pt(II) ions: synthesis, spectral, DFT, magnetic, electrochemical and biological activity studies," *Inorganica Chimica Acta*, vol. 394, pp. 436–445, 2013.
[30] M. Sircoglou, S. Bontemps, M. Mercy et al., "Transition-metal complexes featuring Z-type ligands: agreement or discrepancy between geometry and d^n configuration?," *Angewandte Chemie International Edition*, vol. 46, no. 45, pp. 8583–8586, 2007.
[31] J. Carpenter and F. Weinhold, "Analysis of the geometry of the hydroxymethyl radical by the "different hybrids for different spins" natural bond orbital procedure," *Journal of Molecular Structure*, vol. 169, pp. 41–62, 1988.
[32] G. Frenking, "Understanding the nature of the bonding in transition metal complexes: from Dewar's molecular orbital model to an energy partitioning analysis of the metal–ligand bond," *Journal of Organometallic Chemistry*, vol. 635, no. 1-2, pp. 9–23, 2001.
[33] N. M. Kostic and R. F. Fenske, "Molecular orbital study of bonding, conformations, and reactivity of transition-metal complexes containing unsaturated organic ligands. Electrophilic and nucleophilic additions to acetylide, vinylidene, vinyl, and carbene ligands," *Organometallics*, vol. 1, no. 7, pp. 974–982, 1982.

4

Bioremoval of Different Heavy Metals by the Resistant Fungal Strain *Aspergillus niger*

Ismael Acosta-Rodríguez,[1] Juan F. Cárdenas-González,[2] Adriana S. Rodríguez Pérez,[2] Juana Tovar Oviedo,[1] and Víctor M. Martínez-Juárez[3]

[1]*Universidad Autónoma de San Luis Potosí, Facultad de Ciencias Químicas, Centro de Investigación y de Estudios de Posgrado, Laboratorio de Micología Experimental, Av. Dr. Manuel Nava No. 6, Zona Universitaria, C.P. 78320, San Luis Potosí, Mexico*
[2]*Unidad Académica Multidisciplinaria Zona Media, Universidad Autónoma de San Luis Potosí, Carretera San Ciro de Acosta Km. 4, Ejido Puente del Carmen. C.P. 79617, Río Verde, San Luis Potosí, Mexico*
[3]*Área Académica de Medicina Veterinaria y Zootecnia, Instituto de Ciencias Agropecuarias, Universidad Autónoma Del Estado de Hidalgo, Zona Universitaria, Rancho Universitario Km 1. C.P. 43600, Tulancingo de Bravo Hidalgo, Mexico*

Correspondence should be addressed to Ismael Acosta-Rodríguez; iacosta@uaslp.mx

Guest Editor: Raja Sulaiman Azarudeen

The objective of this work was to study the resistance and removal capacity of heavy metals by the fungus *Aspergillus niger*. We analyzed the resistance to some heavy metals by dry weight and plate: the fungus grew in 2000 ppm of zinc, lead, and mercury, 1200 and 1000 ppm of arsenic (III) and (VI), 800 ppm of fluor and cobalt, and least in cadmium (400 ppm). With respect to their potential of removal of heavy metals, this removal was achieved for zinc (100%), mercury (83.2%), fluor (83%), cobalt (71.4%), fairly silver (48%), and copper (37%). The ideal conditions for the removal of 100 mg/L of the heavy metals were 28°C, pH between 4.0 and 5.5, 100 ppm of heavy metal, and 1 g of fungal biomass.

1. Introduction

Heavy metals are ubiquitous contaminants that have accompanied the man from the earliest ancient times, and unlike other environmental pollutants, heavy metals are chemical elements that man does not create or destroy. The role that man plays in the environmental presence of metals is to introduce into the environment these elements as a result of different human activities, and on the other hand, to alter the chemical or biochemical form in which they are. Metals are naturally subjected to biogeochemical cycles that determine their presence and concentration in different natural environments such as soil, groundwater and surface, air, and living beings. Human intervention can greatly modify the concentration of metals in these environments and facilitate their distribution from the mineral reserves in which metals are naturally confined [1]. From the toxicological point of view, metals often present a marked multiplicity of toxic effects. The specific chemical species of the metal strongly influences these effects, as well as the toxicokinetic variables of absorption, distribution, and excretion. The toxicological significance of heavy metals, taking into account their ubiquity, the extent of their industrial and domestic uses, as well as their environmental persistence, which has to be evaluated based on the characteristics of the chemical compound of which the metal is part, and which determine its mobility environmental and its bioavailability [2].

In Mexico, there are reports of the presence of heavy metals in rivers, lakes, crops, soils, and air urban areas, as well as in coastal environments and marine ecosystems, where the accumulation of toxic metals in fish and shellfish tissues of human consumption is seen [3, 4], and mining is one of the main causes of environmental pollution by heavy metals, mainly due to the inadequate management of their so-called mine tailings. There are reports of widespread contamination in states of the Mexican Republic such as Zacatecas, San Luis Potosí, Guerrero, and Sonora. [4–6]. Reports indicate that in Mexico there could be million tons of pull at the national level, of which they are still unknown the conditions and their potential to affect the environment [1, 7]. The most common example is the soil contamination, which occurs during the extraction of gold and silver, commonly made by amalgamation with mercury and cyanidation. In neither case, there is a total recovery of the compounds or added elements, so it is common to find them in the process residues (mining sludges) in the soluble form [1, 4, 7], so that, the "tailings" contain a great quantity of residual metals that derive from a process of extraction that is not 100% efficient, so they exceed the maximum permissible limits of these metals for soils and waters in Mexico, established in the NOM-147-SEMARNAT/SSA1-2004 [8] and NOM-001-SEMARNAT-1996 [9], respectively. In humans, heavy metals can become very toxic when introduced into the organism. At high concentrations, these can cause skin rashes, stomach upset (ulcers), respiratory problems, weakening of the immune system, damage to the kidneys and liver, hypertension, alteration of genetic material, cancer, neurological disorders, and even death [10]. Despite existing legislation on disposal and waste management, it is clear that the problem persists. The foregoing, together with the waste from mining, aggravates the situation of the contamination by heavy metals in Mexico. Different public institutions have developed investigations to establish the magnitude of this problem and have proposed strategies to contribute to the solution of the same, focusing on the use of biological alternatives that result in a lower alteration of the environment, specifically through the use of microorganisms for the removal of heavy metals or biosorption [11]. There are many reports of the isolation of resistant microorganisms to heavy metals and the use of microbial biomass for the removal of heavy metals, from industrial wastewater and/or contaminated water: the resistance and removal of *Rhizopus stolonifer* to lead, cadmium, copper, and zinc [11], the tolerance and removal mechanisms of heavy metals (lead, cadmium, and chromium), by the fungus *Pleurotus ostreatus* HAAS [12], *Bacillus megaterium* nickel resistance and her capacity of removal [13], heavy metal susceptibility and removal potential (mercury, copper, and lead) of *Rhodotorula mucilaginosa* [14], the resistance of *Alcaligenes* sp. BAPb.1 to lead (II), copper (II), zinc (II), nickel (II), and chromium (VI), and his capacity for removal of lead (II) [15], the isolation and identification of fungi and yeast resistant to lead (II) [16], the resistance and removal of chromium (VI) by *Aspergillus niger* [17], the removal of different heavy metals by *A. niger* [18], the removal of lead, cadmium, copper, and nickel by *A. niger* [19], the removal of aluminum, iron, lead, and zinc by *A. niger* during the bioleaching process [12], and the removal of copper (II), manganese (II), zinc (II), nickel (II), iron (III), lead (II), and cadmium (II) by immobilized cells of *A. niger* [20], with highly satisfactory results. This work reports the removal of different heavy metal in an aqueous solution by a strain of *A. niger* which is highly resistant to some heavy metals.

2. Experimental

2.1. Microorganisms and Heavy Metals Resistant Tests. The fungal strain of *A. niger* was isolated from the polluted air in a fuel station, near to the Faculty of Chemical Science, belonging to the Autonomous University of San Luis Potosí (San Luis Potosi, Mexico) [17], and this was used for the screening. In addition to the above, this fungus was conditioned for years under conditions of biological stress and was inoculated in culture media containing between 0 and 500 ppm of different heavy metals such as chromium, lead, cadmium, arsenic, etc. For the isolation, growth, and pH calibration, we carried out the methodology of Acosta-Rodríguez et al. [21] as follows: on Petri dishes containing modified Lee's minimal medium (LMM) (with 0.25% KH_2PO_4, 0.20% $MgSO_4$, 0.50% $(NH_4)_2SO_4$, 0.50% NaCl, 0.25% glucose, and 2% agar) supplemented with 500 mg/L of K_2CrO_4. The pH of the medium was adjusted with a pH meter Corning Pinnacle 540 and maintained at 5.3 with 100 mmol/L of citrate phosphate buffer. The plates were incubated at 28°C for 7 days. Fungal cultures grown in thioglycolate broth were used as primary inoculums. Heavy metals resistant tests of the isolated strain, the fungi *A. niger*, were performed on liquid LMM containing the appropriate nutritional requirements and different concentrations of heavy metals (as salt), and the dry weight was determined.

2.2. Heavy Metal Resistance Assay. For the resistance test, we followed the methods of Acosta-Rodríguez et al. [21], where Petri dishes were prepared with Sabouraud Dextrose Agar, added with different heavy metals salts. The prepared plates were inoculated with 1×10^6 spores/mL, uniformly spread throughout the dishes, and incubated at 28°C for 7 days, and the growth of the plates was compared with a control.

2.3. Obtaining the Fungal Biomass. For their propagation, 1000 mL Erlenmeyer flasks containing 600 mL of thioglycolate broth (8 g/L) were used. The prepared flasks were inoculated with 1×10^6 spores/mL and were incubated at 28°C, with constant stirring (100 rpm) [21]. After 5 days of incubation, the cells were filtered in Whatman paper No. 1, washed twice with trideionized water, and then dried at 80°C for 12 h in an oven. Finally, the fungal biomass was milled and stored in an amber bottle at room temperature until their use.

2.4. Preparation of Iron Oxide-Coated Biomass. 80 mL of 2 M $Fe(NO_3)_3 \cdot 9H_2O$ was prepared and 1.0 mL of 10 M·NaOH was added to this solution and mixed thoroughly. 20 g of the fungal biomass powder was taken in a porcelain pot, and a mixture of iron oxide and NaOH solution was added to the porcelain pot and homogenized, kept in an oven for 3 h at 80°C. Afterwards, the oven temperature was raised to 110°C and continued for 24 h. The coated biomass powder was separated by crushing with mortar and pestle [22].

2.5. Biosorption Tests for Heavy Metals by Fungal dry Cells. Solutions of heavy metals for analysis were prepared by diluting 1 g/L of stock metal solution. The concentration range of heavy metals solutions was 1–100 mg/L. The pH of each solution was adjusted to the required value by adding 1 M·H_2SO_4 solution before mixing with the fungus with an analyzer Corning Pinnacle 540. The biosorption of the metals by fungal dry cells was determined at different concentrations of 100 mL heavy metal solution, with 1 g of fungal biomass, at 100 rpm, and the sample was filtered, and the supernatant was analyzed for residual heavy metals at different times after a contact period: zinc (II), lead (II), mercury (II), cadmium (II), spectrophotometrically with a Genesys 10S Vis and the dithizone method [23], cobalt (II) by methyl isobutyl ketone [24], fluorine (I) by specific ion, and copper (II), arsenic (III), arsenic (V), and silver (I), by atomic absorption with spectrophotometer Varian SpectrAA 220 [25]. Moreover, biosorption to the contaminated water was examined. To six Erlenmeyer glass flasks add 5 g of fungal biomass and 95 mL of water (263 mg/L of lead (II), 183 mg/L of Mercury (II), and 250 mg/L of cobalt (II)), from the farmland of the "Tanque Tenorio" (which is southeast of the city, in the municipality of Soledad de Graciano Sánchez, S.L.P., Mexico, and is a catchment lagoon of wastewater, of which 60% and 40% are from urban and industrial origin, respectively) (it should be noted that the industrial zone of San Luis Potosí has more than 520 companies, among which are the mining-metallurgists, textiles, and chemicals) [25], and they were incubated during 7 days, stirred at 100 rpm, and filtered in Whatman filter paper No. 1, and the concentration of lead (II), Mercury (II), and zinc (II) of the filtrate was analyzed.

3. Results and Discussion

3.1. Isolation and Identification of a Fungal Strain Resistant to Heavy Metals. The fungal strain was kept in culture medium like LMM containing different concentrations of heavy metals for many years, which caused mechanisms of adaptation and resistance to these metals, causing the fungus to not die intoxicated and could remove several of them. This indicates that these fungi developed the heavy metal tolerance and/or resistance, and they were identified by their macroscopic and microscopic characteristics [26]. In a previous study, it has been reported that the fungus grew in 2000 mg/L (42 μg of dry weight) of chromium (VI), and it presents very good adsorption capacity of chromium (VI) in different conditions [17]. Also, the strain grew on LMM supplemented with different concentrations of heavy metals, about 37.6%, 24.6%, and 13.5%, of zinc (II), mercury (II), and lead (II), respectively, of growth relative to control without metal, and, therefore, probably is resistant to the metals, although, it grew a 16% with 1.4 g/L of arsenic (III), and it is very sensitive to cobalt (II) (12.8% with 600 mg/L) (Figure 1). On the other hand, in plate-resistant testing, the fungus grows in 2000 mg/L of zinc (II), lead (II), mercury (II), and chromium (VI), 1200 g/L of Arsenic (III), 600 mg/L of cobalt (II), and 400 mg/L of cadmium (II) (Table 1), showing that the fungus has the ability to grow at very high concentrations of these different toxic metals, and it can present different mechanisms of resistance and/or adaptation to toxic metals.

Different microorganisms that are heavy metals resistant have been isolated from different contaminated sites: screening the resistance to lead, cadmium, copper, and zinc of five fungal species isolated from soils: *Emericella quadrilineata*, *A. niger*, *Macrophomina phaseolina*, *R. stolonifer*, and *Aspergillus fumigatus*, and the most resistant fungal species (1 g/L of metals) was *R. stolonifer* followed by *M. phaseolina* which showed resistance with all the metals, while *A. niger*, *A. fumigatus*, and *E. quadrilineata* were more sensitive to these heavy metals [11], the fungus *P. ostreatus* HAAS grew very well in 500 mg/L of lead, and concentrations of 30 mg/L of cadmium and 200 mg/L of chromium appeared to inhibit the growth of the fungus [12], *B. megaterium* strain MNSH1-9K-1 tolerate up to 200 ppm of each nickel and vanadium [13], *Alcaligenes* sp. BAPb.1, grow in the presence of 1000 mg/L of lead (II), 600 mg/L of copper (II), 600 mg/L of zinc (II), 400 mg/L of nickel (II) and chromium(VI) [15], *Penicillium* sp., *Trichoderma* sp., and *Alternaria* sp., isolated from the farmland of the "Tanque Tenorio", grow with 500–2000 mg/L of lead (II) [16], *A. niger* has been growing in the presence of different concentrations of metals like nickel, cobalt, iron, magnesium, and manganese [27], and the yeast *Candida tropicalis*, isolated from wastewater from industrial area of Sheikhupura, which is grown in 2.5 g/L of cadmium (II), zinc (II) (1.4 g/L), nickel (II) (1 g/L), Mercury (II) (1.4 g/L), copper (II) (1 g/L), chromium (VI) (1.2 g/L), and lead (II) (1 g/L) [28].

3.2. Removal of Different Heavy Metals by Fungal Biomass of A. niger. On the other hand, we analyzed the capacity of heavy metals removal by dry cell of the fungus. The results are shown in Table 2. The fungus removed efficiently most of the heavy metals analyzed: zinc (II) (100%), mercury (II) (83.2%), fluor (I) (83%), and cobalt (II) (71.4%), and less efficiently: silver (I) (48%) and copper (37%). Dead fungal cells can be effective metal accumulators, and there is evidence that some biomass-based cleanup processes are economically viable [2, 4]. The tolerance of some fungal species to heavy metals, as well as the physiological response to them, has been also determined [2, 11–13]. The removal of heavy metal ions, using fungus as biosorbents, was previously investigated [2, 4, 12, 17–20]. Our results confirm the capacity of the microorganisms biomass for the removal of heavy metals with different effectivity, like bacteria, fungus, yeast, and algae-based microbiological decontamination of heavy metals contaminated soils of different places [2–5]: *M. phaseolina* and *R. stolonifer* for the removal of lead, cadmium, copper, and zinc, from soil [11], the removal of lead, cadmium, and chromium, in liquid culture with *P. ostreatus* HAAS [12], the removal of lead, cadmium, copper, and nickel, with *A. niger* [19], the elimination of copper, cadmium, lead, and zinc in dried soil residues with *A. niger* during the bioleaching process [12], the removal of copper (II) and cadmium (II) in batch systems by immobilized cells of *A. niger* [20], the removal of 90% of chromium (VI) by

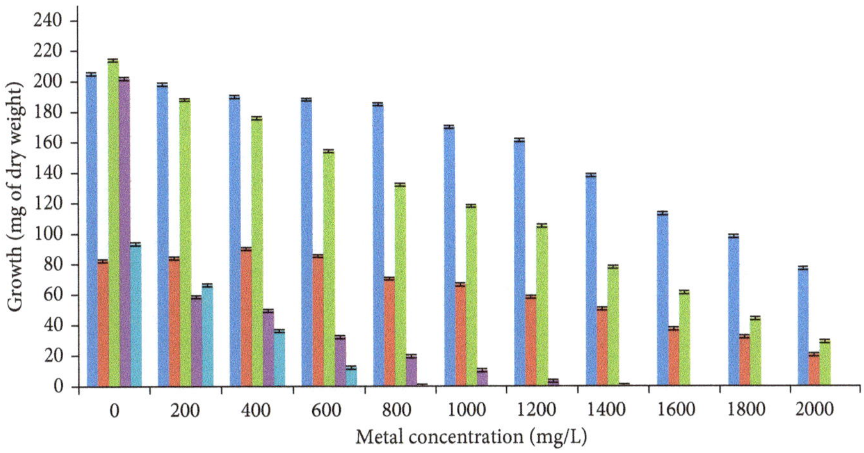

FIGURE 1: Growth in dry weight of *Aspergillus niger* with different heavy metals concentrations: 1×10^6 spores/mL; 28°C, 7 days of incubation; 100 rpm.

TABLE 1: Growth in LMM in plate of *Aspergillus niger* with different heavy metals: 1×10^6 yeast/mL, 28°C, and 7 days of incubation.

Heavy metal	Growth Heavy metal concentration (mg/L)
Zinc (II)	2000
Lead (II)	2000
Mercury (II)	2000
Chromium (VI)	2000
Arsenic (III)	1200
Arsenic (V)	1000
Copper	1000
Silver	1000
Fluor	800
Cobalt	600
Cadmium	400

TABLE 2: Removal of different heavy metals by fungal biomass of *A. niger*: 28°C, 1 g of fungal biomass, 100 rpm, 24 h.

Heavy metals	pH	Initial concentration (mg/L)	% removal
Chromium (VI)	1.0	50	100*
Zinc (II)	5.0	100	100**
Mercury (II)	5.5	100	83.2
Fluor (I)	6.0	10	83.0
Cobalt (II)	5.0	100	71.4
Arsenic (V)	6.0	1.0	69***
Arsenic (III)	6.0	1.0	66***
Lead (II)	4.0	100	59.0
Cadmium (II)	6.0	5.0	57.0
Silver (I)	6.0	100	48.0
Copper (I)	5.0	100	37.0

*30 minutes [17]. **165 minutes. ***Fungal biomass modified with Fe $(NO_3)_3 \cdot 9H_2O$.

NaOH-pretreated *A. niger* biomass, and that heavy metal uptake by live *A. niger* biomass for cadmium (II) and for zinc (II) [29], yeasts isolated from water, soil, and plant environments [30], and other studies with other species of fungi [4, 18, 19, 31–34]. According to our results, we can assume the surface of the biomass coated with iron oxide is partially ionized, causing the pH to approach neutrality; apparently, the OH^- groups present compete for the binding sites with the heavy metals and the biomass, decreasing the removal of metals, while at acid pH, there is a better removal of the heavy metals [35].

3.3. Removal of Heavy Metals in Industrial Wastes with Fungal Biomass. For analyzing the possible use and the ability of *A. niger* biomass to removal of lead (II), cobalt (II), and mercury (II), from wastewater, a removal assay was mounted in an aqueous solution in the presence of 5 g biomass, with 95 mL of nonsterile water contaminated (from "Tanque Tenorio") with 263 mg/L of lead (II), 183 mg/L of mercury (II), and 250 mg/L of cobalt (II), at pH 5.0 (adjusted), 28°C and stirring at 100 rpm. It was observed that at 7 days of incubation, 71%, 69%, and 96.4%, of the heavy metals present in the water contaminated were removal, respectively (Figure 2). The metal removal capability by the biomass of *A. niger* is equal to or greater than the other biomasses that have been studied, like the removal of mercury, cadmium, an copper (4.79%, 10.25%, and 5.49%, respectively), using *R. mucilaginosa* planktonic cells during 48 hours [17], the metal removals during two-step process using *A. niger* reached 84.3%, 84.4%, 25%, and 14.4% for copper, cadmium, lead, and zinc, respectively [12], the removal of cadmium (II) (95%), lead (II) (88%), iron (III) (70%), copper (II) (60%), nickel (II) (48.9%), manganese (II) (37.7%), and zinc (II) (15.4%), from industrial wastewater in batch systems by immobilized cells of *A. niger* [20], the use of the extracellular media of *Alternaria alternata*-containing organic acids and siderophores for the metal leaching (vanadium, aluminum, molybdenum, magnesium, iron, nickel, arsenic, and chromium) [32], the

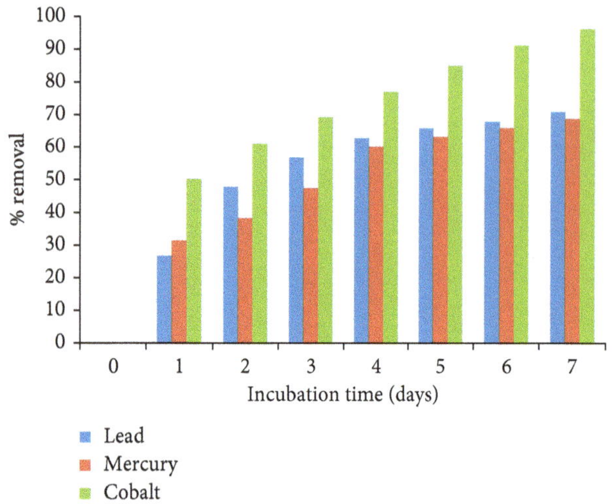

Figure 2: Removal of heavy metals from industrial wastes incubated with 5 g of fungal biomass and 95 mL of water contaminated with 263 mg/L of lead (II), 183 mg/L of mercury (II), and 250 mg/L of cobalt (II), at pH 5.0 (adjusted), 28°C, and 100 rpm.

removal of 67% of arsenic (III) from samples of groundwater contaminated with 1 mg/L from the metalloid, coming from Zimapan, Hidalgo's state, Mexico [36], the 99.35% removal of copper with pure and modified chitosan hydrogels from shrimp shell, from copper leachate [37], *Saccharomyces cerevisiae* and *Torulaspora delbrueckii* decrease in 98.1%, 83.0%, 60.7%, 60.5%, and 54.2% for turbidity, sulphates, BOD, phosphates, and COD, respectively, of the tannery effluent [38], *C. tropicalis* removed 40% of cadmium (II) from the wastewater after 6 days and was also able to remove 78% from the wastewater after 12 days [28], and *S. cerevisiae* "wild-type" (WT) parental strain BY4741, very efficient in removing manganese (II), copper (II), and cobalt (II) from synthetic effluents containing 1-2 mM cations [39]. Industrial effluents often contain more than one type of metal ion; these may interfere in the removal/recovery of the metal ion of interest. Limited information about the effect of cocations is available in the literature. The presence of other cations (cocations) can affect the sorption of metal ions (primary cation) to the biomass, and in some cases, it may affect the removal efficiency [40].

4. Conclusion

We isolated a fungus, which was identified such as *A. niger*, which grow with different heavy metals in LMM, and probably is resistant to the metals. The dead fungal biomass removed efficiently different heavy metals (chromium (VI) and zinc (II) (100%), mercury (II) (83.2%), and fluorine (I) (83%)) at different pH conditions (4.0 for lead (II), 5.0 for zinc (II), and cobalt (II), and 5.5 for mercury (II)), 28°C, and 1 g of fungal biomass. Finally, these results suggest the potential applicability of this fungus for the remediation of heavy metals from polluted soils and waters.

Acknowledgments

This research work was carried out without funds from any institution, but nevertheless it was done with the collaboration of Professors of the Autonomous University of San Luis Potosi, the Multidisciplinary Academic Unit Central Zone-UASLP, and the Autonomous University of the State of Hidalgo, where each teacher uses own and particular resources.

References

[1] L. Albert, "Gestión de los productos químicos," in *En: México Tóxico*, L. A. Albert and M. Jacott, Eds., pp. 21–37, XXI Siglo veintiuno editores, Mexico, 1st edition, 2015.

[2] J. Zaidi and A. Pal, "Review on heavy metal pollution in major lakes of India: remediation through plants," *African Journal of Environmental Science and Technology*, vol. 11, no. 6, pp. 255–265, 2017.

[3] I. Mena Mejia, A. Bustamante González, S. Vargas López, J. I. Olvera Hernández, and J. A. Méndez Espinosa, "Evaluación de la condición ecológica del río Zahuapan," *Revista Internacional de Contaminación Ambiental*, vol. 33, no. 1, pp. 7–19, 2017.

[4] S. A. Covarrubias and J. J. Peña Cabriales, "Contaminación ambiental por metales pesados en México: problemática y estrategias de fitorremediación," *Revista Internacional de Contaminación Ambiental*, vol. 33, no. 2, pp. 7–21, 2017.

[5] E. V. Cortés-Jiménez, V. Mugica-Álvarez, M. C. González-Chávez, R. Carrillo-González, M. Martínez-Gordillo, and M. Vaca-Mier, "Natural revegetation of alkaline tailing heaps at Taxco, Guerrero, Mexico," *International Journal of Phytoremediation*, vol. 15, pp. 127–141, 2013.

[6] E. Ortíz-Ortíz, E. García-Nieto, L. Juárez-Santacruz, M. A. Gómez-Camarillo, E. García-Gallegos, and P. Limón-Huitrón, "Lead exposure: pottery impact in Tlaxcala, México," *Revista Internacional de Contaminación Ambiental*, vol. 33, no. 1, pp. 57–64, 2017.

[7] Y. R. Ramos-Arroyo and C. D. Siebe-Grabach, "Estrategia para identificar jales con potencial de riesgo ambiental en un distrito minero: estudio de caso en el Distrito de Guanajuato, México," *Revista Mexicana de Ciencias Geológicas*, vol. 23, pp. 54–74, 2006.

[8] SEMARNAT, Norma Oficial Mexicana NOM- 001-SEMARNAT 1996. Que establece los límites máximos permisibles de contaminantes en las descargas de aguas residuales en aguas y bienes nacionales. Secretaría de Medio Ambiente, Recursos Naturales y Pesca. Diario Oficial de la Federación, 2003.

[9] SEMARNAT, Norma Oficial Mexicana. NOM- 147-SEMARNAT/SSA1-2004. Que establece los criterios para determinar las concentraciones de remediación de suelos contaminados por arsénico, bario, berilio, cadmio, cromo hexavalente, mercurio, níquel, plata, plomo, selenio, talio y/o vanadio. Secretaría de Medio Ambiente, Recursos Naturales y Pesca. Diario Oficial de la Federación. 2 de marzo de, 2007.

[10] C. Nava Ruíz and M. Méndez Armenta, "Efectos neurotóxicos de metales pesados (cadmio, plomo, arsénico y talio)," *Archivos de Neurociencias (México)*, vol. 16, no. 3, pp. 140–147, 2011.

[11] E. M. Fawzy, F. F. Abdel-Motaal, A. Soad, and S. A. El-Zayat, "Biosorption of heavy metals onto different eco-friendly substrates," *Journal of Bioremediation and Biodegradation*, vol. 8, no. 3, pp. 1-7, 2017.

[12] J. Yang, Q. Wang, Q. Luo, Q. Wang, and T. Wud, "Biosorption behavior of heavy metals in bioleaching process of MSWI fly ash by *Aspergillus niger*," *Biochemical and Engineering Journal*, vol. 46, pp. 294–299, 2009.

[13] A. M. Rivas-Castillo, M. E. Guatemala-Cisneros, and N. G. Rojas-Avelizapa, "Effect of aluminum in *Bacillus megaterium* nickel resistance and removal capability," *Mexican Journal of Biotechnology*, vol. 2, no. 2, pp. 206–220, 2017.

[14] S. Grujić, S. Vasić, I. Radojević, L. Čomić, and A. Ostojić, "Comparison of the *Rhodotorula mucilaginosa* biofilm and planktonic culture on heavy metal susceptibility and removal potential," *Water Air Soil Pollution*, vol. 228, no. 73, pp. 1–8, 2017.

[15] Y. Jin, S. Yu, C. Teng et al., "Biosorption characteristic of *Alcaligenes* sp. BAPb.1 for removal of lead (II) from aqueous solution," *3 Biotech*, vol. 7, no. 123, pp. 1–12, 2017.

[16] M. G. Moctezuma Zárate, A. Robles Galván, J. F. Cárdenas González et al., "Isolation and identification of fungi and yeast resistant to lead (II)," *Journal of Multidisciplinay Engineering Science and Technology*, vol. 4, no. 6, pp. 5273–5278, 2017.

[17] I. Acosta-Rodríguez, J. F. Cárdenas-González, M. G. Moctezuma-Zárate, A. Rodríguez Pérez, and V. M. Martínez-Juárez, "Hexavalent chromium (VI) removal by *Aspergillus niger*," in *Metal-Microbe Interactions and Bioremediation: Principles and Applications for Toxic Metals*, pp. 673–688, S. Das, Ed., CRS Press Book, Taylor & Francis, Boca Raton, FL, USA, 2017.

[18] M. Mukhopadhyay, S. B. Noronha, and G. K. Suraishkumar, "A review on experimental studies of biosorption of heavy metals by *Aspergillus niger*," *Canadian Journal of Chemical Engineering*, vol. 89, pp. 889–900, 2011.

[19] A. Kapoor, T. Viraraghavan, and D. R. Cullimore, "Removal of heavy metals using the fungus *Aspergillus niger*," *Bioresource Technology*, vol. 70, pp. 95–104, 1999.

[20] K. Tsekova, D. Todorova, and S. Ganeva, "Removal of heavy metals from industrial wastewater by free and immobilized cells of *Aspergillus niger*," *International Biodeterioration and Biodegradation*, vol. 64, pp. 447–451, 2010.

[21] I. Acosta-Rodríguez, J. F. Cárdenas-González, V. M. Martínez-Juárez, A. Rodríguez Pérez, M. G. Moctezuma-Zárate, and N. C. Pacheco-Castillo, "Biosorption of heavy metals by *Candida albicans*," in *Advances in Bioremediation and Phytoremediation*, N. Shiomi, Ed., INTECH, 2018.

[22] D. Pokhrel and T. Viraraghavan, "Arsenic removal from an aqueous solution by a modified fungal biomass," *Water Research*, vol. 40, no. 3, pp. 549–552, 2006.

[23] A. E. Greenberg, L. S. Clesceri, and A. D. Eaton, *Standard Methods for the Examination of Water and Wastewater*, American Public Health Association, Washington, DC, USA, 18th edition, 1992.

[24] G. Charlot, *Colorimetric Determination of Elements*, Elsevier Publishing Company, Amsterdam, London, 1964.

[25] SEDECO, *Directorio de empresas que operan en las zonas y parques industriales de la ciudad de San Luis Potosí*, Secretaría de Desarrollo Económico, San Luis Potosí, México, 2014.

[26] M. P. Kirk, F. P. Cannon, C. J. David, and A. J. Stalpers, *Dictionary of the fungi*, CABI Publishing, UK, 2001.

[27] M. Valix and L. O. Loon, "Adaptive tolerance behavior of fungi in heavy metals," *Mineral Engineering*, vol. 16, pp. 193–198, 2003.

[28] A. Rehman and M. S. Anjum, "Multiple metal tolerance and biosorption of cadmium by *Candida tropicalis* isolated from industrial effluents: glutathione as detoxifying agent," *Environmental Monitoring Assessment*, vol. 174, pp. 585–595, 2011.

[29] Y. G. Liu, T. Fan, G. M. Zeng et al., "Removal of cadmium and zinc ions from aqueous solution by living *Aspergillus niger*," *Transactions of Nonferrous Metals Society of China*, vol. 16, pp. 681–686, 2006.

[30] R. Vadkertiova and E. Slavikova, "Metal tolerance of yeasts isolated from water, soil and plant environments," *Journal of Basic Microbiology*, vol. 46, pp. 145–152, 2006.

[31] N. G. Rojas-Avelizapa, J. Otamendi-Valdez, and M. Gómez-Ramírez, "Metal leaching from a spent catalyst by *Alternaria alternata*," *Mexican Journal of Biotechnology*, vol. 2, no. 2, pp. 221–231, 2017.

[32] A. Ali, G. Di, A. Mahar et al., "Mycoremediation of potentially toxic trace elements. A biological tool for soil cleanup: a review," *Pedosphere*, vol. 27, no. 2, pp. 205–222, 2017.

[33] M. Beltrán and A. Gómez, "Metales pesados (Cd, Cr y Hg) su impacto en el ambiente y posibles estrategias biotecnológicas para su remediación," *Revista I3+*, vol. 2, no. 2, pp. 82–112, 2015.

[34] L. F. Sala, S. I. García, J. C. González et al., "Biosorción para la eliminación de metales pesados en aguas de desecho," *Real Sociedad Española Química*, vol. 106, no. 2, pp. 114–120, 2010.

[35] A. Aragon, B. Thomson, and J. Chwirka, *Rapid Small-Scale Column Testing for Arsenic Adsorption*, 2004, http://ipec.utulsa.edu/Ipec/conf2002/aragon_thomson_chwirka_50.pdf.

[36] E. E. Santos Domínguez, J. F. Cárdenas, M. E. Torre, V. M. Martínez, A. Rodríguez, and E. I. Acosta, "Bioadsorción de Arsénico (III) en solución acuosa por la Biomasa Modificada de *Aspergillus niger*," *Avances en Ciencias e Ingeniería*, vol. 8, no. 2, pp. 1–10, 2017.

[37] R. G. Sánchez-Duarte, M. R. Martínez-Macias, M. A. Correa-Murrieta, J. Saldival-Cabrales, D. I. Sánchez-Machado, and J. López-Cervantes, "Síntesis de hidrogeles de quitosano a partir de cáscara de camarón para ensayos de bioadsorción de cobre," *Revista Internacional de Contaminación Ambiental*, vol. 33, pp. 93–98, 2017.

[38] M. L. Nguyen and J. Ruey-Shin, "Modification of cross linked chitosan beads with histidine and *Saccharomyces cerevisiae* for enhanced Ni(II) biosorption," *Journal of Taiwan Institute of Chemical Engineering*, vol. 56, pp. 96–102, 2015.

[39] L. Ruta, C. Paraschivescu, M. Matache, S. Avramescu, and I. C. Farcasanu, "Removing heavy metals from synthetic effluents using "kamikaze" *Saccharomyces cerevisiae* cells," *Applied Microbiology and Biotechnology*, vol. 85, pp. 763–771, 2010.

[40] M. J. Tobin, D. G. Cooper, and R. J. Neufeld, "The effects of cation competition on metal adsorption by *Rhizopus arrhizus* biomass," *Biotechnology and Bioengineering*, vol. 31, pp. 282–286, 1988.

5

Synthesis and Cytotoxic Evaluation of Steroidal Copper (Cu (II)) Complexes

Yanmin Huang,[1] Erbin Kong,[1] Junyan Zhan,[1] Shuang Chen,[1] Chunfang Gan,[1] Zhiping Liu,[1] Liping Pang,[1] and Jianguo Cui[1,2]

[1]*Key Laboratory of Beibu Gulf Environment Change and Resources Utilization, College of Chemistry and Material Science, Guangxi Teachers Education University, Nanning 530001, China*
[2]*Guangxi Colleges and University Key Laboratory of Beibu Gulf Oil and Natural Gas Resource Effective Utilization, Qizhou University, Qizhou, China*

Correspondence should be addressed to Jianguo Cui; cuijgl954@126.com

Academic Editor: Giovanni Natile

Using estrone and pregnenolone as starting materials, some steroidal copper complexes were synthesized by the condensation of steroidal ketones with thiosemicarbazide or diazanyl pyridine and then complexation of steroidal thiosemicarbazones or steroidal diazanyl pyridines with Cu (II). The complexes were characterized by IR, NMR, and HRMS. The synthesized compounds were screened for their cytotoxicity against HeLa, Bel-7404, and 293T cell lines in vitro. The results show that all steroidal copper (II) complexes display obvious antiproliferative activity against the tested cancer cells. The IC_{50} values of complexes **5** and **12** against Bel-7404 (human liver carcinoma) are 5.0 and 7.0 μM.

1. Introduction

Metal-based antitumor drugs play a relevant role in antiblastic chemotherapy [1, 2]. Cisplatin is regarded as one of the most effective drugs [3–8], even if severe toxicities and drug resistance phenomena limit its clinical use [9]. Therefore, in recent years, there has been a rapid expansion in research and development of novel metal-based anticancer drugs in order to improve clinical effectiveness, reduce general toxicity, and broaden the spectrum of activity [10–12].

Copper (Cu) is a transition metal that can exist in oxidised and reduced states. This allows it to participate in redox and catalytic chemistry, making it a suitable cofactor for a diverse range of enzymes and molecules. Cu deficiency or toxicity is implicated in a variety of pathological conditions.

Steroid hormones play an important role in the biochemistry of many cancers; a number of steroidal complexes connected to a metal pharmacophore had been designed and synthesized by many research groups, and their physiological activities were evaluated [13, 14].

Thiosemicarbazones have received considerable attention since the discovery of their cytotoxic activity against cancer cells and bacteriostatic effects [15]. As the disruption of copper homeostasis is a pathological feature of cancer cells, copper complexes had been investigated for their potential applications as anticancer drugs [16]. Cu complexes of thiosemicarbazone (TSC) compounds had been explored as antimalarial, antifungal, antinociceptive, and antibacterial agents [17–19]. Cu complexes of bis(thiosemicarbazones) (CuII(btsc)s) had also been investigated as metallodrugs and diagnostic agents [20]. More recently, Adsule et al. [13] investigated the bioactivity of some new steroidal thiosemicarbazones Cu (II) metal complexes and discovered that some compounds had better antineoplastic activity.

In the present study, some steroidal copper complexes were synthesized by the condensation of steroidal ketones with thiosemicarbazide or diazanyl pyridine and then complexation of steroidal thiosemicarbazones or steroidal diazanyl pyridines with Cu (II). The synthesized compounds were screened for their cytotoxicity against HeLa, Bel-7404, and 293T cell lines in vitro.

2. Materials and Methods

2.1. Materials. The sterols were purchased from Sinopharm Chemical Reagent Co., Ltd., Shanghai, China. All chemicals and solvents are of analytical grade from commercial sources. All solvents were used without further purification unless otherwise specified.

2.2. Instrumentation and Methods. Melting points were determined on an X4 apparatus (Beijing Tech Instrument Co. Ltd., Beijing, China) and were uncorrected. The ^1H and ^{13}C NMR spectra were recorded in $CDCl_3$ on a Bruker AV-600 spectrometer at working frequencies of 600 and 150 MHz and a Bruker AV-300 spectrometer at working frequencies of 300 and 75 MHz, respectively. Chemical shifts are expressed in parts per million (δ) values and coupling constants (J) in Hertz. Infrared spectra were measured with a Thermo Scientific Nicolet IS-10 Spectrophotometer (Thermo Scientific, USA). HREIMS was measured on an Agilent 6210 TOFMS instrument (Agilent Technologies, USA). The cell proliferation assay was undertaken by a MTT method using 96-well plates on a MLLTISKAN MK3 analysis spectrometer (Thermo Scientific, Shanghai, China).

Compounds **1** (L^1) and **2** (L^2) were prepared according to the method of [21].

2.3. Synthesis

2.3.1. General Procedure for Preparation of Steroidal Thiosemicarbazones. Steroidal ketone (0.38 mmol) was dissolved in 40 mL 95% ethanol. After the solution was heated to 65°C, a few drops of glacial acetic acid were added to adjust pH to 3–5, and thiosemicarbazide (1.70 mmol) was added. The mixture was stirred at 60–70°C for 6 h (the progress of the reaction was monitored by TLC, $V_{\text{ethyl acetate}} : V_{\text{petroleum ether}} = 1:2$). Then, the reaction was terminated and majority of solvent was evaporated under reduced pressure. Suitable amount of water was added to the reaction mixture, and the product was extracted with CH_2Cl_2. The combined extract was washed with saturated $NaHCO_3$ solution, water, and saturated brine, dried with anhydrous sodium sulfate, and evaporated under reduced pressure. The resulting residue was separated by the column chromatography using a mixture of ethyl acetate : petroleum ether (1 : 2) to give target products.

*3β-Hydroxypregnenolone-20-semicarbazone (**3**, L^3).* White solid, Yield: 78.0%; m.p. 239–241°C; IR (KBr) ν/cm^{-1}: 3427, 3362, 3219, 3162, 2914, 2888, 1605, 1515, 1433, 1282, 1254; ^1H NMR ($CDCl_3$, 300 MHz) δ: 0.60 (3H, s, 18-CH_3), 0.65 (3H, d, J = 6.6, 21-CH_3), 1.21 (3H, s, 19-CH_3), 3.56–3.45 (1H, m, C_3-H), 5.37 (1H, brs, C_6-H), 6.238 (1H, s, -NH_2), 7.22 (1H, s, -NH_2), 8.51 (1H, s, -NH-); ^{13}C NMR ($CDCl_3$, 75 MHz) δ: 179.0 (-C=S), 153.2 (-C=N), 140.8 (5-C), 121.4 (6-C), 71.1 (3-C), 59.0 (14-C), 56.5 (17-C), 56.4 (9-C), 50.1 (13-C), 42.9 (4-C), 48.8 (12-C), 42.2 (8-C), 37.2 (1-C), 36.5 (10-C), 31.8 (7-C), 31.6 (16-C, 2-C), 24.1 (15-C), 21.0 (11-C), 19.4 (21-C), 17.3 (19-C), 13.2 (18-C); HREIMS: m/z 390.2577 $[M+H]^+$ (calcd for $C_{22}H_{36}N_3OS$, 390.2579).

*3β-Acetyloxypregnenolone-20-semicarbazone (**4**, L^4).* White solid, Yield: 74.5%; m.p. 254-255°C; IR (KBr) ν/cm^{-1}: 3414, 1726, 1589, 1506, 1437, 1369, 1252, 846; ^1H NMR (300 MHz, $CDCl_3$) δ: 0.58 (s, 3H, 18-CH_3), 1.02 (s, 3H, 19-CH_3), 1.91 (s, 3H, 21-CH_3), 2.03 (s, 3H, $COCH_3$), 4.65–4.55 (m, 1H, C3-H), 5.38 (d, 1H, J = 4.5, C5-H), 6.52 (br s, 1H, -NH_2), 7.22 (br s, 1H, -NH_2), 8.63 (s, 1H, -NH); ^{13}C NMR (75 MHz, $CDCl_3$) δ: 13.3 (19-C), 17.5 (18-C), 19.3 (11-C), 21.0 (CH_3CO), 21.4 (21-C), 23.3 (15-C), 24.1 (16-C), 27.7 (2-C), 31.7 (8-C), 32.0 (7-C), 36.6 (10-C), 37.0 (1-C), 38.1 (4-C), 38.8 (12-C), 44.2 (13-C), 50.0 (9-C), 56.4 (17-C), 59.0 (14-C), 73.8 (3-C), 122.3 (6-C), 139.7 (5-C), 153.5 (20-C), 170.6 (C=O), 178.9 (C=S); HREIMS: m/z 432.2633 $[M+H]^+$ (calcd for $C_{24}H_{38}N_3O_2S$, 432.2685).

2.3.2. General Procedure for Preparation of Steroidal Diazanyl Pyridine. A mixture of steroidal ketone (1 mmol) and diazanyl pyridine (1 mmol) in 95% ethanol (30 mL) was stirred at 70–80°C for 6 h. After completion of the reaction, the majority of solvent was evaporated and some water was added to this solution. The mixture was extracted with $CH_3COOC_2H_5$ and the extract was washed with saturated brine, dried with anhydrous sodium sulfate, and evaporated under reduced pressure. The resulting residue was chromatographed on a column of silica gel with mixture of petroleum ether/ethyl acetate (1 : 1) to give steroidal diazanyl pyridine.

*3β-Hydroxyoestrone-17-(2′-diazanyl)pyridine (**9**, L^9).* Yellow solid, Yield: 56.0%; m.p. 269–271°C; IR (KBr) ν/cm^{-1}: 3361, 2935, 1601, 1576, 1444, 995, 871, 768; ^1H NMR (600 MHz, DMSO) δ: 0.85 (3H, s, 18-CH_3), 2.36–2.30 (2H, m, C6-H and C9-H), 2.54 (1H, dd, J = 18.6, 9.0, C6-H), 2.70-2.69 (2H, m, C16-H), 6.45 (1H, d, J = 2.4, C4-H), 6.52 (1H, dd, J = 8.4, 2.4, C2-H), 6.67 (1H, t, J = 6.0, 5′-Py-H), 7.06 (1H, d, J = 8.4, C1-H), 7.07 (1H, d, J = 8.4, 3′-Py-H), 7.56 (1H, td, J = 8.4, 1.8, 4′-Py-H), 8.04 (1H, d, J = 3.6, 6′-Py-H), 8.98 (1H, s, -NH), 9.04 (1H, s, -OH); ^{13}C NMR (150 MHz, DMSO) δ: 17.3 (18-C), 23.0 (11-C), 26.1 (15-C), 26.9 (16-C), 29.2 (7-C), 34.4 (6-C), 38.0 (12-C), 40.1 (8-C), 43.8 (9-C), 44.2 (13-C), 52.2 (14-C), 106.4 (3′-Py-C), 112.8 (2-C), 114.3 (4-C), 115.0 (5′-Py-C), 126.1 (1-C), 130.3 (10-C), 137.2 (4′-Py-C), 137.6 (5-C), 147.5 (6′-Py-C), 155.0 (3-C), 158.3 (2′-Py-C), 162.9 (17-C); HREIMS: $[M+H]^+$ 362.2250 (calcd for $C_{23}H_{28}N_3O$, 362.2232).

*3β-Hydroxypregnenolone-20-(2′-diazanyl)pyridine (**10**, L^{10}).* Yellow solid, Yield: 78.8%; m.p. 234-235°C; IR(KBr) ν/cm^{-1}: 3406, 2932, 1599, 1574, 1442, 838, 768; ^1H NMR (600 MHz, DMSO) δ: 0.55 (3H, s, 18-CH_3), 0.94 (3H, s, 19-CH_3), 1.89 (3H, s, 20-CH_3), 3.35–3.20 (1H, m, C3-αH), 4.64 (1H, br s, NH), 5.27 (1H, s, C6-H), 6.69 (1H, t, J = 6.6, 5′-pyridine-H), 7.06 (1H, d, J = 7.2, 3′-pyridine-H), 7.57 (1H, t, J = 7.2, 4′-pyridine-H), 8.05 (1H, d, J = 6.6, 6′-pyridine-H), 9.07 (s, 1H, -OH); HREIMS: m/z 408.3024 $[M+H]^+$ (calcd for $C_{26}H_{38}N_3O$, 408.3015).

2.3.3. General Procedure for Preparation of Copper Complexes. Steroidal ligand (0.1 mmol) and 0.1 mmol $CuCl_2{\cdot}2H_2O$ were added to 8 mL of methanol. The mixture was stirred for 5 hour at 70°C. The reaction was terminated when large precipitant emerged. The resulting suspension was filtered, washed with ethyl acetate and water, and dried in a desiccator over phosphorus pentoxide to give target products.

*[CuL^1Cl_2] (Compound **5**)*. Compound **5** is a mixture of (*S*)- and (*R*)-configuration isomer (**5**-S : **5**-R = 1.7 : 1, ^{1}H NMR data). Gray yellow solid, Yield: 55%; m.p. 245–247°C; IR (KBr) ν/cm^{-1}: 3441, 1604, 1541, 1409, 1452, 1280, 811, 616; ^{1}H NMR (300 MHz, DMSO) δ: 0.81 (s, 1.07H, 18-CH_3, *R*-), 0.86 (s, 1.71H, 18-CH_3, *S*-), 6.44 (s, 1H, C4-H), 6.50 (d, 1H, *J* = 4.5, C2-H), 7.05 (d, 1H, *J* = 4.5, Cl-H), 7.77 (s, 0.19H, -NH_2, *R*-), 8.00 (s, 0.31H, -NH_2, *S*-), 8.65 (s, 0.36H, -NH_2, *S*-), 8.85 (s, 0.25H, -NH_2, *R*-), 9.03 (s, 1H, -OH), 10.30 (s, 0.34H, -NH-, *S*-), 10.64 (s, 0.20H, -NH-, *R*-).

*[CuL^2Cl_2] (Compound **6**)*. Compound **6** is a mixture of (*S*)- and (*R*)-configuration isomer (**6**-S : **6**-R = 1.5 : 1, ^{1}H NMR data). Gray solid, Yield: 66.7%; m.p. 189-190°C; IR (KBr) ν/cm^{-1}: 3416, 2927, 1604, 1534, 1496, 1447, 876, 816, 751; ^{1}H NMR (600 MHz, DMSO): 0.85 (s, 1.85H, 18-CH_3, *S*-), 1.01 (s, 1.25H, 18-CH_3, *R*-), 2.21 (s, 3H, $COCH_3$), 6.79 (s, 1H, C4-H), 6.83 (d, 1H, *J* = 6.0, C2-H), 7.29 (s, 1H, *J* = 6.0, Cl-H), 7.73 (s, 0.38H, -NH_2), 8.04 (s, 0.45H, -NH_2, *S*-), 8.68 (s, 0.48H, -NH_2, *S*-), 8.84 (s, 0.33H, -NH_2, *R*-), 9.023 (s, 1H, -OH), 10.31 (s, 0.40H, -NH-, *S*-), 10.63 (s, 0.33H, -NH-, *R*-); ^{13}C NMR (150 MHz, DMSO) δ: 16.5 (18-C), 20.6 (CH_3CO), 22.3 (11-C), 26.1 (15-C), 26.4 (16-C), 28.6 (7-C), 28.8 (6-C), 33.5 (12-C), 37.1 (8-C), 37.5 (8-C), 43.1 (9-C), 44.6 (13-C), 51.6 (14-C), 118.6 (2-C), 121.2 (4-C), 126.0 (1-C), 136.8 (5-C), 137.4 (10-C), 148.0 (3-C), 154.7 (17-C), 169.1 ($COCH_3$), 170.1 (C=S), 171.3 (C=S).

*[CuL^3Cl_2] (Compound **7**)*. Compound **7** is a mixture of (*S*)- and (*R*)-configuration isomer (**7**-S : **7**-R = 1 : 1.5, ^{1}H NMR data). Gray solid, Yield: 60.2%; m.p. 189–191°C; IR (KBr) ν/cm^{-1}: 3319, 1604, 1531, 1434, 1369, 1290, 1050, 950; ^{1}H NMR (600 MHz, DMSO) δ: 0.51 (s, 1.2H, 18-CH_3, *S*-), 0.68 (s, 1.8H, 18-CH_3, *R*-), 0.91 (s, 3H, 19-CH_3), 1.96 (s, 1.2H, 21-CH_3, *S*-), 2.00 (s, 1.8H, 21-CH_3, *R*-), 3.23–3.12 (m, 1H, C3-H), 4.61 (s, 1H, OH), 5.24 (s, 1H, C6-H), 7.75 (s, 0.6 H, -NH_2, *R*-), 8.01 (s, 0.4H, -NH_2, *S*-), 8.73 (s, 0.4H, -NH_2, *S*-), 8.95 (s, 0.6H, -NH_2, *R*-), 10.29 (s, 0.4H, -NH-, *S*-), 10.89 (s, 0.6H, -NH-, *R*-); ^{13}C NMR (150 MHz, DMSO) δ: 12.7 (19-C), 18.7 (18-C), 20.3 (11-C), 20.8 (21-C), 21.9 (15-C), 23.7 (16-C), 27.0 (2-C), 31.0 (8-C), 31.1 (7-C), 35.8 (1-C), 36.6 (10-C), 37.4 (4-C), 41.9 (12-C), 43.0 (13-C), 49.2 (9-C), 55.7 (17-C), 58.2 (14-C), 69.7 (3-C), 121.7 (6-C), 139.2 (5-C), 140.9 (20-C), 169.5 (C=S); HREIMS: *m/z* 521.1197 [M–H]$^-$ (calcd for $C_{22}H_{34}Cl_2CuN_3OS$, 521.1196).

*[CuL^4Cl_2] (Compound **8**)*. Compound **8** is a mixture of (*S*)- and (*R*)-configuration isomer (**8**-S : **8**-R = 1.3 : 1, ^{1}H NMR data). Gray solid, Yield: 75%; m.p. 163–165°C; IR (KBr) ν/cm^{-1}: 3424, 1723, 1596, 1534, 1432, 1364, 1040; ^{1}H NMR (600 MHz, DMSO) δ: 0.50 (s, 1.7H, 18-CH_3, *S*-), 0.671 (s, 1.3H, 18-CH_3, *R*-), 0.94 (s, 3H, 19-CH_3), 1.95 (s, 1.7H, 21-CH_3, *S*-), 1.98 (s, 1.3H, 21-CH_3, *R*-), 2.24 (s, 3H, $COCH_3$), 7.754 (s, 0.43H, -NH_2, *R*-), 7.962 (s, 0.49H, -NH_2, *R*-), 8.666 (s, 0.51H, -NH_2, *S*-), 8.929 (s, 0.52H, -NH_2, *S*-), 10.276 (s, 0.56H, -NH, *S*-), 10.881 (s, 0.40H, -NH, *R*-); ^{13}C NMR (150 MHz, DMSO) δ: 12.7 (19-C), 18.7 (18-C), 20.3 (11-C), 20.8 (21-C), 21.9 (15-C), 23.7 (16-C), 27.0 (2-C), 31.0 (8-C), 31.1 (7-C), 35.8 (1-C), 36.6 (10-C), 37.4 (4-C), 41.9 (12-C), 43.0 (13-C), 49.2 (9-C), 55.7 (17-C), 58.2 (14-C), 69.7 (3-C), 121.7 (6-C), 139.2 (5-C), 140.9 (20-C), 169.5 (C=S).

*[CuL^9Cl_2] (Compound **11**)*. Gray solid, Yield: 62.3%; m.p. 210–212°C; IR (KBr) ν/cm^{-1}: 3441, 3324, 2962, 1654, 1611, 1559, 1534, 1501, 1442, 916, 848, 744; ^{1}H NMR (600 MHz, DMSO): 1.51 (3H, s, 18-CH_3), 2.84–2.73 (3H, m, C16-H and C6-H), 6.49 (1H, s, C4-H), 6.56 (1H, br s, C2-H), 7.12 (1H, d, *J* = 6.0, 5′-Py-H), 7.45 (1H, br s, 3′-Py-H), 7.86 (1H, br s, 4′-Py-H), 8.08 (1H, d, *J* = 8.4, Cl-H), 8.74 (1H, br s, 6′-Py-H), 9.11 (1H, s, -NH); ^{13}C NMR (150 MHz, DMSO) δ: 16.1 (18-C), 20.6 (15-C), 20.8 (11-C), 25.9 (7-C), 26.1 (6-C), 29.3 (12-C), 35.9 (16-C), 38.3 (8-C), 42.4 (13-C), 65.0 (14-C), 113.1 (2-C), 114.7 (4-C), 115.3 (3′-Py-C), 117.2 (5′-Py-C), 125.2 (4′-Py-C), 126.2 (1-C), 129.3 (10-C), 133.5 (6′-Py-C), 137.0 (5-C), 143.9 (2′-Py-C), 146.2 (17-C), 155.3 (3-C).

*[$CuL^{10}Cl_2$] (Compound **12**)*. Green solid, Yield: 30.7%; m.p. 224–225°C; ^{1}H NMR (600 MHz, DMSO) δ: 0.96 (3H, s, 18-CH_3), 1.00 (3H, s, 19-CH_3), 2.08 (3H, s, 20-CH_3), 3.37–3.18 (1H, m, C3-αH), 4.63 (1H, s, OH), 5.24 (1H, s, C6-H), 6.21 (0.36H, t, *J* = 4.8), 6.33 (0.33H, d, *J* = 6.6, 5′-Py-H), 7.32 (0.33H, br s, 4′-Py-H), 7.69 (0.32H, dd, *J* = 24.6, 6.6, 3′-Py-H), 8.49 (0.60H, s, 6′-Py-H), 9.07 (0.60H, s), 9.23 (0.60H, d, *J* = 6.6), 9.79 (0.57H, s, -NH); ^{13}C NMR (150 MHz, DMSO) δ: 14.2 (21-C), 18.9 (18-C), 19.7 (11-C), 19.8 (19-C), 20.3 (15-C), 22.2 (16-C), 30.7 (8-C), 31.0 (2-C), 31.1 (7-C), 35.8 (10-C), 36.6 (1-C), 37.8 (12-C), 41.9 (13-C), 42.8 (4-C), 49.2 (17-C), 55.4 (9-C), 57.8 (14-C), 69.7 (3-C), 117.8 (3′-Py-C), 119.8 (5′-Py-C), 122.0 (6-C), 129.2 (4′-Py-C), 140.9 (5-C), 141.0 (6′-Py-C), 149.9 (20-C), 151.9 (2′-Py-C).

2.4. Cytotoxicity Assay. The antiproliferative activity of all Cu(II) metal complexes and steroidal thiosemicarbazones on Bel-7404 (human liver carcinoma), HeLa (human cervical carcinoma), and HEK-293T (normal kidney epithelial) cell lines was determined by using the MTT method and cisplatin as a positive control. The detailed procedure had been reported in our previous work [22].

3. Results and Discussion

3.1. Synthesis and Characterization. The synthetic route and the structures of complexes **5–8** are outlined in Scheme 1. The steroidal thiosemicarbazones **1–4** were obtained by reacting estrone and pregnenolone or their ester with thiosemicarbazide and then the reaction of compounds **1–4** with $CuCl_2{\cdot}2H_2O$ gave steroidal copper (Cu (II)) complexes **5–8** as a mixture of (*R*)- and (*S*)-configuration isomers, respectively. The structures of **5–8** are confirmed by analysis of IR, NMR, and HRMS. Compared with the signal of -NH- in ^{1}H NMR for ligand **3**, the proton chemical shift of -NH- for compound **7** migrates to δ 10.29 (s, 0.4H) and 10.89 (s, 0.6H) ppm of downfield from δ 8.51 ppm of upfield due to the effect of Cu (II) and demonstrates the formation of L^3-Cu (II) complex. The resonances showing of 10.29 and 10.89 ppm belongs to the chemical shift of (*S*)- and (*R*)-**7**, respectively, and illustrates

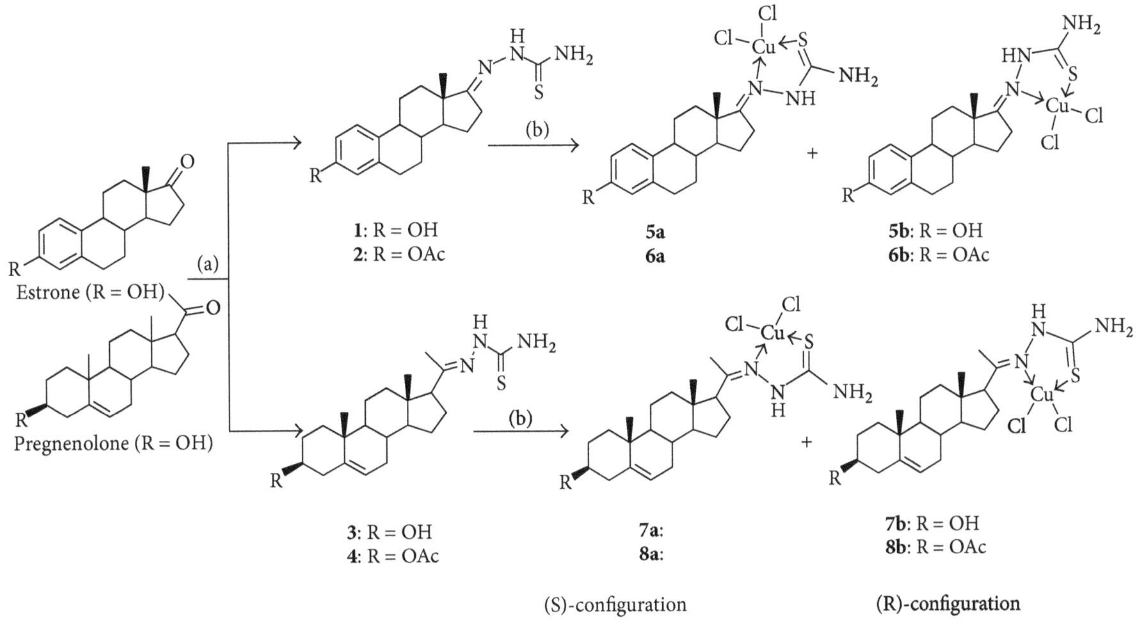

SCHEME 1: Synthesis of complexes **5–8**. Reagents and conditions: (a) thiosemicarbazide, acetic acid, and ethanol; (b) $CuCl_2 \cdot 2H_2O$, $CH_3OH/CHCl_3 = 1:1$.

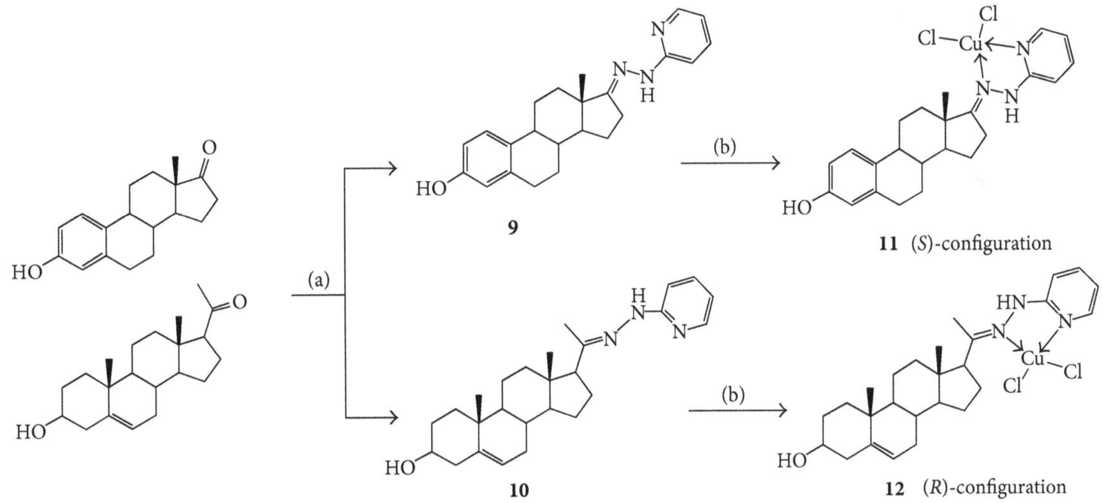

SCHEME 2: Synthesis of complexes **11-12**. Reagents and conditions: (a) 2-hydrazinopyridine, acetic acid, and ethanol; (b) $CuCl_2 \cdot 2H_2O$, $CH_3OH/CHCl_3 = 1:1$.

further that compound **7** is the mixture of (*S*)- and (*R*)-configuration isomers (7-S : 7-R = 1 : 1.5, ^1H NMR data) from the chemical shift of 18-CH_3 and 21-CH_3 (18-CH_3: 0.51 (1.2H, *S*-), 0.68 (1.8H, *R*-) ppm; 21-CH_3: 1.96 (1.2H, *S*-), 2.00 (1.8H, *R*-) ppm).

In order to investigate the effect of different ligand on the antiproliferative activity of complexes, 3*β*-hydroxyoestrone-17-(2′-diazanyl)pyridine-Copper(II) **11** and 3*β*-Hydroxypregnenolone-20-(2′-diazanyl) pyridine-Copper (II) **12** were synthesized according to Scheme 2. Ligands **9** and **10** were obtained as a (*E*)-configuration by reacting estrone or pregnenolone with 2-hydrazinopyridine. Furthermore, the reaction of compounds **9** and **10** with $CuCl_2 \cdot 2H_2O$ gave steroidal copper (Cu (II)) complexes **11** and **12** as (*S*)- and (*R*)-configuration, respectively. The structures of **11** and **12** were confirmed by analysis of IR, NMR, and HRMS.

3.2. Cytotoxic Activity In Vitro. The antiproliferative activities of all steroidal Cu(II) metal complexes were determined in vitro on Bel-7404 (human liver carcinoma), HeLa (human cervical carcinoma), and 293T (normal kidney epithelial) cell lines. The MTT method was used to assay the antiproliferative activity and cisplatin was used as a positive control. The results are summarized as IC_{50} values in μM in Table 1.

From the data shown in Table 1, all steroidal copper (Cu (II)) complexes show an obvious antiproliferative activity

Table 1: Cytotoxicity[a] of steroidal thiosemicarbazone and its Cu-complexes in vitro (IC_{50}: μM)[b].

Compounds	Bel-7404	HeLa	293T
1	19	34	ND
2	>200	42	>200
5	5.0	11	27
6	13	7.7	11
7	9.5	6.8	15.3
8	14.1	10.6	9.1
11	14	ND	14
12	7.0	ND	9.0
Cisplatin	23.2	10.1	10.3

[a]Cytotoxicity as IC_{50} for each cell line is the concentration of compound which reduced by 50% the optical density of treated cells with respect to untreated cells using the MTT assay. [b]Data represent the mean values of the three independent determinations. ND: not determined.

against the tested cancer cells. The compounds **5**–**7** display a better activity to Bell-7404 and HeLa cells compared to that of cisplatin. Comparing the antiproliferative activity of steroidal thiosemicarbazone ligands with that of their copper (II) complexes, we can see that steroidal thiosemicarbazone copper (II) complexes show a better inhibiting activity compared to their homologous ligands (**1** versus **5** and **2** versus **6**). Particularly, complexes **5** and **7** show an excellent antiproliferative activity against Bel-7404 cells with the IC_{50} values of 5.0 and 9.5 μM, and complexes **6** and **7** possess IC_{50} values of 7.7 and 6.8 μM against HeLa cells.

Comparing compound **7** with compound **8**, we can observe that after 3-hydroxyl group of **7** was converted into 3-acetoxy group (compound **8**), the antiproliferative activity of the compound was remarkably decreased and the cytotoxicity to normal cells 293T was increased. The result shows that 3-hydroxyl of the compound to the antiproliferative activity plays an important role.

Unfortunately, these steroidal copper (Cu (II)) complexes to normal kidney epithelial cells (293T) show similar cytotoxicity except for compound **5** which exhibits a smaller activity to 293T cells compared to cisplatin (27 μM versus 10.3 μM).

4. Conclusion

In conclusion, using estrone and pregnenolone as starting materials, through different chemical methods, some steroidal copper (II) complexes were synthesized and characterized by IR, NMR, and HRMS. Their antiproliferative activities were assayed by MTT method. The results show that all steroidal copper (II) complexes display obvious antiproliferative activity against the tested cancer cells, and compounds **5**–**7** show better cytotoxicity compared to a positive control, cisplatin. Among them, complexes **5** and **12** show an excellent antiproliferative activity against Bel-7404 cells with the IC_{50} values of 5.0 and 7.0 μM, and complexes **6** and **7** possess IC_{50} values of 7.7 and 6.8 μM against HeLa cells. The result may be useful for the design of novel chemotherapeutic drugs.

Acknowledgments

The authors acknowledge the financial support of the National Natural Science Foundation of China (no. 21462009), Natural Science Fund of Guangxi province (no. 2014GXNSFAA118052), and the Guangxi Colleges and Universities Key Laboratory of Synthetic and Natural Functional Molecular Chemistry.

References

[1] B. Rosenberg, L. Van Camp, and T. Krigas, "Inhibition of cell division in *Escherichia coli* by electrolysis products from a platinum electrode," *Nature*, vol. 205, no. 4972, pp. 698-699, 1965.

[2] B. Rosenberg, L. Van Camp, E. B. Grimley, and A. J. Thomson, "The inhibition of growth or cell division in *Escherichia coli* by different ionic species of platinum(IV) complexes," *Journal of Biological Chemistry*, vol. 242, no. 6, pp. 1347–1352, 1967.

[3] S. M. Sbovata, F. Bettio, M. Mozzon et al., "Cisplatinum and transplatinum complexes with benzyliminoether ligands; synthesis, characterization, structure-activity relationships, and in vitro and in vivo antitumor efficacy," *Journal of Medicinal Chemistry*, vol. 50, no. 19, pp. 4775–4784, 2007.

[4] D. Kovala-Demertzi, A. Galani, N. Kourkoumelis, J. R. Miller, and M. A. Demertzis, "Synthesis, characterization, crystal structure and antiproliferative activity of platinum(II) complexes with 2-acetylpyridine-4-cyclohexyl-thiosemicarbazone," *Polyhedron*, vol. 26, no. 12, pp. 2871–2879, 2007.

[5] X. Liu, H. Shen, H. Zhu, K. Cui, and S. Gou, "In vitro cytotoxicity study on platinum(II) complexes with epoxysuccinates as leaving groups," *Bioorganic and Medicinal Chemistry Letters*, vol. 17, no. 14, pp. 3831–3834, 2007.

[6] J. Zhang and X. Zhao, "Synthesis, cytotoxicity and DNA-binding levels of ammine/propylamine platinum(II) complexes with carboxylates," *European Journal of Medicinal Chemistry*, vol. 42, no. 2, pp. 286–291, 2007.

[7] A. S. Abu-Surrah and M. Kettunen, "Platinum group antitumor chemistry: design and development of new anticancer drugs complementary to cisplatin," *Current Medicinal Chemistry*, vol. 13, no. 11, pp. 1337–1357, 2006.

[8] G. Momekov, A. Bakalova, and M. Karaivanova, "Novel approaches towards development of non-classical platinum-based antineoplastic agents: design of platinum complexes characterized by an alternative DNA-binding pattern and/or tumor-targeted cytotoxicity," *Current Medicinal Chemistry*, vol. 12, no. 19, pp. 2177–2191, 2005.

[9] K. Barabas, R. Milner, D. Lurie, and C. Adin, "Cisplatin: a review of toxicities and therapeutic applications," *Veterinary and Comparative Oncology*, vol. 6, no. 1, pp. 1–18, 2008.

[10] J. Shao, Z.-Y. Ma, A. Li et al., "Thiosemicarbazone Cu(II) and Zn(II) complexes as potential anticancer agents: syntheses, crystal structure, DNA cleavage, cytotoxicity and apoptosis induction activity," *Journal of Inorganic Biochemistry*, vol. 136, pp. 13–23, 2014.

[11] G. Tamasi, C. Bernini, G. Corbini et al., "Synthesis, spectroscopic and DFT structural characterization of two novel

ruthenium(III) oxicam complexes. In vivo evaluation of anti-inflammatory and gastric damaging activities," *Journal of Inorganic Biochemistry*, vol. 134, pp. 25–35, 2014.

[12] B. Božić, J. Rogan, D. Poleti, N. Trišović, B. Božić, and G. Ušćumlić, "Synthesis, characterization and antiproliferative activity of transition metal complexes with 3-(4,5-diphenyl-1,3-oxazol-2-yl)propanoic acid (oxaprozin)," *Chemical and Pharmaceutical Bulletin*, vol. 60, no. 7, pp. 865–869, 2012.

[13] S. Adsule, S. Banerjee, F. Ahmed, S. Padhye, and F. H. Sarkar, "Hybrid anticancer agents: isothiocyanate-progesterone conjugates as chemotherapeutic agents and insights into their cytotoxicities," *Bioorganic and Medicinal Chemistry Letters*, vol. 20, no. 3, pp. 1247–1251, 2010.

[14] G. A. G. Santos, A. P. Murray, C. A. Pujol, E. B. Damonte, and M. S. Maier, "Synthesis and antiviral activity of sulfated and acetylated derivatives of $2\beta,3\alpha$-dihydroxy-5α-cholestane," *Steroids*, vol. 68, no. 2, pp. 125–132, 2003.

[15] H. Huang, Q. Chen, X. Ku et al., "A series of α-heterocyclic carboxaldehyde thiosemicarbazones inhibit topoisomerase IIα catalytic activity," *Journal of Medicinal Chemistry*, vol. 53, no. 8, pp. 3048–3064, 2010.

[16] C. Duncan and A. R. White, "Copper complexes as therapeutic agents," *Metallomics*, vol. 4, no. 2, pp. 127–138, 2012.

[17] D. X. West and A. E. Liberta, "Thiosemicarbazone complexes of copper(II): structural and biological studies," *Coordination Chemistry Reviews*, vol. 123, no. 1-2, pp. 49–71, 1993.

[18] H. Beraldo and D. Gambino, "The wide pharmacological versatility of semicarbazones, thiosemicarbozones and their metal complexes," *Mini-Reviews in Medicinal Chemistry*, vol. 4, no. 1, pp. 31–39, 2004.

[19] F. Tisato, C. Marzano, M. Porchia, M. Pellei, and C. Santini, "Copper in diseases and treatments, and copper-based anticancer strategies," *Medicinal Research Reviews*, vol. 30, no. 4, pp. 708–749, 2010.

[20] J. L. Dearling, J. S. Lewis, G. E. Mullen, M. J. Welch, and P. J. Blower, "Copper bis(thiosemicarbazone) complexes as hypoxia imaging agents: structure-activity relationships," *Journal of Biological Inorganic Chemistry*, vol. 7, no. 3, pp. 249–259, 2002.

[21] Y. Huang, E. Kong, C. Gan, Z. Liu, Q. Lin, and J. Cui, "Synthesis and antiproliferative activity of steroidal thiosemicarbazone platinum (Pt(II)) complexes," *Bioinorganic Chemistry and Applications*, vol. 2015, Article ID 742592, 7 pages, 2015.

[22] C. Gan, J. Cui, S. Su et al., "Synthesis and antiproliferative activity of some steroidal thiosemicarbazones, semicarbazones and hydrozones," *Steroids*, vol. 87, pp. 99–107, 2014.

6

Synthesis and Characterization of Oxidovanadium (IV) Complexes of 2-((E)-(6-Fluorobenzo [d]thiazol-2-ylimino) methyl)-6-methoxyphenol and their Antimicrobial, Antioxidant, and DNA-Binding Studies

K. Savithri and H. D. Revanasiddappa

Department of Studies in Chemistry, University of Mysore, Manasagangotri, Mysuru, Karnataka 570 006, India

Correspondence should be addressed to H. D. Revanasiddappa; hdrevanasiddappa@yahoo.com

Academic Editor: Jacopo Sgrignani

Two novel oxidovanadium(IV) complexes with a new bidentate (O- and N-) imine-based ligand 2-((E)-(6-fluorobenzo[d]thiazol-2-ylimino)methyl)-6-methoxyphenol (HL) were synthesized under in situ experimental condition where $VOSO_4$ acts as a kinetic template in the ratio 2 : 1 (L : M) and mixed ligand complex using 1,10-phenanthroline (phen) in 1 : 1 : 1 (L : M : phen) ratio. The synthesized compounds were structurally characterized by microanalysis, magnetic susceptibility, FTIR, electronic spectra, TG/DTA, ESR, and molar conductance studies. Based on the spectral studies, the complexes have the general composition [VO $(L)_2$] (C_1) and [VO(L)phen] (C_2) in a square pyramid geometrical fashion. The synthesized compounds were primarily screened for their *in vitro* growth inhibiting activity against different strains of bacteria, namely, *E. coli*, *B. subtilis*, *S. aureus*, and *P. aeruginosa* by the disc diffusion method. Also, the antifungal activity was determined against *C. albicans* and *A. niger* by the Bateman poisoned technique. The *in vitro* antioxidant activity of all the compounds was determined by DPPH free radical-scavenging assay. Intercalative mode of DNA-binding properties of the oxidovanadium(IV) complexes with calf-thymus DNA (CT-DNA) was investigated using UV, fluorescence spectra, and viscosity measurements.

1. Introduction

Vanadium compounds, especially the coordination chemistry of oxidovanadium(IV), have aroused significant interest for several main reasons in recent years especially in the designing of long-acting drugs in metabolism. Research interest in oxidovanadium chemistry derives from its utility in several biological and industrial processes like antimicrobial, spermicidal, antioxidant, antitumor, DNA binding [1], and recently as insulin mimetic [2]. The vital role of vanadium in different chemical and biological systems has motivated the development of vanadium chemistry. The coordination chemistry of nitrogen and oxygen donor ligand is an active area of research. It is evident that the –HC=N linkage in azomethine derivatives is an essential structural requirement for biological activity [3, 4]. The geometry, coordination number, and biological efficacy of oxidovanadium(IV) are highly ligand dependent. It is also known that VO^{2+} is less toxic than the vanadate ion (VO_4^{3-}) and as shown above, with all these characteristics, the complexes exhibit potential therapeutic applications [5].

In addition to that imine bases containing heterocyclic nucleus have potent biological activity.

The *o*-hydroxy imine bases have emerged as an antimicrobial agent of immense interest because of their comprehensive spectrum of *in vitro* and *in vivo* chemotherapeutic action [3, 6]. The metal complexes from bidentate ligand have often been studied recently because of their technical applications in an enhancement of drug action. Transition metal complexes of vanadium with imine bases have been amongst the most widely studied of coordination compounds in the past few years since they are found to be important as analytical and antimicrobial agents. The complexes formed by 3d metals with ligand using both oxygen and nitrogen have been paid close attention in this area.

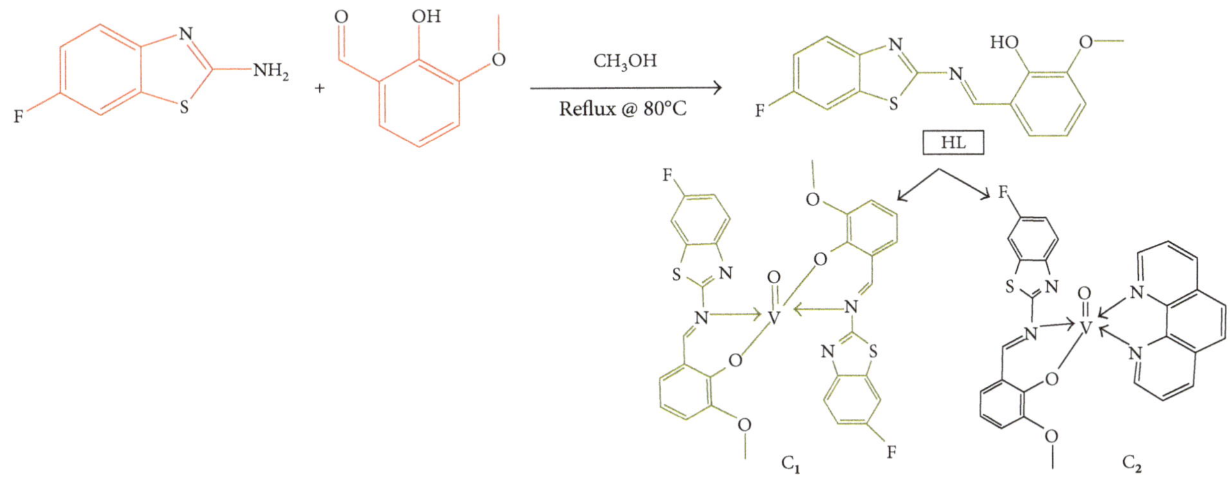

Scheme 1: Synthetic route of ligand HL and its complexes.

Antioxidants interfere with the oxidative processes by scavenging free radicals, chelating free catalytic metals, and by acting as electron donors [7]. Interaction study of small molecules with DNA attracts attention for importance in cancer therapy and molecular biology [8]. DNA is involved in many kinds of life processes, and consequently, it is considered to be the prime target of drugs. The transition metal complexes of 1,10-phenanthroline have been widely employed in DNA studies due to their utility in biomedicinal chemistry and in the design of stereospecific DNA-binding drugs [9]. In this perspective, it is essential to know the antioxidant and DNA binding properties of bioactive compounds for their use in the pharmaceutical and therapeutic field.

The current study is structured to provide a synthesis and characterization of a new imine-based ligand (HL) and its oxidovanadium(IV) complexes. The new compounds have been subjected to characterize by different analytical techniques and also exploit some biological applications of oxidovanadium complexes.

2. Experimental

2.1. Physical Measurements. All the solvents and reagents procured commercially and were used without prior purification. 2-Amino-6-fluorobenzothiazole, 1,10-phenanthroline monohydrate, and *o*-vanillin were purchased from Sigma-Aldrich chemicals, Germany. $VOSO_4{\cdot}6H_2O$ and all other solvents are procured from Merck Specialties Private Limited, Mumbai. The melting point (temperature 0–350°C) of the compounds was determined on an ELICO-3210 apparatus and is uncorrected. FTIR spectra of the compounds were recorded on a PerkinElmer 783 spectrophotometer via attenuated total reflection (FT-ATR) in the range of 4000–400 cm^{-1}. Elemental analyses (CHNO) were performed with a model 240 PerkinElmer elemental analyzer. UV-visible spectra of the complexes in the region 200–800 nm were measured using a UV-Vis spectrophotometer (DU 730 "Life Science" M/S Beckman coulter, USA). Molar conductance measurements were performed for the prepared complexes using an Elico CM-180 conductometer with a cell having a cell constant 1. TGA analysis was performed on a Shimadzu AT-50 thermal analyzer from 30 to 900°C with a heating rate at 10°C/min under N_2 atmosphere. Electron spin resonance (ESR) spectra were recorded using a JEOL JES-TE100 ESR spectrometer in DMSO solution at a liquid-nitrogen temperature on X-band at a frequency of 9.13 GHz under the magnetic field of 300 mT. Ascorbic acid, CT-DNA, and ethidium bromide (EB) were purchased from Sigma Corp., and they were used as received.

2.2. Methodology

2.2.1. Synthesis of Imine-Based Ligand (HL)

(1) Classical Method. The ligand HL was synthesized by the condensation of 2-amino-6-fluorobenzothiazole with *o*-vanillin in 1 : 1 ratio. To an absolute methanolic solution of 2-amino-6-fluorobenzothiazole (10 mmol, 1.103 g), the methanolic aromatic aldehyde, that is, *o*-vanillin (10 mmol, 1.521 g) was added and the reaction mixture was then stirred under reflux at 80°C for 12 h. The product formed after slow evaporation of the solvent was then isolated by filtration, washed, and recrystallized from methanol. Yield = 78–79%, and synthetic pathway of ligand HL and its complexes is displayed in Scheme 1.

(2) Microwave Method. The title compound HL was synthesized by placing an equimolar ratio of the mixture of 2-amino-6-fluorobenzothiazole, and 2-hydroxy-*m*-anisaldehyde was mixed thoroughly in a grinder. The reaction flask was then irradiated in a microwave oven for 7 min using dry methanol (3–4 mL) as a solvent. TLC was used to assess the completion of the reaction progress and to preliminary check on the purity of the product, and the reaction mixture was allowed to attain room temperature (25 ± 2°C). The yellow solid was obtained after recrystallization from methanol, and yield = 87–89%.

(3) HL: 2-((E)-(6-Fluorobenzo[d]thiazol-2-ylimino)methyl)-6-methoxyphenol. Mp = 170–172°C; R_f 0.55, Anal. (%) for $C_{15}H_{11}FN_2O_2S$: Found (calc.): C 54.80 (54.82), H 2.68 (2.63), N 9.11 (9.13), O 5.21 (5.22); *m/z* 302.05 Da; FTIR (υ_{max}/cm^{-1}); (υOH) 3253, (HC=N) azomethine 1662; ^{1}H NMR (295 K/δ

in ppm/DMSO-d_6 400 MHz): HC=N 9.776 (1H, s); phenolic OH 11.45 (1H, s); Ar-H 7.0–7.9 (6H, m); ^{13}C NMR (295 K/δ(ppm)/DMSO-d_6 400 MHz): 164.176, 123.880, 109.423, 158.848, 109.150, 148.414, 135.710, 161.261, 119.463, 159.63, 115.714, 135.323, 124.494, 129.215.

2.2.2. Synthesis of Complexes

(1) Preparation of Complex 1 [C_1]. The metal salt ($VOSO_4 \cdot 6H_2O$) (2 mmol) was dissolved in 10 mL of aqueous alcoholic solution, ligand HL (4 mmol) was mixed in 15 mL ethanol in the ratio of 1 : 2, and the reaction mixture was refluxed for 9 h on a water bath. After a specified period of reaction, the solvent was evaporated slowly by allowing standing at room temperature. The resultant yellow color complex was used for further characterization.

*(2) Preparation of Complex **2** [C_2].* Complex **2** was prepared by mixing an equimolar ratio of metal salt ($VOSO_4 \cdot 6H_2O$) in the aqueous medium and an ethanolic solution of ligand HL, and the mixture was stirred for 10 min. 1,10-Phenanthroline monohydrate dissolved in ethanol was added at the same temperature. It was then kept under reflux for 8 h on a water bath. Then, the solvent was evaporated at RT, and the obtained green color complex was used for further analysis.

(3) C_1: [VO $(L_1)_2]_2$. Yield = 65%, yellow color, Anal. (%) for $C_{30}H_{20}VF_2N_4O_5S_2$; Found (Calc.) C 53.81 (53.02), H 3.01 (3.12), V 7.61 (7.59), N 8.37 (8.21), O 11.95 (12.05); mp ≥ 300°C. *m/z*: 669.02 Da; IR (υ_{max}/cm^{-1}); (HC=N) azomethine 1640; M-O 542; M-N 463; 362, H_2O 3387.

(4) C_2: [VO(L_1)(phen)]. Yield = 78%, green color, Anal. (%) for $C_{27}H_{18}VFN_4O_3S$: Found (Calc.) C 59.13 (58.71), H 3.31 (3.14), V 9.29 (9.12), N 10.22 (10.55), O 8.75 (8.99); mp ≥ 300°C. *m/z*: 548.461 Da; IR (υ_{max}/cm^{-1}); (HC=N) azomethine 1633; M-O 576; M-N 468.

2.3. Bioassay Studies

2.3.1. Antimicrobial Activity. The biological activities of the compounds have been investigated to display the possibility of their uses in the medicinal field. The *in vitro* antimicrobial screening effects of the synthesized compounds were tested against four bacterial strains, namely, *E. coli*. (MTCC 443), *B. subtilis* (MTCC 441), *S. aureus* (MTCC 96), and *P. aeruginosa* (MTCC 424) and two fungal species (*A. niger* MTCC 282 and *C. albicans* MTCC 227). The disc diffusion method using the nutrient agar medium for antibacterial activity [10] and Czapek's agar nutrient medium was used for antifungal activity by the Bateman poisoned technique [11]. The bacteria and fungi were subcultured in the sterile autoclaved medium and were incubated up to 24 h for bacteria (37 ± 2°C) and 72 h for fungi at 22 ± 1°C. Standard antibacterial drug chloramphenicol and antifungal drug fluconazole were used for comparison. The stock solution (1 mg/mL) of the test compounds was prepared in DMSO. The antibiotic activity of each compound was evaluated at three different concentrations on dilution of stock solution, namely, 10, 50, and 100 μg/mL. The discs having a diameter of 5 mm were soaked in the test solutions of the above three concentrations and were placed on an appropriate nutrient medium previously seeded with organisms in the Petri dishes and stored in an incubator at the abovementioned period. In order to clarify any effect of DMSO on the biological screening, separate studies were carried out with solutions alone of DMSO and they showed no activity against any microbial strains. The colony diameter of the test organism was measured with the mm scale after a week. The percentage inhibition of the growth of the test organisms was calculated by the Vincent equation:

$$\text{growth inhibition percentage} = \frac{100\,(C_d - T_d)}{C_d}, \quad (1)$$

where C_d = colony diameter of the control plate and T_d = colony diameter of microbial growth on the test plate.

Biocidal activity was assessed by measuring minimum inhibitory concentrations (MICs). The MIC value of each compound was determined at its lowest concentration (highest dilution) required to arrest the growth of microbes by the standard broth dilution assay. Compounds showing promising antibacterial/antifungal activity were selected for minimum inhibitory concentration (MIC) studies.

2.3.2. Time-Dependent Killing Kinetics. Time-dependent killing kinetic analyses of compounds against *E. coli* strain were performed at 0.5×, 1×, and 2× the MIC, as determined previously, and an initial stock culture with a concentration of 1×10^6 CFU per well. After incubation at the time interval of 6, 12, 18, and 24 h at different MICs of the compounds, the absorbance was measured at 625 nm. The culture treated without the compound was kept as the control. The experiment was performed in triplicate. The rate and extent of killing were expressed as mean $\log_{10} \cdot$CFU$\cdot$mL^{-1} against time [12].

2.3.3. Antioxidant Activity. DPPH has an odd electron number with a strong absorption band at 517 nm. When this electron becomes paired off, the absorption decreases stoichiometrically with respect to the number of electrons taken up [7, 13]. Such a change in the absorbance produced in this reaction has been widely applied to test the capacity of various molecules to act as free radical scavengers. Therefore, DPPH is usually used as a substrate to evaluate antioxidant activity of other antioxidants. The present oxidovanadium complexes are in the +4 oxidation state and act as an electron donor compound. It can reduce the DPPH radical to DPPH (α,α-diphenyl-β-picrylhydrazine) compound, and the VO (IV) ion will be oxidized to its +5 state.

The free radical-scavenging activity of the compounds was measured in terms of hydrogen donating or radical-scavenging ability using the stable radical DPPH described by the Blois method [14]. Briefly, stock solutions of the samples (0.001 g/mL) were prepared by dissolving in DMSO. Different concentrations (20–100 μL) of the stock solution were made up to 3 mL of methanol. A solution of DPPH

(0.0001 mol) in methanol was prepared, and 1 mL of this solution was added to each of the above test solutions. 0.1 M sodium acetate buffer was added. The mixture was shaken vigorously and incubated for 30 min, and then, the absorbance was measured at 517 nm. All the tests were carried out in triplicate and expressed as the mean ± standard deviation (SD). Ascorbic acid (AA) was used as a standard or positive control, parallel to the test compound and in the absence of the test compound/standard used as the negative control. The capability to scavenge the DPPH radical was calculated using the following equation:

$$I(\%) = \left(\frac{A_c - A_{sample}}{A_c}\right) \times 100, \quad (2)$$

where the absorbance of the control reaction mixture excluding the test compounds is termed as A_c and the absorbance of the test compounds is A_{sample}.

2.3.4. DNA-Binding Experiments

(1) Electronic Absorption Spectral Titration. The stock solution of CT-DNA was prepared in 5 mM Tris-HCl/50 mM NaCl buffer at pH 7, which gave $A_{260}/A_{280\,nm}$ of *ca.* 1.8-1.9, indicating that the DNA was sufficiently devoid of protein [15], and the concentration of DNA per nucleotide was determined from its absorption intensity at 260 nm with a known molar absorption coefficient ($\varepsilon_{260} = 6600\ dm^3 \cdot M^{-1} \cdot cm^{-1}$) [16]. DNA stock solutions were kept stored at 4°C and used within four days. The binding constant (K_{bin}) values of the complexes to CT-DNA were obtained by monitoring the changes in the absorption of the spectral band intensity with increasing DNA concentration. After equilibrium had been achieved, the spectra were recorded at 200–550 nm against an analogous solution of DNA as a reference. UV spectral data were fitted into the Wolfe–Shimmer equation (3) [17] to obtain the intrinsic binding constant (K_{bin})

$$\frac{[DNA]}{\varepsilon_a - \varepsilon_b} = \frac{[DNA]}{\varepsilon_b - \varepsilon_f} + \frac{1}{K_{bin}(\varepsilon_b - \varepsilon_f)}, \quad (3)$$

where [DNA] is the concentration of the DNA in base pairs, ε_a is the apparent extinction coefficient ($A_{obs}/[MC]$), ε_b is the extinction coefficient of the metal complex in the fully bound form, and the extinction coefficient for the free metal complex [MC] is ε_f. The value of intrinsic K_{bin} was calculated as the ratio of the slope to intercept. The due correction was made for the absorbance of DNA itself.

(2) EB Competition Assay. The competitive binding activity of oxidovanadium complexes with CT-DNA was studied by the fluorescence spectral method using emission intensity of ethidium bromide (EB) with varying concentrations of DNA. EB emits intense fluorescence light in the presence of DNA at λ_{ex} 600 nm (546 nm) due to strong stacking intercalation between adjacent DNA base pairs. The fluorescence emission intensity can be quenched when the second molecule (synthesized compounds) was added to the CT-DNA bound EB solution. In the fluorescence quenching study, [DNA]/[EB] = 1.13 is kept constant and varied with the compound concentration. The fluorescence emission spectra of the reaction mixture were recorded between 400 and 850 nm at excitation wavelength 350 nm. The fluorescence quenching efficiency is evaluated according to the following classical Stern–Volmer equation:

$$\frac{I_o}{I} = 1 + k_{sv}Q, \quad (4)$$

where the fluorescence intensities without and with the quencher (complexes) are termed I_o and I, respectively, "Q" is the concentration of the quencher, and K_{sv} is the linear Stern–Volmer quenching constant. The measured fluorescence intensities were corrected for the dilution and the inner filter effect in the entire DNA-binding experiments.

(3) Viscosity Measurements. Viscosity experiments were carried on the Ubbelohde viscometer at a static temperature (26 ± 0.1°C). Compounds were added to the DNA solution ($10\ \mu mol \cdot L^{-1}$) with a micropipette. Flow time was recorded for the increasing concentration of the compounds (0–10 μM). Data were graphically presented as $(\eta/\eta_o)^{1/3}$ versus the binding ratio of the concentration of compounds to DNA (r), where η was the viscosity of the DNA in the presence of the compound and η_o was the viscosity of the DNA alone. The relative viscosity η was calculated using the following equation [18]:

$$\eta = \frac{(t - t_o)}{t_o}, \quad (5)$$

where t and t_o represent the flow time of the buffer solution through the capillary and the observed flow time for the DNA in the presence and absence of the compounds, respectively. The mean of replicate measurements was used to evaluate the viscosity of the samples.

3. Results and Discussion

3.1. Chemistry. All the synthesized compounds are stable in the atmospheric condition, nonhygroscopic, insoluble in water but slightly soluble in alcohol and readily soluble in DMF and DMSO. Microwave-assisted synthesis of ligand HL was easy, convenient, short reaction period with a good yield (87–89%) and superior to the conventional method (78–79%). Oxidovanadium complexes are highly colored, fine amorphous powder, stable in the solid phase at the ambient temperature in the light. The solutions were stable on keeping for an extended period in the light. The analytical and physical data of the new compounds are summarized in the experimental section. Analytical data show that the complexes have the stoichiometry to be 1 : 2 (metal : ligand) for complex **1**, whereas 1 : 1 : 1 (metal : ligand : phen) for complex **2**. Physical property and analytical data of newly synthesized imine base and its complexes are summarized in the experimental section. The molar conductance values at the $10^{-3} \cdot M$ concentration are too low to account for any dissociation of the complex in DMF and are presented in Table 1. Hence, the synthesized VO(IV) complexes may be regarded as nonelectrolytes. Structural elucidation of the complexes by multispectroscopic techniques reveals that the vanadium center in complex **1** is coordinated to two bidentate imine-based ligands with the

Table 1: Electronic spectra and magnetic moments of the metal complexes.

Compounds	μ_{eff} (BM)	Λ ($Ohm^{-1}\cdot cm^2\cdot mol^{-1}$)	Electronic spectral bands (nm)	
			MLCT	*d-d* transition
C_1	1.75	13.9	425	555
C_2	1.79	10.4	435	565

two nitrogen and two phenolate oxygen atoms in equatorial positions and oxooxygen in the axial position to complete the square pyramidal N_2O_3 coordination sphere. Similarly, N_3O_2 coordination sphere can be seen around vanadium in complex **2**.

3.1.1. Molar Conductivity Measurements and Magnetic Susceptibility Measurements. The metal complexes were dissolved in DMSO, and the molar conductivities of their solutions at 25 ± 2°C were measured. It is concluded from the results obtained in Table 1, that the complexes have molar conductance value of 10.4–13.9 $ohm^{-1}\cdot mol^{-1}\cdot cm^{-1}$, indicating that all the complexes are nonelectrolytes. The magnetic susceptibility measurements of the complexes were carried out to find out the effective magnetic moment per each metal in the complexes. The number of unpaired electrons possessed by the metal ion can be determined from the effective magnetic moment of the metal ion. The magnetic susceptibility studies from the literature show that VO(IV) complexes have paramagnetic behavior. Values of magnetic susceptibility (μ_{eff}) for the VO(IV) complexes are given in Table 1. A mononuclear oxidovanadium(IV) structure is assigned to both complexes **1** and **2**, and the μ_{eff} values for these complexes are found to be 1.75 and 1.79 BM, respectively. These values are well suited for complexes of oxidovanadium(IV), d^1 system [19, 20].

3.1.2. Electronic Spectra. The electronic absorption spectra of both the complexes were recorded for the freshly prepared solution in DMSO at room temperature over the range 200–800 nm, and the spectral data are listed in Table 1. The electronic spectra of the complexes in which intense absorptions in the region of 328–413 nm are assigned to the spin allowed intraligand (π–π^* and n–π^*) and ligand-to-metal charge transfer (LMCT) O(π) → V(d) transitions. The weak shoulder appeared around 490 (486–518) and ~580 nm are due to the d-d transitions $^2B_2$ ($d_{xy} \rightarrow d_{xz}, d_{yz}$) → ^2E and $^2B_2$ ($d_{xy} \rightarrow d_{x^2-y^2}$) → $^2B_1$ in vanadium complexes which are consistent with the square pyramidal geometry around the V^{4+} ion [21, 22]. The third absorption is not observed as it might bury beneath the high-intensity charge transfer band. The expected *d-d* transition band could not be determined, possibly due to a low concentration of the vanadium(IV) ion in the samples.

3.1.3. IR Spectra. The characteristic IR spectral bands of ligand and its complexes are presented in Table 2. The IR spectrum of the ligand shows a strong band at 1037 cm^{-1} region which is assigned to the cyclic–OCH_3 group of *o*-vanillin. The band for stretching frequency of phenolic OH occurred at 1267 cm^{-1}, whereas in IR spectrum of these complexes, this band is shifted to different frequency showing a strong band at around 1258–1271 cm^{-1} region. It indicates that there existed M-O bond formation and it involved in coordination with the metal ion. In the spectrum of the Schiff base, the strong bands at 1666 cm^{-1} region are attributed to -C=N groups in the free ligand. On chelation, due to the possible drift of lone pair electron density towards the metal ion, the azomethine -C=N band is expected to absorb at a lower frequency in the complexes. The band shifted to 1658–1637 cm^{-1} region indicating the coordination of azomethine nitrogen to the metal. Spectra of the complexes also show new peaks at around 413–452 cm^{-1} and 493–552 cm^{-1} region due to the formation of M-O and M-N bands, respectively. In addition to other bands, the oxidovanadium complex shows its characteristic V=O frequency at 980 and 967 cm^{-1} regions for complexes **1** and **2** [23].

3.1.4. Thermogravimetric (TG)/Differential Thermogravimetric Analysis (DTA). of the complexes was carried out under argon flow in the temperature range of 30–900°C. The thermal data of compounds are listed out in Table 3. The thermal stability curve of C_1 (Figure 1) where decomposition starts with loss of two coordinated organic ligands in C_1 in the range of 300–513°C and the associated weight loss found (46.8–97%) is consistent with the calculated values. There is an exothermal peak on the DTA curve in the range of 450–600°C corresponding to the significant weight loss of the ligand moiety of C_1. Similarly, compound C_2 showed two-step of condensation in the thermogram where phenanthroline decomposition occurred at a temperature (105–126°C) lower than that of the ligand (321–394°C). The percentage of the final metal oxide (V_2O_5) residue of both the complexes was calculated from the weight of the ash obtained.

3.1.5. ESR Spectra. The complexes C_1 and C_2 were also characterized by ESR spectral measurement recorded at liquid nitrogen temperature in the DMSO solution. The X-band ESR spectra of one representative complex in DMSO at liquid nitrogen temperature are given in Figure 2, whereas each spectrum shows a hyperfine pattern of typical eight equidistant lines originating from the interaction of an unpaired electron with ^{51}V, $I = 7/2$ These characteristic spectral features are indicative of monomeric VO(IV) complexes.

The calculated ESR parameters from the spectra of both the complexes are listed in Table 4.

g_{avg} corresponds to the isotropic g_{iso} value was calculated to be 1.952 and 1.958 for C_1 and C_2, respectively. The parameters showing the orders of $g_{||} < g_{\perp}$ and $A_{||} >> A_{\perp}$ indicate that an unpaired electron is present in the $(d_{xy})^1$ orbital with a square pyramidal geometry around the oxidovanadium(IV) complexes in C_{4v} symmetric fashion [5, 24]. Within the limits of errors, the experimental $A_{||}$ agrees with the calculated one, suggesting that the solution structure at low temperature does not significantly deviate from the structure in the solid state.

TABLE 2: Important infrared frequencies (cm^{-1}) of HL and its complexes.

Complexes	ν (C=N)	ν (pH-OH)	ν (OCH_3)	ν (V=O)	ν (M-N)	ν (M-O)
HL	1666	1267	1019	—	—	—
Complex **1**	1658	1271	1083	980	544	448
Complex **2**	1653	—	1025	967	507	452

TABLE 3: TGA analysis of synthesized compounds.

Serial number	Compounds	Stages	Temperature range (TG) (°C)	Probable composition of expelled group/s	Weight loss found (%) Cal.	Weight loss found (%) Expt.	No. of molecules	% residue	Nature
1	C_1	I	302–513	HL	91.2	93.3	2	8.7	V_2O_5
2	C_2	I	105–126	1,10-phen	24.4	23.6	1	12.6	V_2O_5
		II	321–394	HL	47.0	46.2	1		

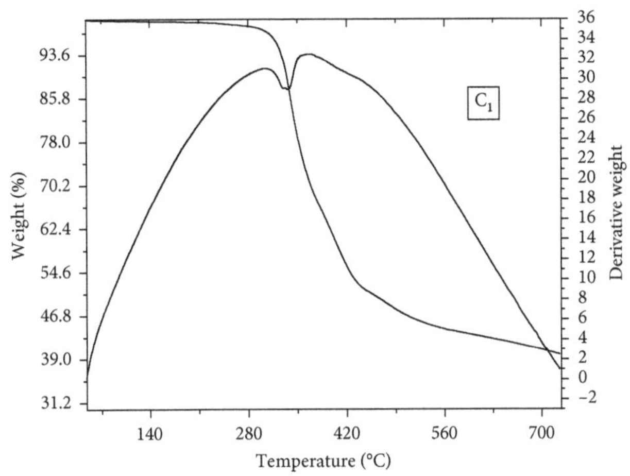

FIGURE 1: TGA-DTA thermogram of complex **1**.

The preceding observations further support the composition of the complexes, which is compatible with the results, obtained by each other methods.

3.2. Biology

3.2.1. Antimicrobial Activity. The antimicrobial activities of the HL and its complexes have been investigated to show their impressive level of antimicrobial activity of MIC values against newly emerging highly drug-resistant pathogens. The results of *in vitro* study of the microbicidal activity of the afresh synthesized complexes against four bacterial strains (Gram negative: *E. coli* and *B. subtilis* and Gram positive: *S. aureus* and *P. aeruginosa*) and two fungal strains (*A. niger* and *C. albicans*) are reported in Table 5.

The free ligand HL and its VO(IV) complexes show varying degree of inhibition effects toward all the organisms, bacteria, and fungi. It is evident that the metal coordination tends to make the ligands strong bacteriostatic and fungistatic agents and efficiently inhibits both bacterial cell and mycelial growth of drug-resistant microbes more than the parent ligand. The oxidovanadium(IV) complex C_2 of mixed ligands showed good antifungal activity rather than antibacterial than the free ligand and is almost nearer to the standard drugs used.

The minimum inhibitory concentration (MIC) of compounds, which showed significant activity against bacterial and fungal species, was also determined (Table 6). The MIC of these compounds varies from 10 to 100 mg/mL. The results indicated that these compounds were the most active in inhibiting the growth of the tested species at 10 mg/mL concentration.

These results show the structure activity relationship between the organism and the substituted saline moiety. The inhibition values indicate that most complexes have higher percentage towards Gram-negative bacteria compared to Gram positive is due to the metal ions shared with the donor atoms (N and O) of the ligand and the π-electron delocalization over the chelate ring increases electron density over C=N nitrogen leading to strong interactions with cell constituents. Such increased activity of the metal chelates can be explained on the basis of the oxidation state of the metal ion and chelation theory (Tweedy's and Overtone's concept) [20, 25].

Mechanism of action seems that the structural components possessing additional azomethine linkage with heteroatoms inhibit enzyme activity due to their deactivation by metal coordination. This permits their efficient membrane permeabilizing ability through the lipoid layer of organisms and destroys their activity [26]. The enhanced antimicrobial potency of the complexes may also be explained by their solubility, conductivity, fitness of the particles, cell permeability, size of the metal ion, and bond length between metal and ligand.

Membrane-targeting antibiotics will exhibit charge-based selectivity toward the bacterial membrane. In general, Gram-positive bacteria have a simpler cell wall than Gram-negative bacteria, and the primary constituent is peptidoglycan, a polysaccharide. In contrast, Gram-negative bacteria have an additional outer bilayer membrane composed of lipopolysaccharides and phospholipids [27, 28].

(1) Time-Dependent Kinetics. Killing kinetics exhibited marked effect made by complexes than free ligand on bacterial strain chosen. The control showed no noticeable bactericidal activity throughout 24 h incubation for all the compounds performed. The ligand HL (Figure 3(a)) at 0.5× MIC showed no significant effect on *E. coli* up to 18 h, and after that, it showed a slow decline up to 24 h. At 1× and 2× the MICs, the rate of killing increased with increasing incubation time

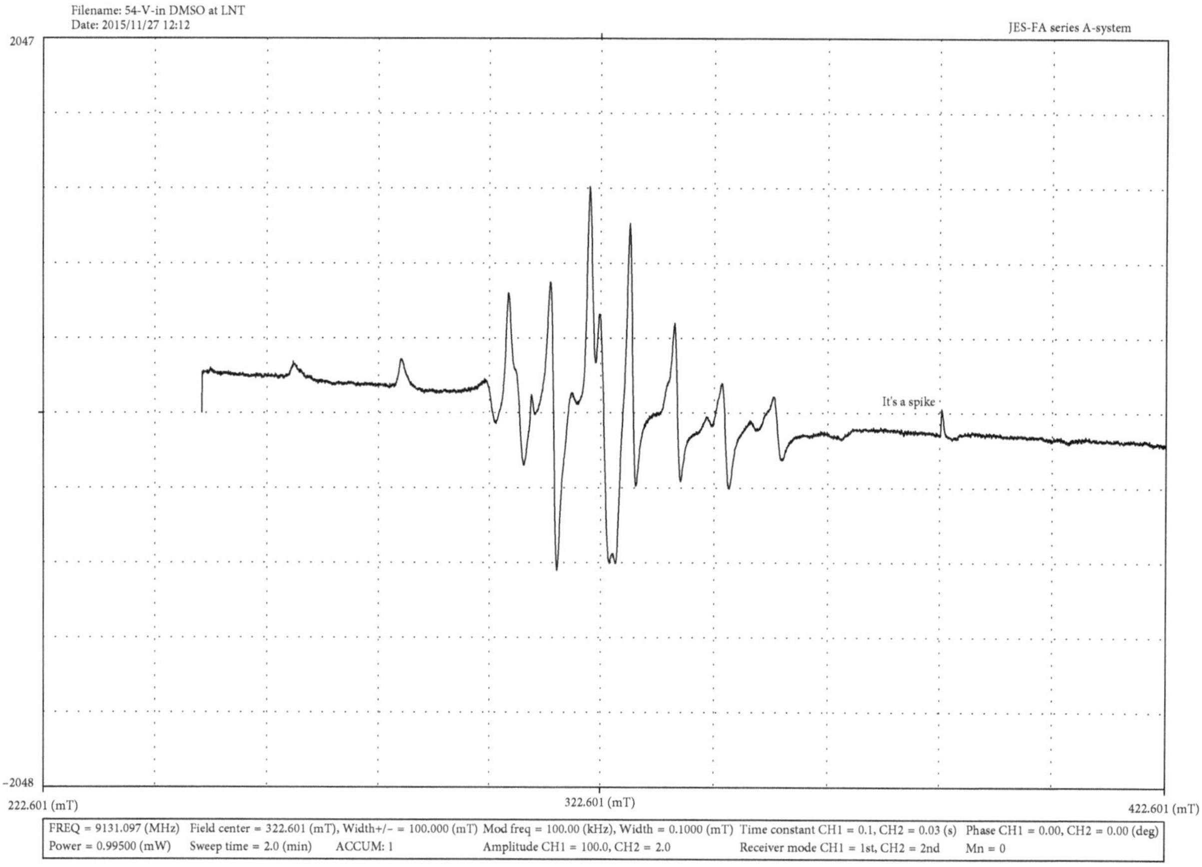

FIGURE 2: ESR spectrum of C_2 in DMSO at liquid nitrogen temperature.

TABLE 4: ESR spectral data of oxidovanadium(IV) complexes in DMSO at liquid nitrogen temperature.

Complexes	$g_{\parallel}$	$g_{\perp}$	g_{avg}	$A_{\parallel}$ ($\times 10^4 \cdot cm^{-1}$)	$A_{\perp}$ ($\times 10^4 \cdot cm^{-1}$)
C_1	1.9302	1.9736	1.952	158.8	53
C_2	1.9310	1.9864	1.958	154.6	54

and reached $\sim 1 \times 10^3$ and $\sim 1 \times 10^2$ CFU·mL^{-1}, respectively. Time-kill curve for complex C_2 is displayed in Figures 3(b) and 3(c) against *E. coli*. When a mild activity was observed at 0.5×, the rate of killing was gradually increased with time and complete killing occurred at 2× the MIC. Complexes C_1 and C_2 showed a more or less similar kind of killing kinetics which showed modest activity at 1× and 2× the MIC.

3.2.2. Antioxidant Activity. Oxidovanadium(IV) complexes exhibit good DNA-binding affinity, and it is considered worthwhile to study the antioxidant activity of these compounds. Free radicals play a vital role in the pathogenesis of various human diseases and aging. In food products, free radicals also cause damage, resulting in diminishing taste and shelf life. Antioxidants are therefore protecting against free radicals and save health. Mostly, strong antioxidant activity is shown by polyphenols. Our study aimed at exploring the most potent antioxidant and examining the factors that give a picture and establish the antioxidant activity with frequent comparison to different concentrations of HL and its VO(IV) complexes (Figure 4). The results revealed that ligand showed low-scavenging potential (~20.32–49.78%) in all the screened concentrations; however, upon complexation with oxidovanadium ions, antioxidant activity has been enhanced significantly (~70–73%) through oxidation of $V^{(IV)}O^{2+}$ to $V^{(V)}O^{+}$ using water as a medium in the reaction mixture since assay was run in the methanolic system [29]. It is believed that the large conjugated system and chelation of the organic molecules with the metal ions by nitrogen and phenolic hydroxyl group benefits the trapping of free radicals. It is further supported by the observed discoloration from purple DPPH radical solution to yellow solution showing scavenging of the DPPH radicals by hydrogen donation.

3.2.3. DNA Binding Study

(1) Electronic Absorption Spectral Features of CT-DNA Binding. Electronic absorption spectroscopy is one of the most useful techniques in DNA-binding studies since the observed changes in the spectra may indicate the mode of interaction.

The UV–Vis titration of both complexes in the absence and in the presence of CT-DNA has been studied in Tris-buffer medium. The absorption spectra of complexes (at a constant concentration of complexes and a wavelength

Table 5: Antimicrobial results of HL and its metal complexes.

Serial number	Compounds (30 μg/disc)	% growth inhibition in μg/mL concentration																	
		Gram-negative bacteria						Gram-positive bacteria						Fungi					
		E. coli			*P. aeruginosa*			*B. subtilis*			*S. aureus*			*C. albicans*			*A. niger*		
		100	50	10	100	50	10	100	50	10	100	50	10	100	50	10	100	50	10
1	HL	16.3	14.8	9.3	48.6	37.1	28.2	47.3	35.3	27.6	23.6	16.7	10.5	40.2	36.6	34.3	41.3	35.5	30.2
2	C_1	67.2	52.2	42.1	66.7	51.8	38.2	68.6	54.6	42.2	67.4	51.1	46.3	79.3	57.5	36.5	62.1	54.3	39.4
3	C_2	73.8	46.3	36.9	61.3	45.2	32.5	77.5	63.2	36.4	73.7	43.7	32.6	70.2	53.6	41.1	87.6	42.5	41.8
4	Ciprofloxacin	96.4	88.2	69.6	85.8	72.9	50.8	92.5	74.6	66.5	91.6	79.5	61.2		—			—	
5	Fluconazole		—			—			—			—		98.6	81.6	72.3	88.2	80.3	66.3
6	Negative control (DMSO)		100			100			100			100			100			100	

Each value was replicated thrice.

Table 6: Minimum inhibitory concentration (mg/mL) of compounds against bacteria and fungi.

Serial number	Compounds	*E. coli*	*P. aeruginosa*	*B. subtilis*	*S. aureus*	*C. albicans*	*A. niger*
1	HL	15	25	30	15	30	22
2	C_1	20	15	10	10	20	30
3	C_2	10	20	15	20	10	10

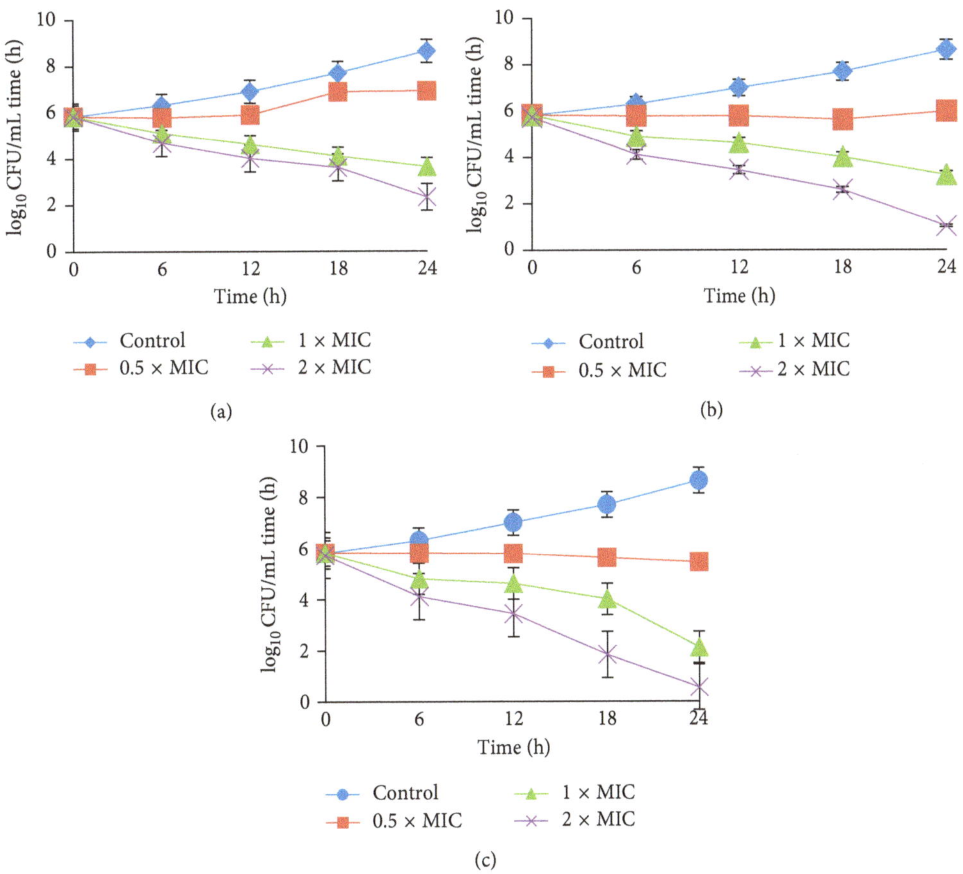

Figure 3: Time-course kinetics of the bactericidal activity of (a) HL, (b) C_1, and (c) C_2 on *E. coli* after exposure to 0.5×, 1×, and 2× the MIC and the control group treated with normal saline.

range of 800–200 nm) are given in Figure 5. An increase in the concentration of CT-DNA causes hypochromism for the high-energy absorption intensity band of complex **1** at 218 nm, and on the contrary, the low energy band at 269 nm shows the same with DNA addition. A similar pattern has been observed for the complex **2** at ~218–231 nm (218, 221, 224, and 231 nm), which exhibits hypochromism and a significant bathochromism of about 12 nm. These spectral

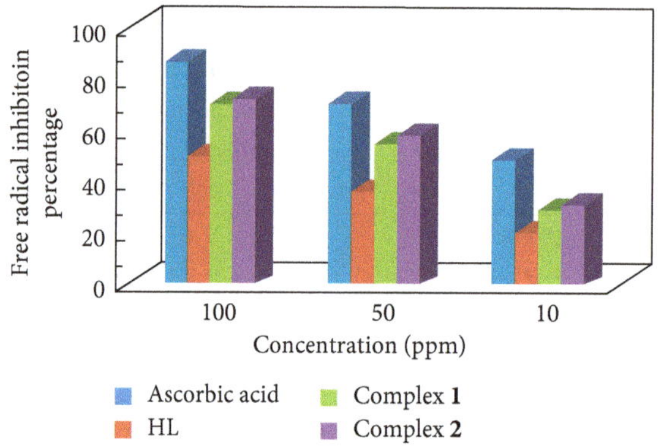

FIGURE 4: Antioxidant activity of the imine-based ligand (HL) and its metal complexes.

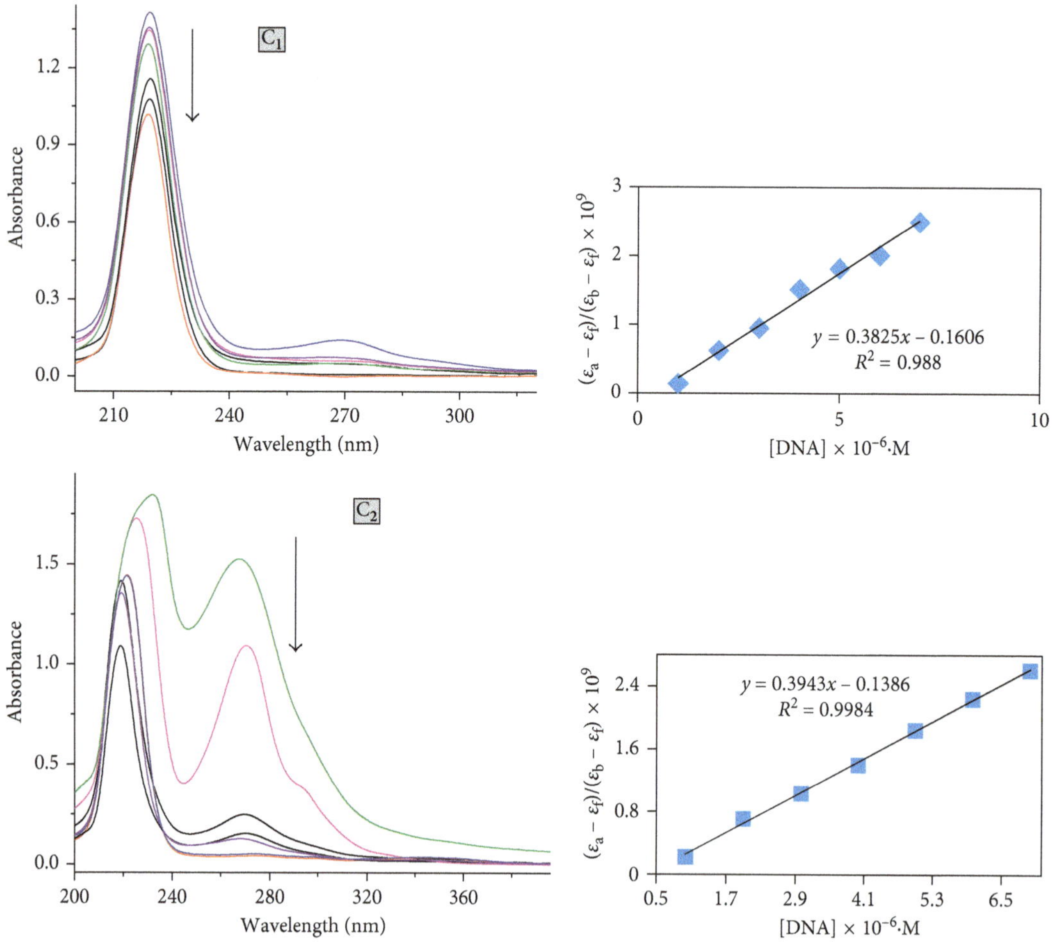

FIGURE 5: The electronic absorption spectra of complex C_1 and C_2 (20 μM) in the absence and presence of DNA in Tris/NaCl buffer (pH = 7.2). The arrows show the absorbance changes upon increasing concentration of DNA (0–25 μM) at RT. The plot shows $[DNA]/(\varepsilon_a - \varepsilon_f) \times 10^{-5}$ versus [DNA].

characteristics are typical complexes to have higher binding propensity with DNA most likely through a stacking mode of interaction between the planar aromatic chromophore and the DNA base pairs. To further elucidate the binding strength of the complexes, intrinsic binding constants K_{bin} were calculated by monitoring the changes of absorbance in the MLCT bands with increasing amounts of CT-DNA. The binding constant (K_{bin}) values of complexes are quantified from (1), and the values found to be 5.95×10^5, 9.86×10^6 and 5.61×10^3 for complexes C_1, C_2, and HL, respectively. The obtained K_{bin} values of the complexes are almost parallel to those reported for the typical classical intercalator ethidium bromide (K_{bin}, 10^6–$10^7 \cdot M^{-1}$) and suggests that the binding mechanism of oxidovanadium complexes with DNA is going through the intercalative mode, while the intercalative ability of nonplanar ligand to DNA appears weak and

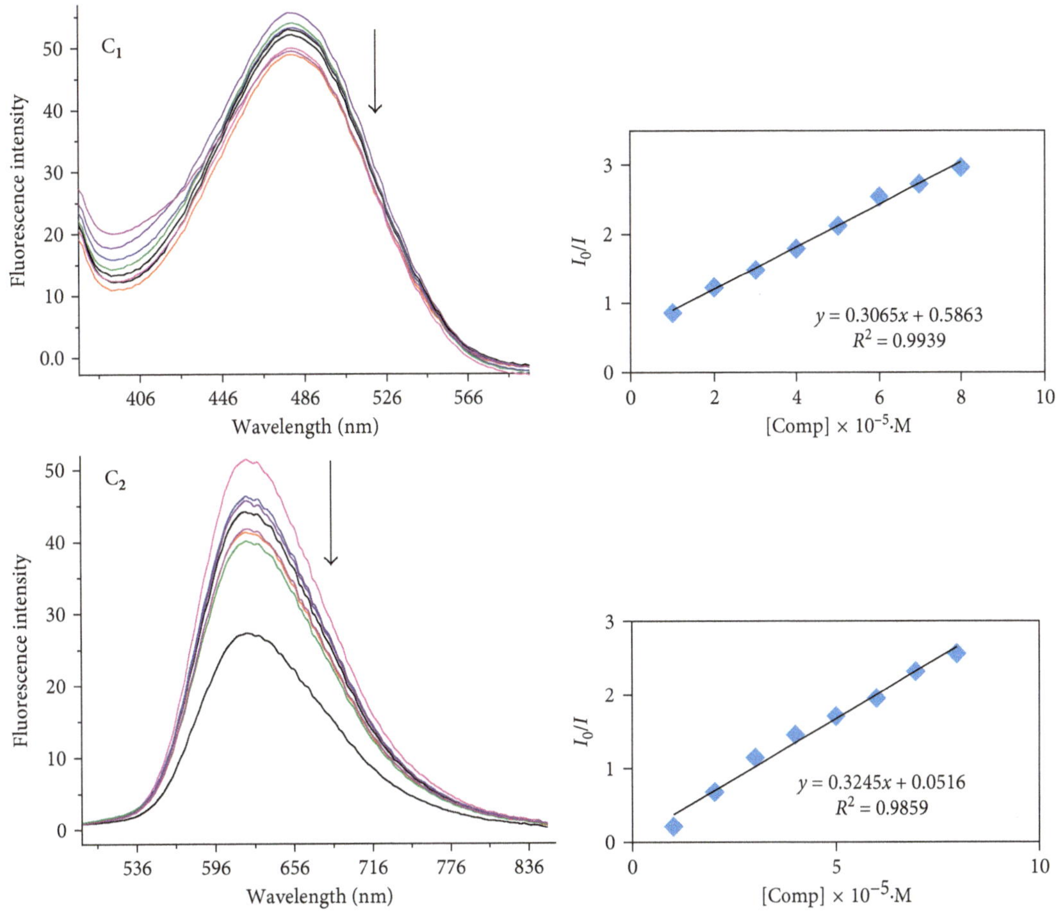

Figure 6: Fluorescence quenching curves of EB-DNA in the presence of complex (C_1 and C_2) at the concentration 0–35 μM at the interval of 5 in Tris-HCl/NaCl buffer (pH 7.2). The arrow indicates the effect of increasing the complex concentration on the fluorescence emission of ethidium bromide bound CT-DNA. The plot shows the linear fit of F_o/F versus [complex].

suggests that ligand is involved in DNA groove binding. The relative K_{bin} value and pronounced hyperchromism of complexes C_2 in comparison to C_1 with respect to CT-DNA were owing to the planar and larger surface area of the 1,10-phenanthroline ring appended deeply in the complex cause enlargement of size of the DNA.

It also suggested that the DNA-binding affinities of these oxidovanadium complexes are associated with the electronic effects of the substituents introduced on the benzene ring of the ligands. The imine-based ligand would encourage partial intercalation of compounds in the minor groove of DNA helices by engaging in hydrogen-bonding interactions between coordinated -NH- and -OH bands with the functional groups positioned on the edge of DNA bases [30].

(2) Fluorescence Emission Titrations for DNA Interaction. Fluorescence quenching tendency is further examined to ascertain the relative binding mode of oxidovanadium complexes with CT-DNA due to high sensitivity and selectivity of this method [31]. The oxidovanadium complexes either in aqueous solution or the presence of CT-DNA do not exhibit any fluorescence emission. So, the interaction of the complexes to DNA has been studied by the competitive ethidium bromide (EB) binding assay.

Successive addition of the DNA solution into a constant concentration of complexes led to changes in the fluorescence intensity of complexes. Quenching of fluorescence intensity of compounds was observed and is displayed in Figure 6. The fluorescence spectra of complex **1** showed the emission band at λ_{em} 486 nm, and in complex **2**, the emission band was observed at 656 nm when excited at wavelength λ_{ex} 350 nm and 620 nm, respectively. The intense fluorescence of the DNA-EB complex system can be quenched by the addition of a third molecule which can bind DNA by the intercalative mode by displacing EB. This significant decrease in fluorescence intensity affords strong support in favor of an intercalative binding pattern of the complex into the DNA double helix.

Furthermore, the fluorescence quenching study in the presence of the complexes was analyzed by the Stern–Volmer quenching constant (K_{sv}). The obtained SV plots are linear (Figure 6), and the values are found to be 1.09 (±0.02) × 10^4·M^{-1} for complex **1** and 3.25 (±0.01) × 10^4·M^{-1} at 37°C. Hence, the fluorescence quenching has static mechanism as the fluorophore, and the quencher collides together in the ground state. The values of K_q obtained for DNA was about 10^{12}·M^{-1}·S^{-1}. This again proves that the fluorescence quenching of the biomolecule occurs by the

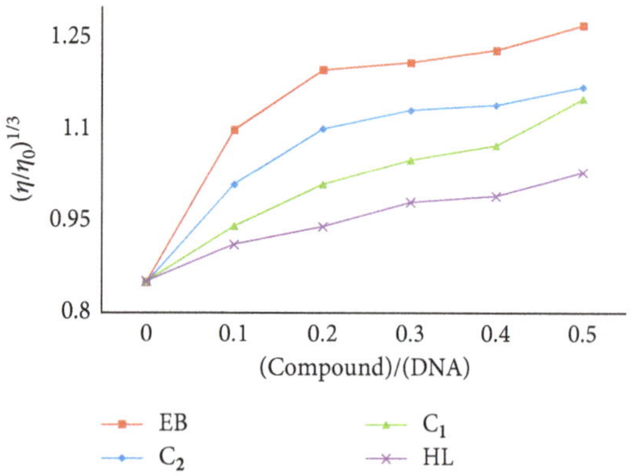

FIGURE 7: Effect on relative viscosity (±0.1) of CT-DNA under the influence of increasing amount of compounds at 26 ± 0.1°C in 5 mM Tris-HCl buffer (pH 7.2).

static mode since k_q values are higher than limiting diffusion constant of the diffusional quenching for biopolymers ($2*10^{11} \cdot M^{-1} \cdot S^{-1}$).

(3) Viscosity Measurements. The most critical, least ambiguous, and effective means to analyze the binding mode of compounds to DNA is the viscometry technique in solution can be carried out in the absence of X-ray structural data. A classical intercalation probe (EB) between the two strands of DNA results in significant increase in viscosity of the DNA solution due to an increase in separation of base pairs at the intercalation sites and hence, an increase in overall DNA length. In contrast, partial, nonclassical intercalation of ligand could crook the DNA helix, resulting in shortening of DNA length and, concomitantly, reducing its viscosity [32]. The thickness of CT-DNA increases with increase in the ratio of complexes to CT-DNA is associated with the specific intercalation binding mode [28]. Also, the effects of relative viscosity of CT-DNA under the influence of increasing amounts of the complexes at 26 ± 0.1°C (Figure 7) resemble the binding mode of ethidium bromide to DNA. It can be deduced that the obtained results were parallel to the above UV-spectroscopic data such as hypochromism and red-shift of complexes in the presence of DNA.

4. Conclusion

In the present investigation, the synthesis, spectral characterization, *in vitro* biopotency studies of VO(IV) complexes of imine-based ligand (HL) were carried out. Elemental analysis confirms the stoichiometry in complexes. Physical properties like molar conductivity, IR, magnetic susceptibility, and electronic spectral properties gave satisfactory analytical data coupled with ESR, and TGA studies suggested the nature of coordination for both the oxidovanadium(IV) complexes to be square pyramidal geometry with point group C_{4V}. The synthesized compounds had a broad spectrum of antimicrobial activity against a panel of highly resistant G −ve, G +ve, and fungal pathogens which showed that the complexes have better antibiotic action than its parent ligand and standard drugs used. The antioxidant activity of the complexes found to be as a good scavenger of DPPH radical comparing to free ligand. Spectral titration assessed that the metal complexes strongly bind with DNA relative to the ligand, and the results are corroborated with each other techniques.

Acknowledgments

One of the authors K. Savithri is thankful to the University Grants Commission, New Delhi, for the award of Rajiv Gandhi National Senior Research fellowship and also wishes to thank IOE Vigyan Bhavan, University of Mysore, for spectral measurements.

References

[1] D. C. Crans, J. J. Smee, E. Gaidamauskas, and L. Yang, "The chemistry and biochemistry of vanadium and the biological activities exerted by vanadium compounds," *Chemical Reviews*, vol. 104, no. 2, pp. 849–902, 2004.

[2] J. Korbecki, I. Baranowska-Bosiacka, I. Gutowska, and D. Chlubek, "Biochemical and medical importance of vanadium compounds," *Acta biochimica Polonica*, vol. 59, no. 2, pp. 195–200, 2012.

[3] G. Nizami and R. Sayyed, "Antimicrobial, electrochemical and thermodynamic studies of Schiff base complexes and their potential as anticarcinogenic and antitumor agents : a review," *Journal of Chemical Technology & Biotechnology*, vol. 10, no. 10, pp. 40–51, 2017.

[4] A. Jamshidvand, M. Sahihi, V. Mirkhani, and M. Moghadam, "Studies on DNA binding properties of new Schiff base ligands using spectroscopic, electrochemical and computational methods : influence of substitutions on DNA-binding," *Journal of Molecular Liquids*, vol. 253, pp. 61–71, 2018.

[5] E. Bagdatli, E. Altuntas, and U. Sayin, "Synthesis and structural characterization of new oxovanadium(IV) complexes derived from azo-5-pyrazolone with prospective medical importance," *Journal of Molecular Structure*, vol. 1127, pp. 653–661, 2017.

[6] M. K. Prashanth, M. Madaiah, H. D. Revanasiddappa, and K. N. Amruthesh, "Synthesis, characterization, and BSA binding studies of some new benzamides related to Schiff base," *ISRN Organic Chemistry*, vol. 2013, Article ID 791591, 2013.

[7] M. M. Ibrahim, G. A. M. Mersal, A. M. M. Ramadan, S. Y. Shaban, M. A. Mohamed, and S. Al-Juaid, "Synthesis, characterization and antioxidant/cytotoxic activity of oxovanadium(IV) complexes of methyliminodiacetic acid and ethylenediaminetetracetic acid," *Journal of Molecular Structure*, vol. 1137, pp. 742–755, 2017.

[8] B. Jing, J. Dong, J. Li, T. Xu, and L. Li, "Synthesis, crystal structure, and DNA interaction of an oxovanadium(IV) complex containing L-valine Schiff base and 1,10-phenanthroline," *Journal of Coordination Chemistry*, vol. 66, no. 3, pp. 520–529, 2013.

[9] U. Saha and K. K. Mukherjea, "DNA binding and nuclease activity of an oxovanadium valinato-Schiff base complex," *International Journal of Biological Macromolecules*, vol. 66, pp. 166–171, 2014.

[10] M. Balouiri, M. Sadiki, and S. K. Ibnsouda, "Methods for *in vitro* evaluating antimicrobial activity: a review," *Journal of Pharmaceutical Analysis*, vol. 6, no. 2, pp. 71–79, 2016.

[11] H. M. Vinusha, K. Shivaprasad, S. Chandan, and M. Begum, "Imino-4-methoxyphenol thiazole derived Schiff base ligands: synthesis, spectral characterization and antimicrobial activity," *Journal of Chemical Sciences*, vol. 6, no. 3, pp. 3–6, 2015.

[12] T. M. Chandra Babu, S. S. Rajesh, B. V. Bhaskar et al., "Molecular docking, molecular dynamics simulation, biological evaluation and 2D QSAR analysis of flavonoids from Syzygium alternifolium as potent anti-Helicobacter pylori agents," *RSC Advances*, vol. 7, no. 30, pp. 18277–18292, 2017.

[13] B. Ilhan-Ceylan, E. Tuzun, Y. Kurt, M. Acikgoz, S. Kahraman, and U. Bahri, "Oxovanadium(IV) complexes based on S-alkyl-thiosemicarbazidato ligands. Synthesis, characterization, electrochemical, and antioxidant studies," *Journal of Sulfur Chemistry*, vol. 36, no. 4, pp. 434–449, 2015.

[14] M. Blois, "Antioxidant determinations by the use of a stable free radical," *Nature*, vol. 181, no. 4617, pp. 1199-1200, 1958.

[15] K. Zheng, F. Liu, Y. Li, Z. Wu, and C. Yan, "Synthesis and structure elucidation of new μ-oxamido-bridged dicopper(II) complexes showing in vitro anticancer activity : evaluation of DNA/protein-binding properties by experiment and molecular docking," *Journal of Inorganic Biochemistry*, vol. 156, pp. 75–88, 2016.

[16] L. Shivakumar, K. Shiva Prasad, and H. D. Revanasiddappa, "Synthesis, characterization, antioxidant, antimicrobial, DNA binding and cleavage studies of mononuclear Cu(II) and Co(II) complexes of 3-hydroxy-N′-(2-hydroxybenzylidene)-2-naphthohydrazide," *European Journal of Chemistry*, vol. 2, no. 3, pp. 394–403, 2011.

[17] A. Wolfe, G. H. Shimer, and T. Meehan, "Polycyclic aromatic hydrocarbons physically intercalate into duplex regions of denatured DNA," *Biochemistry*, vol. 26, no. 20, pp. 6392–6396, 1987.

[18] L. H. Abdel-Rahman, A. M. Abu-dief, E. F. Newair, and S. K. Hamdan, "Some new nano-sized Cr(III), Fe(II), Co(II), and Ni(II) complexes ligand : structural characterization, electrochemical, antioxidant, antimicrobial, antiviral assessment and DNA interactions," *Journal of Photochemistry and Photobiology B: Biology*, vol. 160, pp. 18–31, 2016.

[19] V. Singh, P. Bora, and H. S. Yadav, "Oxovanadium(IV) complexes with ligands derived by condensation of 1, 2-diacetylbenzene with 2-aminobenzamide and β-diketones," *Acta metallomica-MEEMB*, vol. 11, no. 2–4, pp. 171–179, 2014.

[20] M. K. Sahani, S. K. Pandey, O. P. Pandey, and S. K. Sengupta, "A series of novel oxovanadium(IV) complexes: synthesis, spectral characterization and antimicrobial study," *Journal of Molecular Structure*, vol. 1074, pp. 401–407, 2014.

[21] H. Wang, Y. Liu, M. Li et al., "Multifunctional TiO_2 nanowires-modified nanoparticles bilayer film for 3D dye-sensitized solar cells," *Optoelectronics and Advanced Materials: Rapid Communications*, vol. 4, no. 8, pp. 1166–1169, 2010.

[22] S. B. Gajera, J. V. Mehta, and M. N. Patel, "DNA interaction, cytotoxicity, antibacterial and antituberculosis activity of an oxovanadium(IV) complexes derived from fluoroquinolones and 4-hydroxy-5-((4-hydroxyphenyl)diazenyl)thiazole-2(3H)-thione Sanjay," *RSC Advances*, vol. 5, no. 5, pp. 21710–21719, 2015.

[23] R. C. Maurya, J. Chourasia, D. Rajak et al., "Oxovanadium(IV) complexes of bioinorganic and medicinal relevance: synthesis, characterization and 3D molecular modeling of some oxovanadium(IV) complexes involving O, N-donor environment of salicylaldehyde-based sulfa drug Schiff bases," *Arabian Journal of Chemistry*, vol. 9, pp. S1084–S1100, 2016.

[24] E. Garribba, G. Micera, A. Panzanelli, and D. Sanna, "Electronic structure of oxovanadium(iv) complexes of α-hydroxycarboxylic Acids," *Inorganic Chemistry*, vol. 42, no. 13, pp. 3981–3987, 2003.

[25] M. Rizzotto, "Metal complexes as antimicrobial agents," *A Search for Antibacterial Agents*, 2012.

[26] M. A. Bonilla, M. L. Cerrada, and M. F. Garcia, "Antimicrobial activity of chitosan in food, agriculture and biomedicine," in *Polymeric Materials with Antimicrobial Activity: From Synthesis to Applications, RSC Polymer Chemistry Series No. 10*, RSC Publishing, Cambridge, UK, pp. 22–53, 2014.

[27] L. Chen, Y. Zhu, D. Yang, R. Zou, J. Wu, and H. Tian, "Synthesis and antibacterial activities of antibacterial peptides with a spiropyran fluorescence probe," *Scientific Reports*, vol. 4, no. 1, pp. 1-7, 2014.

[28] Z. Mandegani, Z. Asadi, M. Asadi, H. R. K. Heidarib, and B. Rastegari, "Synthesis, characterization, DNA binding, cleavage activity, cytotoxicity and molecular docking of new nano water-soluble $[M(5\text{-}CH_2PPh_3\text{-}3,4\text{-}salpyr)](ClO_4)_2$ (M = Ni, Zn) complexes," *Dalton Transactions*, vol. 45, no. 15, pp. 6592–6611, 2016.

[29] D. Sanna, V. Ugone, A. Fadda, G. Micera, and E. Garibba, "Behavior of the potential antitumor $V^{IV}O$ complexes formed by flavonoid ligands. 3. Antioxidant properties and radical production capability," *Journal of Inorganic Biochemistry*, vol. 161, pp. 18–26, 2016.

[30] V. Rajendiran, R. Karthik, M. Palaniandavar et al., "Mixed-ligand copper(II)-phenolate complexes : effect of coligand on enhanced DNA and protein binding, DNA cleavage, and anticancer activity," *Inorganic Chemistry*, vol. 46, no. 20, pp. 8208–8221, 2007.

[31] J. Deng, G. Su, P. Chen, Y. Du, Y. Gou, and Y. Liu, "Evaluation of DNA binding and DNA cleavage of nickel(II) complexes with tridentate α-N-heterocyclic thiosemicarbazones ligands," *Inorganica Chimica Acta*, vol. 471, pp. 194–202, 2018.

[32] W. Hong, F. Huang, T. Huan, X. Xu, Q. Han, and G. Wang, "Comparative studies on DNA-binding and in vitro antitumor activity of enantiomeric ruthenium(II) complexes," *Journal of Inorganic Biochemistry*, vol. 180, pp. 54–60, 2018.

7

A QSAR Study based on SVM for the Compound of Hydroxyl Benzoic Esters

Li Wen,[1] Qing Li,[2] Wei Li,[1] Qiao Cai,[1] and Yong-Ming Cai[3]

[1] *School of Public Health, Guangdong Pharmaceutical University, Guangzhou Higher Education Mega Centre, Guangzhou 510006, China*
[2] *Guangdong Provincial Center for Disease Control and Prevention, Guangzhou 511430, China*
[3] *School of Medical Information Engineering, Guangdong Pharmaceutical University, Guangzhou Higher Education Mega Centre, Guangzhou 510006, China*

Correspondence should be addressed to Yong-Ming Cai; cym@gdpu.edu.cn

Academic Editor: Konstantinos Tsipis

Hydroxyl benzoic esters are preservative, being widely used in food, medicine, and cosmetics. To explore the relationship between the molecular structure and antibacterial activity of these compounds and predict the compounds with similar structures, Quantitative Structure-Activity Relationship (QSAR) models of 25 kinds of hydroxyl benzoic esters with the quantum chemical parameters and molecular connectivity indexes are built based on support vector machine (SVM) by using R language. The External Standard Deviation Error of Prediction ($SDEP_{ext}$), fitting correlation coefficient (R^2), and leave-one-out cross-validation (Q^2_{LOO}) are used to value the reliability, stability, and predictive ability of models. The results show that R^2 and Q^2_{LOO} of 4 kinds of nonlinear models are more than 0.6 and $SDEP_{ext}$ is 0.213, 0.222, 0.189, and 0.218, respectively. Compared with the multiple linear regression (MLR) model (R^2 = 0.421, RSD = 0.260), the correlation coefficient and the standard deviation are both better than MLR. The reliability, stability, robustness, and external predictive ability of models are good, particularly of the model of linear kernel function and eps-regression type. This model can predict the antimicrobial activity of the compounds with similar structure in the applicability domain.

1. Introduction

1.1. Conceptual Framework. QSAR [1, 2] is used to research the relationship between the molecular structure and biological activity and physicochemical characteristics, reveal the quantitative relationship, predict the activity of unknown compounds, and direct the synthesis of new materials [3–5]. QSAR is considered as one of the promising technologies and is widely used at present because of making up the loss of experimental data, reducing the cost of testing, and achieving high throughput prediction and screening [6]. Many international organizations and regulatory agencies have supported and promoted the use of QSAR and thought that QSAR can be used as an alternative to animal experiments. Health Canada, the United States of Food and Drug Administration (FDA), Environmental Protection Agency (EPA), the European Union, and the Organization for Economic Cooperation and Development (OECD) apply QSAR to identify potential health hazards, screening, and priority [7]. After recent years of development, QSAR has become a frontier topic in medicinal chemistry, environmental chemistry, life science, analytical chemistry, computer chemistry, and even pesticide [8–11].

Hydroxyl benzoic esters are important kinds of preservatives, which are widely used in medicine, food, cosmetics, pesticides, and other fields [12]. At present, there are about 60 kinds of food preservatives in the world [13]. The benzoic acid and sorbic acid are productive in China, but the usage is little because of the high toxicity of benzoic acid and the high price of sorbic acid. Hydroxyl benzoic esters have high efficiency, low toxicity, compatibility, and other advantages; the performance of antibacterial is stronger than benzoic acid and sorbic acid because it has a phenolic hydroxyl [14]. So it is of great significance to study and apply the antibacterial activity of hydroxyl benzoic esters.

1.2. Research Status of SVM in QSAR. SVM is a machine learning algorithm based on statistical learning theory proposed by Cortes et al. [15–17]. SVM can be used for pattern recognition, regression analysis and function fitting, and so forth because it possesses favorable mathematical properties, such as the uniqueness of the solution, nondependence on the dimension of the input space, and so forth. The optimal solution of SVM is superior to the traditional learning methods. In recent years, SVM is applied to the study of QSAR of the compound. Hou et al. [18] investigated the QSAR of the antimalarial activity of PfDHODH inhibitors by generating four computational models using a multiple linear regression (MLR) and a SVM based on a dataset of 255 PfDHODH inhibitors. Sharma et al. [19] drew support from SVM and MLR studying the activity of HIV-1 capsid inhibitors. SVM model was found more efficient in prediction. Khuntwal et al. [20] used MLR and SVM to develop QSAR models for a dataset of 34 tetrahydrobenzothiophene derivatives. Zhiming et al. [21] by using ridge regression (RR) and SVM built QSAR models of bitter tasting thresholds (BTT) and cytotoxic T lymphocyte (CTL) and predicted independent test data. Results showed that the fitting, LOOCV, and external prediction accuracies were superior to the reported results of the existing literature. Zhang et al. [22] took the benzene compounds as the research object, combining the molecular structure of the quantitative description with MLR or nonlinear regression statistical methods SVM, to build successfully the acute toxicity QSAR models and mutagenic QSAR models of benzene compounds. By comparing the linear and nonlinear QSAR models, Zhang Xiao-Long discovered that the stability and prediction ability of nonlinear QSAR models are better than those of multiple linear QSAR models. In the literature, there are very few researches about QSAR of the hydroxyl benzoic esters. Jiang et al. [23] used MLR to build the model of QSAR and it can well predict the MIC and $t_{0.5}$ in the range of atomic number (the number of C among 1–4 on the ester chain of MIC and 1–3 of $t_{0.5}$). Qiu et al. [24] optimized the molecular structures of eleven kinds of p-hydroxyl benzoic esters by using density functional theory (DFT) B3LYP method of quantum chemistry and then used stepwise multiple linear regression to select the descriptors and to generate the best prediction model that relates the structural features to inhibitory activity. The QSAR results showed that the lowest unoccupied molecular orbit E_{LUMO} and the increase of dipole moment μ were the main independent factors contributing to the antifungal activity of the compounds. SVM has shown obvious advantages in the QSAR research, but QSAR study of the compound of hydroxyl benzoic esters is confined to the linear model at present; there is no literature on the nonlinear QSAR analysis of the system.

In this paper, we use the quantum chemical parameters and molecular connectivity indexes to analyze the antibacterial activity of the hydroxyl benzoic esters. The QSAR model is established by the SVM algorithm in the R software. We obtain the structure-activity relationship between the molecular structural parameters and the antibacterial activity of *Escherichia coli* under the most stable configuration, which provides a basis of predicting the antibacterial activity of similar compounds.

2. Method

2.1. Data Preparation

2.1.1. Basic Information of Hydroxyl Benzoic Esters. This paper took the 25 hydroxyl benzoate group compounds as the research object, including 10 o-hydroxyl benzoic esters, 2 m-hydroxyl benzoic esters, and 13 p-hydroxyl benzoic esters. Their details are shown in Table 1.

2.1.2. Terminal Value. The antimicrobial half-life ($t_{1/2}$) (h) at the condition of minimum inhibition concentration of 25 hydroxyl benzoic esters was collected from the literature [23], in the form of logarithm ($\mathrm{lgt}_{1/2}$) to express its antibacterial activity. The results are shown in Table 2.

2.2. Calculation and Selection of Molecular Descriptors. The quantum chemical parameters [25] and molecular connectivity indexes [26] can well explain the antibacterial activity of compounds and have good correlation between them; therefore, this paper selects them with a clear physical meaning as the descriptor.

2.2.1. The Quantum Chemical Parameters. In this paper, the quantum chemical parameters are calculated by the latest Gaussian 09 software [27] that is a quantum chemistry software of semiempirical calculation and ab initio calculation of United States Gaussian company. Gaussian 09 in the calculation can carry out the molecular structure through the View Gauss 5 software directly and create the input files of molecular structures. In the calculation, Gaussian 09 software calls directly the input file and translates it into the form of redundant internal coordinates automatically. The results of the calculation are output by the text. Each time before calculation, a suitable chemistry model (computational method) should be established for the system in order to achieve balance in terms of computational cost and accuracy [27, 28]. The method of this paper is B3LYP/6-31G DFT/(d). Because all the molecular configurations are optimal configurations and the geometry optimization is convergent and there is no virtual frequency by the frequency analysis, therefore, all the data are true and reliable. Find out the useful quantum chemical parameters from the output file. The values are shown in Table 3.

2.2.2. The Molecular Connectivity Indexes. Molecular connectivity indexes which mainly reflect the number of atoms in molecules, valence bond and branch information, and so forth are the constants that are calculated according to the molecular structure. Each order index has a different meaning. Many studies show that ${}^{5}\mathbf{X}^{\mathbf{v}}{}_{\mathbf{P}}$ can characterize a lot of information, which has a great significance in explaining the influence of structure on biological activity [29, 30]. So, this

TABLE 1: The basic information of hydroxyl benzoate esters.

ID	Compound	Abbreviation
1	Methyl o-hydroxyl benzoate esters	M-o-HB
2	Ethyl o-hydroxyl benzoate esters	E-o-HB
3	Propyl o-hydroxyl benzoate esters	P-o-HB
4	Isopropyl o-hydroxyl benzoate esters	IP-o-HB
5	Butyl o-hydroxyl benzoate esters	B-o-HB
6	Isobutyl o-hydroxyl benzoate esters	IB-o-HB
7	Isoamyl o-hydroxyl benzoate esters	IA-o-HB
8	Octyl o-hydroxyl benzoate esters	O-o-HB
9	Benzyl o-hydroxyl benzoate esters	Be-o-HB
10	Phenyl o-hydroxyl benzoate esters	Ph-o-HB
11	Methyl m-hydroxyl benzoate esters	M-m-HB
12	Ethyl m-hydroxyl benzoate esters	E-m-HB
13	Methyl p-hydroxyl benzoate esters	M-p-HB
14	Ethyl p-hydroxyl benzoate esters	E-p-HB
15	Propyl p-hydroxyl benzoate esters	P-p-HB
16	Isopropyl p-hydroxyl benzoate esters	IP-p-HB
17	Butyl p-hydroxyl benzoate esters	B-p-HB
18	Isobutyl p-hydroxyl benzoate esters	IB-p-HB
19	Amyl p-hydroxyl benzoate esters	A-p-HB
20	Isoamyl p-hydroxyl benzoate esters	IA-p-HB
21	Heptyl p-hydroxyl benzoate esters	H-p-HB
22	Octyl p-hydroxyl benzoate esters	O-p-HB
23	Isooctyl p-hydroxyl benzoate esters	IO-p-HB
24	Nonyl p-hydroxyl benzoate esters	N-p-HB
25	Benzyl p-hydroxyl benzoate esters	Be-p-HB

study selects 8 molecular connectivity indexes, including ${}^{0}\mathbf{X}^{v}{}_{P}$, ${}^{1}\mathbf{X}^{v}{}_{P}$, ${}^{2}\mathbf{X}^{v}{}_{P}$, ${}^{3}\mathbf{X}^{v}{}_{P}$, ${}^{4}\mathbf{X}^{v}{}_{P}$, ${}^{5}\mathbf{X}^{v}{}_{P}$, ${}^{3}\mathbf{X}^{v}{}_{C}$, and ${}^{4}\mathbf{X}^{v}{}_{PC}$. The results are shown in Table 4.

2.3. Establishment of Models

2.3.1. Partition of Dataset. The rational division of datasets is a very hot research topic in the field of QSAR. There are a variety of methods. In this paper, Random Sampling (RS) [31] is used to divide the raw data into training set (22 kinds) and test set (3 kinds, o-hydroxyl benzoic esters, m-hydroxyl benzoic esters, and p-hydroxyl benzoic esters). The training set is used to establish the SVM nonlinear models, and the test set tests the external prediction ability of the models.

2.3.2. Modeling Method. Through the R software program, the training set with 22 compounds is used to build the nonlinear models by SVM algorithm based on the selected descriptors. Firstly, we standardize the data and then establish 4 models of kernel for radial, linear, eps-regression, and nu-regression type, respectively.

2.4. Model Validation. Model validation is very important for QSAR research, which consists of two aspects: internal validation to test the fitting ability and robustness of models and external validation to test the model's predictive ability. Both internal and external validations are equally important [32].

2.4.1. Internal Validation. There are many methods to estimate a model's stability, robustness, and internal predictive ability, such as the fitting correlation coefficient, cross-validation, random model test, Y random, and various residual errors (like Root Mean Squared Errors (RMSEs), standard residual error, etc.) [33]. In this paper, the fitting correlation coefficient (R^2) between the experimental and predicted values of the training dataset and leave-one-out cross-validation (Q^2_{LOO}) are used to test the reliability, robustness, stability, and whether the models are overfitting or not.

2.4.2. External Validation. A very important purpose of the QSAR models is to predict the related activity data of new or even nonsynthetic compounds, in order to guide the design and synthesis of compounds with desirable activity, or to screen the compounds. This requires that the model has good predictive ability and generalization ability; however, cross-validation can only explain the internal predictive ability of models and good internal prediction ability does not mean the excellent external prediction ability [34–36]; that is, good cross-validation Q^2_{cv} is a necessary but nonsufficient condition for the high external predictive ability [35]. The only way

TABLE 2: Antimicrobial half-life ($t_{1/2}$) (h) at the condition of minimum inhibition concentration of hydroxyl benzoic esters.

ID	$t_{1/2}$	$lgt_{1/2}$
1	66.0	1.819
2	74.0	1.869
3	138.0	2.139
4	115.0	2.060
5	17.0	1.230
6	15.0	1.176
7	20.5	1.312
8	66.0	1.819
9	42.0	1.623
10	12.0	1.079
11	43.0	1.633
12	56.0	1.748
13	76.0	1.880
14	94.0	1.973
15	176.0	2.245
16	156.0	2.193
17	61.5	1.789
18	29.0	1.462
19	34.5	1.537
20	28.0	1.447
21	28.0	1.447
22	39.8	1.599
23	31.0	1.491
24	36.8	1.565
25	29.0	1.462

to evaluate the external predictive ability of the model is to test the model with the new compound (namely, external test set that is not involved in the process of descriptor selection and model establishment). The parameters of evaluation model's external predictive ability include R^2_{ext}, external Q^2_{ext}, and $SDEP_{ext}$. In this paper, the test set is used to predict the corresponding $lgt_{1/2}$ and external predictive ability of the models is evaluated by $SDEP_{ext}$.

2.5. Extraction of Key Descriptors. We use principal component analysis to extract the most critical molecular descriptors of the hydroxyl benzoic esters for antibacterial half-life.

3. Results

3.1. Internal Prediction and Scatter Plot. Four nonlinear SVM models based on the selected descriptors are established by using training set. Experimental values and internal prediction results of $lgt_{1/2}$ are shown in Table 5 and scatter plot in Figure 1.

3.2. Parameters of Internal Validation. See Table 6.

3.3. Results of External Validation. $lgt_{1/2}$ of the test set is predicted, respectively, by 4 SVM models and the results are shown in Table 7. $SDEP_{ext}$ of the models and the residual between experimental values and the predicted results of $lgt_{1/2}$ are displayed in Table 8. Scatter plots of experimental values and prediction results by 4 SVM models of 25 compounds of $lgt_{1/2}$ are shown in Figure 2.

3.4. Results of Principal Component Analysis. See Tables 10 and 11.

4. Discussion and Conclusion

The degree of freedom and the speed of the preservative molecule determine the effective collision between the central atom of reactivity and the group or atom of microbial molecular activity. As a result, the antimicrobial property of the preservative is essentially determined by the electronic behavior of the preservative and the microorganism, that is, the quantum biochemical characterization of preservative. Therefore, from the perspective of quantum chemistry to study the relationship between the structure and properties of compound, the effective antimicrobial groups of preservative can be explained in essence [37]. Jiang et al. [23] use multiple linear regression to establish the linear model of 25 kinds of hydroxyl benzoic esters. The parameters are shown in Table 9. Results showed that R^2 was only 0.421, but the equation had good linear relationship when the number of C atoms was less than 4. When the number of C atoms in the ester group is more than 4, the influencing factors become more complex and cannot be described by simple linear relationship and may be in nonlinear or diversified relationship. So we use the R language to write the program and establish 4 kinds of nonlinear models through the SVM machine algorithm for 25 hydroxyl benzoic esters and predict $lgt_{1/2}$. Predicted results of training set are shown in Table 5. The scatter plot of experimental and predicted $lgt_{1/2}$ is drawn by using R software. Figure 1 shows that the predicted and experimental values are in good agreement and the linearity is obvious. According to literatures, if the value of R^2 is greater than 0.6 [35, 38] and Q^2 is greater than 0.5, the model is good, and model is excellent when the values are more than 0.9 [39]. Tropsha et al. [6] recommend R^2 and Q^2 to be greater than 0.6. Table 6 shows that both R^2 and Q^2_{LOO} are greater than 0.6 and R^2 and Q^2_{LOO} of two models with linear kernel function are close to 0.75, so we may think that the stability, robustness, and internal predicted ability of the 4 models are better and the models are not overfitting because R^2 is larger than Q^2_{LOO} by no more than 25%. By RS extracting, the para-, ortho-, and metacompound from 25 hydroxyl benzoic esters make up external test set to test the models, and the prediction results are shown in Table 7. The parameters from Table 8 show that the residual values of $lgt_{1/2}$ of the test set are in the range of -0.037244~0.322733 and $SDEP_{ext}$ is 0.213, 0.222, 0.189, and 0.218, respectively. The results indicate that the 4 models have high external predictive ability among themselves; in particular the model of the linear kernel function and eps-regression type is better than the other 3 models. Scatter plots of experimental values and prediction results by 4 SVM models of 25 compounds of $lgt_{1/2}$ are shown in Figure 2. The results show that

TABLE 3: Quantum chemical parameters of hydroxyl benzoic group at B3LYP/6-31G(d) level.

ID	E (hartree)	ZPE (KJ/mol)	E_{HOMO} (e V)	E_{LUMO} (e V)	ΔE (e V)	$\mu\ (T_D)$ $\mu\ (D_X)$	$\mu\ (D_Y)$ $\mu\ (D_Z)$	P
1	−535.4	390.3	−0.2384	−0.06313	0.17524	2.736 2.560	0.964 0.000	105.3
2	−574.7	464.7	−0.2370	−0.06142	0.17557	3.026 −2.648	−1.464 0.000	117.8
3	−614.0	539.2	−0.2367	−0.06093	0.17573	3.155 2.808	−1.440 0.000	130.0
4	−614.0	537.7	−0.2359	−0.05991	0.17596	3.091 −2.862	1.167 0.000	130.7
5	−653.3	614.2	−0.2364	−0.06066	0.17576	3.182 1.294	−2.906 0.000	142.3
6	−653.3	613.0	−0.2368	−0.06100	0.17578	3.169 −1.990	2.434 −0.393	142.2
7	−692.6	688.3	−0.2363	−0.06060	0.17574	3.165 2.136	2.334 0.047	154.6
8	−810.6	912.9	−0.2362	−0.06035	0.17580	3.276 2.254	−2.377 0.000	191.1
9	−766.4	603.6	−0.2381	−0.05969	0.17842	1.302 −1.093	−0.182 0.755	165.9
10	−727.1	526.4	−0.2421	−0.06882	0.17325	0.922 0.529	0.755 0.000	165.9
11	−574.7	461.5	−0.2408	−0.05736	0.18346	3.350 0.594	3.297 0.000	94.45
12	−535.4	387.3	−0.2419	−0.05896	0.18293	3.284 −1.208	3.053 0.0006	84.22
13	−535.4	387.8	−0.2450	−0.05006	0.19497	1.278 0.280	−1.247 0.000	106.0
14	−574.7	462.1	−0.2439	−0.04860	0.19528	1.194 −1.182	−0.169 0.000	118.4
15	−614.0	537.0	−0.2436	−0.04820	0.19540	1.201 −1.178	0.235 0.000	130.6
16	−614.0	535.2	−0.2433	−0.04748	0.19577	1.054 1.004	0.319 0.000	131.3
17	−653.3	611.3	−0.2434	−0.04798	0.19545	1.122 −0.519	−0.995 0.000	142.9
18	−653.3	610.6	−0.2439	−0.04838	0.19550	1.288 −0.785	−1.012 −0.131	142.8
19	−692.6	686.3	−0.2433	−0.04789	0.19544	1.1777 0.7889	−0.8744 0.0000	155.2
20	−692.6	685.6	−0.2434	−0.04795	0.19547	1.086 −0.589	−0.912 0.023	155.2
21	−771.3	836.0	−0.2433	−0.04777	0.19550	1.169 −1.155	−0.179 0.000	179.6
22	−810.6	911.2	−0.2432	−0.04772	0.19552	1.089 −0.258	−1.058 0.000	191.7
23	−810.6	910.2	−0.2433	−0.04780	0.19550	1.201 0.414	1.113 −0.179	191.7
24	−849.9	984.9	−0.2432	−0.04770	0.19553	1.169 1.046	−0.522 0.000	203.9
25	−776.4	601.9	−0.2442	−0.04966	0.19452	1.260 −0.916	−0.697 −0.512	174.3

TABLE 4: Molecular connectivity indexes of hydroxyl benzoic group.

ID	${}^0X^v_P$	${}^1X^v_P$	${}^2X^v_P$	${}^3X^v_P$	${}^4X^v_P$	${}^5X^v_P$	${}^3X^v_C$	${}^4X^v_{PC}$
1	6.073	3.117	2.009	1.226	0.608	0.259	0.190	0.235
2	6.780	3.705	2.238	1.316	0.706	0.311	0.190	0.223
3	7.487	4.205	2.653	1.478	0.770	0.381	0.190	0.223
4	7.650	4.100	2.971	1.436	0.775	0.349	0.426	0.389
5	8.194	4.705	3.007	1.772	0.885	0.426	0.190	0.223
6	8.358	4.561	3.509	1.588	0.818	0.421	0.599	0.629
7	9.065	5.061	3.836	1.969	0.963	0.415	0.599	0.512
8	11.020	6.705	4.421	2.772	1.623	0.904	0.190	0.223
9	9.167	5.262	3.290	2.157	1.187	0.607	0.308	0.407
10	8.460	4.824	3.014	1.898	1.008	0.425	0.258	0.327
11	6.073	3.111	2.042	1.087	0.647	0.255	0.213	0.321
12	6.780	3.699	2.271	1.177	0.780	0.320	0.213	0.309
13	6.073	3.111	2.038	0.904	0.412	0.173	0.213	0.292
14	6.780	3.699	2.267	0.994	0.508	0.237	0.213	0.280
15	7.487	4.199	2.683	1.155	0.572	0.305	0.213	0.280
16	7.650	4.094	3.001	1.114	0.576	0.281	0.449	0.392
17	8.194	4.699	3.036	1.449	0.686	0.350	0.213	0.280
18	8.358	4.555	3.539	1.266	0.620	0.353	0.622	0.446
19	8.902	5.199	3.390	1.699	0.894	0.431	0.213	0.280
20	9.065	5.055	3.865	1.646	0.765	0.384	0.622	0.568
21	10.310	6.199	4.097	2.199	1.248	0.703	0.213	0.280
22	11.020	6.699	4.451	2.449	1.425	0.828	0.213	0.280
23	11.180	6.555	4.637	2.173	1.374	0.792	0.622	0.568
24	11.730	7.199	4.804	2.699	1.601	0.953	0.213	0.280
25	9.1670	5.256	3.609	1.792	0.823	0.467	0.331	0.464

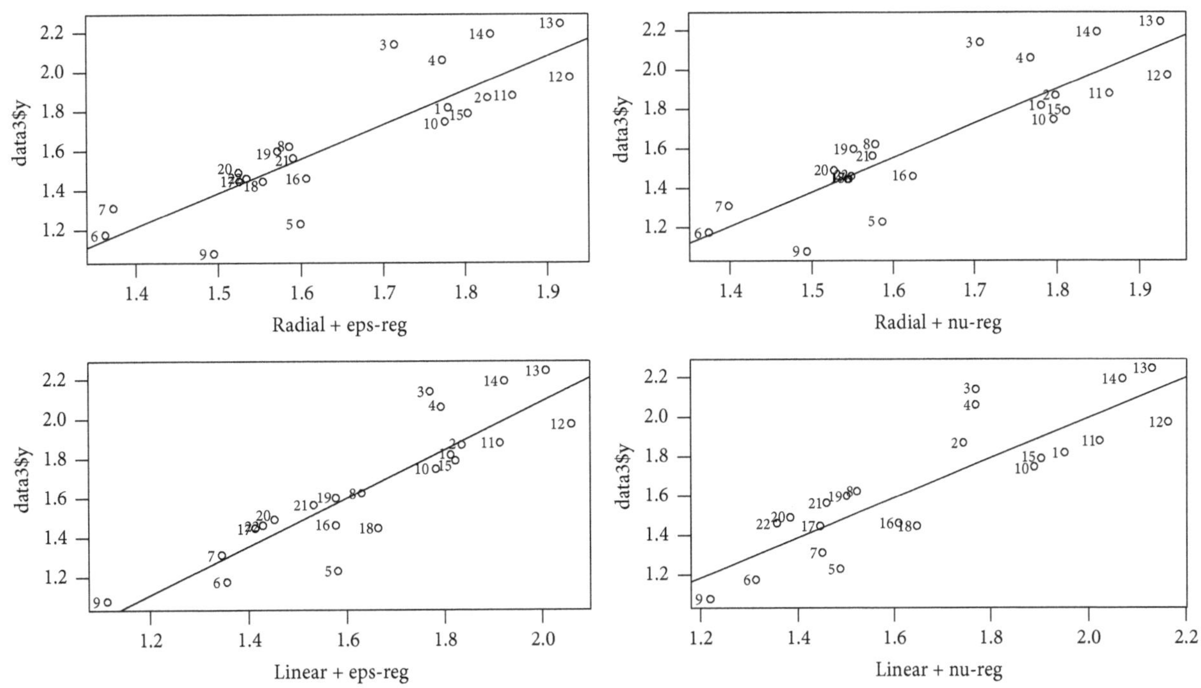

FIGURE 1: Scatter plot of experimental values and 4 SVM models' internal prediction results of lgtl/2. *Note.* The horizontal coordinates, respectively, represent the predicted values of lgtl/2 of 4 SVM models, and the longitudinal coordinates express the experimental results.

Table 5: Experimental values and 4 SVM models' internal prediction results of lgtl/2.

ID	$lgt_{1/2}$	Radial + eps-reg	Radial + nu-reg	Linear + eps-reg	Linear + nu-reg
1	1.819	1.779216	1.781302	1.812285	1.951165
2	1.869	1.827756	1.798580	1.835602	1.744102
3	2.139	1.713926	1.707563	1.769820	1.770619
4	2.060	1.772371	1.768500	1.792442	1.770267
5	1.230	1.599024	1.587027	1.580694	1.487888
6	1.176	1.363148	1.374539	1.355817	1.313377
7	1.312	1.373201	1.398522	1.345323	1.450414
9	1.623	1.585821	1.578657	1.629972	1.523181
10	1.079	1.494724	1.494087	1.112267	1.219324
12	1.748	1.775344	1.796207	1.781397	1.889308
13	1.880	1.858219	1.863383	1.913434	2.022421
14	1.973	1.927762	1.934197	2.059377	2.163645
15	2.245	1.916552	1.925729	2.008084	2.130922
16	2.193	1.831546	1.848498	1.922720	2.069817
17	1.789	1.803731	1.811289	1.822127	1.903807
18	1.462	1.606335	1.624347	1.576873	1.609259
20	1.447	1.525688	1.545616	1.413420	1.445952
21	1.447	1.554127	1.544298	1.663491	1.646921
22	1.599	1.571473	1.551537	1.576665	1.501430
23	1.491	1.524709	1.527338	1.452021	1.384236
24	1.565	1.590526	1.575234	1.531546	1.459150
25	1.462	1.534525	1.549003	1.428528	1.357148

Note. Radial + eps-reg, radial + nu-reg, linear + eps-reg, and linear + nu-reg, respectively, represent the 4 SVM models where kernel function is radial and linear and type is eps-regression and nu-regression.

Table 6: R^2 and Q^2_{LOO} of 4 SVM models.

Parameters	Radial + eps-reg	Radial + nu-reg	Linear + eps-reg	Linear + nu-reg
R^2	0.614	0.613	0.756	0.740
Q^2_{LOO}	0.611	0.608	0.747	0.731

Note. Radial + eps-reg, radial + nu-reg, linear + eps-reg, and linear + nu-reg, respectively, represent the 4 SVM models where kernel function is radial and linear and type is eps-regression and nu-regression.

Table 7: Experimental values and prediction results of test set of $lgt_{1/2}$.

ID	8	11	19
Compound	O-o-HB	M-m-HB	A-p-HB
$lgt_{1/2}$	1.819000	1.633000	1.537000
Radial + eps-reg	1.501886	1.762810	1.675167
Radial + nu-reg	1.496267	1.776929	1.691083
Linear + eps-reg	1.559133	1.670244	1.732065
Linear + nu-reg	1.535459	1.711716	1.772964

Table 8: Experimental and predicted values of $lgt_{1/2}$ residual and $SDEP_{ext}$.

ID	8	11	19	$SDEP_{ext}$
Radial + eps-reg	0.317114	−0.129810	−0.138167	0.213
Radial + nu-reg	0.322733	−0.143929	−0.154083	0.222
Linear + eps-reg	0.259867	−0.037244	−0.195065	0.189
Linear + nu-reg	0.283541	−0.078716	−0.235964	0.218

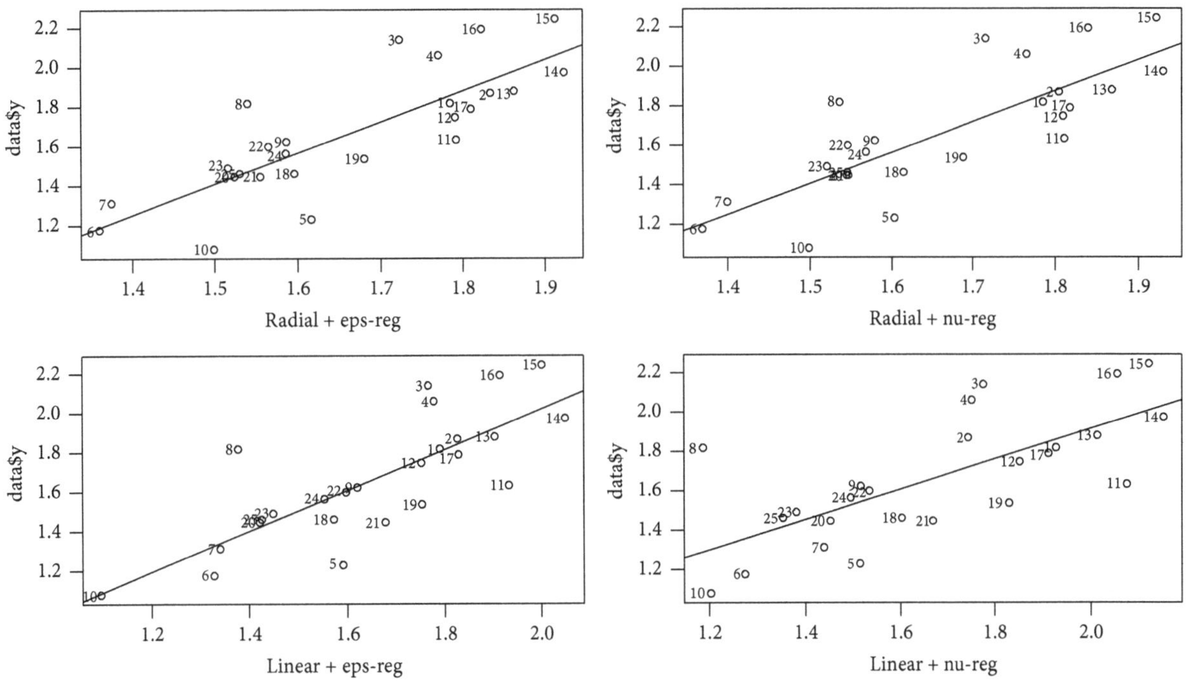

FIGURE 2: Scatter plot of experimental values and 4 SVM models' prediction results of lgtl/2. *Note.* The horizontal coordinates, respectively, represent the predicted values of lgtl/2 of 4 SVM models, and the longitudinal coordinates express the experimental results.

TABLE 9: The parameters of the MLR model established by Wang Deng-Ju.

Method	n	R	R^2	R_{ad}	RSD
MLR	25	0.649	0.421	0.311	0.260
MLR	8 ($C < 4$)	0.969	0.938	0.856	0.076

TABLE 10: Contribution rate of the first three principal components.

	Comp. 1	Comp. 2	Comp. 3
Standard deviation	188.7974802	38.08956667	4.3318149940
Proportion of variance	0.9602859	0.03908591	0.0005055312
Cumulative proportion	0.9602859	0.99937176	0.9998772898

the overall prediction of the 4 SVM models is better and, particularly, the linear relationship between predictive and experimental value of the model, where kernel function is linear and type is eps-regression, is the best.

In Table 10, the principal component analysis shows that the proportion of variance of the first principal component reaches 96.03%; therefore, the first principal component is taken only. Table 11 shows that the first principal component includes E (total energy), ZPE (zero-point vibrational energy), and p (polarizability). We consider that E, ZPE, and p are the key factors for antibacterial half-life of hydroxyl benzoic esters. p is a kind of structural parameter characterized by molecular deformation tensor under the action of external electric field. It is the most important property that p is related to the volume of the molecule and p contains information about the molecular interaction that is able to characterize the properties of the molecule as an electron acceptor. Since the coefficients of p and ZPE are negative, this indicates that the value of p and ZPE is greater and the antibacterial half-life of hydroxyl benzoic esters is shorter but E is just the opposite because the coefficient is positive.

TABLE 11: Loadings of the first three principal components.

	Comp. 1	Comp. 2	Comp. 3
E	0.467	0.838	0.277
ZPE	−0.871	0.491	
$\mu\ (D_Y)$			−0.152
p	−0.152	−0.238	0.944

In summary, QSAR nonlinear model obtained by quantum chemical parameters and molecular connectivity indexes can better predict the antibacterial activity of hydroxyl benzoic esters. The introduction of SVM algorithm solves the problem of poor correlation of QSAR and complex nonlinear relationship between the molecular descriptors when formula weight is large, which provides a basis for the prediction of the antibacterial activity of compounds with similar structure.

Therefore, the main conclusions of this paper are as follows:

(1) The establishment of the 4 kinds of nonlinear models using 25 hydroxyl benzoic acid esters by SVM method, through internal and external validation, the stability, and robustness, and internal and external predictive ability of 4 kinds of models are good; that

is, the models are available and may predict new compounds in the applicability domain.

(2) The model of linear kernel function and eps-regression type has the largest R^2 and $Q^2{}_{LOO}$, the minimum $SDEP_{ext}$, and the optimal linear relationship between predictive and experimental value of $lgt_{1/2}$ in 4 kinds of SVM models, which is the optimal model.

(3) SVM algorithm is a good method to solve the problem of multicollinearity and complex nonlinear relationship between molecular descriptors in QSAR modeling.

(4) E, ZPE, and p are the key factors for antibacterial half-life of hydroxyl benzoic esters.

Acknowledgments

This study was supported the Natural Science Foundation of Guangdong Province (Grant no. 2014A030313585), the Natural Sciences Funds, China (Grant no. 81473588, 2014), and Guangdong Province Science and Technology New Drug R&D Key Project (Grant no. 2013A022100041).

References

[1] P. R. Duchowicz, E. A. Castro, and F. M. Fernández, "Alternative algorithm for the search of an optimal set of descriptors in QSAR-QSPR studies," *Communications in Mathematical and in Computer Chemistry*, vol. 55, no. 1, pp. 179–192, 2006.

[2] G. Mu, H. Liu, Y. Wen, and F. Luan, "Quantitative structure-property relationship study for the prediction of characteristic infrared absorption of carbonyl group of commonly used carbonyl compounds," *Vibrational Spectroscopy*, vol. 55, no. 1, pp. 49–57, 2011.

[3] H. Zhu, I. Rusyn, A. Richard, and A. Tropsha, "Use of cell viability assay date improves the prediction accuracy of conventional quantitative structure-activity relationships models of animal carcinogenicity," *Environmental Health Perspectives*, vol. 116, no. 4, pp. 506–513, 2008.

[4] J. C. Drosos, M. Viola-Rhenals, and R. Vivas-Reyes, "Quantitative structure-retention relationships of polycyclic aromatic hydrocarbons gas-chromatographic retention indices," *Journal of Chromatography A*, vol. 1217, no. 26, pp. 4411–4421, 2010.

[5] A. A. D'Archivio, M. A. Maggi, P. Mazzeo, and F. Ruggieri, "Quantitative structure-retention relationships of pesticides in reversed-phase high-performance liquid chromatography based on WHIM and GETAWAY molecular descriptors," *Analytica Chimica Acta*, vol. 628, no. 2, pp. 162–172, 2008.

[6] A. Tropsha, "Best practices for QSAR model development, validation, and exploitation," *Molecular Informatics*, vol. 29, no. 6-7, pp. 476–488, 2010.

[7] T. Steger-Hartmann and S. Boyer, *Computer-Based Prediction Models in Regulatory Toxicology*, vol. 1, Springer-Verlag, Berlin, Germany, 2014.

[8] V. Ruusmann, S. Sild, and U. Maran, "QSAR DataBank - an approach for the digital organization and archiving of QSAR model information," *Journal of Cheminformatics*, vol. 6, no. 1, pp. 1–17, 2014.

[9] T. W. Schultz, M. T. D. Cronin, J. D. Walker, and A. O. Aptula, "Quantitative structure-activity relationships (QSARS) in toxicology: a historical perspective," *Journal of Molecular Structure: THEOCHEM*, vol. 622, no. 1-2, pp. 1–22, 2003.

[10] B. Ma, H. Chen, M. Xu, T. Hayat, Y. He, and J. Xu, "Quantitative structure-activity relationship (QSAR) models for polycyclic aromatic hydrocarbons (PAHs) dissipation in rhizosphere based on molecular structure and effect size," *Environmental Pollution*, vol. 158, no. 8, pp. 2773–2777, 2010.

[11] P. Y. Lee and C. Y. Chen, "Toxicity and quantitative structure–activity relationships of benzoic acids to *Pseudokirchneriella subcapitata*," *Journal of Hazardous Materials*, vol. 165, no. 1–3, pp. 156–161, 2009.

[12] C. Charnock and T. Finsrud, "Combining esters of para-hydroxy benzoic acid (parabens) to achieve increased antimicrobial activity," *Journal of Clinical Pharmacy and Therapeutics*, vol. 32, no. 6, pp. 567–572, 2007.

[13] A. L. Hoyt, D. Bushman, N. Lewis, and R. Faber, "Developing a modified preservative efficacy testing approach as a predictive tool for the evaluation of preservative systems in liquid home care products under variable test conditions," *Journal of AOAC International*, vol. 95, no. 1, pp. 203–205, 2012.

[14] R. Mathammal, K. Sangeetha, M. Sangeetha, R. Mekala, and S. Gadheeja, "Molecular structure, vibrational, UV, NMR, HOMO-LUMO, MEP, NLO, NBO analysis of 3,5 di tert butyl 4 hydroxy benzoic acid," *Journal of Molecular Structure*, vol. 1120, pp. 1–14, 2016.

[15] C. Cortes and V. Vapnik, "Support-vector networks," *Machine Learning*, vol. 20, no. 3, pp. 273–297, 1995.

[16] C. Zhang, Y. Tian, and N. Deng, "The new interpretation of support vector machines on statistical learning theory," *Science China. Mathematics*, vol. 53, no. 1, pp. 151–164, 2010.

[17] M. Ha, C. Wang, and J. Chen, "The support vector machine based on intuitionistic fuzzy number and kernel function," *Soft Computing*, vol. 17, no. 4, pp. 635–641, 2013.

[18] X. Hou, X. Chen, M. Zhang, and A. Yan, "QSAR study on the antimalarial activity of Plasmodium falciparum dihydroorotate dehydrogenase (PfDHODH) inhibitors," *SAR and QSAR in Environmental Research*, vol. 27, no. 2, pp. 101–124, 2016.

[19] N. Sharma, K. R. Ethiraj, M. Yadav et al., "Identification of LOGP values and electronegativities as structural insights to model inhibitory activity of HIV-1 capsid inhibitors—a SVM and MLR aided QSAR studies," *Current Topics in Medicinal Chemistry*, vol. 12, no. 16, pp. 1763–1774, 2012.

[20] K. Khuntwal, M. Yadav, A. Nayarisseri, S. Joshi, D. Sharma, and S. Suhane, "Credential role of van der Waal volumes and atomic masses in modeling hepatitis C virus NS5B polymerase inhibition by Tetrahydrobenzo-thiophenes using SVM and MLR aided QSAR studies," *Current Bioinformatics*, vol. 8, no. 4, pp. 465–471, 2013.

[21] W. Zhiming, N. Han, and Y. Zheming, "Feature selection for high-dimensional data based on ridge regression and SVM and its application in peptide QSAR modeling," *Journal of Physical Chemistry*, vol. 03, pp. 498–507, 2013.

[22] X. L. Zhang, Z. X. Zhou, Y. H. Liu et al., "Predicting acute toxicity of aromatic amines by linear and nonlinear regression methods," *Chinese Journal of Structural Chemistry*, vol. 33, no. 2, pp. 244–252, 2014.

[23] C.-C. Jiang, R.-J. Zhou, D.-J. Wang et al., "A QSAR study on antibacterial activity of p-hydroxybenzoate esters," *Modern Food Science and Technology*, vol. 30, no. 6, pp. 98–102, 2014.

[24] S. Qiu, C. Jiang, R. Zhou, and D. Wang, "A QSAR study on antibacterial activity of p-hydroxybenzoate esters," *Modern Food Science and Technology*, vol. 30, no. 6, pp. 98–102, 2014.

[25] A. Y. Musa, W. Ahmoda, A. A. Al-Amiery, A. A. H. Kadhum, and A. B. Mohamad, "Quantum chemical calculation for the inhibitory effect of compounds," *Journal of Structural Chemistry*, vol. 54, no. 2, pp. 301–308, 2013.

[26] A. Mozrzymas, "On the spacer group effect on critical micelle concentration of cationic gemini surfactants using molecular connectivity indices," *Combinatorial Chemistry and High Throughput Screening*, vol. 19, no. 6, pp. 481–488, 2016.

[27] A. Frisch, M. J. Frisch, F. R. Clemente et al., *Gaussian 09 User's Reference*, Gaussian Company Press, 2011.

[28] Anonymous, "Gaussian 09 Software Ported to 64-bit MAC OS Using PGI Compilers," *Computer Workstations*, vol. 23, no. 6, article 17, 2010.

[29] M. Atabati and R. Emamalizadeh, "The hydrogen perturbation in molecular connectivity indices and their application to a QSPR study," *Journal of Solution Chemistry*, vol. 41, no. 11, pp. 1922–1936, 2012.

[30] A. Mozrzymas, "Modelling of the critical micelle concentration of cationic gemini surfactants using molecular connectivity indices," *Journal of Solution Chemistry*, vol. 42, no. 11, pp. 2187–2199, 2013.

[31] A. Yasri and D. Hartsough, "Toward an optimal procedure for variable selection and QSAR model building," *Journal of Chemical Information & Computer Sciences*, vol. 41, pp. 1218–1227, 2001.

[32] P. Gramatica and A. Sangion, "A Historical excursus on the statistical validation parameters for qsar models: a clarification concerning metrics and terminology," *Journal of Chemical Information and Modeling*, vol. 56, no. 6, pp. 1127–1131, 2016.

[33] E. M. de Haas, T. Eikelboom, and T. Bouwman, "Internal and external validation of the long-term QSARs for neutral organics to fish from ECOSAR™," *SAR and QSAR in Environmental Research*, vol. 22, no. 5-6, pp. 545–559, 2011.

[34] P. Gramatica, "Principles of QSAR models validation: internal and external," *QSAR and Combinatorial Science*, vol. 26, no. 5, pp. 694–701, 2007.

[35] G. Alexander and T. Alexander, "Beware of q2!," *Journal of Molecular Graphics and Modelling*, vol. 20, no. 4, pp. 269–276, 2002.

[36] J. Polanski, R. Gieleciak, and A. Bak, "Probability issues in molecular design: Predictive and modeling ability in 3D-QSAR schemes," *Combinatorial Chemistry and High Throughput Screening*, vol. 7, no. 8, pp. 793–807, 2004.

[37] B. Kocjan and J. Sliwiok, "Chromatographic and spectroscopic comparison of the hydrophobilici-ty of vitamins D_2 and D_3," *Journal of Planar Chromatography. Modern TLC*, vol. 7, no. 4, pp. 327-328, 1994.

[38] A. Tropsha, P. Gramatica, and V. K. Gombar, "The importance of being earnest: validation is the absolute essential for successful application and interpretation of QSPR models," *QSAR and Combinatorial Science*, vol. 22, no. 1, pp. 69–77, 2003.

[39] L. Eriksson, J. Jaworska, A. P. Worth, M. T. D. Cronin, R. M. McDowell, and P. Gramatica, "Methods for reliability and uncertainty assessment and for applicability evaluations of classification- and regression-based QSARs," *Environmental Health Perspectives*, vol. 111, no. 10, pp. 1361–1375, 2003.

8

Study on Adsorption of Cu and Ba from Aqueous Solutions using Nanoparticles of *Origanum* (*OR*) and *Lavandula* (*LV*)

Ghadah M. Al-Senani and Foziah F. Al-Fawzan

Department of Chemistry, College of Science, Princess Nourah Bint Abdulrahman University, Riyadh, Saudi Arabia

Correspondence should be addressed to Ghadah M. Al-Senani; gmalsnany@pnu.edu.sa

Academic Editor: Raja Sulaiman Azarudeen

Wild herbs (*Origanum* (*OR*) and *Lavandula* (*LV*)) were used as environment-friendly adsorbents in this study. The adsorbents were used for adsorption of Cu and Ba from water. The adsorption of heavy metals onto *OR* and *LV* was dependent on particle size, dose, and solution pH. The diameter of adsorbent particles was less than 282.8 nm. The adsorption follows second-order kinetics. Langmuir and Freundlich models have been applied to describe the equilibrium data, and the thermodynamic parameters, the Gibbs free energy, ΔG°, enthalpy, ΔH°, and entropy, ΔS°, have been determined. The positive value of ΔH° suggests that the adsorption of heavy metals by the wild herbs is endothermic. The negative values of ΔG° at all the studied temperatures indicate that the adsorption is a spontaneous process. It can be concluded that *OR* and *LV* are promising adsorbents for the removal of heavy metals from aqueous solutions over a range of concentrations.

1. Introduction

The adsorption of toxic substances resulting from the accumulation of industrial wastes is important and one of the most dangerous challenges facing the environment and society nowadays. One of the most significant contaminants affecting water resources is heavy metals. These metal ions present a significant risk to animals and humans because of their high toxicity at both low and high concentrations in soil and water. The search for new techniques to remove these contaminants has involved both chemical and biological methods.

Contaminating materials have become increasingly dangerous with increasing technological development, and need for diverse heavy metals, for example, in ore processing and other modern industries, prompted the organizations concerned with the preservation of the environment to develop restrictions and laws for laboratories and concerning the treatment industrial waste before its release into the environment so that toxic materials do not exceed the allowable limits. Researchers have, thus, devoted efforts to find effective methods for the removal of contaminants from waste, but these processes are often economically expensive; therefore, we must find ways to bypass conventional and high-cost advanced adsorption technologies.

The adsorption of contaminants on the surfaces of solids is an effective method for heavy metal remediation. Activated carbon is considered to be an efficient, competitive material for this task. However, the cost of production is still high; thus, many researchers have begun to search for alternative adsorbents made from local natural materials. Plants are one type of alternative material that can be used to remove heavy metal ions from water systems and soil.

The current proliferation of technology and development of science has been an enormous boon for humans. They have become dependent on technology and scientific development in various aspects such as daily activities, trade, industry, and work. Heavy metals are structural elements such as lead, zinc, arsenic, cadmium, copper, titanium, cobalt, lithium, aluminum, and mercury and can be in the form of metals or dissolved salts.

These metals are present in the environment in air, water, and soil. For example, factory chimneys release metal oxides into the air, thus transmitting heavy metal pollution to humans, animals, and plants. In addition, car exhausts release lead oxides, resulting from the combustion of tetraethyl lead, into the atmosphere, and this is one of the most widespread routes to leading contamination of marine organisms with metals and the transit of these contaminants via sea fishing to humans and animals. Furthermore, agricultural soil is one of the most important sources of food polluted by heavy metals, which arises through the irrigation of crops with polluted water or the use of pesticides. In this situation, the metals are transmitted through the vascular system of plants and fruits. Therefore, field crops irrigated with drainage water polluted by heavy metals are one of the most important and most dangerous sources for the entry of toxic heavy metals into the human body [1].

There are many methods for adsorption of heavy metals from the environment, both chemical and physical. However, some of these are not economically feasible. Therefore, it is necessary to investigate low-cost, effective alternatives. The adsorption of heavy metals by the adsorption technology is a good alternative, and it is used in the treatment of wastewater and soil. To compare the adsorbent substances, the cost, as well as effectiveness, must be considered. Activated carbon is a highly effective alternative used for the adsorption of heavy metals from wastewater, but it is soluble under extremely acidic conditions [2].

Consequently, there is an increased interest from researchers into low-cost, effective alternative adsorbents. Many natural materials used in the adsorption of heavy metals (cadmium, copper, chromium, lead, nickel, cobalt, and lithium), such as *Diplotaxis harra*, *Glebionis coronaria*, coffee grounds, banana peel, fruit and vegetable peels, cactus, rice straw, wheat straw, and salvinia plant, have been discovered [3–8].

The aim of this work is to study the adsorption of Cu and Ba from water using microparticles of *Origanum* (*OR*) and *Lavandula* (*LV*), which are considered environmentally safe and low-cost. In this study, we use thermodynamic, kinetic, equilibrium, and adsorption isotherms in the analysis of the adsorption behavior.

2. Experimental

2.1. Materials. All the chemicals used in this study were of analytical grade. $Cu(Cl)_2{\cdot}4H_2O$ (98%) and $Ba(Cl_2){\cdot}4H_2O$ (90%) were purchased from Sigma-Aldrich (Germany).

2.2. Preparation of the Adsorbents. Origanum (Figure 1) is a leafy plant used as a kind of aromatic spices and flavorings in foods, as well as the benefits of medical oregano. It also contains a variety of antioxidant compounds, such as rosmarinic acid. The chemicals in oregano are those that give it distinctive aroma and flavor. It includes thymol, albinin, limonene, carvacrol, and carophyllene [9].

Lavandula (*LV* Figure 1) is a plant that grows and prefers sunny open areas. Lavender is used for insomnia, nervousness, and depression [10].

The wild herbs used in this study were purchased from a perfumer. The dried herbs were then cut into small pieces and milled to a powder using a laboratory planetary ball mill (DECO-PBM-V-0.4 L). The powders were sieved into particles of 50, 100, 200, 300, and 400 μm using an Octagon D200 digital sieve shaker. The adsorbents were stored in glass bottles for further use without any pretreatment.

2.3. Methods. For the adsorption study, distilled water was used to prepare various solutions at the desired concentrations from the stock solution. The adsorption experiments were performed in a series of flasks containing 100 mL solutions of metal ions at desired concentration and mass of adsorbent herbs. The mixtures were shaken for 16 h at 120 rpm using a shaker ("Rotaterm" orbital and linear shaker). The mixtures were filtered, and the heavy metals concentrations were determined by inductively coupled plasma (ICP) mass spectrometry. The effect of adsorbent doses from 0.1 to 2 g on the metal adsorption was studied. The initial concentration was 10 mg/L. The effect of pH on the heavy metal adsorption process was investigated at pH values from 3 to 12, adjusted either with 1 M·HCl or 1 M·NaOH using a pH meter, to monitor the change. In addition, the effect of the adsorbent particle size was investigated from 400 to less than 50 μm, and the contact time was varied between 1 and 3 h. The effect of temperature was studied from 25° to 60°C for 3 h using 0.5 g of the adsorbent at an initial metal concentration of 10 mg/L.

The adsorption equilibrium was investigated for different metal concentrations between 1 and 100 mg/L. In addition, kinetic experiments were conducted with 10 mg/L metal and 0.5 g of the adsorbent at room temperature with stirring at 120 rpm for 6 h.

The adsorption at equilibrium, q_e (mg/g), was calculated by using the following equation:

$$q_e = \frac{(C_0 - C_e)V}{m}, \tag{1}$$

where C_0 and C_e (mg/L) are the liquid-phase heavy metal concentrations initially and at equilibrium, respectively; V is the volume of the solution (L); and m is the mass of the dry adsorbent (g).

The percentage of metal adsorption ($Ads_{HM}\%$) from the solution was calculated as follows:

$$Ads_{HM}\% = \frac{(C_0 - C_e)}{C_0} \times 100. \tag{2}$$

The pseudo-first-order and pseudo-second-order adsorption kinetic models were applied to describe the behavior of the adsorbent.

The equilibrium data were then fitted using the Langmuir and Freundlich isotherm models.

The thermodynamic parameters calculated to describe the adsorption process onto *OR* and *LV* include the changes in standard enthalpy (ΔH°), standard entropy (ΔS°), and standard free energy (ΔG°).

(a) (b)

FIGURE 1: (a) *Origanum* (*OR*) and (b) *Lavandula* (*LV*).

2.4. Characterization of the Adsorbents. Fourier-transform (FT-IR) analysis was used to determine the functional groups in the adsorbents. The particle size of adsorbents was measured using an XRD. The initial metal concentrations of the adsorbents were determined for Cu and Ba from ICP measurements.

3. Results and Discussion

3.1. Characterization of the Adsorbents. Figure 2 shows the FT-IR absorption bands of the adsorbent. The peaks indicate the presence of functional groups containing O-H and N-H at 3300 cm^{-1} wavelength and carboxylic acid (–COOH) groups at 1600 cm^{-1} and 1150 cm^{-1} wavelengths, which play a major role in the adsorption process and efficiently adsorb metal ions via ion exchange [11].

The mineral composition of biomass ash commonly shows a lowest amount of Cu and Ba (not exceeding 0.072 mg/L) in *OR*, as does *LV*, and hence the Cu and Ba were selected for further study (Figure 3).

The smallest size obtained after milling the wild herbs was 50 μm. Figure 4 shows the size of adsorbent particles that had been ground to less than 50 μm. The diameter of the particle is less than 282.8 nm, which is smaller than the size reached by milling, and smaller particles can increase the effectiveness of adsorption.

The small particle size was characterized by fine porosity and a large internal surface area, while the powder form has larger diameter pores and smaller internal surface area. A smaller pore size results in stronger and greater adsorption capacity because the small adsorbent particle size reduces the pathway for both mass transport and internal diffusion of the adsorbent inside the adsorbents [12].

3.2. Effect of Particle Sizes of Adsorbent. The experimental effect of particles on metal adsorption was investigated for five sizes (400 to less than 50 μm) at room temperature and initial pH (5 and 7 for Cu and Ba, respectively). Figure 5 indicates that metal adsorption increased with decreasing particle size for *OR* and *LV*. This behavior can be attributed to the increased internal surface area with decreasing particle size [12].

This behavior arises because most of the internal surface of these particles can be used for adsorption. The smaller particle size gives higher adsorption rates because metal ions have a short path for transfer inside the pores of the small particle adsorbents [13].

However, it is expected that the use of the smaller particle size should give greater removal rate because of the increased surface area, and as the particle size increases, the number of small pores increases. The particle dimensions determine the propagation distance, because the dimensions of the adsorbent particles obtained by standard sieves depend on the particle length and width; thus, this explains the decreased adsorption of metals with increasing adsorbent particle size [14].

3.3. Effect of Adsorbent Dosage. The effect of the adsorbent doses on heavy metal adsorption by *OR* and *LV* was studied. As shown in Figure 6, the adsorption percentage of Cu and Ba increased with increase of adsorbent doses. The increase in adsorption percentage with increasing adsorbent dose arises because of the increased availability of exchange sites or surface areas at higher concentrations of adsorbents [15]. Similar behavior has been observed in previous studies and arises from the effect of interactions between mineral ions and adsorbent. The increased dose is consistent with the greater area and larger number of adsorption sites [16].

3.4. Effect of pH. As shown in Figure 7, the percent adsorption of Cu and Ba increased significantly with increasing pH from 3 to 5, and then stabilized until reaching pH 7; after that, the adsorption was a slight decrease above pH 7 for both adsorbents *OR* and *LV* [15].

4. Adsorption Kinetic Model

To understand the rate and type of adsorption on adsorbents, we studied several kinetic models.

4.1. Pseudo-First-Order Kinetic Model. The pseudo-first-order model is expressed by the following equation:

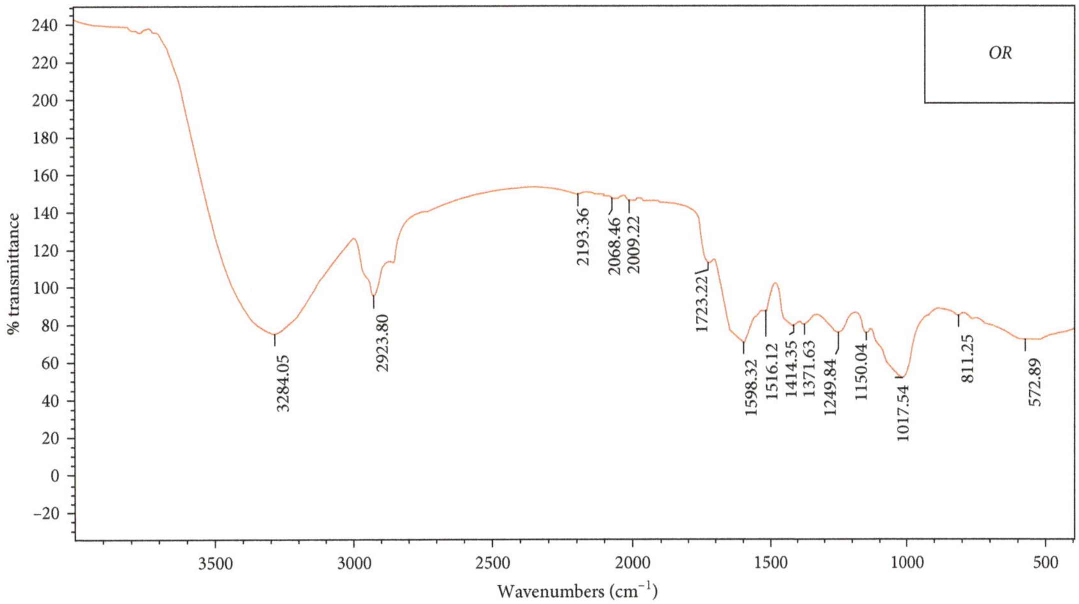

(a)

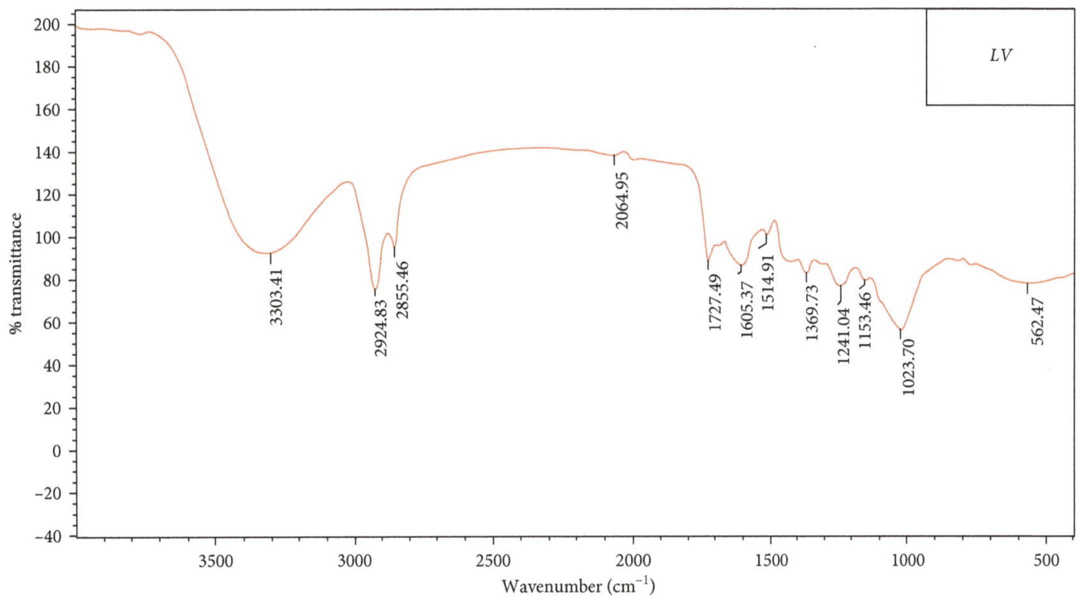

(b)

FIGURE 2: FT-IR spectra of the functional groups in (a) *Equisetum*, *OR*, and (b) *Teucrium*, *LV*.

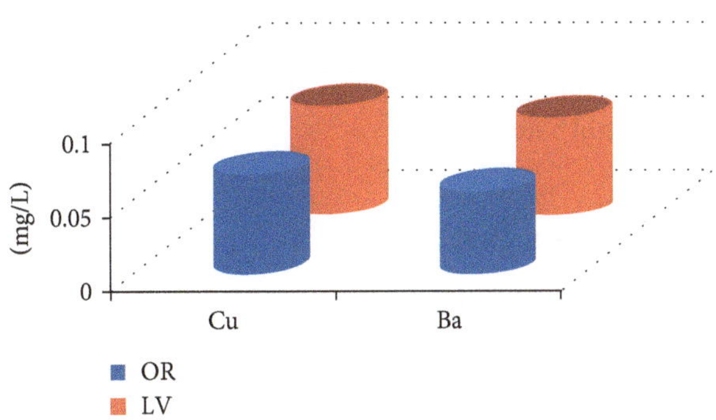

FIGURE 3: Concentration of heavy metals in the adsorbent.

$$\log(q_e - q_t) = \log q_e - \frac{K_1 t}{2.303}, \quad (3)$$

where q_e and q_t represent the amount of metal ions adsorbed at equilibrium and at time, t, respectively and K_1 is pseudo-first-order rate constant.

As shown in Figure 8 and Table 1, the regression coefficient ($R^2 \le 0.96$) indicates that the experimental data accurately support the first-order model to describe the adsorption kinetics of Cu ions, but do not support it with Ba ions. However, we found that the typical values of q_m are lower than the experimental values for Cu ions. This indicates that both metal ions and the adsorbent are involved

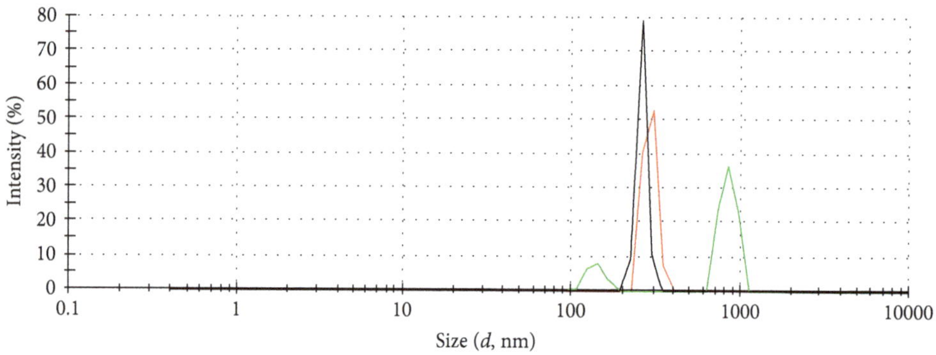

FIGURE 4: XRD of the particle size of adsorbents.

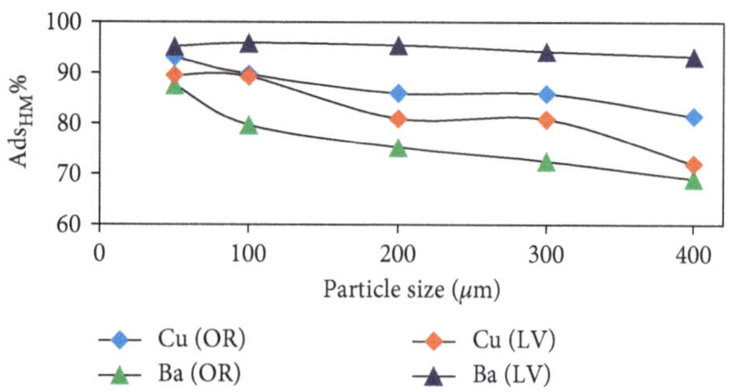

FIGURE 5: The effect of particle size on the metal ion adsorption by *OR* and *LV* (initial pH, 120 rpm, 10 mg/L of metals, 0.5 g of adsorbents, and 25°C for 16 h).

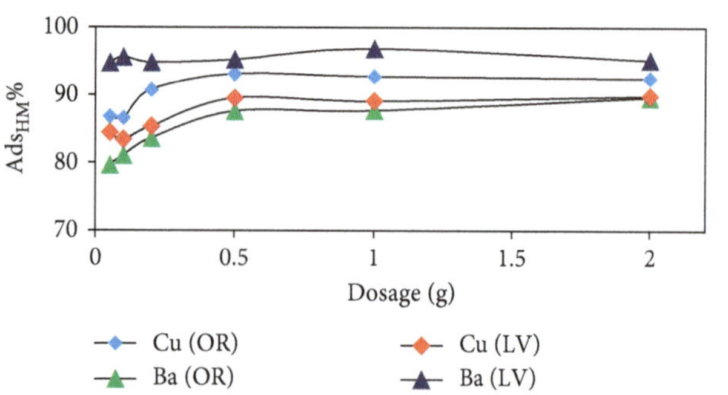

FIGURE 6: The effect of dosage on the metal ion adsorption by *OR* and *LV* (initial pH, 120 rpm, 10 mg/L of metals, 50 μm size of adsorbents, and 25°C for 16 h).

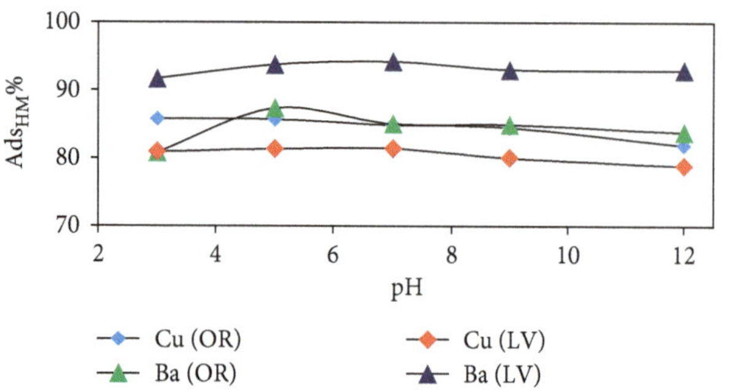

FIGURE 7: The effect of pH on the metal ion adsorption by *OR* and *LV* (120 rpm, 10 mg/L of metals, 50 μm size and 0.5 g of adsorbents, and 25°C for 16 h).

in the adsorption process [17]. Therefore, the first-order model may not be suitable for explaining the adsorption kinetics of ions on the adsorbents. Similar results were obtained for the kinetics of adsorption of different metal ions on other adsorbents [17–19].

4.2. Pseudo-Second-Order Kinetic Model. The pseudo-second-order model is expressed by the following equation:

$$\frac{t}{q_t} = \frac{1}{K_2 {q_e}^2} + \frac{t}{q_e}, \tag{4}$$

where K^2 is the pseudo-second-order rate constant.

The experimental data showed the best fit with the highest correlation coefficients ($R^2 \geq 0.996$) for the pseudo-second-order model for each of the metal ions. In addition, the experimental values were well matched with the calculated data (Table 1). Therefore, the experimental data accurately support the pseudo-second-order kinetic model for the adsorption of metal ions.

This finding indicates that the rate-limiting step was not the resistance of the boundary layer [8]. Thus, the rate-limiting step is chemisorption involving valence forces through the exchange of electrons between the metal ions and the different functional groups in the adsorbent [20]. This is consistent with the observed rapid adsorption.

As shown in Figure 8, the kinetic analysis for the adsorption of metal ions on the catalysts revealed one step: the rate of adsorption in the equilibrium state. This phenomenon indicates that the chemical adsorption is the dominant step on the adsorption process [11, 21]. The adsorption half-time ($t1/2$) for the adsorption of half the metal ions in minutes was between 13.45 and 21.80 min, indicating significant affinity of the adsorbent molecules for the adsorbate [8].

5. Adsorption Isotherm

To understand the mechanism of metal ion adsorption from an aqueous solution by the adsorbents, we tested several adsorption isotherm models. The Langmuir adsorption isotherm was applied in accordance with the following equation:

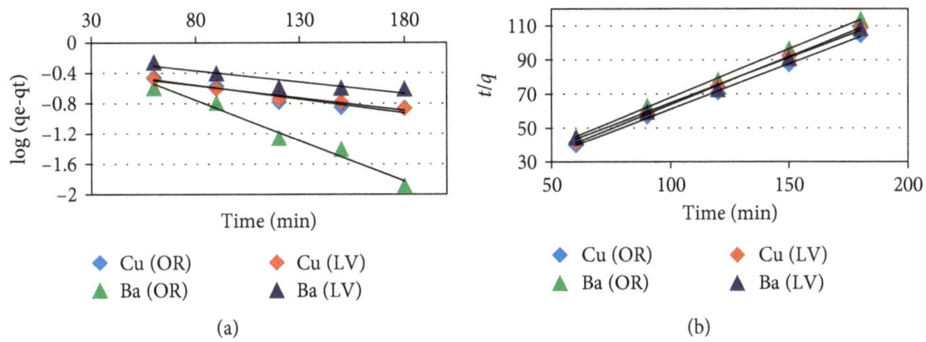

FIGURE 8: The plot of (a) pseudo-first order (POF) and (b) pseudo-second order (POS) for the metal ion adsorption onto *OR* and *LV* (initial pH, 120 rpm, 10 mg/L of metals, 50 μm size and 0.5 g of adsorbents, and 25°C at time 60, 90, 120, 150, and 180 min).

TABLE 1: Kinetic model parameters for the metal ion adsorption onto *OR* and *LV*.

Kinetic model	Parameter	Adsorbent (*OR*)		Adsorbent (*LV*)	
		Cu	Ba	Cu	Ba
	$q_{e,exp}$ (mg/g)	1.86	1.75	1.79	1.90
Pseudo-first order	$q_{e,calc}$ (mg/g)	0.54	0.52	0.50	0.74
	$K_1 \times 10^3$(min^{-1})	8.52	13.36	7.60	6.91
	R^2	0.904	0.669	0.962	0.823
Pseudo-second order	$q_{e,calc}$ (mg/g)	1.88	1.75	1.78	1.88
	$K_2 \times 10^3$ (g/mg·min)	36.96	30.92	41.81	24.37
	h (mg/g·min)	0.13	0.09	0.13	0.09
	$t1/2$ (min)	14.43	18.52	13.45	21.80
	R^2	0.999	0.999	0.999	0.996

$$\frac{C_e}{q_e} = \frac{1}{K_L q_m} + \frac{C_e}{q_m}, \quad (5)$$

where C_e is the adsorbate equilibrium concentration, q_e is the observed adsorption capacity at equilibrium, q_m is the maximum adsorption capacity, and K_L is the adsorption equilibrium constant. The Langmuir adsorption isotherm showed a linear relationship between C_e/q_e and C_e for the adsorption of metals on *OR* and *LV* (Figure 9), and the slope and R^2 data are presented in Table 2. The plot exhibited good correlation coefficient and linearity. The R^2 values are very close to unity, indicating strong correlation with the Langmuir adsorption isotherm. The maximum adsorption capacity of the adsorbent, q_m (mg/g), for metal ions is high, which can be explained by their noncovalent interaction with ions. Another reason may be the presence of carboxylic acid-containing sugars in the herbs, which bind the metal ions [22]. So, the values of R_L were calculated from the following equation:

$$R_L = \frac{1}{1 + K_L q_m} \log C_e. \quad (6)$$

$R_L \leq 1$, indicating that the Langmuir isotherm is appropriate for the adsorption process. For a single adsorption system, C_0 is usually the highest concentration of the liquid phase encountered [23].

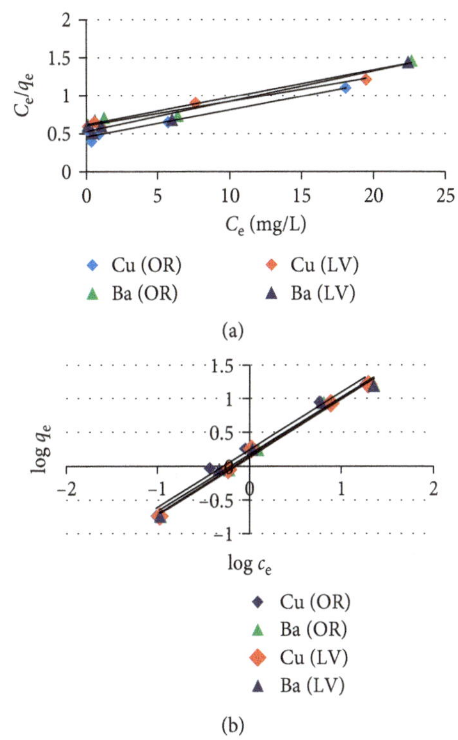

FIGURE 9: The (a) Langmuir and (b) Freundlich isotherms for the metal ion adsorption onto *OR* and *LV* (initial pH, 120 rpm, 50 μm size of adsorbents, and 25°C for 16 h).

TABLE 2: Adsorption isotherm parameters for the metal ion adsorption onto *OR* and *LV*.

Adsorption isotherm	Parameter	Adsorbent (*OR*)		Adsorbent (*LV*)	
		Cu	Ba	Cu	Ba
Langmuir	q_m (mg/g)	28.17	27.62	31.25	25.06
	$K_L \times 10^3$ (L/mg)	77.8	58.80	52.84	75.53
	R_L	0.56	0.63	0.65	0.57
	R^2	0.963	0.965	0.975	0.976
Freundlich	K_F (mg/g) (L/mg)$^{1/n}$	1.74	1.35	1.40	1.50
	n	1.17	1.16	1.16	1.18
	R^2	0.983	0.988	0.994	0.982

The Freundlich adsorption isotherm is given by the following equation:

$$\log q_e = \log K_F + \frac{1}{n}\log C_e, \quad (7)$$

where n is the Freundlich equilibrium coefficient; values of $\log q_e$ are plotted against $\log C_e$ (Figure 9). The slope and R^2 data are presented in Table 2. In the slope of the Freundlich plot, $1/n$ is greater than 1, which indicates that adsorption is cooperative. Higher values of K_F and n indicate a higher affinity for the adsorption of metals [4]. Values of $1/n < 1$ indicate the favorable adsorption of all metals, and values of n within the range 1–10 represent good adsorption and a favorable physical process. Higher values of K_F indicate a high adsorption intensity [24].

From the Langmuir and Freundlich isotherms, it is clear the metal ions were strongly adsorbed in the pores of the herbs.

6. Thermodynamic Parameters

In order to understand the nature of adsorption, the thermodynamic parameters such as free energy change (ΔG°), enthalpy change (ΔH°), and entropy change (ΔS°) could be calculated using the following equations:

$$\Delta G^\circ = -\mathrm{RT}\ln K_{eq},$$

$$\ln K_{eq} = \frac{\Delta S^\circ}{R} - \frac{\Delta H^\circ}{\mathrm{RT}}, \quad (8)$$

$$\Delta G^\circ = \Delta H^\circ - \mathrm{T}\Delta S^\circ.$$

The values of the thermodynamic parameters are shown in Table 3. The linearized van 't Hoff plots of $\ln K_{eq}$ versus $1/T$ are shown in Figure 10.

The values of ΔH° and ΔS° were determined from the slope and intercept of the linear plot.

The parameters indicate that the adsorption process is spontaneous and endothermic. The positive value of ΔS° indicates an increase in the degrees of freedom (or disorder) of the adsorbed species for Cu and Ba ions.

The negative values of ΔG° indicate that the thermodynamic process was spontaneous and feasible for Cu and Ba ions [8]. Moreover, the increase in negative ΔG° values with increasing temperature shows an increased probability of adsorption at higher temperatures, which is consistent with an earlier report [25].

TABLE 3: Thermodynamic parameters for the metal ion adsorption onto *OR* and *LV*.

Parameter	Adsorbent (*OR*)		Adsorbent (*LV*)	
	Cu	Ba	Cu	Ba
ΔH° (J/mol)	1867.07	9096.35	3748.28	10161.4
ΔS° (J/mol K)	14.52	33.14	17.16	45.89
ΔG° (kJ/mol) at 298K	−2.46	−0.78	−1.36	−3.51
ΔG° (kJ/mol) at 333K	−2.97	−1.94	−1.96	−5.12
R^2	0.966	0.964	0.968	0.957

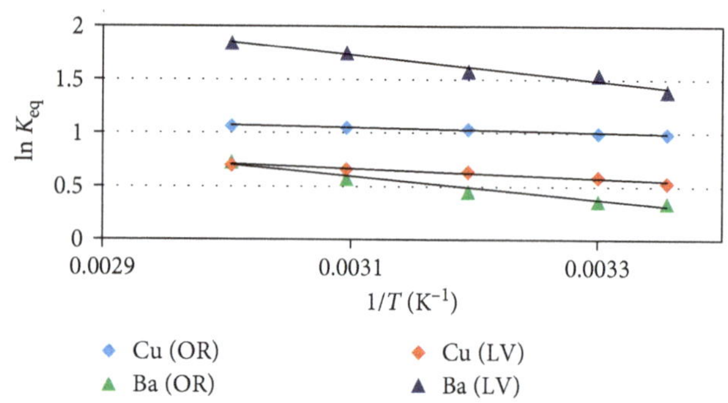

FIGURE 10: van 't Hoff plots for the metal ion adsorption onto *OR* and *LV* (initial pH, 120 rpm, 10 mg/L of metals, and 50 μm size and 0.5 g of adsorbents for 16 h).

7. Conclusion

In this study, the adsorption of Cu and Ba onto *OR* and *LV* was investigated with respect to the particle size, dosage of the adsorbent, and pH. Characterization of the adsorbents revealed the presence of carboxylic groups that were involved in the adsorption of the metal ions. Adsorption was maximum with 50 μm particle size and 0.5 g of the adsorbent. Metal ion adsorption onto wild herbs follows the pseudo-first-order and pseudo-second-order kinetic models. The adsorption isotherm data for the Cu and Ba ions fitted well with the Langmuir and Freundlich models. The thermodynamic parameters show that the adsorption process was feasible, spontaneous, and endothermic in nature for Cu and Ba. Hence, the wild herbs can be used as a low-cost adsorbent for the removal of heavy metal ions from aqueous solutions. Further study is warranted to evaluate the potential of the adsorbents for heavy metal removal from aqueous solutions.

Acknowledgments

The researchers would like to thank the Deanship of Scientific Research at Princess Nourah Bint Abdulrahman University for providing the funding for this study under Project no. 216-38-ص.

References

[1] K. M. Al-Qahtani, "Assessment and treatment some of the local seeds then used in removal of heavy metals," *Journal of American Science*, vol. 11, no. 11, pp. 198–203, 2015.

[2] K. L. Wasewar, "Adsorption of metals onto tea factory waste: a review," *International Journal of Recent Research and Applied Studies*, vol. 3, no. 3, pp. 303–322, 2010.

[3] T. Derbe, H. Dargo, and W. Batu, "Cactus potential in heavy metal (Pb and Cd) removal in water sample collected from rural area around Adigrat town," *Chemistry and Materials Research*, vol. 7, no. 3, pp. 84–92, 2015.
[4] B. Dhir and R. Kumar, "Adsorption of heavy metals by *Salvinia* biomass and agricultural residues," *International Journal of Environmental Research*, vol. 4, no. 3, pp. 427–432, 2010.
[5] M. Hossain, H. Hao Ngo, W. Guo, and T. Nguyen, "Removal of copper from water by adsorption onto banana peel as bioadsorbent," *International Journal of GEOMATE*, vol. 2, no. 4, pp. 227–234, 2012.
[6] N. Jain, "Removal of heavy metal by using different fruit peels, vegetable peels and organic waste-a review," *International Journal of Advanced Research*, vol. 3, no. 11, pp. 916–920, 2015.
[7] J. Seniunait, R. Vaiškunait, and V. Bolutien, "Coffee grounds as an adsorbent for copper and lead removal form aqueous solutions," in *Proceedings of 9th Conference "Environmental Engineering"*, Vilnius, Lithuania, May 2014.
[8] H. Tounsadi, A. Khalidi, M. Abdennour, and N. Barka, "Biosorption potential of *Diplotaxis harra* and *Glebionis coronaria* L. biomasses for the removal of Cd(II) and Co(II) from aqueous solutions," *Journal of Environmental Chemical Engineering*, vol. 3, no. 3, pp. 822–830, 2015.
[9] Bedoor al-Agha, http://capsuleh.com/oregano-benefits, 2017.
[10] http://www.nabataty.com/%D9%84%D8%A7%D9%81%D9%86%D8%AF%D8%B1, 2016.
[11] H. Nacke, A. Gonçalves, M. Campagnolo et al., "Adsorption of Cu (II) and Zn (II) from water by *Jatropha curcas* L. as biosorbent," *Open Chemistry*, vol. 14, no. 1, pp. 103–117, 2016.
[12] M. A. Al-Anber, "Thermodynamics approach in the adsorption of heavy metals," in *Thermodynamics-Interaction Studies-Solids, Liquids and Gases*, J. C. Moreno-Pirajan, Ed., IntechOpen, London, UK, 2011.
[13] M. A. Al-Anber, "Removal of high-level $Fe3^{+}$ from aqueous solution using Jordanian inorganic materials: bentonite and quartz," *Desalination*, vol. 250, no. 3, pp. 885–891, 2010.
[14] I. Osu Charles and S. Odoemelam, "Studies on adsorbent dosage, particle sizes and Ph constraints on biosorption of Pb (II) and Cd(II) ions from aqueous solution using modified and unmodified *Crasstrotrea gasar* (Bivalve) biomass," *International Archive of Applied Sciences and Technology*, vol. 1, no. 1, pp. 62–68, 2010.
[15] K. M. Al-Qahtani, "Water purification using different waste fruit cortexes for the removal of heavy metals," *Journal of Taibah University for Science*, vol. 10, no. 5, pp. 700–708, 2016.
[16] A. Samindika, "A comprehensive study of Cd (II) removal from aqueous solution via adsorption and solar photocatalysis," *Electronic Thesis and Dissertation Repository*, vol. 19, p. 1783, 2013.
[17] M. Arshadi, M. Amiri, and S. Mousavi, "Kinetic, equilibrium and thermodynamic investigations of Ni(II), Cd(II), Cu(II) and Co(II) adsorption on barley straw ash," *Water Resources and Industry*, vol. 6, pp. 1–17, 2014.
[18] M. Meitei and M. Prasad, "Adsorption of Cu (II), Mn (II) and Zn (II) by *Spirodela polyrhiza* (L.) Schleiden: equilibrium, kinetic and thermodynamic studies," *Ecological Engineering*, vol. 71, pp. 308–317, 2014.
[19] J. Caz´on, M. Viera, E. Donati, and E. Guibal, "Zinc and cadmium removal by biosorption on *Undaria* pinnatifida in batch and continuous processes," *Journal of Environmental Management*, vol. 129, pp. 423–434, 2013.
[20] M. S. Rahman and K. V. Sathasivam, "Heavy metal adsorption onto *Kappaphycus* sp. from aqueous solutions: the use of error functions for validation of isotherm and kinetics models," *BioMed Research International*, vol. 2015, Article ID 126298, 13 pages, 2015.
[21] N. Feng, X. Guo, S. Liang, Y. Zhu, and J. Liu, "Biosorption of heavy metals from aqueous solutions by chemically modified orange peel," *Journal of Hazardous Materials*, vol. 185, no. 1, pp. 49–54, 2011.
[22] M. Lasheen, N. Ammar, and H. Ibrahim, "Adsorption/desorption of Cd(II), Cu(II) and Pb(II) using chemically modified orange peel: equilibrium and kinetic studies," *Solid State Sciences*, vol. 14, no. 2, pp. 202–210, 2012.
[23] S. Yalcin, "The mechanism of heavy metal biosorption on green marine macroalga *Enteromorpha linza*," *Data in CLEAN-Soil, Air, Water*, vol. 42, no. 3, pp. 251–259, 2014.
[24] L. Tofan, R. Wenkert, C. Paduraru, and P. Carmen, "Natural and waste materials as green sorbents for Cd(ll) removal from aqueous effluents," *Environmental Engineering and Management Journal*, vol. 15, no. 5, pp. 1049–1058, 2016.
[25] C. Massocatto, E. Paschoal, N. Buzinaroet et al., "Preparation and evaluation of kinetics and thermodynamics studies of lead adsorption onto chemically modified banana peels," *Desalination and Water Treatment*, vol. 51, no. 28–30, pp. 5682–5691, 2013.

9

Bioactive Potential of 3D-Printed Oleo-Gum-Resin Disks: *B. papyrifera*, *C. myrrha*, and *S. benzoin* Loading Nanooxides—TiO_2, P25, Cu_2O, and MoO_3

Diogo José Horst,[1] Sergio Mazurek Tebcherani,[2] Evaldo Toniolo Kubaski,[3] and Rogério de Almeida Vieira[4]

[1] *Federal University of Technology-Paraná, 84016-210 Ponta Grossa, PR, Brazil*
[2] *Department of Production Engineering, Federal University of Technology-Paraná, 84016-210 Ponta Grossa, PR, Brazil*
[3] *Department of Materials Science, State University of Ponta Grossa, 84030-900 Ponta Grossa, PR, Brazil*
[4] *Department of Earth and Exact Sciences, Federal University of São Paulo, 89972-270 São Paulo, SP, Brazil*

Correspondence should be addressed to Diogo José Horst; diogohorst@gmail.com

Academic Editor: Francesco P. Fanizzi

This experimental study investigates the bioactive potential of filaments produced via hot melt extrusion (HME) and intended for fused deposition modeling (FDM) 3D printing purposes. The oleo-gum-resins from benzoin, myrrha, and olibanum in pure state and also charged with 10% of metal oxide nanoparticles, TiO_2, P25, Cu_2O, and MoO_3, were characterized by ultraviolet-visible (UV-Vis) and Fourier transform infrared (FTIR) spectroscopy, energy-dispersive X-ray microanalysis (EDXMA), scanning electron microscopy (SEM), and differential scanning calorimetry (DSC). Disks were 3D-printed into model geometries (10×5 mm) and the disk-diffusion methodology was used for the evaluation of antimicrobial and antifungal activity of materials in study against the clinical isolates: *Staphylococcus aureus, Pseudomonas aeruginosa, Escherichia coli*, and *Candida albicans*. Due to their intrinsic properties, disks containing resins in pure state mostly prevent surface-associated growth; meanwhile, disks loaded with 10% oxides prevent planktonic growth of microorganisms in the susceptibility assay. The microscopy analysis showed that part of nanoparticles was encapsulated by the biopolymeric matrix of resins, in most cases remaining disorderly dispersed over the surface of resins. Thermal analysis shows that plant resins have peculiar characteristics, with a thermal behavior similar to commercial available semicrystalline polymers, although their structure consists of a mix of organic compounds.

1. Introduction

Infections caused by pathogenic microorganisms are of great concern in many fields [1]. Hospital-acquired infections are one of the major problems increasing mortality and morbidity. Microbial contamination may occur from various sources and invasive interventions [2]. *Staphylococcus aureus* is a Gram-positive major human pathogen that causes skin and soft-tissue infections, life-threatening infections such as pneumonia and sepsis, and toxinoses including toxic shock syndrome [3]. *Pseudomonas aeruginosa* is a Gram-negative bacillus that is rapidly becoming one of the major causes of opportunistic and nosocomial infections which have become a worldwide problem [4]. *Escherichia coli* is a Gram-negative pathogen which has been regarded as an important indicator bacterium; it causes severe diseases, such as hemolytic uremic syndrome (HUS), hemorrhagic colitis [5], and thrombotic thrombocytopenic purpura which can be fatal in some cases [6, 7]. *Candida albicans* is renowned as the leading fungal pathogen of oral candidosis, which manifests in a variety of clinical guises ranging from common denture associated infections in otherwise healthy individuals to systemic infections in human immunodeficiency virus disease [8, 9].

Although bacteria have shown the ability to acquire resistance to many antibiotics, however, in nature, there are several examples of antibiotics to which resistance has not yet developed [10, 11]. The history of medicine and pharmacy is well known for using plant oleo-gum-resins and extracts in curing diseases; these are known to have analgesic, antioxidant, antifungal, antiseptic, antibacterial, astringent, sedative, and stimulant therapeutic properties, among others [12–21]. The bioactivity of oils and extracts obtained species *Commiphora myrrha, Styrax benzoin*, and *Boswellia papyrifera* has been investigated by several researches [22–31]; these aromatic resins basically consist of monoterpenes (C10H16), triterpenes (C30H48), and sesquiterpenes (C15H24) with unique combinations, besides benzoic, myrrholic, and boswellic acids, respectively [32–37]; the demonstration of the presence of secondary metabolites in medicinal plants oils, extracts, and resins provides a scientific validation for the popular use of these plants [38–42].

Besides that, a greener approach for the biosynthesis of colloidal metal nanoparticles and dispersion/encapsulation of drugs using natural oleo-gum-resins has been suggested as being effective and more environmental friendly [43–49]. Moreover, the development of multifunctional nanocomposite materials with enhanced mechanical and antimicrobial properties has been studied [50–52]. Indeed, nanoparticles (NPs) are widely used in the field of healthcare, presenting numerous advantages in medical and biotechnological applications [53, 54] and increasingly attracting researchers due to their unique properties, such as submicrometer size (1–100 nm), large surface-to-volume ratio, and advanced reactivity [55, 56].

It is imperative to improve the applicability of 3D printing for pharmaceutical purposes by searching novel materials. The investigation of different physicochemical properties and adequate processing parameters of such materials is important for successful additive manufacturing of personalized geometries [57]. In the last years, the use of 3D printing for the development of drug delivery systems, medical devices, bone tissue engineering, and antimicrobial materials has shown promising results with a large possibility of applications [58–61]. Beyond that, the use of hot melt extrusion (HME) in the fabrication of novel antimicrobial filaments for pharmaceutical application has steadily increased [62–67]. Nature can combine brittle minerals and organic molecules into hybrid composites that are highly organized to achieve exceptional properties [68, 69]; organic-inorganic hybrid nanostructures and materials on their basis are promising class of multifunctional advanced materials [70, 71]. Within this context, the purpose of this study was to evaluate the potential of hybrid engineered materials intended for fused deposition modeling (FDM) 3D printing, testing its bioactivity against clinical pathogenic organisms including Gram-positive, Gram-negative bacteria, and fungus.

2. Materials and Methods

2.1. Materials and Reagents. Oleo-gum-resins from benzoin *(Styrax benzoin)* harvested in Singapore, olibanum *(Boswellia papyrifera)* originally from Ethiopia, and myrrh *(Commiphora myrrha)* from Somalia were purchased from Mountain Rose Herbs (Eugene, Oregon, USA). Titanium dioxide (TiO_2), anatase (P25), molybdenum trioxide (MoO_3), and copper I (Cu_2O) oxide were purchased from Plasmachem (GmbH, Germany); the nanoparticles have sizes between 10 nm to 1 nm. Prior to testing, all materials were sterilized using UV radiations. Clinical isolates of *Staphylococcus aureus* ATCC 6538; *Pseudomonas aeruginosa* ATCC 9027; *Escherichia coli* ATCC 8739; and *Candida albicans* ATCC 2091 were purchased from the American Type Culture Collection (ATCC, Manassas, VA, USA). Mueller-Hinton agar and Sabouraud dextrose agar were purchased from Kasvi (Curitiba, PR, Brazil).

2.2. Preparation of Materials. Resins in powder form in pure state and also loaded with 10% (w.t) oxides were added to a hot melt desktop screw extruder (Filastruder, GA, USA) forming printable filaments measuring 1.75 mm diameter; the extrusion was performed at temperatures within 70–85°C and the materials were cooled at ambient conditions; the extrusion speed was maintained at 20 rpm. In sequence, disks (10 × 5 mm) were manufactured using a FDM 3D printer (Prusa Mendel–I3, USA). The printing temperature was maintained at 80°C and the heating of table was maintained at 60°C; the printing feeding speed was maintained at 10 mm/min; the output measure of the hot end is 0.4 mm.

2.3. Susceptibility Assay In Vitro. Mueller-Hinton agar (MHA) was used to determine the antimicrobial activity of materials in study against bacteria and Sabouraud dextrose agar (SDA) was prepared for fungi; the agar media were prepared by following manufacturer instructions. The plates were autoclaved, in sequence; holes were made for insertion of sampling disks. Using a sterile transfer loop, suspensions containing 5×10^6 CFU/mL^{-1} were inoculated in the agar media. Then, the plates were incubated at bacteriological greenhouse during 48 hours at 34°C, with growth checks performed at every 6 hours. The bioactivity of materials was evaluated by comparing the inhibition zone (diameter in mm) and also number of colony forming units (CFU) in relation to the control plates with positive growth. The semiquantitative K-B disk-diffusion method was used to determine the antimicrobial and antifungal activity of materials [72, 73]. To obtain a more accurate counting of the viable cells, the software ImageJ with the plugin automated colony counting was used [74–76].

2.4. Scanning Electron Microscopy (SEM) and Energy-Dispersive X-Ray Microanalysis (EDXMA). Scanning electron microscopy (SEM) analysis was performed using equipment Jeol (JSM-6610LV). Samples of resins in pure state and resins loading metal oxides were placed on a carbon tape and subjected to a vacuum and electron beam. The structural identification of the samples was performed by X-rays diffraction measurements; the equipment used was a Bruker diffractometer (D8 Advance) equipped with a lynx-eye detector. We used a copper X-ray generator tube with radiation Kalfa 1 = 1.5406 A. The power was adjusted to 1600 W (40 kV and 40 mA) for evaluating the signals diffracted in the region between 20° and 120° 2 theta in step 0.025°/s. Samples were prepared in order to avoid any preferential orientation of hkl planes in a standard circular sample holder with a diameter of approximately 2.5 cm.

2.5. UV-Visible (UV-Vis) Absorption and Fourier Transform Infrared (FTIR) Spectroscopy. Spectroscopy analysis in the ultraviolet-visible region (UV-Vis) was carried out utilizing Shimadzu equipment (UV-1800) calibrated at a bandwidth of 1 nm; the wavelength range was maintained from 300 to 900 nm. The Fourier transformed infrared (FTIR) spectroscopy measurements were performed using a Shimadzu Spectrometer Prestige 21 (Shimadzu Corporation, Koyoto, Japan) with a resolution of 2 cm^{-1}. The measurements were carried out on KBr pellets which have transparency in the infrared region 400–4000 cm^{-1}. In this context, the powder samples resins and resins oxides (1 mg) were ground with KBr (300 mg, spectroscopic, high purity). To form the tablets, the mixture was placed in a hydraulic press applying approximately 10 tones, while the air was extracted by a mechanical pump. The FTIR spectra of the samples were recorded at ambient temperature and spectral band between 4000 and 300 cm^{-1}.

2.6. Differential Scanning Calorimetry (DSC). In the calorimetric analysis, samples were weighed (3.0 ± 0.5 mg) and hermetically sealed in aluminum crucibles being placed in a Shimadzu calorimeter model DSC-60 under an atmosphere of nitrogen, flow 50 ml min^{-1}; the heating ratio was maintained from 20°C min^{-1} to 550°C. The heating rate was maintained at 10°/min and the nitrogen flow was 100 ml/min. The equipment was calibrated for temperature with indium standard (156.6 ± 0.3°C) through their melting peak. The enthalpy and heat flow were calibrated using the heat of fusion of indium (28.59 J/g ± 0.30) using the same conditions as the samples. The correction factor was calculated in accordance with the procedures and specifications from Shimadzu.

2.7. Statistical Analysis. For data treatment of the susceptibility assay in vitro, the Experimental Design in Contextualized Blocks was used [77, 78]. In this research, the blocks are the 16 samples (raw data) and the treatments are 15 types of material + control plates, therefore totalizing 16 essays; the assay was carried out in quadruplicate for tested microorganisms; for data treatment, the statistical software SASM-Agri-8.1 was used [79].

2.8. List of Abbreviations. In text, the resins in pure state were identified as follows: *Styrax benzoin* (B pure), *Commiphora myrrha* (M pure), and *Boswellia papyrifera* (P pure). The oxides were identified as titanium dioxide (TiO_2); anatase oxide (P25); copper I (Cu_2O); and molybdenum trioxide (MoO_3), respectively.

3. Results and Discussion

3.1. Susceptibility Assay In Vitro. According to Figure 1(a) during the initial period of 6 hours, the growth of *C. albicans* was inhibited by materials B (pure), B + TiO_2, B + MoO_3, B + Cu_2O, B + P25, P + MoO_3, P + Cu_2O, P + P25, M + MoO_3, M + Cu_2O, B + MoO_3, B + Cu_2O, B + P25, P + Cu_2O, and M + Cu_2O ($p = 0.05$).

In 12-hour testing period, its growth of was limited by resin B (pure), B + TiO_2, B + MoO_3, B + Cu_2O, B + P25, P + TiO_2, P + MoO_3, P + P25, M pure, M + TiO_2, M + P25, B + P25, P + P25, and M + TiO_2. During 24-hour period, its growth was hampered by all materials, with one exception M + MoO_3; at this period, the most efficient materials were B + Cu_2O, B + P25, and P + Cu_2O. At 36 hours, B + Cu_2O, P + Cu_2O, M + Ti_2O, and M + Cu_2O were the most efficient materials. In the test period of 48 hours, its growth was restrained by all materials with one exception in P + P25 and the most efficient materials were B (pure), B + TiO_2, B + MoO_3, B + Cu_2O, P + MoO_3, P + Cu_2O, M (pure), M + TiO_2, and M + MoO3. In short, all materials tested (numerically) impede the proliferation of *C. albicans* ($p = 5\%$), highlighting B + TiO_2, B + MoO_3, B + Cu_2O, P + MoO_3, P + Cu_2O, M (pure), and M + MoO_3 which were the most efficient bioactive materials during the 48-hour assay.

According to Figure 1(b), during initial 6-hour testing period, the most effective materials were B + Cu_2O, P + TiO_2, M (pure), M + Cu_2O, and M + P25. Already during 12-hour testing period, the materials B (pure), B + MoO_3, P + MoO_3, and M + MoO_3 were the most effective. During 24-hour assay, the materials B + TiO_2, B + MoO_3, B + Cu_2O, P + TiO_2, P + Cu_2O, M + TiO_2, M + Cu_2O, and M + P25 were the most efficient and, during 36-hour period, the materials B + MoO_3, P + TiO_2, P + Cu_2O, M + TiO_2, and M + Cu_2O were the most effective. At the final period of 48-hour assay, all materials showed significant effectiveness in relation to the control plate, with only one exception P + P25 ($p = 5\%$). In sum, all materials under study (numerically) prevent the proliferation of *E. coli*, highlighting B + MoO_3, P + Cu_2O, P + TiO_2, and M + TiO_2, which stood out as the most bioactive materials.

In relation to Figure 1(c), during the initial period of 6 hours, among the materials that inhibited the growth of this bacterium, B + MoO_3, B + Cu_2O, P + Cu_2O, and M + P25 stood out. However, in the period of 12 hours, all materials tested showed no significant difference concerning the control plates, being considered ineffective during this period of assay. In 24-hour period, the materials B + Cu_2O, B + TiO_2, B + MoO_3, B + Cu_2O, B + P25, P + MoO_3, M + TiO_2, M + MoO_3, and M + Cu_2O were the most effective in detaining its growth. In the period of 36 hours, B + MoO_3, B + Cu_2O, P + Cu_2O, M + TiO_2, and M + Cu_2O were the most efficient. At the final period of 48 hours, the materials B + MoO_3, B + P25, P + MoO_3, M + TiO_2, and M + MoO_3 were the most effective in inhibiting its growth ($p = 5\%$). In short, all materials (numerically) were effective against the bacteria in question, highlighting B + MoO_3, P + MoO_3, and M + TiO_2 and presenting the higher inhibition rates.

According to Figure 1(d) during the initial period of 6 hours, in all cases, there was no significant difference between the materials and the control plate. However, during 12-hour period, significant differences were noted, highlighting B + MoO_3 and M + MoO_3 which were the most effective in restraining its growth. In the period of 24 hours, in most cases, there was no significant difference between the

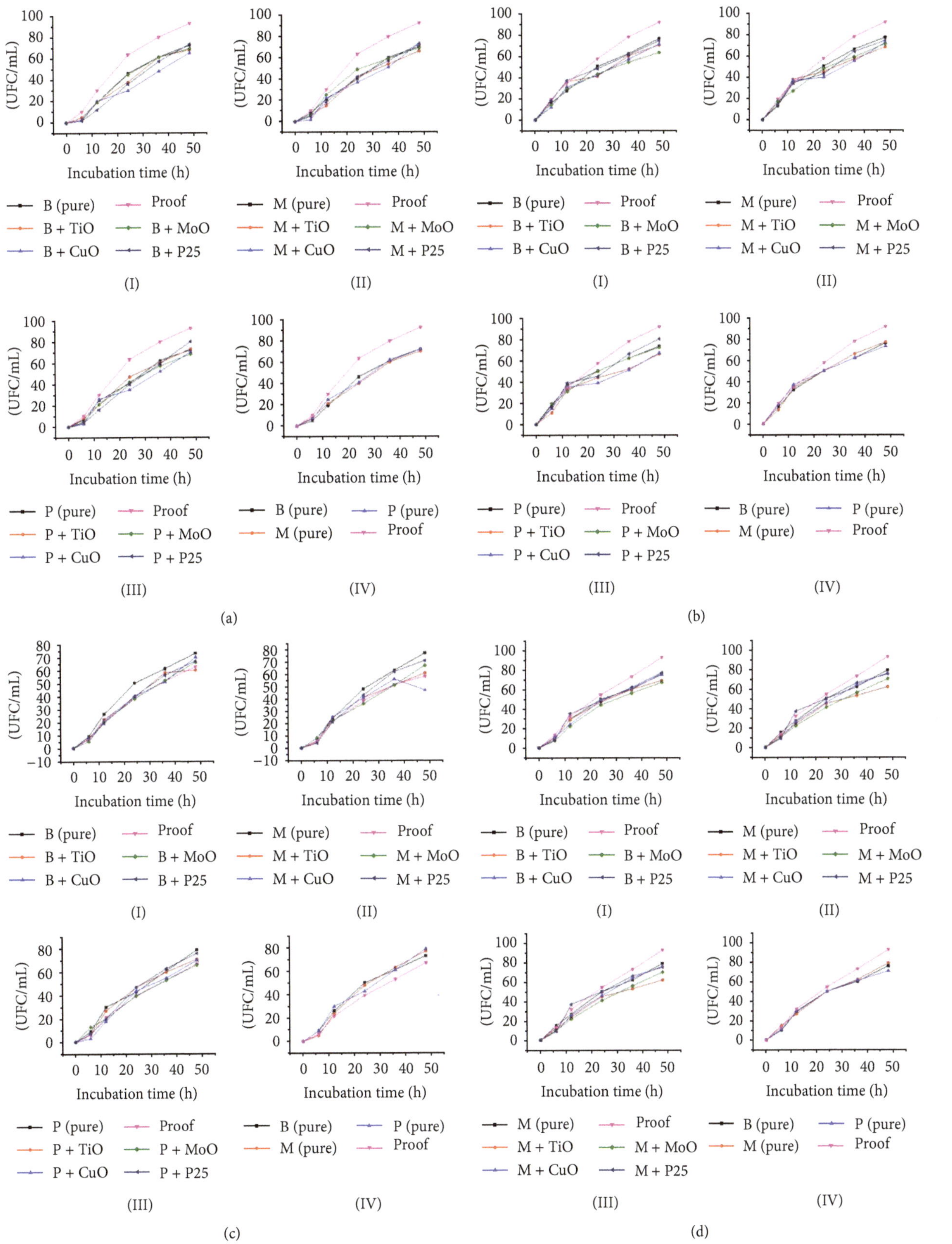

FIGURE 1: Antibiogram of clinical isolates: (a) refers to *C. albicans*; (b) refers to *E. coli*; (c) refers to *S. aureus*; and (d) refers to *P. aeruginosa* microorganisms.

TABLE 1: Antibacterial activity of materials against selected pathogenic strains.

Oleo-gum-resin	Inhibition zone (mm)			
Nanooxide	*S. aureus*	*E. coli*	*P. aeruginosa*	*C. albicans*
M (pure)	1.5 ± 1.5	3.0 ± 0.4	5.5 ± 0.6	2.4 ± 3.2
P (pure)	2.0 ± 1.0	3.1 ± 0.5	4.7 ± 0.88	4.4 ± 2.4
B (pure)	1.1 ± 1.1	4.0 ± 0.4	4.0 ± 2.5	1.8 ± 1.8
M + TiO_2	5.2 ± 0.18	8.8 ± 0.32	7.2 ± 0.28	6.8 ± 0.15
M + P25	6.0 ± 0.97	6.4 ± 0.11	6.1 ± 0.35	6.1 ± 0.26
M + MoO_3	5.5 ± 0.22	7.0 ± 0.20	6.6 ± 0.32	6.7 ± 0.92
M + Cu_2O	5.0 ± 0.28	6.5 ± 0.91	6.4 ± 1.57	5.5 ± 0.36
P + TiO_2	5.5 ± 1.55	8.0 ± 0.14	7.7 ± 1.90	6.0 ± 0.21
P + P25	5.8 ± 0.10	5.2 ± 0.26	6.2 ± 0.88	6.0 ± 1.74
P + MoO_3	5.5 ± 0.23	6.5 ± 0.36	7.0 ± 0.30	6.9 ± 1.59
P + Cu_2O	5.6 ± 1.81	8.0 ± 0.53	7.4 ± 1.39	6.6 ± 2.13
B + TiO_2	6.0 + 0.38	7.7 ± 0.77	5.0 ± 1.80	6.8 ± 1.27
B + P25	7.6 ± 0.53	6.2 ± 0.65	4.5 ± 0.36	6.1 ± 2.05
B + MoO_3	6.1 ± 1.70	8.3 ± 0.40	5.2 ± 1.24	6.5 ± 2.46
B + Cu_2O	7.5 ± 1.15	7.0 ± 0.88	4.8 ± 0.36	7.1 ± 2.15

Note. The experiments were done in quadruplicate and the results were interpreted in terms of standard deviation of mean diameter of zone of inhibition.

materials and control plates, with just one exception M + MoO_3. In the time interval of 36 hours, in most cases, there were significant differences; it is noteworthy that B + MoO_3, M + TiO_2 and M + MoO_3 were the most bioactive. In the period of 48 hours, in all cases, there were no significant differences between the materials and control plate, with just one exception M + TiO_2, highlighting that B + MoO_3, M + MoO_3, and M + TiO_2 were the most efficient biocides in inhibiting the proliferation of *S. aureus*.

Table 1 shows the results of the antibacterial and antifungal activity of materials against the tested pathogenic microorganisms; the results are exposed as the mean values obtained after 48 hours of assay.

The results from Table 1 show that the bioactivity of plant resins in pure state was effective and the addition of 10% w.t of oxides nanoparticles increased the efficiency of materials, as expected. Regarding the resin of *C. myrrha*, the antibiogram of bacteria and fungus under test corroborates the results obtained by Omer et al. [80] and in Alhussaini et al. [81]; concerning the resin of *B. papyrifera*, the results present in this study are in accordance with Camarda et al. [25], Abdallah et al. [82], Abdalah and Khalid [83], and de Rapper et al. [84]; and in relation to the resin of *S. benzoin* the results showed here are in accordance with the findings previously obtained by Dahni et al. [85] and De Rapper et al. [86].

3.2. SEM and EDXMA. Figure 2 shows the results of scanning electronic microscopy and energy-dispersive X-ray microanalysis.

According to Figure 2, the SEM analysis showed that part of oxides nanoparticles was encapsulated by the matrix of resins, and part remained heterogeneously dispersed over the surface of resins. Therefore, better ways of addition of oxides need to be studied, aiming for homogeneous nanostructure of the materials in the best way possible. One option is to increase the shearing forces during extrusion or premixing the content could improve the dispersant size and homogeneity in the dispersion by fractioning large agglomerates [87]. Another approach could be to feed the nanoparticles in suspension into the extrusion line [88]. During the printing of sample disks, an incomplete dispersion of the colloidal oxides in the disks was observed in the form of microsized domains of agglomerated nanoparticles; such agglomeration has previously been reported as a general problem for ceramic nanoparticles dispersed in polymers [89].

As exposed in Figure 3, the EDXMA analysis revealed information about the crystalline structure of samples.

According to Figure 3, the peaks identified at $2\theta = 15^\circ$ confirm the crystalline structure of the *B. papyrifera* [90]; the peaks found at $2\theta = 35^\circ$ and 45° confirm the molecular structure of *C. myrrha* [91]; the peaks at $2\theta = 35^\circ$ and 40° match with the chemical structure of *S. benzoin* [92]. No secondary phase was found samples, so it was possible to index the monoclinic phase of titanium oxide (PDF number 65-5714) and P25 anatase + rutile (PDF number 21-1272 and 21-1276, respectively). Those peaks at scattering angles of 25.26°, 36.94°, 48.05°, 53.89°, 55.06°, and 62.681° correspond to the reflections from the (101), (004), (200), (105), (211), and (204) crystalline planes of anatase (P25) and TiO_2 oxides [93]; it was possible to index the monoclinic phase of copper I (PDF number 89-5898) and molybdenum oxide (PDF number 13-345); the diffraction peak found at $2\theta = 26^\circ$ found in sample B + P25 is attributed to the hexagonal structure of graphite (002) [94]; the diffraction peaks found at 2θ = 29,6°, 36,7°, and 42,5° are attributed to Cu_2O crystalline planes (110), (200), and (220), respectively [95]. Lastly, the diffraction peaks at 2θ around 23.20°, 25.51°, 27.18°, 33.66°; 38.76°; 52.64°; 58.79°; and 67.28° correspond to (110), (040), (021), (111), (131), (211), (081), and (261); XRD standard data planes indexed to pure monocyclic and orthorhombic structure of MoO_3 [96].

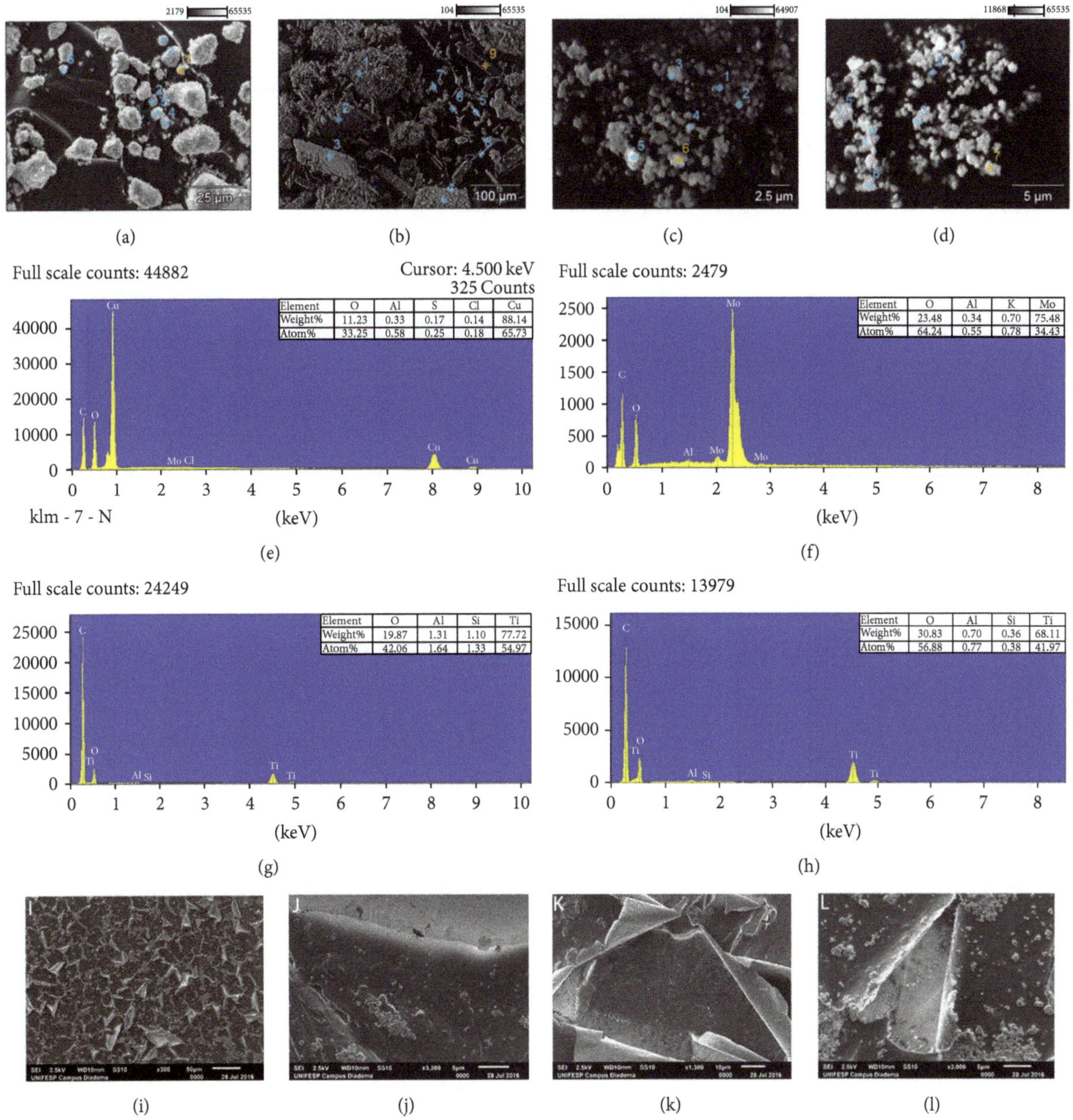

Figure 2: EDXMA analysis of metal oxides: Cu_2O (a); MoO_3 (b); P25 (c); TiO_2 (d); Cu_2O (e); MoO_3 (f); P25 (g); TiO_2 (h). SEM analysis of materials: B + P25 (i); M + TiO_2 (j); P + Cu_2O (k); B + MoO_3 (l).

3.3. UV-Vis and FTIR Spectroscopy. Figure 4 shows the UV-Visible spectrograms of samples.

According to Figure 4, the peaks at 350 nm to 420 nm confirm the molecular vibration of oleo-gum-resin of olibanum [97]; the peak found near of 390 nm with another prominent peak at 480 nm confirms the molecular vibration of myrrh extract [98]; the peak found at 350 nm with vibration extending to another peak at 450 nm confirms molecular vibration of benzoin resin [99]; the intensity of spectra obtained at wavelengths 350 nm is typical for the crystalline structure of anatase + rutile (P25) and titanium dioxide (TiO_2) [100]; the spectra showing intensity at wavelengths >300 nm typical for diametric and/or oligomeric species confirm the molecular vibration of MoO_3 [101]. Lastly, distinct peaks observed at 600 nm with stretching band until 800 nm confirm the vibration Cu_2O [102, 103].

Figure 5 shows FTIR spectrograms.

According to Figure 5, the FTIR spectrum of olibanum shows a peak at 3422.37 cm^{-1} indicating the presence of -OH- group, and the peak at 1705.33 cm^{-1} indicates the presence of carbonyl group The stretching band at 3428 cm^{-1} (O–H) and at 2930 cm^{-1} (C–H) and the bending bands of C–H appear at 1455 and 1378 cm^{-1} and the stretching band of C=O in carboxyl group appears at 1717 cm^{-1}, the stretching bands of C–O in carboxyl group are identified at 1243 cm^{-1}, and the stretching band of C=O at 1737 cm^{-1} indicates the presence of esters in olibanum resin [90].

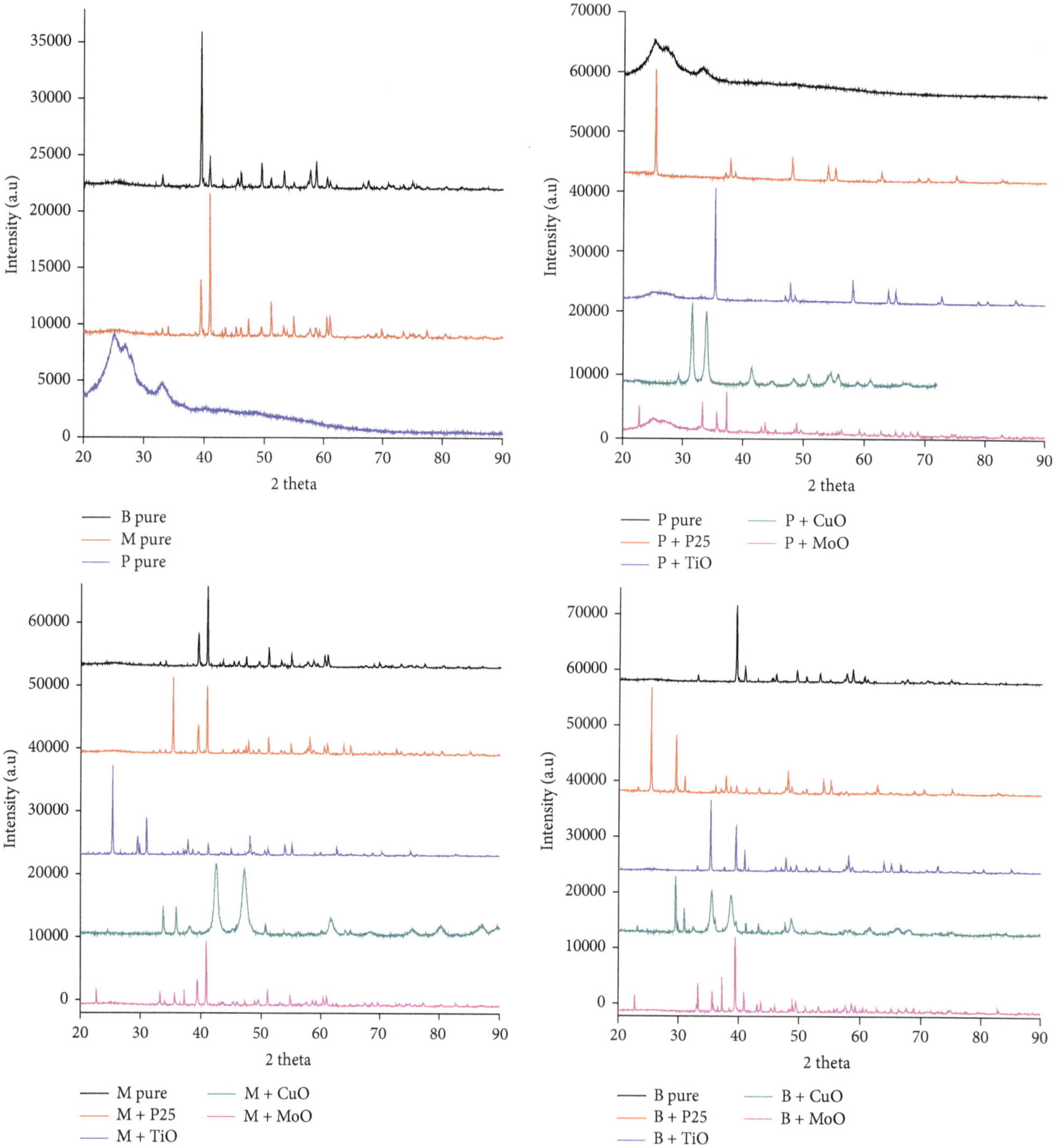

FIGURE 3: X-ray diffractograms of samples in pure state and also doped with metal oxides nanoparticles.

The FTIR spectrum of benzoin shows stretching band of carboxyl group (C=O) at 1719 cm^{-1}, and aromatic skeletal bands at 1601, 1516, and 1451 cm^{-1}, stretching band of C–O in carboxyl group at 1273 cm^{-1}, and a bending band at 712 cm^{-1} show a phenyl group (Ph–H) peaks identified at 1207 to 1441 cm^{-1} and 1376 to 1450 cm^{-1} confirming the presence of coniferyl benzoate [104]; the peak found near to 1650 cm^{-1} evidences an aliphatic unsaturation with strong C=C bonds, the peak near to 1610 cm^{-1} shows weak aromatic unsaturation [105], the peaks found at around 2872 and 2923 cm^{-1} can be attributed to C–H asymmetric and symmetric stretching vibration of methylene [106], also the band observed at around 1450 cm^{-1} is due to C–H stretching vibration of methylene bridge, a peak at 1560 cm^{-1} can be assigned to stretching vibration of a carboxylate group (-COOH) [107], the broad peaks at 3500 cm^{-1} 3.420 cm^{-1} can be assigned to

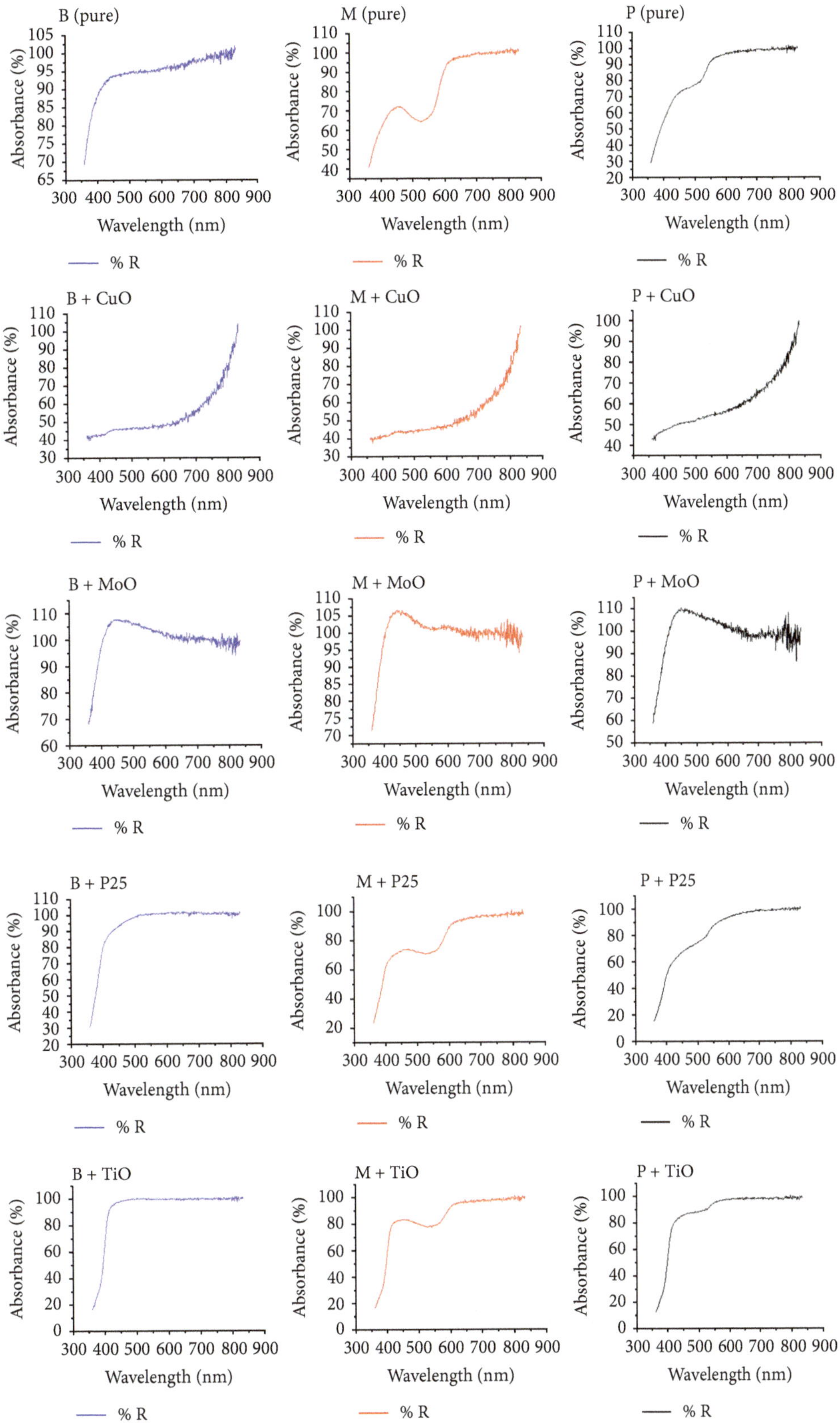

FIGURE 4: UV-Vis absorption spectra of samples in pure state and also doped with metal oxides nanoparticles.

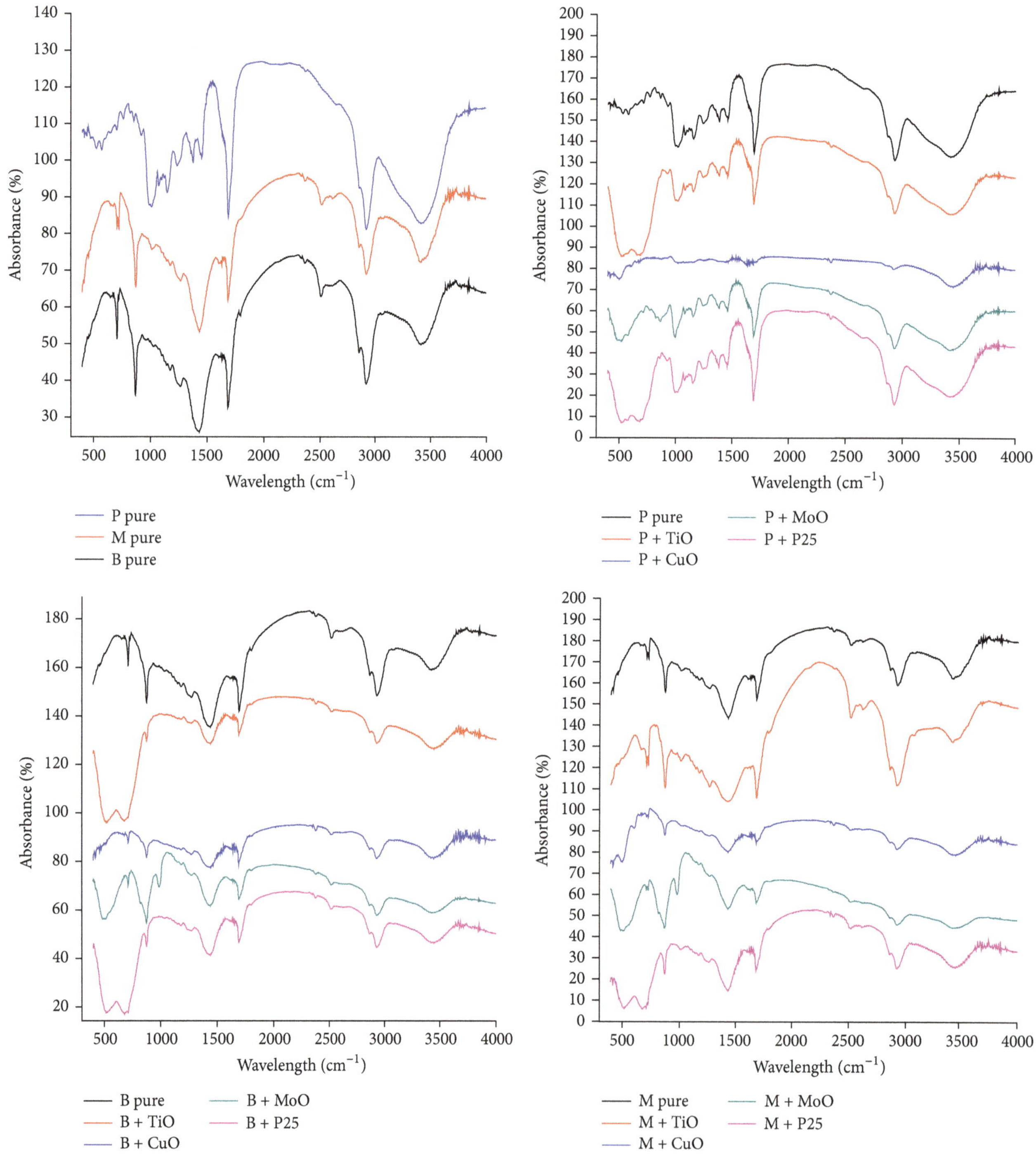

FIGURE 5: FTIR spectra of samples in pure state and also doped with metal oxides nanoparticles.

stretching of functional groups O–H [108], additionally peaks at 1580 to 1590 cm^{-1} correspond to stretching vibration of C=C–C aromatic rings [109], and lastly, a band observed at around 1400 cm^{-1} refers to C–H stretching vibration of vinyl [110] and broad peaks at 1070 cm^{-1} can be associated with stretching vibration of C–O–C [111].

The spectrum of myrrh shows the presence of broad bands located at 3.450, 1.630, and 1.550 cm^{-1} which are attributed to stretching of O–H, –COOH, and C=C and the shift of C=O vibration (symmetric stretching) of –COOH groups; an intense band at 1630 cm^{-1} confirms the molecular vibration of myrrh resin [105]; the stretching band identified at 1025 to 1200 cm^{-1} corresponds to the C–O stretching, while the weak bands at 1340 to 1450 cm^{-1} can be attributed to aliphatic hydrocarbons (CH_2 and CH_3) groups, groups of aldehydes (–CHO) and ketones (C=O), and the bending modes of bonds in alcohols (O–H), phenols (-OH), and carboxylic acids (-COOH); the bands at 1620 and 1650 cm^{-1} correspond to aromatic rings, while the bands around 2920 to 2930 cm^{-1} are analogous to the asymmetric stretching of the C–H bonds [106]; the strong broad band appearing at 3440 cm^{-1} can be assigned to the stretching vibrations of various groups in alcohols (O–H) and phenols (-OH) [105, 112].

The FTIR spectra of resins doped with titanium oxide nanoparticles reveal a small peak at 1640 cm^{-1} and a large broad peak between 3450 and 3200 cm^{-1}, corresponding to the stretching vibrations of absorbed water, as well as hydroxyl (OH) groups present in the surface of TiO_2–P25 nanopowder [113]; the broad peak between 600 and 400 cm^{-1} can be assigned to the presence of Ti–O–Ti bonds [114]. Concerning the copper I oxide addition, the peaks found at 529 and 602 cm^{-1} denotes the (Cu–O) stretching vibration of monoclinic CuO phase; the broad peak at about 490–620 cm^{-1} (central at 548 cm^{-1}) was due to an overlap between Cu–O stretching vibration of Cu_2O/CuO and (-OH) hydroxyl vibrations at 490–510 cm^{-1} [115]; the peak found at 620 cm^{-1} is related to Cu_2O crystals [116], and the stretching bands found at 298 to 620 cm^{-1} match the crystalline structure of Cu_2O [117, 118].

Regarding molybdenum nanopowder, two peaks found at 876.5 and 595.8 cm^{-1} are assigned to MoO_3 phase [119]; the peak at 996 cm^{-1} was associated with the terminal stretching vibration of molybdenum in its oxidized form (MoO) that is an indication of layered MoO_3 phase; the bands at 867 cm^{-1} and 558 cm^{-1} are assigned to stretching vibrations and bending vibrations of the Mo–O–Mo units [120]; the bands at around 3435 cm^{-1} and 1614 cm^{-1} can be attributed to the stretching and bending vibrations of (O–H) hydroxyl groups in the adsorbed water [121].

It is important to emphasize that FTIR absorption bands of different compounds in wood resin exudates may be discriminated according to their responses to a given thermal stimulation, because different compounds usually respond in different ways to the same stimulation. For example, the spectral bands of a volatile compound will decrease synchronously if this compound evaporates when the sample is heated [122]; since plant samples usually are complex mixtures, signal-resolving methods are necessary to find the spectral features of compounds of interest in the signal-overlapped IR spectra.

3.4. Differential Scanning Calorimetry (DSC). Figure 6 shows the DSC curves of materials under analysis.

According to Figure 6, the thermophase diagram of myrrh in pure state presents characteristics of a semicrystalline polymer. The glass transition temperature (T_g) of this resin occurs at 75°C; from ambient temperature to this point, the material presents glassy behavior; it is hard, inflexible, and brittle, from 75°C to 300°C and this resin exhibits rubbery behavior, at this range; it is soft and flexible, from 300°C to 500°C, and the material presents viscoelastic state; the resin reaches its melting temperature (T_m) at 500°C. The peak at 150°C indicates a primary crystallization and the baseline change with a prominent peak at 300°C indicates a secondary crystallization, so the crystallization temperature (T_c) of this material occurs between 150°C and 300°C. In comparison to other available polymers, this resin shows similar characteristics to commercial polyurea.

The resin of benzoin also presents characteristics of a semicrystalline polymer such as polyurea. From ambient temperature till reaching 80°C (T_g), this biopolymers exhibits glassy state; from 80°C to 275°C, it presents rubbery state, with rigid crystalline phase and amorphous mobile phase; two exothermic peaks at 350°C and 420°C indicate that crystallizations occur at this temperature range (T_c), from 275°C to 500°C; the resin presents viscoelastic state; finally, this resin reaches its melting temperature (T_m) at 500°C.

Similarly, the resin of olibanum presents characteristics of semicrystalline polymers, such as polyester. Its glass transition temperature (T_g) occurs at 80°C; from this point forward until 300°C, this biopolymer presents rubbery state; two peaks at 300°C and 400°C indicate that crystallizations occur at this temperature range (T_c); lastly at 500°C, this biopolymer reaches its melting temperature (T_m).

When the temperature of resins reaches its T_m, the melting of crystallites occurs; at this point, the system power reaches the level needed to win the secondary intermolecular forces between the chains of the crystalline phase, destroying the regular packing structure, thereupon changing from rubbery state to viscous state. This transition only occurs in crystalline phase, so this interpretation only makes sense if it is applied to semicrystalline polymers [123].

It is evident that the addition of oxides nanoparticles will interfere in activation energy of particles by breaking existing chemical bonds between the atoms of each substance, thus favoring the occurrence of other chemical bonds and synthesis of a new substances, and also will influence the thermal-mechanical properties of samples (Ehrenstein et al., 2006) [124]. It will also depend on the degree of crystallinity, since higher crystallinity will result in a harder and more thermally stable but also more brittle material, whereas the amorphous regions provide certain elasticity and impact resistance [125, 126].

As exposed in the DSC signal, the addition of TiO_2 and P25 reveals peaks at 300°C and 500°C; at high temperatures, TiO_2 nanoparticles dehydrate and coarsen, the final stable phase upon grain growth being always rutile [127]. The overall process (phase transformation, water loss, and coarsening) is irreversible; the TiO_2 and P25 samples contain adsorbed water which has the properties of bulk liquid water and a small fraction bound very tightly, probably in the form of hydroxyl groups [128]. The addition of Cu_2O shows peaks at 300°C and 450°C, related to the removal of water from the surface [129] and the corrosion of copper nanoparticles [130], concerning the addition of MoO_3. The occurrence of peaks at 300°C and 500°C is due to the removal of water and recrystallizations of phases present in the nanopowder [131].

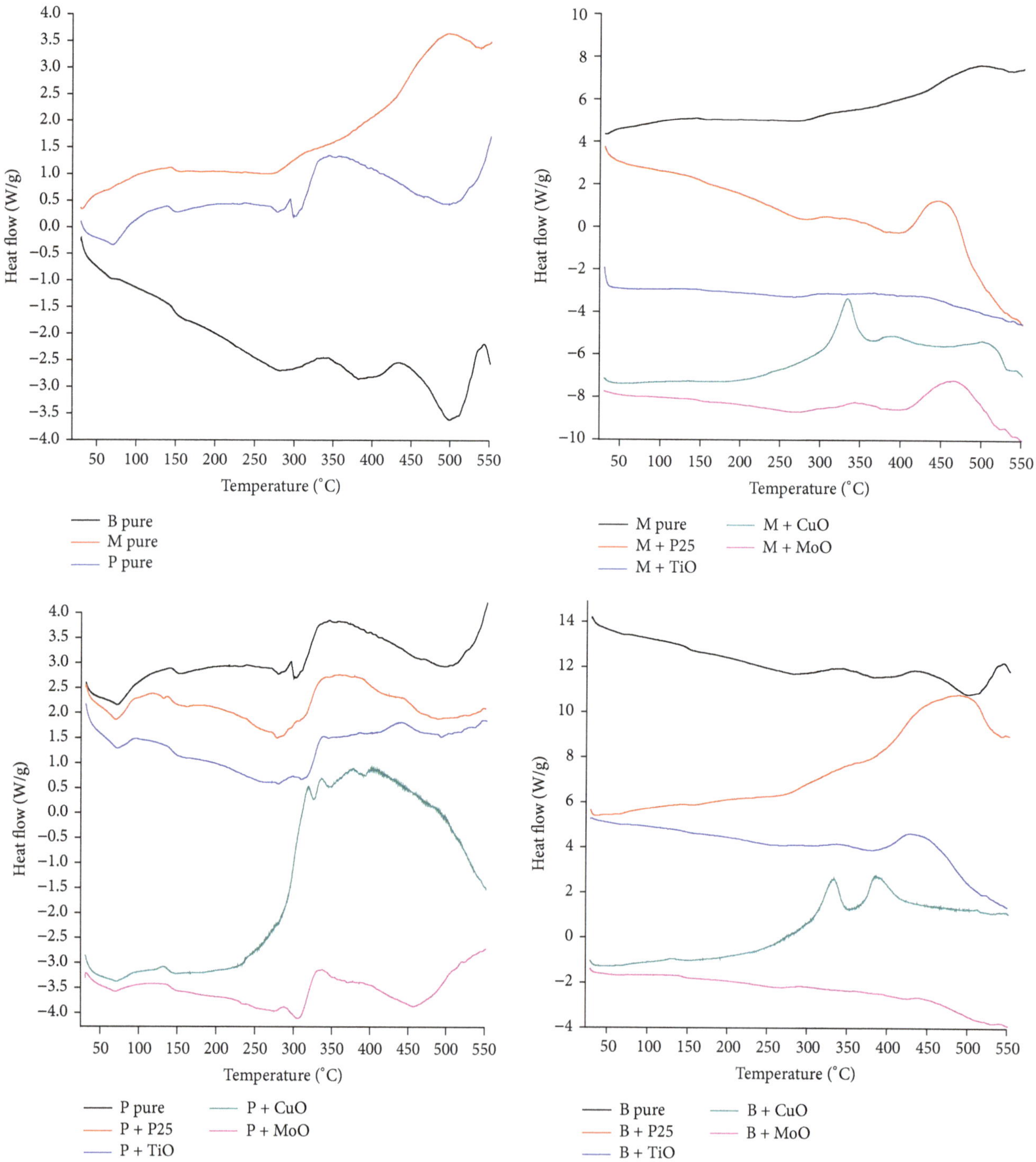

FIGURE 6: DSC phase diagram of samples in pure state and also doped with metal oxides nanoparticles.

In short, resins showed thermal behavior inherent to semicrystalline polymers such as polyester and polyurea; at some point, the molecules disposed in amorphous matrix obtain enough freedom of motion to spontaneously rearrange themselves into crystalline forms. This transition from amorphous solid to crystalline solid was evidenced by distinct exothermic peaks, as the temperature increases to 500°C samples, eventually reaching its melting point.

4. Conclusion

The biopolymers tested showed inherent characteristics of commercial available semicrystalline polymers; in most cases, the materials inhibited the proliferation of clinical pathogens under study, and, as expected, the addition of oxides nanoparticles increased the bacteriostatic effect. Although their addition was not well structured during the production of filaments and disks, nanoparticles remained disorderly dispersed over the matrix of resins, in most cases being encapsulated by the same.

Acknowledgments

The authors acknowledge the financial support from the Coordination of Improvement of Higher Level Personnel (CAPES, Brazil) through the Ph.D. scholarship granted.

References

[1] A. Muñoz-Bonilla and M. Fernández-García, "Polymeric materials with antimicrobial activity," *Progress in Polymer Science*, vol. 37, no. 2, pp. 281–339, 2012.

[2] O. Cuvas Apan, T. Z. Apan, and A. Apan, "In vitro antimicrobial activity of commonly used vasoactive drugs," *Journal of Clinical Anesthesia*, vol. 34, 2016.

[3] T.-W. Wan, Y. Tomita, N. Saita et al., "Emerging ST121/agr4 community-associated methicillin-resistant Staphylococcus aureus (MRSA) with strong adhesin and cytolytic activities: Trigger for MRSA pneumonia and fatal aspiration pneumonia in an influenza-infected elderly," *New Microbes and New Infections*, vol. 13, pp. 17–21, 2016.

[4] L. Gao, Y. Zhang, Y. Wang et al., "Reduction of PCN biosynthesis by NO in Pseudomonas aeruginosa," *Redox Biology*, vol. 8, pp. 252–258, 2016.

[5] E. Almaas, B. Kovács, T. Vicsek, Z. N. Oltvai, and A.-L. Barabási, "Global organization of metabolic fluxes in the bacterium *Escherichia coli*," *Nature*, vol. 427, no. 6977, pp. 839–843, 2004.

[6] A. M. Ibekwe, P. M. Watt, C. M. Grieve, V. K. Sharma, and S. R. Lyons, "Multiplex fluorogenic real-time PCR for detection and quantification of Escherichia coli O157:H7 in dairy wastewater wetlands," *Applied and Environmental Microbiology*, vol. 68, no. 10, pp. 4853–4862, 2002.

[7] X. Sun, C. Lei, L. Guo, and Y. Zhou, "Separable detecting of Escherichia coli O157H:H7 by a giant magneto-resistance-based bio-sensing system," *Sensors and Actuators, B: Chemical*, vol. 234, pp. 485–492, 2016.

[8] D. Williams and M. Lewis, "Pathogenesis and treatment of oral Candidosis," *Journal of Oral Microbiology*, vol. 3, no. 2011, article 5771, 2011.

[9] A. N. B. Ellepola, L. P. Samaranayake, and Z. U. Khan, "Extracellular phospholipase production of oral Candida albicans isolates from smokers, diabetics, asthmatics, denture wearers and healthy individuals following brief exposure to polyene, echinocandin and azole antimycotics," *Brazilian Journal of Microbiology*, vol. 47, no. 4, pp. 911–916, 2016.

[10] C. L. Ventola, "The antibiotic resistance crisis: part 2: management strategies and new agentsts," *Pharmacy and Therapeutics*, vol. 40, no. 5, pp. 344–352, 2015.

[11] T. D. Michl, K. E. S. Locock, S. S. Griesser, M. Haeussler, L. Meagher, and H. J. Griesser, "Bio-inspired antimicrobial polymers," in *Biosynthetic Polymers for Medical Applications*, pp. 87–127, Elsevier Science B.V, Amsterdam, Netherlands, 2016.

[12] D. Schillaci, V. Arizza, T. Dayton, L. Camarda, and V. Di Stefano, "In vitro anti-biofilm activity of Boswellia spp. oleogum resin essential oils," *Letters in Applied Microbiology*, vol. 47, no. 5, pp. 433–438, 2008.

[13] M. F. T. Medeiros and U. P. De Albuquerque, "The pharmacy of the Benedictine monks: The use of medicinal plants in Northeast Brazil during the nineteenth century (1823-1829)," *Journal of Ethnopharmacology*, vol. 139, no. 1, pp. 280–286, 2012.

[14] S. A. Sargin, E. Akçicek, and S. Selvi, "An ethnobotanical study of medicinal plants used by the local people of Alaşehir (Manisa) in Turkey," *Journal of Ethnopharmacology*, vol. 150, no. 3, pp. 860–874, 2013.

[15] Z. Hübsch, R. L. Van Zyl, I. E. Cock, and S. F. van Vuuren, "Interactive antimicrobial and toxicity profiles of conventional antimicrobials with Southern african medicinal plants," *South African Journal of Botany*, vol. 93, pp. 185–197, 2014.

[16] H. Kong, J. Yang, Y. Zhang, Y. Fang, K. Nishinari, and G. O. Phillips, "Synthesis and antioxidant properties of gum arabic-stabilized selenium nanoparticles," *International Journal of Biological Macromolecules*, vol. 65, pp. 155–162, 2014.

[17] A. A. Shahat, A. Y. Ibrahim, and M. S. Elsaid, "Polyphenolic content and antioxidant activity of some wild Saudi Arabian asteraceae plants," *Asian Pacific Journal of Tropical Medicine*, vol. 7, no. 7, pp. 545–551, 2014.

[18] A. N. Shikov, O. N. Pozharitskaya, V. G. Makarov, H. Wagner, R. Verpoorte, and M. Heinrich, "Medicinal Plants of the Russian Pharmacopoeia; Their history and applications," *Journal of Ethnopharmacology*, vol. 154, no. 3, pp. 481–536, 2014.

[19] S. Sultana, H. M. Asif, N. Akhtar, and K. Ahmad, "Medicinal plants with potential antipyretic activity: A review," *Asian Pacific Journal of Tropical Disease*, vol. 5, no. 1, pp. S202–S208, 2015.

[20] E. Özdemir and K. Alpınar, "An ethnobotanical survey of medicinal plants in western part of central Taurus Mountains: Aladaglar (Nigde-Turkey)," *Journal of Ethnopharmacology*, vol. 166, pp. 53–65, 2015.

[21] M. Bahmani, K. Saki, S. Shahsavari, M. Rafieian-Kopaei, R. Sepahvand, and A. Adineh, "Identification of medicinal plants effective in infectious diseases in Urmia, northwest of Iran," *Asian Pacific Journal of Tropical Biomedicine*, vol. 5, no. 10, pp. 858–864, 2015.

[22] P. Mendonça Pauletti, A. R. Araújo, M. C. M. Young, A. M. Giesbrecht, and V. Da Silva Bolzani, "nor-Lignans from the leaves of Styrax ferrugineus (Styracaceae) with antibacterial and antifungal activity," *Phytochemistry*, vol. 55, no. 6, pp. 597–601, 2000.

[23] M. A. Saeed and A. W. Sabir, "Antibacterial activities of some constituents from oleo-gum-resin of Commiphora mukul," *Fitoterapia*, vol. 75, no. 2, pp. 204–208, 2004.

[24] R. A. A. Mothana and U. Lindequist, "Antimicrobial activity of some medicinal plants of the island Soqotra," *Journal of Ethnopharmacology*, vol. 96, no. 1-2, pp. 177–181, 2005.

[25] L. Camarda, T. Dayton, V. Di Stefano, R. Pitonzo, and D. Schillaci, "Chemical composition and antimicrobial activity of some oleogum resin essential oils from Boswellia spp. (Burseraceae)," *Annali di Chimica*, vol. 97, no. 9, pp. 837–844, 2007.

[26] S. F. van Vuuren, G. P. P. Kamatou, and A. M. Viljoen, "Volatile composition and antimicrobial activity of twenty commercial frankincense essential oil samples," *South African Journal of Botany*, vol. 76, no. 4, pp. 686–691, 2010.

[27] M. P. Paraskeva, S. F. van Vuuren, R. L. van Zyl, H. Davids, and A. M. Viljoen, "The in vitro biological activity of selected South African Commiphora species," *Journal of Ethnopharmacology*, vol. 119, no. 3, pp. 673–679, 2008.

[28] J. Wanner, E. Schmidt, S. Bail et al., "Chemical composition and antibacterial activity of selected essential oils and some of their main compounds," *Natural Product Communications*, vol. 5, no. 9, pp. 1359–1364, 2010.

[29] T. Shen, G. Li, X. Wang, and H. Lou, "The genus Commiphora: a review of its traditional uses, phytochemistry and pharmacology," *Journal of Ethnopharmacology*, vol. 142, no. 2, pp. 319–330, 2012.

[30] S. Al-Daihan, M. Al-Faham, N. Al-shawi et al., "Antibacterial activity and phytochemical screening of some medicinal plants commonly used in Saudi Arabia against selected pathogenic microorganisms," *Journal of King Saud University - Science*, vol. 25, no. 2, pp. 115–120, 2013.

[31] A. A. Mohamed, S. I. Ali, F. K. EL-Baz, A. K. Hegazy, and M. A. Kord, "Chemical composition of essential oil and in vitro antioxidant and antimicrobial activities of crude extracts of Commiphora myrrha resin," *Industrial Crops and Products*, vol. 57, pp. 10–16, 2014.

[32] F. Wang, Y.-B. Wang, H. Chen, L. Chen, S.-W. Liang, and S.-M. Wang, "Two new triterpenoids from the resin of Styrax tonkinensis," *Journal of Asian Natural Products Research*, vol. 17, no. 8, pp. 823–827, 2015.

[33] L. O. Hanus, T. Rezanka, V. M. Dembitsky, and A. Moussaieff, "Myrrh -*Commiphora* chemistry," *Biomedical Papers*, vol. 149, no. 1, pp. 3–28, 2005.

[34] F. S. Li, D. L. Yan, R. R. Liu, K. P. Xu, and G. S. Tan, *Chemical Constituents of* Boswellia carterii *(Frankincense)*, vol. 8, pp. 25–27, 2007.

[35] D. Poeckel and O. Werz, "Boswellic acids: Biological actions and molecular targets," *Current Medicinal Chemistry*, vol. 13, no. 28, pp. 3359–3369, 2006.

[36] H. P. T. Ammon, "Boswellic acids in chronic inflammatory diseases," *Planta Medica*, vol. 72, no. 12, pp. 1100–1116, 2006.

[37] H. Safayhi and E.-R. Sailer, "Anti-inflammatory actions of pentacyclic triterpenes," *Planta Medica*, vol. 63, no. 6, pp. 487–493, 1997.

[38] K. H. C. Başer, B. Demirci, A. Dekebo, and E. Dagne, "Essential oils of some *Boswellia* spp., myrrh and opopanax," *Flavour and Fragrance Journal*, vol. 18, no. 2, pp. 153–156, 2003.

[39] G. Tayoub, I. Schwob, J.-M. Bessière et al., "Composition of volatile oils of Styrax (Styrax officinalis L.) leaves at different phenological stages," *Biochemical Systematics and Ecology*, vol. 34, no. 9, pp. 705–709, 2006.

[40] J.-J. Filippi, C. Castel, X. Fernandez, M. Rouillard, M. Gaysinski, and S. Lavoine-Hanneguelle, "An unusual acenaphthylene-type sesquiterpene hydrocarbon from Siam and Sumatra benzoin gum," *Phytochemistry Letters*, vol. 2, no. 4, pp. 216–219, 2009.

[41] D. L. Custódio and V. F. Veiga-Junior, "True and common balsams," *Brazilian Journal of Pharmacognosy*, vol. 22, no. 6, pp. 1372–1383, 2012.

[42] P. Burger, A. Casale, A. Kerdudo et al., "New insights in the chemical composition of benzoin balsams," *Food Chemistry*, vol. 210, pp. 613–622, 2016.

[43] S. Kumar and S. K. Gupta, "Rosin, A naturally derived excipient in drug delivery systems," *Polim. Med*, vol. 43, no. 1, pp. 45–48, 2013.

[44] S. K. Bajpai and M. Kumari, "A green approach to prepare silver nanoparticles loaded gum acacia/poly(acrylate) hydrogels," *International Journal of Biological Macromolecules*, vol. 80, pp. 177–188, 2015.

[45] C. Dong, X. Zhang, H. Cai, and C. Cao, "Facile and one-step synthesis of monodisperse silver nanoparticles using gum acacia in aqueous solution," *Journal of Molecular Liquids*, vol. 196, pp. 135–141, 2014.

[46] A. J. Kora and L. Rastogi, "Green synthesis of palladium nanoparticles using gum ghatti (Anogeissus latifolia) and its application as an antioxidant and catalyst," *Arabian Journal of Chemistry*, 2015.

[47] M. Bernela, M. Ahuja, and R. Thakur, "Enhancement of anti-inflammatory activity of glycyrrhizic acid by encapsulation in chitosan-katira gum nanoparticles," *European Journal of Pharmaceutics and Biopharmaceutics*, vol. 105, pp. 141–147, 2016.

[48] S. K. Bajpai, M. Jadaun, and S. Tiwari, "Synthesis, characterization and antimicrobial applications of zinc oxide nanoparticles loaded gum acacia/poly(SA) hydrogels," *Carbohydrate Polymers*, vol. 153, pp. 60–65, 2016.

[49] H. Joy Prabu and I. Johnson, "Plant-mediated biosynthesis and characterization of silver nanoparticles by leaf extracts of Tragia involucrata, Cymbopogon citronella, Solanum verbascifolium and Tylophora ovata," *Karbala International Journal of Modern Science*, vol. 1, no. 4, pp. 237–246, 2015.

[50] J. Rodríguez-Hernández, "Antimicrobial micro/nanostructured functional polymer surfaces," in *Nanobiomaterialsin Antimicrobial Therapy. Applications of Nanobiomaterials*, vol. 6, pp. 153–192, Elsevier Science B.V, Amsterdam, Netherlands, 2016.

[51] M. Quaresimin, R. Bertani, M. Zappalorto, A. Pontefisso, F. Simionato, and A. Bartolozzi, "Multifunctional polymer nanocomposites with enhanced mechanical and anti-microbial properties," *Composites Part B: Engineering*, vol. 80, article no. 3615, pp. 108–115, 2015.

[52] M. S. Ganewatta and C. Tang, "Controlling macromolecular structures towards effective antimicrobial polymers," *Polymer (United Kingdom)*, vol. 63, pp. A1–A29, 2015.

[53] D. F. Williams, "On the nature of biomaterials," *Biomaterials*, vol. 30, no. 30, pp. 5897–5909, 2009.

[54] R. K. Jha, P. K. Jha, K. Chaudhury, S. V. Rana, and S. K. Guha, "An emerging interface between life science and nanotechnology: present status and prospects of reproductive healthcare aided by nano-biotechnology," *Nano Reviews*, vol. 5, no. 1, article 22762, 2014.

[55] C. Buzea, I. I. Pacheco, and K. Robbie, "Nanomaterials and nanoparticles: sources and toxicity," *Biointerphases*, vol. 2, pp. MR17–MR71, 2007.

[56] L. Zhang, X. Wang, Y. Miao et al., "Magnetic ferroferric oxide nanoparticles induce vascular endothelial cell dysfunction and inflammation by disturbing autophagy," *Journal of Hazardous Materials*, vol. 304, pp. 186–195, 2016.

[57] J. J. Water, A. Bohr, J. Boetker et al., "Three-dimensional printing of drug-eluting implants: Preparation of an antimicrobial polylactide feedstock material," *Journal of Pharmaceutical Sciences*, vol. 104, no. 3, pp. 1099–1107, 2015.

[58] N. Sandler, I. Kassamakov, H. Ehlers, N. Genina, T. Ylitalo, and E. Haeggstrom, "Rapid interferometric imaging of printed drug laden multilayer structures," *Scientific Reports*, vol. 4, article no. 4020, 2014.

[59] T. M. Rankin, N. A. Giovinco, D. J. Cucher, G. Watts, B. Hurwitz, and D. G. Armstrong, "Three-dimensional printing surgical

instruments: are we there yet?" *Journal of Surgical Research*, vol. 189, no. 2, pp. 193–197, 2014.

[60] J. Yue, P. Zhao, J. Y. Gerasimov et al., "3D-Printable Antimicrobial Composite Resins," *Advanced Functional Materials*, vol. 25, no. 43, pp. 6756–6767, 2015.

[61] A. Ion, E. Andronescu, D. Řadulescu et al., "Biocompatible 3d matrix with antimicrobial properties," *Molecules*, vol. 21, no. 1, Article ID 21010115, 2016.

[62] M. Karl, D. Djuric, and K. Kolter, "Pharmaceutical Excipients for Hot-Melt Extrusion," *Pharmaceutical Technology*, vol. 35, no. 5, pp. 74–82, 2011.

[63] M. Wilson, M. A. Williams, D. S. Jones, and G. P. Andrews, "Hot-melt extrusion technology and pharmaceutical application," *Therapeutic Delivery*, vol. 3, no. 6, pp. 787–797, 2012.

[64] M. Lu, Z. Guo, Y. Li et al., "Application of hot melt extrusion for poorly water-soluble drugs: Limitations, advances and future prospects," *Current Pharmaceutical Design*, vol. 20, no. 3, pp. 369–387, 2014.

[65] M. Stanković, H. W. Frijlink, and W. L. J. Hinrichs, "Polymeric formulations for drug release prepared by hot melt extrusion: application and characterization," *Drug Discovery Today*, vol. 20, no. 7, article no. 1575, pp. 812–823, 2015.

[66] A. Melocchi, F. Parietti, A. Maroni, A. Foppoli, A. Gazzaniga, and L. Zema, "Hot-melt extruded filaments based on pharmaceutical grade polymers for 3D printing by fused deposition modeling," *International Journal of Pharmaceutics*, vol. 509, no. 1-2, pp. 255–263, 2016.

[67] J. Goole and K. Amighi, "3D printing in pharmaceutics: A new tool for designing customized drug delivery systems," *International Journal of Pharmaceutics*, vol. 499, no. 1-2, pp. 376–394, 2016.

[68] G. Mayer, "Rigid biological systems as models for synthetic composites," *Science*, vol. 310, no. 5751, pp. 1144–1147, 2005.

[69] E. Munch, M. E. Launey, D. H. Alsem, E. Saiz, A. P. Tomsia, and R. O. Ritchie, "Tough, bio-inspired hybrid materials," *Science*, vol. 322, no. 5907, pp. 1516–1520, 2008.

[70] J. Yuan and A. H. E. Müller, "One-dimensional organic-inorganic hybrid nanomaterials," *Polymer*, vol. 51, no. 18, pp. 4015–4036, 2010.

[71] C. Sanchez, B. Julián, P. Belleville, and M. Popall, "Applications of hybrid organic-inorganic nanocomposites," *Journal of Materials Chemistry*, vol. 15, no. 35-36, pp. 3559–3592, 2005.

[72] A. W. Bauer, W. M. Kirby, J. C. Sherris, and M. Turck, "Antibiotic susceptibility testing by a standardized single disk method," *The American Journal of Clinical Pathology*, vol. 45, no. 4, pp. 493–496, 1966.

[73] B. Bonev, J. Hooper, and J. Parisot, "Principles of assessing bacterial susceptibility to antibiotics using the agar diffusion method," *Journal of Antimicrobial Chemotherapy*, vol. 61, no. 6, pp. 1295–1301, 2008.

[74] J. Marotz, C. Lübbert, and W. Eisenbeiß, "Effective object recognition for automated counting of colonies in Petri dishes (automated colony counting)," *Computer Methods and Programs in Biomedicine*, vol. 66, no. 2-3, pp. 183–198, 2001.

[75] M. Putman, R. Burton, and M. H. Nahm, "Simplified method to automatically count bacterial colony forming unit," *Journal of Immunological Methods*, vol. 302, no. 1-2, pp. 99–102, 2005.

[76] S. Sieuwerts, F. A. M. De Bok, E. Mols, W. M. De Vos, and J. E. T. Van Hylckama Vlieg, "A simple and fast method for determining colony forming units," *Letters in Applied Microbiology*, vol. 47, no. 4, pp. 275–278, 2008.

[77] M. I. Rodrigues and A. F. Lemma, *Planejamento de experimentos e otimização de processos*, Cárita, Campinas, SP, Brazil, 2nd edition, 2009.

[78] C. A. Mucelin, *Estatística Elementar e experimental aplicada às tecnologias*, Medianeira, PR, Brazil, Valério, 2nd edition, 2006.

[79] M. G. Canteri, R. A. Althaus, J. S. Virgens Filho, E. A. Giglioti, and C. V. Godoy, "SASM - Agri: Sistema para anßlise e separaτπo de mΘdias em experimentos agrφcolas pelos mΘtodos ScoftKnott, Tukey e Duncan," *Revista Brasileira de Agrocomputaτπo*, vol. 1, no. 2, pp. 18–24, 2001.

[80] S. A. Omer, S. E. I. Adam, and O. B. Mohammed, "Antimicrobial activity of commiphora myrrha against some bacteria and candida albicans isolated from gazelles at king khalid wildlife research centre," *Research Journal of Medicinal Plant*, vol. 5, no. 1, pp. 65–71, 2011.

[81] M. S. Alhussaini, A. M. Saadabi, and M. I. Alghonaim, "An evaluation of the antimicrobial activity of commiphora myrrha nees (Engl.) oleo-gum resins from Saudi Arabia," *Journal of Medical Sciences (Faisalabad)*, vol. 15, no. 4, pp. 198–203, 2015.

[82] E. M. Abdallah, A. S. Khalid, and N. Ibrahim, "Antibacterial activity of oleo-gum resins of *Commiphora molmol* and *Boswellia papyrifera* against methicillin resistant *Staphylococcus aureus* (MRSA)," *AnnalidiChimica*, vol. 97, no. 9, pp. 837–844, 2009.

[83] E. M. Abdalah and A. E. Khalid, "A preliminary evaluation of the antibacterial effects of commiphora molmol and boswellia papyrifera oleo-gum resins vapor," *International Journal of Chemical and Biochemical Sciences*, vol. 1, pp. 1–15.

[84] S. de Rapper, S. F. Van Vuuren, G. P. P. Kamatou, A. M. Viljoen, and E. Dagne, "The additive and synergistic antimicrobial effects of select frankincense and myrrh oils—a combination from the pharaonic pharmacopoeia," *Letters in Applied Microbiology*, vol. 54, no. 4, pp. 352–358, 2012.

[85] R. Dahni, G. K. Eri, A. V. S. Gita Samira et al., "Anti-microbial screening of some novel benzoin derivatives," *International Journal of Advances in Pharmaceutical Research*, vol. 2, no. 12, pp. 639–642.

[86] S. De Rapper, G. Kamatou, A. Viljoen, and S. van Vuuren, "The *in vitro* antimicrobial activity of *Lavandula angustifolia* essential oil in combination with other aroma-therapeutic oils," *Evidence-based Complementary and Alternative Medicine*, vol. 2013, Article ID 852049, 10 pages, 2013.

[87] D. Shen, L. Fang, X. Chen, and Y. Tang, "Structure and properties of polyacrylic acid modified hydroxyapatite/liquid crystal polymer composite," *Journal of Reinforced Plastics and Composites*, vol. 30, no. 13, pp. 1155–1163, 2011.

[88] R. Baumgartner, A. Eitzlmayr, N. Matsko, C. Tetyczka, J. Khinast, and E. Roblegg, "Nano-extrusion: A promising tool for continuous manufacturing of solid nano-formulations," *International Journal of Pharmaceutics*, vol. 477, no. 1, pp. 1–11, 2014.

[89] T. Takayama, M. Todo, and A. Takano, "The effect of bimodal distribution on the mechanical properties of hydroxyapatite particle filled poly(L-lactide) composites," *Journal of the Mechanical Behavior of Biomedical Materials*, vol. 2, no. 1, pp. 105–112, 2009.

[90] S. Mohanty and M. G. Krishna, "Proximate analysis and standardization of plant exudates: gum olibanum and gum dikamali," *International Journal of Pharmaceutical Sciences Review and Research*, vol. 24, no. 1, pp. 172–176, 2014.

[91] A. M. Atta, H. A. Al-Lohedan, and S. A. Al-Hussain, "Synthesis of stabilized myrrh-capped hydrocolloidal magnetite nanoparticles," *Molecules*, vol. 19, no. 8, pp. 11263–11278, 2014.

[92] J. Du, H. Singh, and T.-H. Yi, "Antibacterial, anti-biofilm and anticancer potentials of green synthesized silver nanoparticles using benzoin gum (Styrax benzoin) extract," *Bioprocess and Biosystems Engineering*, vol. 39, no. 12, pp. 1923–1931, 2016.

[93] Z. He, W. Que, J. Chen, Y. He, and G. Wang, "Surface chemical analysis on the carbon-doped mesoporous TiO2 photocatalysts after post-thermal treatment: XPS and FTIR characterization," *Journal of Physics and Chemistry of Solids*, vol. 74, no. 7, pp. 924–928, 2013.

[94] J. Hu, Y. Bandog, J. Zhan, C. Zhi, and D. Golberg, "Carbon nanotubes as nanoreactors for fabrication of single-crystalline Mg3N2 nanowires," *Nano Letters*, vol. 6, no. 6, pp. 1136–1140, 2006.

[95] D. Cao, N. Nasori, Z. Wang et al., "Facile surface treatment on Cu2O photocathodes for enhancing the photoelectrochemical response," *Applied Catalysis B: Environmental*, vol. 198, pp. 398–403, 2016.

[96] M. Balaji, J. Chandrasekaran, and M. Raja, "Role of substrate temperature on MoO3 thin films by the JNS pyrolysis technique for P-N junction diode application," *Materials Science in Semiconductor Processing*, vol. 43, pp. 104–113, 2016.

[97] A. J. Kora, R. B. Sashidhar, and J. Arunachalam, "Aqueous extract of gum olibanum (*Boswellia serrata*): a reductant and stabilizer for the biosynthesis of antibacterial silver nanoparticles," *Process Biochemistry*, vol. 47, no. 10, pp. 1516–1520, 2012.

[98] I. M. El-Sherbiny, E. Salih, and F. M. Reicha, "Green synthesis of densely dispersed and stable silver nanoparticles using myrrh extract and evaluation of their antibacterial activity," *Journal of Nanostructure in Chemistry*, vol. 3, no. 8, pp. 1–7, 2013.

[99] M. Hovaneissian, P. Archier, C. Mathe, G. Culioli, and C. Vieillescazes, "Analytical investigation of styrax and benzoin balsams by HPLC-PAD-fluorimetry and GC-MS," *Phytochemical Analysis*, vol. 19, no. 4, pp. 301–310, 2008.

[100] Y. Man, L. Mu, Y. Wang, S. Lin, G. L. Rempel, and Q. Pan, "Synthesis and characterization of rutile titanium dioxide/polyacrylate nanocomposites for applications in ultraviolet light-shielding materials," *Polymer Composites*, vol. 36, no. 1, pp. 8–16, 2015.

[101] J. P. Thielemann, T. Ressler, A. Walter, G. Tzolova-Müller, and C. Hess, "Structure of molybdenum oxide supported on silica SBA-15 studied by Raman, UV-Vis and X-ray absorption spectroscopy," *Applied Catalysis A: General*, vol. 399, no. 1-2, pp. 28–34, 2011.

[102] M. S. M. Suan, M. R. Johan, N. L. Hawari, and H. A. Ching, "Annealing effects on the properties of copper oxide thin films prepared by chemical deposition," *International Journal of Electrochemical Science*, vol. 6, no. 12, pp. 6094–6104, 2011.

[103] G. Varuguese, V. Rini, S. P. Suraj, and K. T. Usha, "Characterization and optical studies of copper oxide nanostructures doped with lanthanum ions," *Advances in Materials Science*, vol. 14, no. 4, pp. 49–60, 2014.

[104] J.-B. Chen, Q. Zhou, and S.-Q. Sun, "Direct chemical characterization of natural wood resins by temperature-resolved and space-resolved Fourier transform infrared spectroscopy," *Journal of Molecular Structure*, vol. 1115, pp. 55–62, 2016.

[105] H. G. M. Edwards and M. J. Falk, *Spectrochimica Acta-Part A: Molecular and Biomolecular Spectroscopy*, vol. 53, Elsevier, Amsterdam, Netherlands, 1997.

[106] M. Xue, X. Zhang, Z.-F. Wu, H. Wang, X. Ding, and X.-Y. Tian, "Preparation and flame retardancy of polyurethane/POSS nanocomposites," *Chinese Journal of Chemical Physics*, vol. 26, no. 4, pp. 445–450, 2013.

[107] F. Jiang, X. Wang, and D. Wu, "Design and synthesis of magnetic microcapsules based on n-eicosane core and Fe3O4/SiO2 hybrid shell for dual-functional phase change materials," *Applied Energy*, vol. 134, pp. 456–468, 2014.

[108] S. N. Kharat and V. D. Mendhulkar, ""synthesis, characterization and studies on antioxidant activity of silver nanoparticles using Elephantopus scaber leaf extract"," *Materials Science and Engineering C*, vol. 62, pp. 719–724, 2016.

[109] C. M. Baicea, V. I. Luntraru, D. I. Vaireanu, E. Vasile, and R. Trusca, "Composite membranes with poly(ether ether ketone) as support and polyaniline like structure, with potential applications in fuel cells," *Central European Journal of Chemistry*, vol. 11, no. 3, pp. 438–445, 2013.

[110] B. Guan, P. A. Latif, and T. Yap, "Physical preparation of activated carbon from sugarcane bagasse and corn husk and its physical and chemical characteristics," *Int. J. Eng. Res. Sci. Technol*, vol. 2, pp. 1–14, 2013.

[111] H. Yang, R. Yan, H. Chen, D. H. Lee, and C. Zheng, "Characteristics of hemicellulose, cellulose and lignin pyrolysis," *Fuel*, vol. 86, no. 12-13, pp. 1781–1788, 2007.

[112] M. Yilmaz, H. Turkdemir, M. A. Kilic et al., "Biosynthesis of silver nanoparticles using leaves of Stevia rebaudiana," *Materials Chemistry and Physics*, vol. 130, no. 3, pp. 1195–1202, 2011.

[113] Q. Chen and N. L. Yakovlev, "Adsorption and interaction of organosilanes on TiO2 nanoparticles," *Applied Surface Science*, vol. 257, no. 5, pp. 1395–1400, 2010.

[114] F. Cheng, S. M. Sajedin, S. M. Kelly, A. F. Lee, and A. Kornherr, "UV-stable paper coated with APTES-modified P25 TiO2 nanoparticles," *Carbohydrate Polymers*, vol. 114, pp. 246–252, 2014.

[115] A. Pendashteh, M. F. Mousavi, and M. S. Rahmanifar, "Fabrication of anchored copper oxide nanoparticles on graphene oxide nanosheets via an electrostatic coprecipitation and its application as supercapacitor," *Electrochimica Acta*, vol. 88, pp. 347–357, 2013.

[116] J.-J. Ruan, Y.-Q. Huo, and B. Hu, "Three-dimensional Ni(OH)2/Cu2O/CuO porous cluster grown on nickel foam for high performance supercapacitor," *Electrochimica Acta*, vol. 215, pp. 108–113, 2016.

[117] T. Yu, X. Zhao, Z. X. Shen, Y. H. Wu, and W. H. Su, "Investigation of individual CuO nanorods by polarized micro-Raman scattering," *Journal of Crystal Growth*, vol. 268, no. 3-4, pp. 590–595, 2004.

[118] M. A. Dar, S. H. Nam, Y. S. Kim, and W. B. Kim, "Synthesis, characterization, and electrochemical properties of self-assembled leaf-like CuO nanostructures," *Journal of Solid State Electrochemistry*, vol. 14, no. 9, pp. 1719–1726, 2010.

[119] Y. Zhu, X. Li, D. Zhang et al., "Tuning the surface charges of MoO3 by adsorption of polyethylenimine to realize the electrophoretic deposition of high-exothermic Al/MoO3 nanoenergetic films," *Materials and Design*, vol. 109, pp. 652–658, 2016.

[120] C. V. Subba Reddy, E. H. Walker Jr., C. Wen, and S.-I. Mho, "Hydrothermal synthesis of MoO3 nanobelts utilizing poly(ethylene glycol)," *Journal of Power Sources*, vol. 183, no. 1, pp. 330–333, 2008.

[121] L. Wang, X. Zhang, Y. Ma, M. Yang, and Y. Qi, "Rapid microwave-assisted hydrothermal synthesis of one-dimensional MoO3 nanobelts," *Materials Letters*, vol. 164, pp. 623–626, 2016.

[122] J.-B. Chen, S.-Q. Sun, and Q. Zhou, "Data-driven signal-resolving approaches of infrared spectra to explore the macroscopic and microscopic spatial distribution of organic and inorganic compounds in plant," *Analytical and bioanalytical chemistry*, vol. 407, no. 19, pp. 5695–5706, 2015.

[123] S. V. Canevarolo Junior, *Polimer Science*, Artliber, São Paulo, SP, Brazil, 2nd edition, 2006.

[124] G. W. Ehrenstein, *Polymericmaterials:Structure, Properties, Applications*, Hanser Gardner Publications, Cincinnat, Ohio, USA, 2001.

[125] G. Menges, E. Haberstroh, W. Michaeli, and E. Schmachtenberg, *Plastics Materials Science*, Carl Hanser Verlag, Munchen, Germany.

[126] C. E. Carraher and R. B. Seymour, *Eymour/Carraher'Spolymerchemistry*, Marcel Dekke, New York, NY, USA, 6th edition, 2003.

[127] C. Marinescu, A. Sofronia, C. Rusti et al., "DSC investigation of nanocrystalline TiO2 powder," *Journal of Thermal Analysis and Calorimetry*, vol. 103, no. 1, pp. 49–57, 2011.

[128] G. Madras, B. J. McCoy, and A. Navrotsky, "Kinetic model for TiO2 polymorphic transformation from anatase to rutile," *Journal of the American Ceramic Society*, vol. 90, no. 1, pp. 250–255, 2007.

[129] N. Topnani, S. Kushwaha, and T. Athar, "Wet synthesis of copper oxide nanopowder," *International Journal of Green Nanotechnology: Materials Science and Engineering*, vol. 1, no. 2, pp. M67–M73, 2009.

[130] Y. Wan, X. Wang, H. Sun, Y. Li, K. Zhang, and Y. Wu, "Corrosion behavior of copper at elevated temperature," *International Journal of Electrochemical Science*, vol. 7, no. 9, pp. 7902–7914, 2012.

[131] A. Lagashetty, V. Havanoor, S. Basavaraja, and A. Venkataraman, "Synthesis of MoO3 and its polyvinyl alcohol nanostructured film," *Bulletin of Materials Science*, vol. 28, no. 5, pp. 477–481, 2005.

10

The Effect of Hydrofluoric Acid Surface Treatment on the Cyclic Fatigue Resistance of K3 NiTi Instruments

Kee-Yeon Kum[1] and Seok Woo Chang[2]

[1]*Department of Conservative Dentistry, Seoul National University Dental Hospital, Seoul Dental Hospital for Disabled, Dental Research Institute, Seoul National University School of Dentistry, Seoul, Republic of Korea*
[2]*Department of Conservative Dentistry, School of Dentistry, Kyung Hee University, Seoul, Republic of Korea*

Correspondence should be addressed to Seok Woo Chang; swc2007smc@gmail.com

Academic Editor: Ian S. Butler

The aim of this study was to investigate the effect of 50% hydrofluoric acid (HF) surface treatment on the cyclic fatigue resistance (CFR) of K3 NiTi instruments. Twenty as-received and twenty HF-treated K3 NiTi instruments were compared in CFR. The surface texture and fracture surface of two instrument groups were examined with a scanning electron microscope (SEM). Additionally, any change of Ni and Ti composition from both instrument groups was investigated using energy dispersive spectrometry. The results were analyzed with t-test. The HF-treated K3 group showed statistically higher cyclic fatigue resistance than as-received K3 group ($P < 0.05$). HF-treated K3 instruments showed smoother and rounded surface compared to as-received K3 under SEM observation. The fracture surfaces of both groups showed typical patterns of cyclic fatigue fracture. There was no difference in surface Ni and Ti composition between two groups. HF treatment of K3 instruments smoothed the file surface and increased the cyclic fatigue resistance, while it had no effect on surface ion composition and the file fracture pattern.

1. Introduction

In modern endodontics, many clinicians use engine driven nickel-titanium (NiTi) rotary instruments, which makes root canal preparation more fast and effective [1]. However, accidental instrument separation is a concern for every clinician who uses NiTi rotary instruments in their practice. The two main fracture mechanisms of NiTi rotary instrument are cyclic fatigue fracture and torsional fracture [2]. Cyclic fatigue fracture has been reported to be affected by many factors such as manufacturing process [2], root canal geometry [3], instrumentation motion [4], cross-sectional configuration, rotational speed [5], and surface treatment method [6]. It is recently focused that surface defects such as microcracks and machining grooves, which were induced in machining process during manufacturing, are the stress concentration point where the initiation of file fracture begins to propagate [5, 7–9]. The grinding procedure across the grains of NiTi wire has been also suggested to deteriorate the mechanical strength of the instrument owing to the machining defects on instrument surface and the residual stress within the internal structure [10–12]. Previous studies supported this postulation in that the ground file showed much lower cyclic fatigue resistance than twisted files [5]. Therefore, various surface treatment methods such as electropolishing [13], ion implantation [14], and chemical and physical vapor deposition [15, 16] have been tried for improving mechanical properties and reducing surface defects of ground NiTi rotary instruments. These methods were reported to be effective in improving the mechanical properties of NiTi rotary instruments, but there have not been firmly established consensus about the effectiveness of such treatment for improving cyclic fatigue resistance of NiTi rotary instruments, yet.

Acid treatment is widely used method during manufacturing process to chemically polish the surface of ground NiTi instruments. Hydrofluoric acid (HF), which is the ingredients of Kroll's reagent, is used for chemical polishing of ground NiTi alloy [17]. In spite of the surface treatment during manufacturing, previous studies clearly showed that the as-received ground NiTi rotary instruments have rough surface which include rolling marks, microcracks, and pitting under a scanning electron microscope (SEM) [5, 18]. In addition, our preliminary study showed that chemical polishing with higher concentrations of HF produced smoother surface of NiTi instruments compared with that of as-received NiTi instruments (unpublished data).

Therefore, in this study, we compared the cyclic fatigue resistances of K3 NiTi rotary instruments before and after surface treatment with HF. In addition, to visually examine any change induced in the instrument surface, the SEM images for as-received and HF-treated K3 instruments were investigated. Furthermore, EDS (energy dispersive spectrometry) analysis was carried out to investigate any change in surface ion composition of K3 files by HF surface treatment. Fractographic analysis of fracture surface was also performed under SEM.

2. Materials and Methods

2.1. Treatment of K3 NiTi Rotary Instruments with Hydrofluoric Acid. The NiTi rotary instruments used in this study were K3 with #30/06 taper (SybronEndo, Orange, CA). Twenty as-received K3 instruments were soaked in 50% HF for 2 minutes and washed with distilled water. Twenty as-received K3 instruments were included as a control group.

2.2. Cyclic Fatigue Resistance Test. The cyclic fatigue resistance was measured with 20 as-received K3 instruments and 20 HF-treated K3 instruments. All the instruments had been previously examined using an OPMI dental surgical microscope (Zeiss, Oberkochen, Germany) for any sign of surface defect or deterioration. A fatigue tester (Denbotix, Bucheon, Korea) was used in this study. This equipment was designed to provide cyclic tensile and compressive stress on the tip area of the instrument, so that it simulates the conditions encountered in clinical situations [5] (Figure 1).

A stainless-steel artificial canal with 1.5 mm inner diameter, 8 mm length (straight portion), 60° angle of curvature, and 5 mm radius of curvature was used in this experiment. To simulate the up and down pecking motion, continuous 6 mm up and down motion was applied at 0.5 cycles per second. Each NiTi instrument was engaged on an electric torque-controlled motor (Aseptico, Woodinville, WA) with a 20 : 1 reduction handpiece and rotated at the speed recommended by manufacturer (300 rpm). During experiment, the artificial canal was filled with RC-prep (Premier Dental Products, Norristown, PA, USA) to minimize the friction and heat generation in artificial canal. When an instrument fractures in the artificial canal, the internal sensor detects the torque change on the canal wall and the electric motor automatically stops. Thereafter, the experimenter visually confirmed instrument fracture. The time elapsed until the instrument fracture was automatically recorded by a digital timer accurate to 0.1 seconds. The time to fracture was multiplied by the number of rotations per minute (rpm), to calculate the number of cycles to failure (NCF) for each NiTi instrument. Using the SPSS statistical package (Version 12, SPSS, Chicago, IL), the NCF of as-received and HF-treated samples were analyzed by one-way analysis of the variance (ANOVA) and Tukey post hoc test. P value less than 0.05 was considered to be statistically significant.

TABLE 1: The results of fatigue resistance test of HF-treated and as-received K3 instruments.

NCF	K3 ($n = 20$)
As-received	79825.5 (±14472.35)[a]
HF-treated	90835.5 (±13170.87)[b]

NCF: number of cycles to fracture; HF: hydrofluoric acid. Different alphabetical letter represents statistically significant difference.

TABLE 2: Ni and Ti composition in as-received and HF-treated samples in K3 (atomic%).

Weight percent	Ni	Ti
As-received ($n = 3$)	55.07 ± 2.50	44.93 ± 2.50
HF-treated ($n = 3$)	54.14 ± 2.70	45.86 ± 2.70

2.3. SEM Examination of As-Received and HF-Treated Surfaces of K3 Instruments. To investigate the effect of HF on the surface texture of K3 instruments, as-received and HF-treated instruments were examined with SEM (Hitachi S-4700, Tokyo, Japan) at a magnification of ×200 and ×10,000.

2.4. Comparison of Ni and Ti Composition in As-Received and HF-Treated K3 Files. To investigate the effect of HF on the change of Ni and Ti ion composition of K3 instrument, energy dispersive spectrometry (EDS) was carried out in as-received and HF-treated K3 instruments ($n = 3$).

2.5. Examination of Fracture Surfaces. After the cyclic fatigue fracture, the fracture surfaces were examined by SEM. Fractographic analysis of all fractured instruments was performed by SEM (Hitachi S-4700, Tokyo, Japan) at magnifications of ×5,000 to determine the modes or patterns of fracture.

3. Results

3.1. Cyclic Fatigue Resistance Test. The results of cyclic fatigue resistance test were shown in Table 1. The cyclic fatigue resistance in HF-treated K3 files was significantly higher than that of as-received K3 files ($P < 0.05$).

3.2. SEM Examination of Surfaces in As-Received and HF-Treated K3 Files. SEM images of as-received K3 instruments clearly showed rough surfaces with metal roll-overs and machining marks (Figures 2(a) and 2(c)). On the contrary, the HF-treated K3 showed smoothening of rough surfaces (Figures 2(b) and 2(d)). It was shown that the flute surface of K3 instrument was smoothened and the sharp edges of milling mark were rounded.

3.3. Comparison of Ni and Ti Composition in As-Received and HF-Treated K3 Files. As-received and HF-treated K3 file had no difference in NiTi composition. The results were illustrated in Table 2 and Figure 3.

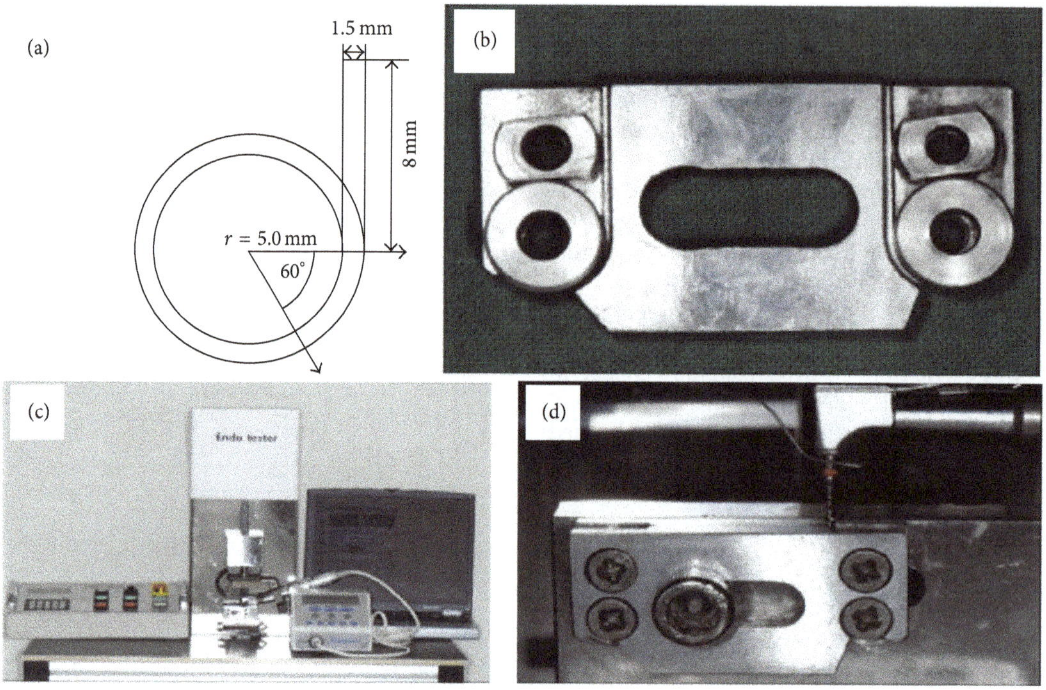

FIGURE 1: Cyclic fatigue resistance measuring equipment and diagram of an artificial canal. (a) Schematic diagram of artificial canal. (b) Metal blocks constituting artificial canal. (c) Setup of cyclic fatigue resistance measuring equipment. (d) Working NiTi instrument inside the artificial canal during experiment.

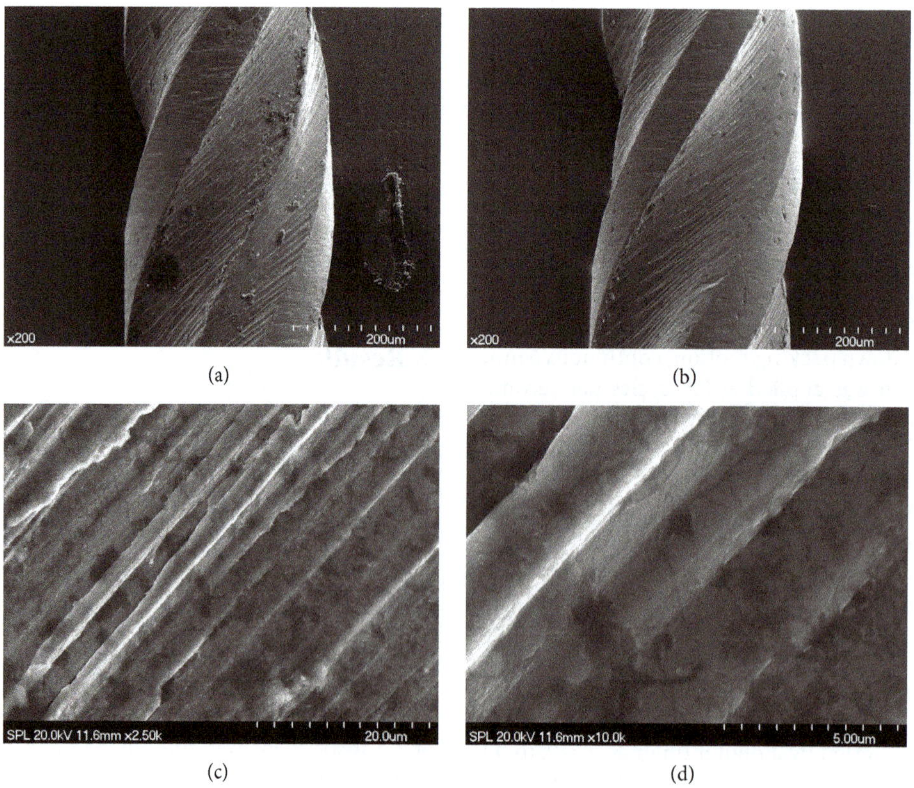

FIGURE 2: SEM images of the surfaces of as-received and HF-treated K3 instruments. As-received K3 files (a, c) revealed rough surfaces with metal roll-overs and machining marks. On the contrary, the HF-treated K3 (b, d) showed smoothening of rough surfaces. (a) As-received K3 instrument (×200). (b) HF-treated K3 instrument (×200). (c) As-received K3 instrument (×2,500). (d) HF-treated K3 instrument (×10,000). SEM: scanning electron microscope; HF: hydrofluoric acid.

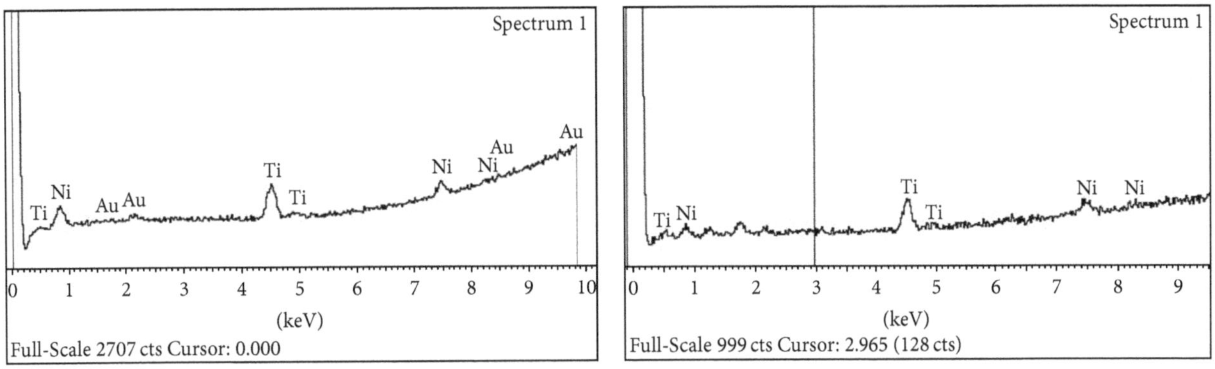

FIGURE 3: EDS analysis of as-received K3 instrument (a) and HF-treated K3 instrument (b). As-received and HF-treated K3 instruments showed no difference in ion composition.

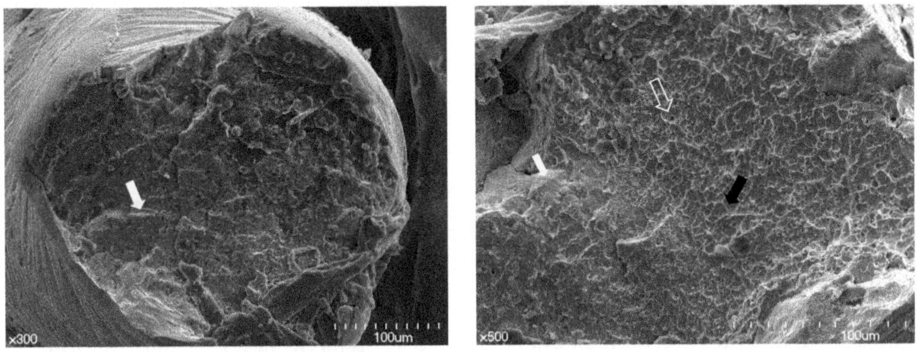

FIGURE 4: Fracture surface of as-received (a) and HF-treated (b) K3 instruments. Both surfaces revealed characteristic patterns of cyclic fatigue fractures with multiple lines (white arrows in (a, b)) and cone and cup appearance (black arrows in (b)). Pore-like structures were also shown (unfilled white arrow in (b)). The multiple lines pointed to by white arrow show the initiation site for the fatigue fracture process.

3.4. Examination of Fracture Surface. Figure 4(a) (as-received K3 instrument) and Figure 4(b) (HF-treated K3 instrument) show the fracture surface of each group of K3 NiTi instruments. Figure 4(a) shows the multiple lines (white arrow) and cone and cup appearance at the same time. Figure 4(b) also shows the multiple lines (white arrow), dimple (black arrow) appearance, and pore-like structures (unfilled white arrow). Multiple lines correspond to the advance of the crack front in successive cycles.

4. Discussion

There are several kinds of surface treatments to improve surface characteristics of ground NiTi rotary instrument. Among these, electropolishing has been relatively much investigated. However, there is a still controversy regarding the effectiveness of electropolishing on cyclic fatigue resistance of NiTi rotary files. Some studies [19–21] demonstrated the positive effect of electropolishing on the fatigue resistance of NiTi rotary files, while other studies [22] did not. This was similar to the case of ion implantation. Gavini et al. [23] reported that the nitrogen ion implantation was effective in increasing the fatigue resistance of NiTi rotary instrument. On the other hand, Wolle et al. [14] reported that argon implantation reduced the fatigue resistance of NiTi rotary instrument.

However, until now, no study has been published on the effect of simple treatment of strong acid on the cyclic fatigue resistance of NiTi rotary instruments. This is the first study that investigated the effect of concentrated HF treatment on the surface smoothness and the fatigue resistance of NiTi rotary instrument.

HF is a component of Kroll's reagent, which is used for chemical polishing of manufactured NiTi rotary instruments [18]. In this experiment, higher concentration (50%) than that of HF in Kroll's reagent was used and two main actions were postulated with this acid treatment. Firstly, as was supported by SEM examination results obtained with K3 instruments, HF surface treatment reduced the rough surface and stress concentrating microdefect, which might improve the cyclic fatigue resistance in K3 instruments. This is in agreement with result of Karn et al. [24]. Secondly, as was suggested by a previous study [25], HF surface treatment might anodize and passivate the titanium oxide layer which was produced during manufacturing, which might also improve cyclic fatigue resistance. The smoothening of rough surfaces of NiTi rotary files could give additional benefit which is related to biocompatibility. The rougher surface and larger surface area tend to have more corrosion than smoother and smaller surface area. The surface corrosion of NiTi alloy could release nickel ion that can evoke allergic reactions [26] in contacting tissues. In this sense, HF treatment which resulted in smoother surface in this experiment is postulated to contribute to increasing the cyclic fatigue resistance of NiTi files as well as lowering the corrosion tendency of them. In this sense, future investigations on the corrosion resistance of HF-treated NiTi instruments are needed.

"Kroll's reagent," the mixture of HF and nitric acid in water, has been widely used as chemical etching agent for alloys used for hip and knee prosthesis [27] and dental implant [28, 29]. "Kroll's reagent" is listed as ASTM (American Society of Testing Materials) #187 etchant which was recommended to be used for microchemical etching of cast irons. Usually, the relative composition of HF in Kroll's solution is approximately 5%. However, in this experiment, 10 times higher concentrations of HF were used for chemical etching, because our preliminary study showed that high concentrations of HF were effective in increasing the fatigue resistance of K3 NiTi files (unpublished data).

The composition of Ni and Ti in NiTi rotary files has been known to be 55 : 45 and the composition of Ni and Ti has been known to affect the mechanical properties of NiTi alloy [30]. The present study showed that the HF treatment did not affect the Ni and Ti composition. Further studies regarding the effect of various surface treatments or corrosion of NiTi rotary instruments on the Ni and Ti composition would be necessary.

This study has several limitations, one of which was that the minute surface changes (e.g., formation of surface oxide) could not be thoroughly investigated; the other is that the investigation of minute surface changes in instruments could be better investigated with more precise devices. A surface sensitive technique (e.g., X-ray photoelectron spectroscopy and Auger Spectroscopy) would be more useful to understand the chemical changes of the surface after HF etching (e.g., changes of Ni/Ti ratio, increase of oxygen concentration, and formation of passivation layers). So the future study using more precise equipment than EDS to investigate exact Ni and Ti ratio might be necessary. Future studies regarding the effect of HF on cutting efficiency might be also necessary. Moreover, future studies on the effect of HF concentration and duration of the treatment on surface characteristics and resistance to fatigue fracture might be helpful to better understand the role of HF treatment.

In conclusion, this study clearly demonstrated that HF surface treatment could be beneficial in improving the surface smoothness and thus increasing the cyclic fatigue resistance of ground NiTi rotary files.

References

[1] C. R. Glosson, R. H. Haller, S. Brent Dove, and C. E. del Rio, "A comparison of root canal preparations using Ni-Ti hand, Ni-Ti engine-driven, and K-Flex endodontic instruments," *Journal of Endodontics*, vol. 21, no. 3, pp. 146–151, 1995.

[2] S.-Y. Park, G. S. P. Cheung, J. Yum, B. Hur, J.-K. Park, and H.-C. Kim, "Dynamic torsional resistance of nickel-titanium rotary instruments," *Journal of Endodontics*, vol. 36, no. 7, pp. 1200–1204, 2010.

[3] P. M. Di Fiore, "A dozen ways to prevent nickel-titanium rotary instrument fracture," *The Journal of the American Dental Association*, vol. 138, no. 2, pp. 196–249, 2007.

[4] M.-C. Avoaka and Y. Haïkel, "Influence of axial movement on fatigue of ProFile® Ni-Ti rotary instruments: an in vitro evaluation," *Bosnian Journal of Basic Medical Sciences*, vol. 10, no. 2, pp. 107–111, 2010.

[5] S.-R. Oh, S.-W. Chang, Y. Lee et al., "A comparison of nickel-titanium rotary instruments manufactured using different methods and cross-sectional areas: ability to resist cyclic fatigue," *Oral Surgery, Oral Medicine, Oral Pathology, Oral Radiology and Endodontology*, vol. 109, no. 4, pp. 622–628, 2010.

[6] C. Praisarnti, J. W. W. Chang, and G. S. P. Cheung, "Electropolishing enhances the resistance of nickel-titanium rotary files to corrosion-fatigue failure in hypochlorite," *Journal of Endodontics*, vol. 36, no. 8, pp. 1354–1357, 2010.

[7] S. B. Alapati, W. A. Brantley, T. A. Svec, J. M. Powers, and J. C. Mitchell, "Scanning electron microscope observations of new and used nickel-titanium rotary files," *Journal of Endodontics*, vol. 29, no. 10, pp. 667–669, 2003.

[8] G. Kuhn, B. Tavernier, and L. Jordan, "Influence of structure on nickel-titanium endodontic instruments failure," *Journal of Endodontics*, vol. 27, no. 8, pp. 516–520, 2001.

[9] U. Inan and N. Gonulol, "Deformation and fracture of mtwo rotary nickel-titanium instruments after clinical use," *Journal of Endodontics*, vol. 35, no. 10, pp. 1396–1399, 2009.

[10] J. H. Yao, S. A. Schwartz, and T. J. Beeson, "Cyclic fatigue of three types of rotary nickel-titanium files in a dynamic model," *Journal of Endodontics*, vol. 32, no. 1, pp. 55–57, 2006.

[11] T. R. Tripi, A. Bonaccorso, and G. G. Condorelli, "Cyclic fatigue of different nickel-titanium endodontic rotary instruments," *Oral Surgery, Oral Medicine, Oral Pathology, Oral Radiology and Endodontology*, vol. 102, no. 4, pp. e106–e114, 2006.

[12] G. Gambarini, "Cyclic fatigue of ProFile rotary instruments after prolonged clinical use," *International Endodontic Journal*, vol. 34, no. 5, pp. 386–389, 2001.

[13] W. Callister, *Materials Science and Engineering. An Introduction*, JohnWilley Sons, 2000.

[14] C. F. B. Wolle, M. A. Z. Vasconcellos, R. Hinrichs, A. N. Becker, and F. B. Barletta, "The Effect of Argon and Nitrogen Ion Implantation on Nickel-Titanium Rotary Instruments," *Journal of Endodontics*, vol. 35, no. 11, pp. 1558–1562, 2009.

[15] T. R. Tripi, A. Bonaccorso, and G. G. Condorelli, "Fabrication of hard coatings on NiTi instruments," *Journal of Endodontics*, vol. 29, no. 2, pp. 132–134, 2003.

[16] T. R. Tripi, A. Bonaccorso, E. Rapisarda et al., "Depositions of nitrogen on NiTi instruments," *Journal of Endodontics*, vol. 28, no. 7, pp. 497–500, 2002.

[17] C. L. Chu, C. Y. Chung, Y. P. Pu, and P. H. Lin, "Graded surface structure in chemically polished NiTi shape memory alloy after NaOH treatment," *Scripta Materialia*, vol. 52, no. 11, pp. 1117–1121, 2005.

[18] M. H. Wong, F. T. Cheng, and H. C. Man, "Characteristics, apatite-forming ability and corrosion resistance of NiTi surface modified by AC anodization," *Applied Surface Science*, vol. 253, no. 18, pp. 7527–7534, 2007.

[19] M. E. Anderson, J. W. H. Price, and P. Parashos, "Fracture resistance of electropolished rotary nickel-titanium endodontic instruments," *Journal of Endodontics*, vol. 33, no. 10, pp. 1212–1216, 2007.

[20] A. A. Sinan, S. E. Thiémélé-Yacé, E. Abouattier-Mansilla, K. Vallaeys, and F. Diemer, "Effects of electropolishing on rupture resistance in cyclic fatigue and on the surface condition of endodontic instruments," *Odonto-stomatologie tropicale = Tropical dental journal*, vol. 33, no. 129, pp. 23–28, 2010.

[21] H. P. Lopes, C. N. Elias, V. T. L. Vieira et al., "Odontostomatologie tropicale," *Journal of Endodontics*, vol. 36, no. 10, pp. 1653–1657, 2010.

[22] T. B. Bui, J. C. Mitchell, and J. C. Baumgartner, "Effect of electropolishing profile nickel-titanium rotary instruments on cyclic fatigue resistance, torsional resistance, and cutting efficiency," *Journal of Endodontics*, vol. 34, no. 2, pp. 190–193, 2008.

[23] G. Gavini, O. F. Pessoa, F. B. Barletta, M. A. Z. Vasconcellos, and C. L. Caldeira, "Cyclic fatigue resistance of rotary nickel-titanium instruments submitted to nitrogen ion implantation," *Journal of Endodontics*, vol. 36, no. 7, pp. 1183–1186, 2010.

[24] T. A. Karn, F. C. Kelly, F. C. Eichmiller, and L. S. W. Spångberg, "Fracture behavior of Ni-Ti rotary endodontic instruments," *Journal of Dental Research*, vol. 83, no. 1, supplement 2684, 2004.

[25] F. T. Cheng, P. Shi, G. K. H. Pang, M. H. Wong, and H. C. Man, "Microstructural characterization of oxide film formed on NiTi by anodization in acetic acid," *Journal of Alloys and Compounds*, vol. 438, no. 1-2, pp. 238–242, 2007.

[26] M. L. Walsh, V. H. Smith, and C. M. King, "Type 1 and type IV hypersensitivity to nickel," *Australasian Journal of Dermatology*, vol. 51, no. 4, pp. 285–286, 2010.

[27] V. Amigó, M. D. Salvador, F. Romero, C. Solves, and J. F. Moreno, "Microstructural evolution of Ti-6Al-4V during the sintering of microspheres of Ti for orthopedic implants," *Journal of Materials Processing Technology*, vol. 141, no. 1, pp. 117–122, 2003.

[28] J. M. Martínez González, F. García Sabán, J. Ferrándiz Bernal, J. C. Gonzalo Lafuente, J. Cano Sánchez, and C. Barona Dorado, "Removal torque and physico-chemical characteristics of dental implants etched with hydrofluoric and nitric acid. An experimental study in Beagle dogs," *Medicina Oral, Patologia Oral y Cirugia Bucal*, vol. 11, no. 3, pp. E281–E285, 2006.

[29] R. d. Cravero Marta and I. J. Carlos, "Assessing double acid-etched implants submitted to orthodontic forces and used as prosthetic anchorages in partially edentulous patients," *The Open Dentistry Journal*, vol. 2, no. 1, pp. 30–37, 2008.

[30] K. Mehrabi, H. Bahmanpour, A. Shokuhfar, and A. Kneissl, "Influence of chemical composition and manufacturing conditions on properties of NiTi shape memory alloys," *Materials Science and Engineering A*, vol. 481-482, no. 1-2 C, pp. 693–696, 2008.

11

Biomimetic Approach to CO_2 Reduction

Ilaria Gamba

Universidad de La Laguna, San Cristóbal de La Laguna, Santa Cruz de Tenerife, Spain

Correspondence should be addressed to Ilaria Gamba; ilaria.gamba@gmail.com

Academic Editor: Spyros P. Perlepes

The development of artificial photosynthetic technologies able to produce solar-fuels from CO_2 reduction is a fundamental task that requires the employment of specific catalysts being accomplished. Besides, effective catalysts are also demanded to capture atmospheric CO_2, mitigating the effects of its constantly increasing emission. Biomimetic transition metal complexes are considered ideal platforms to develop efficient and selective catalysts to be implemented in electrocatalytic and photocatalytic devices. These catalysts, designed according to the inspiration provided by nature, are simple synthetic molecular systems capable of mimic features of the enzymatic activity. The present review aims to focus the attention on the mechanistic and structural aspects highlighted to be necessary to promote a proper catalytic activity. The determination of these characteristics is of interest both for clarifying aspects of the catalytic cycle of natural enzymes that are still unknown and for developing synthetic molecular catalysts that can readily be applied to artificial photosynthetic devices.

1. Introduction

The employment of devices based on artificial photosynthetic technologies may be the most convincing strategy for generating enough clean energy to satisfy the needs of the expanding world [1, 2]. However, the development of such devices is not an easy task. Different fundamental components, each one committed to a specific function, must be matched together to mimic the ultimate functionality of the natural photosynthetic apparatus [3]. Thus, inspired by natural photosynthesis, the crucial components of an artificial photosynthetic device should be a light harvester for the absorption of sunlight and the generation of charge-separation (e.g., semiconductor or molecular dye) and a reduction active reaction site and an oxidation active reaction site, where the redox processes occur [4]. Such a design is represented by the so called "hybrid photocatalytic" devices, in which semiconductors are employed as light harvesters and transition metal complexes as molecular catalysts to assist specific redox reactions: the carbon dioxide reduction and the water oxidation [5–7]. Indeed, the setting exploited for hybrid photocatalytic devices presents significant advantages. The use of semiconductors as light harvesters, in fact, ensures a broad spectral absorption associated with a good robustness of the materials. Also, the involvement of specific transition metal complexes as catalysts assists and facilitates both the occurrence of redox reactions and the generation of charge separation through fast charge transfers between the semiconductor and the molecular complexes. In particular, the coupling between semiconductors and biomimetic complexes (i.e., hydrogenase, formate dehydrogenases, and carbon monoxide dehydrogenases mimics) may reveal to be an efficient strategy for aiding the CO_2 reduction, improving the current solar fuel technologies and stimulating the applications of these devices.

Besides the development of artificial photosynthetic devices, aimed to stimulate "solar fuels" production, the investigation focused on the activation and conversion of CO_2 is also of main interest to improve the strategies nowadays used for the treatment of its constantly increasing emissions [8]. Indeed, the rising of CO_2 emissions has led to serious greenhouse effects and global warming, facing the necessity to adopt CO_2 conversion technologies, able to transform this greenhouse gas into value-added products, to ensure an equilibrate global carbon cycle, and to mitigate the global warming through the reduction of atmospheric CO_2 [9].

However, the CO_2 activation, reduction, or conversion is not easy to accomplish. CO_2 is a thermodynamically stable molecule and kinetically inert. Indeed, the electrocatalytic CO_2 reduction demands high over-potentials to occur and results in poor selectivity of product formation (mixtures of CO and formate are generally obtained). Furthermore, CO_2 reduction is typically characterized by low faradaic efficiency due to the competition with the hydrogen evolution reaction (HER), which takes place at the same range of potentials [10]. In this context, the development of novel, specific, and efficient catalysts to perform CO_2 activation, reduction, or conversion is nowadays a major challenge to be faced. Transition metal compounds can be used as ideal platform to realize these tasks, both considering homogeneous and heterogeneous (thus, associated to an electrode) applications [11–14].

The present review aims to focus the attention on the progresses achieved employing transition metal complexes, designed according to the inspiration devoted to the bio-inorganic approach, as catalysts to promote carbon dioxide (CO_2) two-electron reduction. Furthermore, the active site structures and the mechanistic features associated with the CO_2 reduction activities of natural enzymatic systems (in particular carbon monoxide dehydrogenases, CODH, and formate dehydrogenases, FDH, and enzymes) are herein described, with the aim of highlighting fundamental principles of their efficient catalytic activity. These aspects are of main importance to inspire the design of simple synthetic molecular systems able to reproduce the enzymes functionality.

Moreover, the development of electrocatalysts, able to perform the two-electron reduction of CO_2 to CO and/or formate, closely related to the enzymatic activity, are here listed. Particular attention is paid to the description of proposed reaction mechanisms and the general structural features that can dictate the selectivity of the products generated from CO_2 reduction. On the other hand, photoelectrocatalytic systems, which represent attractive alternatives to develop efficient catalysts, [12, 15], were not included in the present discussion.

2. CO_2 Activation and Reduction by Enzymes and Transition Metal Complexes

Nature is widely known to be a source of efficient catalysts: the enzymes and metalloenzymes, which are natural machines that can be esteemed for performing extremely difficult chemical processes with elegant selectivity, fast rates, and low activation energies. Furthermore, metalloenzymes present the advantage of being constructed exclusively from copious bioavailable materials and typically employ only metals abundant in the Earth's crust, such as first row transition metals [16, 17].

However, when considering applications that require sustainability and scalability, enzymes have the major drawback of being quite unstable and work normally only over the limited range of conditions in which the proteins remain correctly folded. Thus, they are not suitable for use directly in all applications, especially when compared to synthetic compounds that can mimic enzymes structure and functionality. All in all, if not directly considered for applications, enzymes serve as inspiration to discover fundamental principles of fast catalysis that can be executed by simple synthetic molecules. In particular, the enzymes that have inspired the development of biomimetic synthetic catalysts for CO_2 reduction are the carbon monoxide dehydrogenases and the formate dehydrogenases [18, 19].

Key structural and mechanistic features related to these enzymes are herein summarized in the following paragraphs, included in the section dedicated to "carbon dioxide reduction enzymes."

3. Carbon Dioxide Reduction Enzymes

3.1. Carbon Monoxide Dehydrogenases (CODHs). There are two types of known carbon monoxide dehydrogenase (CODH) enzymes, which are the MoCu-CODHs and the Ni-CODHs, respectively, characterized by different metals (MoCu or Ni) contained in their active sites [18]. Both enzymes are able to catalyze the oxidation of CO to CO_2, albeit Ni-CODHs upkeep the process at much higher turnover frequencies (k_{cat} values: $4 \times 10^4\ s^{-1}$) than MoCu-CODHs ($100\ s^{-1}$). Under certain circumstances, CO_2 reduction activity has also been observed in case of Ni-CODHs (although with lower turnover frequencies compared to CO oxidation, approximately $45\ s^{-1}$) but never detected to date for MoCu-CODHs [20, 21].

Structural and mechanistic aspects, related in particular to Ni-CODHs, are of main interests also for the development of synthetic catalysts. However, the extreme sensitivity of Ni-CODHs to the presence of dioxygen (indeed, they are found only in anaerobic bacteria) made difficult mechanistic investigation and structural study. In fact, organisms growth, purification, and all manipulation related to these enzymes require definitely strictly anaerobic and often low temperature conditions [22]. Despite these difficulties, the X-ray structures of five Ni–CODHs have been solved to date [23–33], clearly showing the presence of different types of metal clusters: [$NiFe_3S_4$], called the C-clusters, and [Fe_4S_4], called B- and D-clusters [19]. The reversible CO/CO_2 conversion is believed to occur at the C-clusters [22, 34, 35], whereas the B- and D-cluster would be part of the electrons-transporting chain of the protein system. The structure of these C-clusters can be described as a distorted [Fe_4S_4] cluster with a nickel atom replacing one of the iron atoms and an iron atom pendent to the cluster, called the "dangling iron" (Figure 1). The metal atoms, which coordinate to cysteine residues and inorganic sulfide ligands are ordinated in a cubane structure, like the one described for [Fe_4S_4] clusters. The dangling iron of the [$NiFe_3S_4$] clusters is coordinated by a cysteine and a sulfide ligand of the cubane structure and by a histidine and a hydroxyl ligands (Figure 1) [28].

Four distinct oxidation states of the C-cluster were identified on the basis of spectroscopic studies: C_{ox}, C_{red1}, C_{int}, and C_{red2}. C_{ox} (Ni^{2+}, Fe^{3+}) is an oxidized state of the cluster considered inactive. C_{red1} (Ni^{2+}, Fe^{2+}) and C_{red2} (Ni^0, Fe^{2+}), which have one and three electrons more reduced than C_{ox}, respectively, are proposed to be the active states of the enzyme, directly involved in the catalytic cycle of reversible CO/CO_2 conversion. Finally, C_{int} (Ni^+, Fe^{2+}) is described as the intermediate state between C_{red1} and C_{red2}, thus, two electrons less than C_{ox}, and it seems to be an inactive state as well as C_{ox} [28].

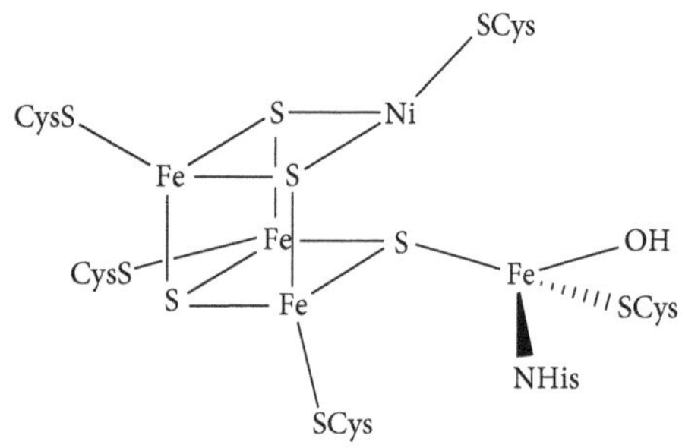

Figure 1: Molecular structure of the C-cluster contained in the Ni-CODH enzymes-active site.

The catalytic mechanism proposed, for the reversible CO/CO_2 conversion assisted by Ni–CODHs, suggests that the reaction site should be located between the Ni atom and the dangling Fe atom. Thus, during the first step of CO oxidation (as depicted in Figure 2), the Ni atom is the center responsible for the initial CO binding, when the cluster is in the C_{red1} state. Upon CO binding, the C-cluster undergoes reduction to the C_{red2} state and then transfers the electrons through the B- and D-clusters in the enzyme, generating the intermediate C_{red1}-CO state. Afterwards, the hydroxyl group, which is coordinated to the dangling Fe atom, attacks the bounded CO molecule at the carbon atom, leading to the C_{red1}-CO_2 form. At this stage, upon binding of water, a CO_2 molecule and a proton are released, reducing the C-cluster at the C_{red2}. The C-cluster starting state, C_{red1}, is recovered upon oxidation of C_{red2} mediated by the B- and D-clusters. Finally, the presence of a histidine residue located in the protein near to the C-cluster is proposed to facilitate the transfer of protons during the reaction and stabilize the intermediate steps providing hydrogen bonds to the coordinated CO and CO_2.

3.2. Formate Dehydrogenase (FDH). The metal-dependent formate dehydrogenase (FDHs) enzymes are able to reversibly catalyze the two-electron reduction of CO_2 to formate. Between different classes of known FDH enzymes, the NAD^+-independent FDHs are the ones recognized as responsible for CO_2 reduction reaction. Like the Ni-CODHs, the NAD^+-independent FDHs are also extremely sensitive to atmospheric dioxygen, and their manipulation also requires strictly anaerobic conditions. The structures of several of these enzymes were resolved (by a combination of both X-ray crystallographic and EXAFS measurements) and show the presence in their active sites of molybdenum or tungsten atoms, displaying common active sites architecture [36]. According to these data, the reaction site can be described by Mo (or W) center in a trigonal-bipyramidal geometry coordinated by two pyranopterins, one Se-cysteine residue, and one sulfido ligand (Figure 3).

The mechanism proposed for the FDH-catalyzed formate reversible oxidation is currently still under discussion. In case of Mo-FDHs, the Mo atom contained in the active site would assume during the catalytic cycle the Mo(VI), Mo (V), and Mo(IV) formal oxidation states, also supported by the contribution of the redox noninnocent ligand pyranopterin. In this scenario, the Mo(VI) state is responsible for formate oxidation and the Mo(IV) state is used for CO_2 reduction [36]. Early attempts to propose a reaction mechanism envisioned the possibility of generating of a vacant coordination site, meant for formate binding, upon Se-cysteine dissociation (Figure 4). Thus, initially formate binds at the Mo(VI) center (oxidized form) in the vacant site generated by the released of the Se-cysteine. Subsequently, the α proton of the formate is removed, probably by the nearby selenide anion. Reduction of the Mo(VI) to the Mo (IV) state is finally associated with CO_2 release. The catalytic cycle closes with the regeneration of the initial Mo(VI) state, upon two-electron oxidation of the reduced Mo(IV) form (electrons are transferred from Mo(IV) via a [4Fe–4S] cluster center to an external electron acceptor) and deprotonation of the Se atom that recover its coordination at the metal center [37].

Another suggested panorama for the formate oxidation by metal-dependent FDHs pictures the direct hydride transfer to Mo(VI), followed by hydride migration to the sulfur [38] (Figure 5, mechanism A).

More recently, Reisner and Hirst have suggested a mechanistic proposal supporting that the formate oxidation occurs through a five-membered transition state. In this scenario, formate coordinate to the Mo(VI) by its oxygen lone pairs and then, upon formate oxidation, CO_2 and Mo (IV)-SH products are generated. The five-membered transition intermediate can either undergo proton-coupled electron-transfer (PCET) reaction or as a hydride-transfer (HT) reaction (Figure 5, mechanism B) [36].

Besides these formulations that picture the dissociation of the Se-cysteine ligand, a scenario, in which the saturated coordination environment at the Mo atom is preserved, has been also considered [39, 40]. In this latter case, the direct hydride transfer of the formate α-hydrogen to the terminal sulfido group of the Mo(VI)=S moiety, without substrate binding at the metal, is proposed (Figure 5, mechanism C) [40].

4. Carbon Dioxide Reduction Mediated by Synthetic Catalysts

4.1. CO_2 Coordination Modes at Metal Centers. The transformation of carbon dioxide mediated by transition metal catalysts is a research area that is captivating the attention of the scientific community [11, 41, 42]. Diverse products, including CO, formate, methanol, and methane, can be derived from CO_2 reduction catalyzed by transition metal complexes. In this context, a recent review published by Grice details and collects the more recent achievements in the field of CO_2 reduction with homogeneous early transition metal complexes (group 3–7). The review focuses the attention on the stoichiometric and catalytic reductions to formate, methanol, and similar products, which have been more frequently observed when early transition metal complexes are employed as catalysts [42]. Besides this report, several publications recently appear highlighting the opportunities to take advantage of transition metal complexes to reduce CO_2 and to convert it in value-added products

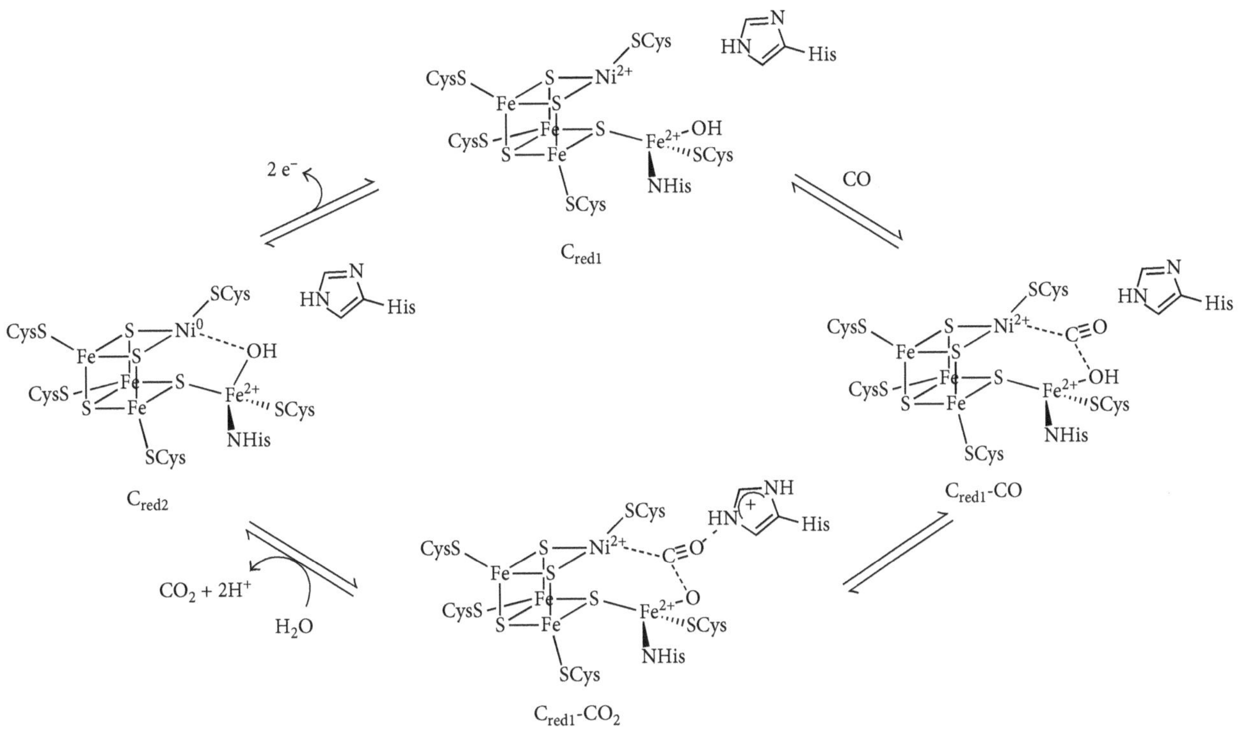

FIGURE 2: Proposed mechanism for the reversible CO oxidation mediated by Ni-CODH enzymes [18, 19, 28]. Reproduced with permission from [18].

[43–47]. For most of these transformations, the CO_2 activation and reduction is initiated by its coordination at a metal center.

The regulation of the CO_2 binding mode at metal complexes is of major importance for the development of catalysts that can selectively reduce CO_2, because it can dictate the distribution of the reaction products [45, 48, 49]. In this regard, a recent reevaluation of known CO_2-metal complex adducts displays the presence of several binding modes, allowing their complete and accurate classification, as well as the rationalization of the oxidation states assumed by carbon and metal atoms involved in the CO_2 binding [50].

The possible CO_2 metal binding modes are categorized according to the number of directly bonded metal centers and depending on the connectivity (how the atoms are attached to one another). The principal adducts derived from the simplest CO_2 monometallic complexes are summarized in Figure 6.

At a mononuclear transition metal center, CO_2 can either coordinate through the C or the O atom, where η^O exhibits a richer variety of coordination possibilities than the η^C, including η^1 and η^2-(C,O) side-on binding fashions.

The η^1-binding mode is represented by the coordination of one of the CO_2 oxygen atom at the metal center. Different types of η^1-binding modes can be distinguished depending on the kind of interaction that determines the nature of the binding. On the basis of these data, it is possible to define the existence of η^1- [1], η^1- [2], and η^1- [3] adducts. In case of η^1- [1], CO_2 datively bound through one of its oxygen atoms to the metal centers. η^1- [2] adducts are characterized by metal-O interaction of electrostatic nature, involving the positively charged metal

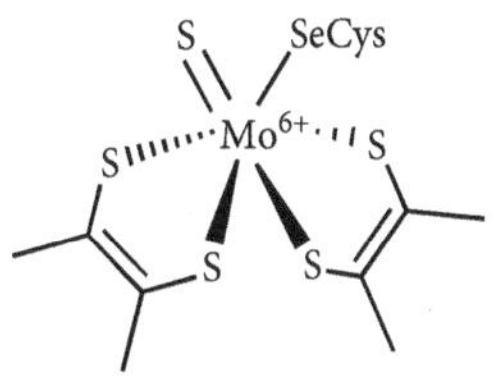

FIGURE 3: Molecular structure of Mo-dependent FDH enzymes-active site.

and the induced negative quadrupole moments at the CO_2 oxygen atom. Finally, η^1- [3] adducts represent the patterns in which the binding situation is better described by the interaction developed between the one-electron oxidize metal center and the reduced CO_2^- ligand, where the metal would have a positive charge and a negative charge would be located at the non-coordinated oxygen atom. The radical carbon would form a double bond to the datively coordinated oxygen atom (Figure 6).

Besides, the CO_2 binding mode of η^2-(C,O) side-on adducts can be described by two limiting situations: one in which CO_2 binds by π-donation to the metal, identified as η^2-(C,O) [1], and another in which the bound order of the double C=O bond is reduced as results of the π-back donation of the metal center, with concomitant formation of metal-C and metal-O bonds, recognized as η^2-(C,O) [2] (Figure 6).

Cases in which the CO_2 binding is shared between two, three, or four metal atoms result in an even wider variety of interaction types. A detailed description of these situations can be found in the recent review published by Paparo and Okuda [50].

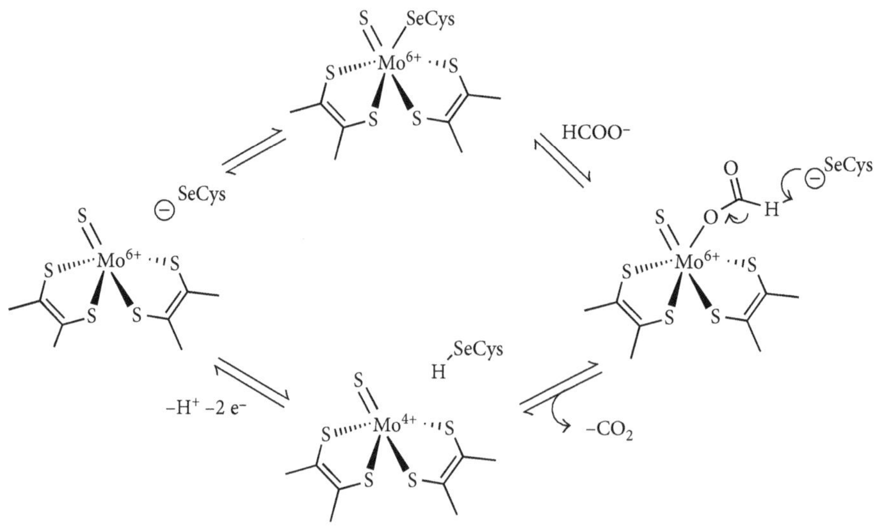

FIGURE 4: Early mechanistic proposal for the FDH-catalyzed reversible formate oxidation [36, 37]. Reproduced with permission from [37].

4.2. Structural Models of CODHs. Attempts to mimic the Ni-CODH active sites have been focused either on the preparation of monometallic Ni and heterobimetallic Ni-Fe complexes or on the construction of [NiFe3S4] clusters.

Few examples appear in the literature describing mononuclear Ni-CO_2 η^2-adducts, which present spectroscopic characteristics in close relation to the Ni center of the C-cluster found in Ni-CODHs enzymes. The first structurally characterized Ni-CO_2 η^2-adduct, $((PCy_3)_2Ni(\eta^2\text{-}CO_2))$, was reported in 1975 by Aresta and co-workers, [51, 52], followed by the more recent example presented by Hillhouse, the (dtbpe)Ni(η^2-CO_2) (dtbpe = 1,2-bis(di-tert-butylphosphino) ethane) adduct [53] (Figures 7(a) and 7(b)). Both the compounds are based on tetracoordinated nickel centers, with two phosphorus donor atoms and one η^2-CO_2 ligand.

More recently, another interesting Ni-CO_2 η^2-adduct was structurally characterized by Lee et al., the ($PP^{Me}P$)Ni (η^2-CO_2) adduct (where $PP^{Me}P$ is the PMe[2-P^iPr_2-C_6H_4]$_2$ ligand, Figure 7(c)) [54]. In this latter case, the Ni center is 5-coordinated by three phosphorus donor atoms and one CO_2 ligand in a η^2-binding mode. Besides the same group, taking advantage of the similar but anionic PNP ligand (PNP^- = -N [2-P^iPr_2-4-Me-C_6H_3]$_2$) reported the synthesis and characterization of novel nickel-carboxylate adducts. The carboxylate ligand, at the Ni center of these compounds, is stabilized by Lewis acids, such as a proton (nickel hydroxycarbonyl, (PNP)NiCOOH), sodium (nickel sodium carboxylate, (PNP) NiCOONa), or another metal ion.

Using this strategy, also a dinuclear nickel-iron carboxylate species, (PNP)Ni-μ-CO_2-κC:$\kappa^2 O,O'$-Fe(PNP), was prepared and characterized [56]. This latter complex represents the first example of nickel-iron hetero-bimetallic complex equipped with a bridging CO_2 ligand, structurally reminiscent of the NiFe-binuclear active site of CODH enzymes.

In parallel to these studies, the synthetic effort spent for mimicking the unique [$NiFe_3S_4$] core structure resulted in the biomimetic compounds reported by Holm and coworkers, which represent the closest synthetic analogues of the C-cluster found in anaerobic CODHs [55, 57–59]. The approach used satisfies fundamental structural requirements of the CODHs-active site modeling, as the incorporation of a Ni atom at the apex of an iron-sulfur cubane [57]. Also, rational synthetic modifications forced the planarization of the tetrahedral nickel center, upon breaking of one Ni-S bond, attaining square planar Ni(II) in a Ni-Fe-S cubanoid cluster, as owned by the enzyme (Figure 7(d)) [55, 58]. Unfortunately, the presence of a pending iron atom in a synthetic complex was not ever realized. Neither of these compounds, however, was found able of promoting CO_2 reduction or CO oxidation activities.

4.3. Functional Models for CODHs: Electrocatalytic CO_2 Reduction to CO Mediated by Molecular Complexes. A great research effort was dedicated to the development of molecular catalysts able to act as CODHs functional models. The transformation of CO_2 in solution and in particular its reduction to CO through the assistance the of transition metal complexes as electrocatalysts have been comprehensively studied and reviewed [42, 45, 60–65]. These compounds have no structural relation with the active site of the CODHs and cannot be considered structural mimics of these enzymes. However, metal electrocatalysts can reproduce the CODHs activity of CO_2 reduction to CO, thus resulting in fundamental interests to elucidate aspects of the catalytic mechanism of these enzymes that are still unknown.

In this context, the catalytic model systems more closely related to the CODHs activity are herein described by detailing in the following paragraphs the most representative examples of transition metal compounds based on phosphine or macrocyclic ligands.

4.3.1. Transition Metal Complexes Bearing Phosphine Ligands. Transition metal phosphine complexes have crucially contributed to the understanding of the possible intermediates involved in the CO_2 reduction reactions. In this context, palladium complexes bearing polydentate phosphine ligands, of the type

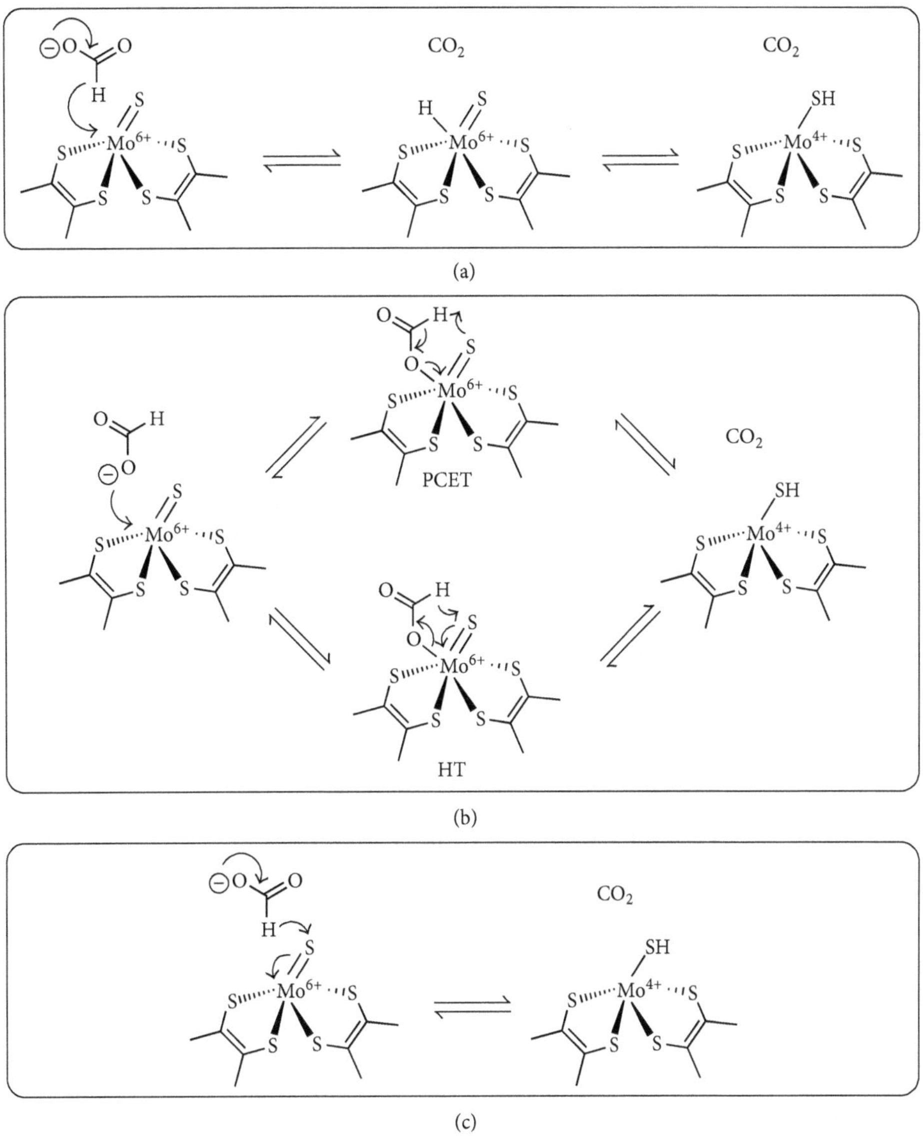

FIGURE 5: Proposed reaction mechanisms by (a) Tiberti et al. [38], (b) Robinson et al. [36], and (c) Niks et al. [40]. Reproduced with permission from [36].

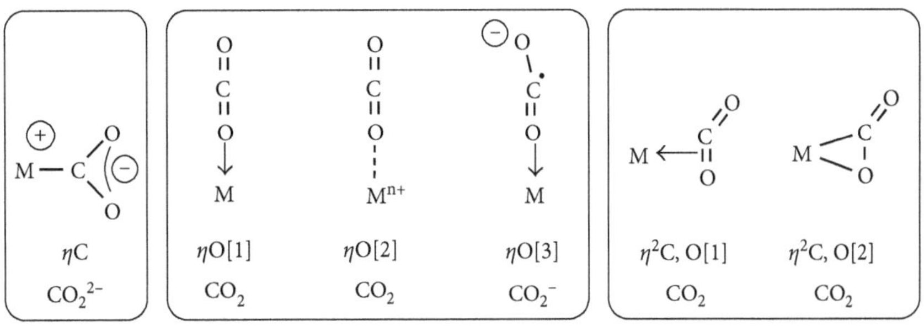

FIGURE 6: Possible binding modes of CO_2-monometallic complexes. Reproduced with permission from [50].

$[Pd(triphosphine)(solvent)]^{2+}$, represent the most extensively studied catalysts for the electrocatalytic CO_2 reduction.

The mechanism of this class of catalysts, firstly reported by DuBois and coworkers [66, 67], has been studied in detail [18, 45, 49, 61, 68, 69].

As represented in Figure 8, the proposed catalytic mechanism began with the Pd(II) **(1)** electrochemical reduction to Pd(I) **(2)**, performing the reaction in DMF and in presence of acid. Following this initial reduction, CO_2 coordinate at the Pd(I) center and the metal-carboxylate

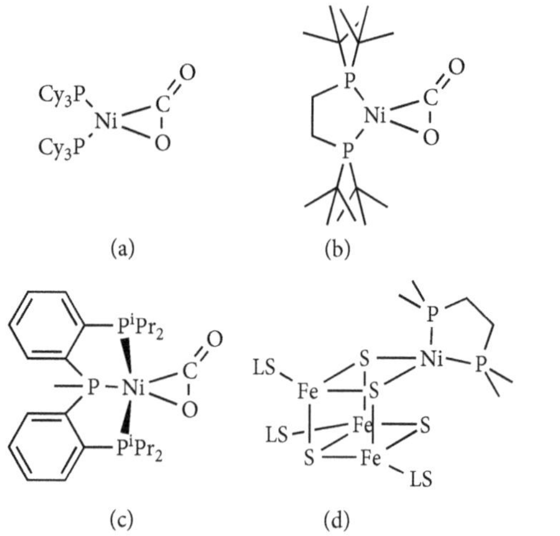

FIGURE 7: Molecular structure of the Ni-based catalysts reported by (a) Aresta et al. [51], (b) Hillhouse et al. [53], (c) Lee et al. [54], and (d) Holm et al. [55].

intermediate **(3)** is generated upon oxidation of Pd(I) to Pd (II). Subsequent protonation of the carboxylate ligand affords a metallocarboxylic acid intermediate **(4)**. The second electroreduction of Pd(II) to Pd(I) **(5)** causes the release of the coordinated solvent molecule **(6)**. Then, additional protonation of the carboxylic acid ligand allows the formation of "dihydroxy carbene" intermediate **(7)**. CO would be generated at this intermediate, by dehydration of the dihydroxy carbene species with concomitant oxidation of Pd (I) to Pd(II) **(8)**. The regeneration of the initial catalysts is achieved by solvent coordination and CO release.

The catalytic current observed was found to depend on the acid concentration and the pH dependence observed is consistent with two protonation steps being involved in the transition state [61].

These studies have highlighted the importance of having a weakly bounded molecule coordinated at the metal center in order to develop a proper catalytic activity. Thus, a vacant coordination site would be required for the reaction to occur. In particular, the production of CO from CO_2, which involves the cleavage of a C-O bond, would be favored by the presence of a vacant coordination site adjacent to the dihydroxy carbene ligand. This specific structural characteristic allows the formation of a metal-O bond upon cleavage of the C-O bond, resulting in the formal migration of a water molecule as depicted in Figures 8 and 9.

The requisite of having a vacant coordination site at a specific position to the CO_2-bounded ligand, featured in case of these transition metal catalysts, can be found also of biological importance to enlighten aspects of the catalytic mechanism proposed for natural enzymes.

On the other hand, inspired by the bimetallic active site of CODHs, dinuclear analogues of such mononuclear [Pd (triphosphine)(solvent)]$^{2+}$ complexes were designed, with the aim of enhancing the CO_2-binding properties through the creation of a bifunctional active site [70]. The activity of homobimetallic catalysts was found much improved compared to their monometallic analogues. However, rapid formation of a Pd-Pd bond inactivates the catalyst, resulting in a relatively low TON. Such an inactivation step is prevented in the native Ni-Fe CODH enzymes due to the difference in redox potentials of Ni and Fe centers, which results in a slight tendency to promote metal-metal bound formation.

4.3.2. Transition Metal Complexes Bearing Macrocyclic Ligands

(1) N_4-Macrocyclic Ligands. The first report about N_4-macrocyclic transition metal catalysts, which efficiently performed the electrocatalytic reduction of CO_2 to CO, appeared in 1980 was published by Eisenberg (Figure 10) [71].

In this work, cobalt and nickel N_4-macrocyclic complexes were demonstrated to be able to reduce CO_2 to CO, albeit at high overpotentials (ranging from −1.3 to −1.6 V versus SCE). Furthermore, at such potentials, these systems suffer the competing production of molecular hydrogen.

More recently, Sauvage and coworkers have extensively studied the reduction of CO_2 mediated by Ni-cyclam (cyclam = 1,4,8,11-tetraazacyclotetradecane) complexes (Figure 11) [48, 72]. Interestingly, CO_2 reduction catalyzed by Ni-cyclam complexes, performed in water, leads exclusively to the formation of CO as product. Whereas, using mercury electrodes, CO_2 conversion mediated by transition metal complexes displays normally the formation of CO together with formate and oxalate as common products [73]. Furthermore, CO_2 reduction catalyzed by Ni-cyclam complexes, performed in water-DMF (dimethylformamide) mixture, leads to the formation of both CO and formate in different ratio depending on the reduction potential applied.

In the 1980s, Collin and coworkers proposed a reaction mechanism to rationalize how both CO and formate can be generate from CO_2 reduction mediated by Ni-cyclam complexes (Figure 11) [74–76].

[Ni(cyclam)]$^+$ generated by one-electron reduction from [Ni(cyclam)]$^{2+}$ is the species responsible for the initial CO_2 binding and the active species of the catalytic cycle. The mechanism proposed envisions that the CO_2 reduction likely proceeds through an inner-sphere electron transfer mechanism. The formation of an adduct, upon CO_2 coordination to the catalyst, should represent the initial step of the catalytic cycle. However, [Ni(CO_2)(cyclam)]$^+$ has never been trapped at sufficiently high concentration to be fully spectroscopically characterized.

In early study by Collin, it was suggested that CO_2 can actually bind to [Ni(cyclam)]$^+$ through three possible coordination modes, in equilibrium between each others.

The reaction pathway followed after the formation of the initial CO_2 adduct would be dictated by the adopted binding mode. Thus, upon uptake of an electron and a proton, η^1-CO_2 species yield to a carboxylate intermediate, [Ni(C(O) OH)(cyclam)]$^+$, that subsequently undergoes heterolytic C–O bond cleavage to generate CO in the presence of a proton. Instead, the η^1-OCO complexes evolve into nickel-formate species, [Ni(OCOH)(cyclam)]$^+$.

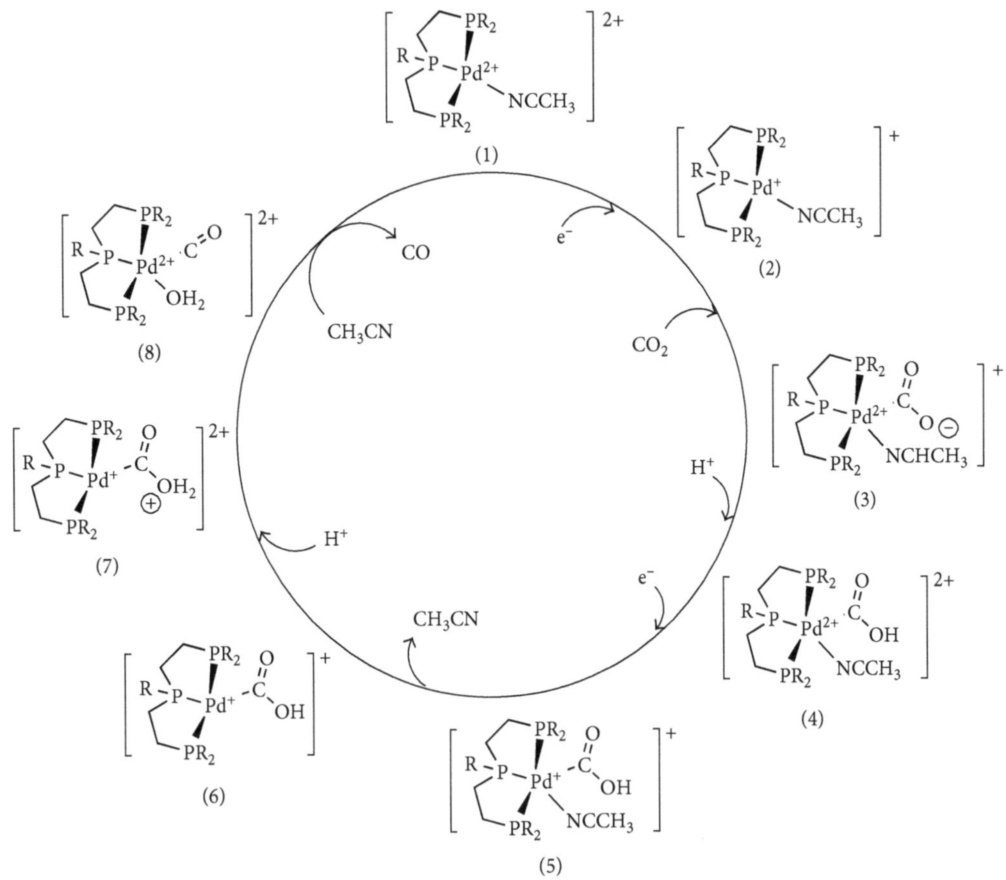

FIGURE 8: Proposed catalytic cycle for the Dubois Pd catalyst [49, 61]. Reproduced with permission from [49].

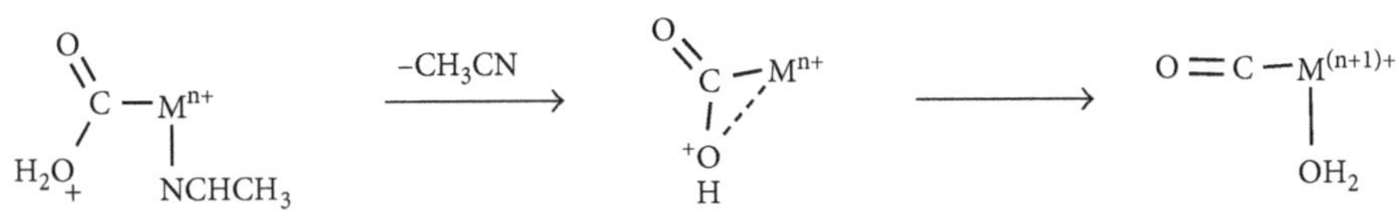

FIGURE 9: Mechanism proposed for the cleavage of the C-O bond at a metal center. Reproduced with permission from [49].

The selectivity for CO generation observed in case of the reaction mediated in water by Ni(cyclam) catalysts was recently elegantly explained by the aid of computational calculations. The formation of η^1-CO_2 adducts was indeed found energetically favored by 14 kcal/mol compare to η^1-OCO. Thus, under the studied conditions, the reaction pathway proposed to result in formate generation would be blocked [48].

(2) Porphyrin Ligands. A remarkable example of macrocyclic metal complex employed as CO_2 reduction catalysts is represented by the iron porphyrin FeTDHPP complex, where TDHPP is the 5,10,15,20-tetrakis(2′,6″-dihydroxylphenyl)-porphyrin ligand (Figure 12(a)). FeTDHPP was reported to catalyze the CO_2 reduction to CO, in acidic dimethylformamide (DMF) solutions, with modest overpotential of less than 0.5 V and 90% faradaic yield for CO production through 50 million

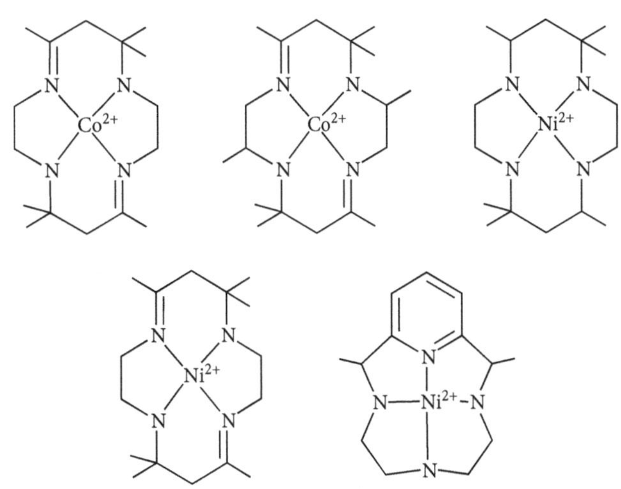

FIGURE 10: Eisenberg catalysts for the reduction of CO_2 to CO [71].

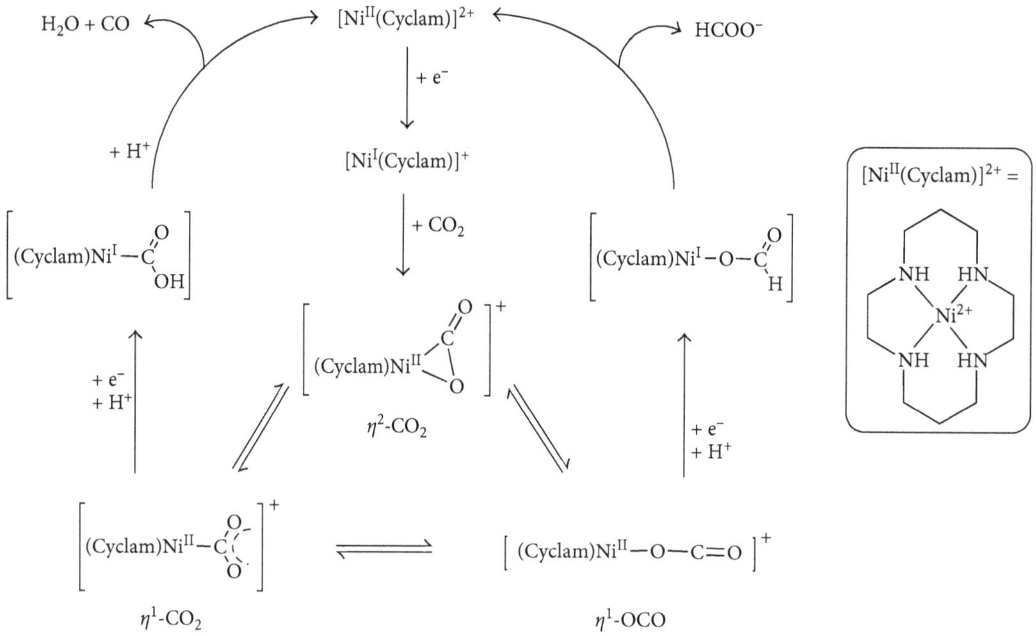

Figure 11: Proposed mechanism for CO_2 reduction mediated by [Ni(II)cyclam]. Reproduced with permission from [48].

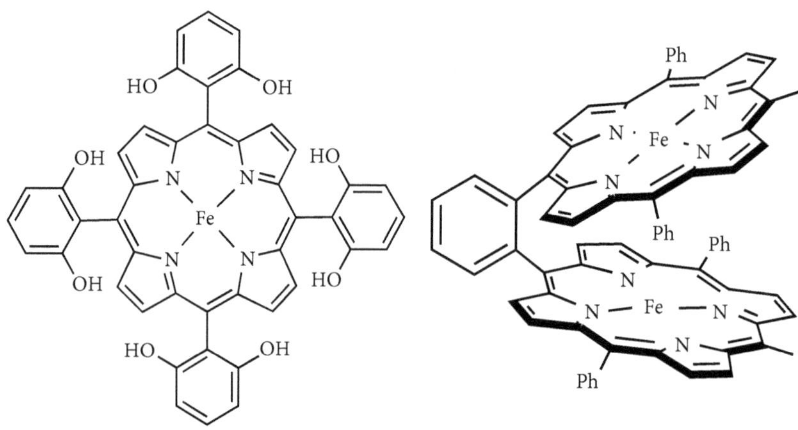

(a) (b)

Figure 12: Molecular structure of (a) FeTDHPP and (b) ortho-Fe_2DTPP complexes.

turnovers over 4 hours ($3500\,s^{-1}$) [77]. The authors suggested that this notable activity could be due to the presence of pendant hydroxyl group that assists the cleavage of the C-O bond by either favoring the proton transfer to the CO_2 bounded at the metal center or stabilizing the Fe-CO_2 adduct by hydrogen bounding.

Inspired by the Ni-Fe containing metalloenzymes, cofacial Fe(0) tetraphenyl porphyrin dimers and ortho- and meta-Fe_2DTPP (Figure 12(b)) were also reported [13, 78]. These compounds were demonstrated to be effective catalysts for CO_2 electroreduction. The catalysts exhibited excellent selectivity for CO generation and improved faradaic yield (95%) and turnover frequencies ($4300\,s^{-1}$), at a moderate overpotential (0.66 V).

(3) N_5-Macrocyclic Ligands. Robert and coworkers recently reported two novel efficient electrocatalysts, based on cobalt and iron complexes bearing N_5-macrocyclic ligands (Figure 13) [79].

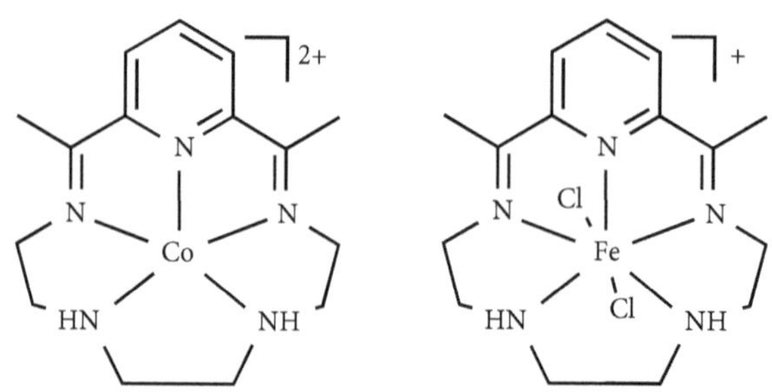

Figure 13: Molecular structure of the cobalt and iron complexes [79].

The electrocatalytic reduction of CO_2 mediated by the cobalt complex produces CO with high efficiency and

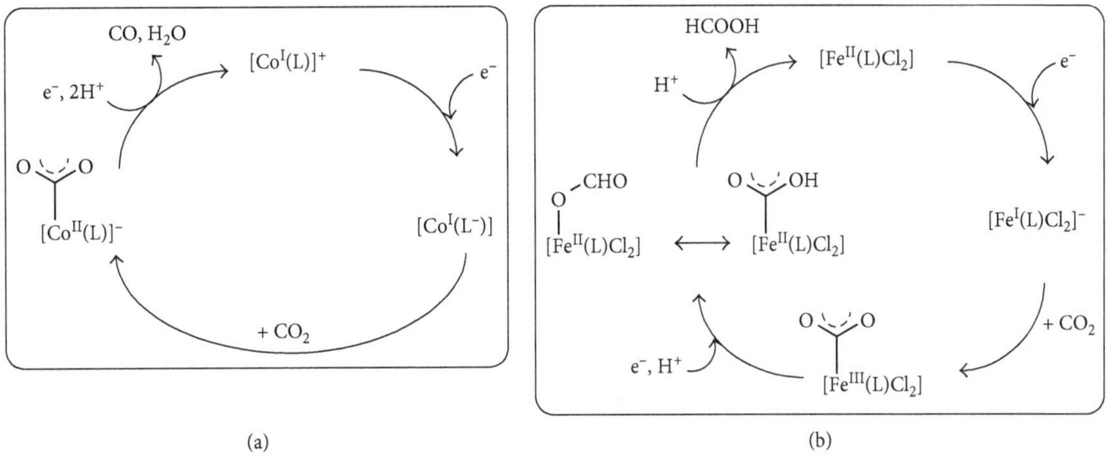

FIGURE 14: Reaction mechanism proposed for the CO_2 reduction mediated by cobalt (a) and iron (b) catalysts. Reproduced with permission from [79].

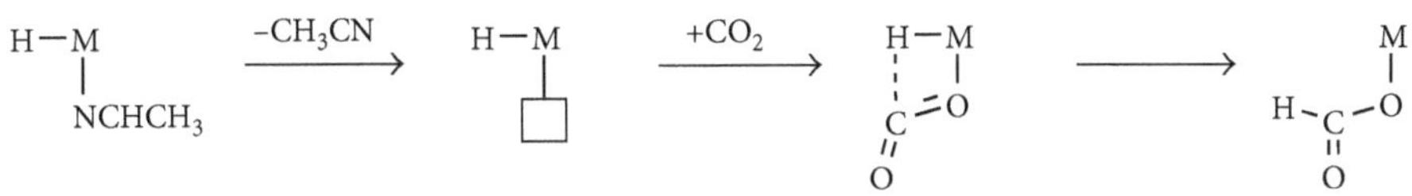

FIGURE 15: Mechanism proposed for the CO_2 insertion into a metal-H bond. Reproduced with permission from [49].

selectivity. On the other hand, the process mediated by the Fe complex generates mainly formic acid.

The mechanisms proposed for the two different reaction pathways are depicted in Figure 14.

The authors suggested that the selectivity of products formation through the two-electron reduction of CO_2, CO, or HCOOH could be attributed to distinct protonation and reduction steps for the reactions mediated by cobalt or iron, respectively. The breaking of the C-O bond would be favored at the cobalt-based catalysts (Figure 14(a)), whereas the iron-based catalysts support the formation of a η^1-OCOH intermediate, which would evolve into formic acid release upon protonation (Figure 14(b)).

4.4. Functional Models for FDHs: Electrocatalytic CO_2 Reduction to Formate Mediated by Molecular Complexes. Homogeneous transition metal catalysts were certainly found able to reduce CO_2 to formate through diverse reaction pathways; for example, as previously mentioned, η^1-OCOH or η^1-OCO intermediates upon protonation would evolve into formic acid or formate release, respectively. In particular, the electrocatalytic systems more closely related to the activity of FDH enzymes actuate the CO_2 reduction to formate following mechanistic paths that involve the CO_2 insertion into a metal-H bond.

An important breakthrough in the field of functional FDH enzymes modeling came in 1981, when Darensbourg et al. reported that anionic metal hydrides react readily with CO_2 to generate metal-formate species [80]. The research of Darensbourg and coworkers showed for the first time that CO_2 can be inserted into a Ni-H bond of a nickel-triphosphine complex, the trans-$[(H)_2Ni(PCy_3)_2]$, to generate trans-$[(H)(HCO_2)Ni(PCy_3)_2]$ species [81].

The presence of two adjacent sites, one destined to coordinate a hydride ligand and the second for CO_2 binding, likely facilitate C-H bond formation and subsequent formate production, as depicted in Figure 15 [49].

Since the finding by Darensbourg and coworkers, numerous research groups have studied this reaction mediated by transition metal catalysts (especially employing late transition metal as Ru, Rh, Ir, Ni, and Pd).

On the basis of these studies, it was highlighted that the insertion of CO_2 into the M-H bond could actually occur following two possible pathways.

Thus, CO_2 can either coordinate in a η^2-fashion to the metal center and receive the migration of the hydride ligand at the carbon atom or develop a weak H···CO_2 interaction, which leads to M-H bond cleavage and formate generation (Figures 16 and 17) [41].

The former mechanistic path is the generally accepted (Figure 16 and Intermediate A in Figure 17) [82], while the latter alternative (Intermediate B, Figure 17) has been supported essentially by theoretical studies [83–85].

The addition of a base to generate formate salt (as highlighted in Figure 16) is aimed at facilitating the catalysts regeneration. In fact, precipitation of formate salts allowed the easy separation of products and catalysts.

5. Final Remarks

The design of biomimetic transition metal complexes can decisively contributed to elucidate aspects that concern enzymes structure and reactivity. The statement of both the

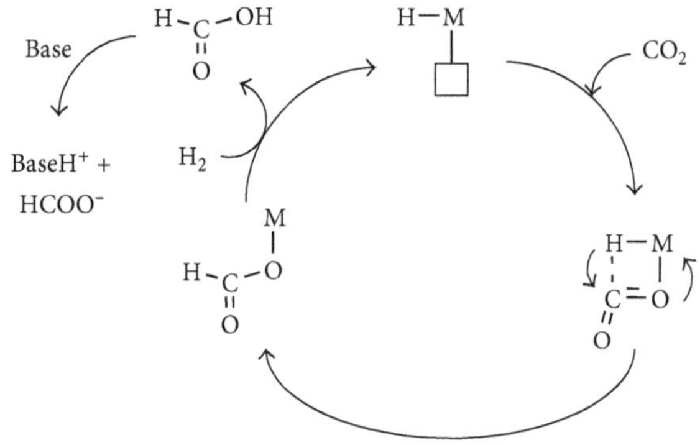

FIGURE 16: General reaction mechanism describing the catalytic CO_2 reduction to formate [41]. Reproduced with permission from [41].

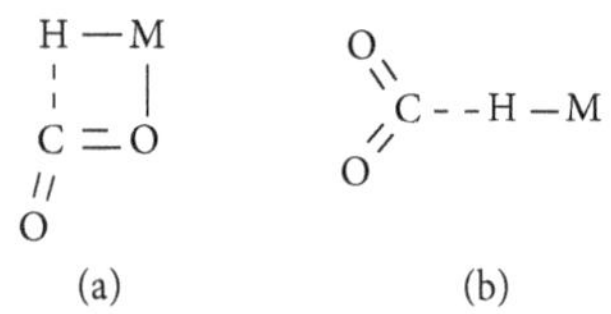

FIGURE 17: Intermediates proposed for CO_2 reduction to formate.

structural and functional characteristics, mandatory to develop a proper enzymatic activity, is critical from a fundamental point of view and may result in key-importance for the construction of synthetic catalysts employable for industrial applications.

As product of this mutual convenience, efficient transition metal catalysts and electrocatalysts, mimicking the enzymatic CO_2 two-electrons reduction to CO and/or formate, have been developed, and the more significant related examples are reviewed herein. Special emphasis is given to define the structural characteristics that have been highlighted to be compulsory for achieving a fruitful catalytic activity. The development of novel compounds that satisfy these features can lead to their readily application for electrocatalytic devices and to a better understanding of the mechanism of action of both natural and synthetic systems.

References

[1] P. A. Owusu, S. Asumadu-Sarkodie, P. Ameyo, and S. Dubey, "A review of Ghana's water resource management and the future prospect," *Cogent Engineering*, vol. 3, no. 1, p. 1167990, 2016.

[2] S. Sorrell, "Reducing energy demand: a review of issues, challenges and approaches," *Renewable and Sustainable Energy Reviews*, vol. 47, pp. 74–82, 2015.

[3] G. O. Larrazábal, A. J. Martín, and J. Pérez-Ramírez, "Building blocks for high performance in electrocatalytic CO_2 reduction: materials, optimization strategies, and device engineering," *Journal of Physical Chemistry Letters*, vol. 8, no. 16, pp. 3933–3944, 2017.

[4] J. Barber, "Photosynthetic energy conversion: natural and artificial," *Chemical Society Reviews*, vol. 38, no. 1, pp. 185–196, 2009.

[5] F. Wen and C. Li, "Hybrid artificial photosynthetic systems comprising semiconductors as light harvesters and biomimetic complexes as molecular cocatalysts," *Accounts of Chemical Research*, vol. 46, no. 11, pp. 2355–2364, 2013.

[6] C. Liu, B. C. Colón, M. Ziesack, P. A. Silver, and D. G. Nocera, "Water splitting-biosynthetic system with CO_2 reduction efficiencies exceeding photosynthesis," *Science*, vol. 352, no. 6290, pp. 1210–1213, 2016.

[7] P. D. Tran, L. H. Wong, J. Barber, and J. S. C. Loo, "Recent advances in hybrid photocatalysts for solar fuel production," *Energy and Environmental Science*, vol. 5, no. 3, pp. 5902–5918, 2012.

[8] D. C. Thomas, "Carbon dioxide mitigation: a challenge for the twenty-first century," in *Carbon Management: Implications for R&D in the Chemical Sciences and Technology: A Workshop Report to the Chemical Sciences Roundtable*, pp. 33–43, N.R.C. U.C.S. Roundtable, National Academies Press (US), Washington, DC, USA, 2001.

[9] T. R. Anderson, E. Hawkins, and P. D. Jones, "CO_2, the greenhouse effect and global warming: from the pioneering work of Arrhenius and Callendar to today's Earth System Models," *Endeavour*, vol. 40, no. 3, pp. 178–187, 2016.

[10] R. Kortlever, J. Shen, K. J. P. Schouten, F. Calle-Vallejo, and M. T. M. Koper, "Catalysts and reaction pathways for the electrochemical reduction of carbon dioxide," *Journal of Physical Chemistry Letters*, vol. 6, no. 20, pp. 4073–4082, 2015.

[11] R. J. Lim, M. Xie, M. A. Sk et al., "A review on the electrochemical reduction of CO_2 in fuel cells, metal electrodes and molecular catalysts," *Catalysis Today*, vol. 233, pp. 169–180, 2014.

[12] P. Sutra and A. Igau, "Emerging Earth-abundant (Fe, Co, Ni, Cu) molecular complexes for solar fuel catalysis," *Current Opinion in Green and Sustainable Chemistry*, vol. 10, pp. 60–67, 2018.

[13] D.-M. Feng, Y.-P. Zhu, P. Chen, and T.-Y. Ma, "Recent advances in transition-metal-mediated electrocatalytic CO_2 Reduction: from homogeneous to heterogeneous systems," *Catalysts*, vol. 7, no. 12, pp. 373–391, 2017.

[14] P. C. Sahoo, M. Kumar, S. K. Puri, and S. S. V. Ramakumar, "Enzyme inspired complexes for industrial CO_2 capture: opportunities and challenges," *Journal of CO_2 Utilization*, vol. 24, pp. 419–429, 2018.

[15] B. Kumar, M. Llorente, J. Froehlich, T. Dang, A. Sathrum, and C. P. Kubiak, "Photochemical and photoelectrochemical reduction of CO_2," *Annual Review of Physical Chemistry*, vol. 63, no. 1, pp. 541–569, 2012.

[16] A. Warshel, P. K. Sharma, M. Kato, Y. Xiang, H. Liu, and M. H. M. Olsson, "Electrostatic basis for enzyme catalysis," *Chemical Reviews*, vol. 106, no. 8, pp. 3210–3235, 2006.

[17] W. J. Albery and J. R. Knowles, "Free-energy profile for the reaction catalyzed by triosephosphate isomerase," *Biochemistry*, vol. 15, no. 25, pp. 5627–5631, 1976.

[18] L. J. Gan, J. Laureanti, and A. K. Jones, "Biomimetic complexes for production of dihydrogen and reduction of CO_2," in *Homo- and Heterobimetallic Complexes in Catalysis. Topics in Organometallic Chemistry*, P. Kalck, Ed., Springer International, Berlin, Germany, 2015.

[19] M. Can, F. A. Armstrong, and S. W. Ragsdale, "Structure, function, and mechanism of the nickel metalloenzymes, CO dehydrogenase, and acetyl-coa synthase," *Chemical Reviews*, vol. 114, no. 8, pp. 4149–4174, 2014.

[20] V. Svetlitchnyi, C. Peschel, G. Acker, and O. Meyer, "Two membrane-associated NiFeS-carbon monoxide dehydrogenases

from the anaerobic carbon-monoxide-utilizing eubacterium carboxydothermus hydrogenoformans," *Journal of Bacteriology*, vol. 183, no. 17, pp. 5134–5144, 2001.
[21] B. Zhang, C. F. Hemann, and R. Hille, "Kinetic and spectroscopic studies of the molybdenum-copper CO dehydrogenase from oligotropha carboxidovorans," *Journal of Biological Chemistry*, vol. 285, no. 17, pp. 12571–12578, 2010.
[22] C. V. Kumar and E. H. Asuncion, "DNA binding studies and site selective fluorescence sensitization of an anthryl probe," *Journal of the American Chemical Society*, vol. 115, no. 19, pp. 8547–8553, 1993.
[23] T. I. Doukov, T. M. Iverson, J. Seravalli, S. W. Ragsdale, and C. L. Drennan, "A Ni-Fe-Cu center in a bifunctional carbon monoxide dehydrogenase/acetyl-CoA synthase," *Science*, vol. 298, no. 5593, pp. 567–572, 2002.
[24] C. Darnault, A. Volbeda, E. J. Kim et al., "Ni-Zn-Fe4-S4 and Ni-Ni-Fe4-S4 clusters in closed and open α subunits of acetyl-CoA synthase/carbon monoxide dehydrogenase," *Nature Structural Biology*, vol. 10, no. 4, pp. 271–279, 2003.
[25] C. L. Drennan, J. Heo, M. D. Sintchak, E. Schreiter, and P. W. Ludden, "Life on carbon monoxide: X-ray structure of Rhodospirillum rubrum Ni-Fe-S carbon monoxide dehydrogenase," *Proceedings of the National Academy of Sciences*, vol. 98, no. 21, pp. 11973–11978, 2001.
[26] H. Dobbek, V. Svetlitchnyi, L. Gremer, R. Huber, and O. Meyer, "Crystal structure of a carbon monoxide dehydrogenase reveals a Ni-4Fe-5S cluster," *Science*, vol. 293, no. 5533, pp. 1281–1285, 2001.
[27] H. Dobbek, V. Svetlitchnyi, J. Liss, and O. Meyer, "Carbon monoxide induced decomposition of the active site Ni–4Fe–5S cluster of CO dehydrogenase," *Journal of the American Chemical Society*, vol. 126, no. 17, pp. 5382–5387, 2004.
[28] J.-H. Jeoung and H. Dobbek, "Carbon dioxide activation at the Ni,Fe-cluster of anaerobic carbon monoxide dehydrogenase," *Science*, vol. 318, no. 5855, pp. 1461–1464, 2007.
[29] T. I. Doukov, L. C. Blasiak, J. Seravalli, S. W. Ragsdale, and C. L. Drennan, "Xenon in and at the end of the tunnel of bifunctional carbon monoxide dehydrogenase/acetyl-CoA synthase," *Biochemistry*, vol. 47, no. 11, pp. 3474–3483, 2008.
[30] Y. Kung, T. I. Doukov, J. Seravalli, S. W. Ragsdale, and C. L. Drennan, "Crystallographic snapshots of cyanide- and water-bound C-clusters from bifunctional carbon monoxide dehydrogenase/acetyl-CoA synthase," *Biochemistry*, vol. 48, no. 31, pp. 7432–7440, 2009.
[31] J.-H. Jeoung and H. Dobbek, "Structural basis of cyanide inhibition of Ni, Fe-containing carbon monoxide dehydrogenase," *Journal of the American Chemical Society*, vol. 131, no. 29, pp. 9922-9923, 2009.
[32] J.-H. D. Jeoung and H. Dobbek, "n-Butyl isocyanide oxidation at the NiFe4S4OH(x) cluster of CO dehydrogenase," *Journal of Biological Inorganic Chemistry*, vol. 17, no. 2, pp. 167–173, 2012.
[33] W. Gong, B. Hao, Z. Wei et al., "Structure of the alpha2epsilon2 Ni-dependent CO dehydrogenase component of the *Methanosarcina barkeri* acetyl-CoA decarbonylase/synthase complex," *Proceedings of the National Academy of Sciences*, vol. 105, no. 28, pp. 9558–9563, 2008.
[34] J. Heo, C. R. Staples, C. M. Halbleib, and P. W. Ludden, "Evidence for a ligand CO that is required for catalytic activity of CO dehydrogenase from *Rhodospirillum rubrum*," *Biochemistry*, vol. 39, no. 27, pp. 7956–7963, 2000.
[35] J. Heo, C. M. Halbleib, and P. W. Ludden, "Redox-dependent activation of CO dehydrogenase from *Rhodospirillum rubrum*," *Proceedings of the National Academy of Sciences*, vol. 98, no. 14, pp. 7690–7693, 2001.
[36] W. E. Robinson, A. Bassegoda, E. Reisner, and J. Hirst, "Oxidation-state-dependent binding properties of the active site in a Mo-containing formate dehydrogenase," *Journal of the American Chemical Society*, vol. 139, no. 29, pp. 9927–9936, 2017.
[37] H. C. A. Raaijmakers and M. J. Romão, "Formate-reduced *E. coli* formate dehydrogenase H: the reinterpretation of the crystal structure suggests a new reaction mechanism," *Journal of Biological Inorganic Chemistry*, vol. 11, no. 7, pp. 849–854, 2006.
[38] M. Tiberti, E. Papaleo, N. Russo, L. De Gioia, and G. Zampella, "Evidence for the formation of a Mo–H intermediate in the catalytic cycle of formate dehydrogenase," *Inorganic Chemistry*, vol. 51, no. 15, pp. 8331–8339, 2012.
[39] L. B. Maia, L. Fonseca, I. Moura, and J. J. G. Moura, "Reduction of carbon dioxide by a molybdenum-containing formate dehydrogenase: a kinetic and mechanistic study," *Journal of the American Chemical Society*, vol. 138, no. 28, pp. 8834–8846, 2016.
[40] D. Niks, J. Duvvuru, M. Escalona, and R. Hille, "Spectroscopic and kinetic properties of the molybdenum-containing, NAD^+-dependent formate dehydrogenase from *Ralstonia eutropha*," *Journal of Biological Chemistry*, vol. 291, pp. 1162–1174, 2016.
[41] M. Cokoja, C. Bruckmeier, B. Rieger, W. A. Herrmann, and F. E. Kühn, "Transformation of carbon dioxide with homogeneous transition-metal catalysts: a molecular solution to a global challenge?," *Angewandte Chemie International Edition*, vol. 50, no. 37, pp. 8510–8537, 2011.
[42] K. A. Grice, "Carbon dioxide reduction with homogenous early transition metal complexes: Opportunities and challenges for developing CO_2 catalysis," *Coordination Chemistry Reviews*, vol. 336, pp. 78–95, 2017.
[43] D. Yu, S. P. Teong, and Y. Zhang, "Transition metal complex catalyzed carboxylation reactions with CO_2," *Coordination Chemistry Reviews*, vol. 293-294, pp. 279–291, 2015.
[44] Q. Liu, L. Wu, R. Jackstell, and M. Beller, "Using carbon dioxide as a building block in organic synthesis," *Nature Communications*, vol. 6, no. 1, pp. 5933–5948, 2015.
[45] E. E. Benson, C. P. Kubiak, A. J. Sathrum, and J. M. Smieja, "Electrocatalytic and homogeneous approaches to conversion of CO_2 to liquid fuels," *Chemical Society Reviews*, vol. 38, no. 1, pp. 89–99, 2009.
[46] W. Zhang, Y. Hu, L. Ma et al., "Progress and perspective of electrocatalytic CO_2 reduction for renewable carbonaceous fuels and chemicals," *Advanced Science*, vol. 5, no. 1, pp. 1700275–1700299, 2018.
[47] P. Braunstein, D. Matt, and D. Nobel, "Reactions of carbon dioxide with carbon-carbon bond formation catalyzed by transition-metal complexes," *Chemical Reviews*, vol. 88, no. 5, pp. 747–764, 1988.
[48] J. Song, E. L. Klein, F. Neese, and S. Ye, "The mechanism of homogeneous CO_2 reduction by Ni(cyclam): product selectivity, concerted proton–electron transfer and C–O bond cleavage," *Inorganic Chemistry*, vol. 53, no. 14, pp. 7500–7507, 2014.
[49] M. Rakowski Dubois and D. L. Dubois, "Development of molecular electrocatalysts for CO_2 reduction and H_2 production/oxidation," *Accounts of Chemical Research*, vol. 42, no. 12, pp. 1974–1982, 2009.

[50] A. Paparo and J. Okuda, "Carbon dioxide complexes: bonding modes and synthetic methods," *Coordination Chemistry Reviews*, vol. 334, pp. 136–149, 2017.

[51] M. Aresta, C. F. Nobile, V. G. Albano, E. Forni, and M. Manassero, "New nickel–carbon dioxide complex: synthesis, properties, and crystallographic characterization of (carbon dioxide)-bis(tricyclohexylphosphine)nickel," *Journal of the Chemical Society, Chemical Communications*, no. 15, pp. 636-637, 1975.

[52] M. Aresta and C. F. Nobile, "(Carbon dioxide)bis(trialkylphosphine)nickel complexes," *Journal of the Chemical Society, Dalton Transactions*, no. 7, pp. 708–711, 1977.

[53] J. S. Anderson, V. M. Iluc, and G. L. Hillhouse, "Reactions of CO_2 and CS2 with 1,2-Bis(di-tert-butylphosphino)ethane complexes of Nickel(0) and Nickel(I)," *Inorganic Chemistry*, vol. 49, no. 21, pp. 10203–10207, 2010.

[54] Y.-E. Kim, J. Kim, and Y. Lee, "Formation of a nickel carbon dioxide adduct and its transformation mediated by a Lewis acid," *Chemical Communications*, vol. 50, no. 78, pp. 11458–11461, 2014.

[55] R. Panda, C. P. Berlinguette, Y. Zhang, and R. H. Holm, "Synthesis of MFe_3S_4 clusters containing a planar MIISite (M = Ni, Pd, Pt), a structural element in the C-Cluster of carbon monoxide dehydrogenase," *Journal of the American Chemical Society*, vol. 127, no. 31, pp. 11092–11101, 2005.

[56] C. Yoo and Y. Lee, "Carbon dioxide binding at a Ni/Fe center: synthesis and characterization of Ni(η1-CO2-κC) and Ni-μ-CO2-κC:κ2O,O′-Fe," *Chemical Science*, vol. 8, no. 1, pp. 600–605, 2017.

[57] J. Wilkie and I. H. Williams, "Transition-state structural variation in a model for carbonyl reduction by lactate dehydrogenase: computational validation of empirical predictions based upon Albery-More O'Ferrall-Jencks diagrams," *Journal of the American Chemical Society*, vol. 114, no. 13, pp. 5423–5425, 1992.

[58] R. Panda, Y. Zhang, C. C. McLauchlan et al., "Initial structure modification of tetrahedral to planar Nickel(II) in a nickel–iron–sulfur cluster related to the C-Cluster of carbon monoxide dehydrogenase," *Journal of the American Chemical Society*, vol. 126, no. 20, pp. 6448–6459, 2004.

[59] J. Sun, C. Tessier, and R. H. Holm, "Sulfur ligand substitution at the Nickel(II) sites of cubane-type and cubanoid $NiFe_3S_4$ clusters relevant to the C-clusters of carbon monoxide dehydrogenase," *Inorganic Chemistry*, vol. 46, no. 7, pp. 2691–2699, 2007.

[60] A. Gennaro, A. A. Isse, and E. Vianello, "Electrochemical reduction of CO_2 catalysed by transition metal complexes," in *Molecular Electrochemistry of Inorganic, Bioinorganic and Organometallic Compounds*, A. J. L. Pombeiro and J. A. McCleverty, Eds., pp. 311-316, Springer, Dordrecht, Netherlands, 1993.

[61] D. L. Dubois, "Development of transition metal phosphine complexes as electrocatalysts for CO_2 and CO reduction," *Comments on Inorganic Chemistry*, vol. 19, no. 5, pp. 307–325, 1997.

[62] J. Shen, R. Kortlever, R. Kas et al., "Electrocatalytic reduction of carbon dioxide to carbon monoxide and methane at an immobilized cobalt protoporphyrin," *Nature Communications*, vol. 6, no. 1, p. 8177, 2015.

[63] D. S. Laitar, P. Müller, and J. P. Sadighi, "Efficient homogeneous catalysis in the reduction of CO_2 to CO," *Journal of the American Chemical Society*, vol. 127, no. 49, pp. 17196-17197, 2005.

[64] H. Yoshio, K. Katsuhei, and S. Shin, "Production of Co and Ch_4 in electrochemical reduction of CO_2 at metal electrodes in aqueous hydrogencarbonate solution," *Chemistry Letters*, vol. 14, no. 11, pp. 1695–1698, 1985.

[65] J. A. Ramos Sende, C. R. Arana, L. Hernandez, K. T. Potts, M. Keshevarz, and H. D. Abruna, "Electrocatalysis of CO_2 reduction in aqueous media at electrodes modified with electropolymerized films of vinylterpyridine complexes of transition metals," *Inorganic Chemistry*, vol. 34, no. 12, pp. 3339–3348, 1995.

[66] D. L. DuBois and A. Miedaner, "Mediated electrochemical reduction of CO_2. Preparation and comparison of an isoelectronic series of complexes," *Journal of the American Chemical Society*, vol. 109, no. 1, pp. 113–117, 1987.

[67] D. L. DuBois, A. Miedaner, and R. C. Haltiwanger, "Electrochemical reduction of carbon dioxide catalyzed by Pd (triphosphine)(solvent)$(BF4)_2$ complexes: synthetic and mechanistic studies," *Journal of the American Chemical Society*, vol. 113, no. 23, pp. 8753–8764, 1991.

[68] A. Majumdar, "Bioinorganic modeling chemistry of carbon monoxide dehydrogenases: description of model complexes, current status and possible future scopes," *Dalton Transactions*, vol. 43, no. 32, pp. 12135–12145, 2014.

[69] J. W. Raebiger, J. W. Turner, B. C. Noll et al., "Electrochemical reduction of CO_2 to CO catalyzed by a bimetallic palladium complex," *Organometallics*, vol. 25, no. 14, pp. 3345–3351, 2006.

[70] J. Hawecker, J.-M. Lehn, and R. Ziessel, "Efficient photochemical reduction of CO_2 to CO by visible light irradiation of systems containing $Re(bipy)(CO)_3X$ or $Ru(bipy)_3{}^{2+}$–Co^{2+} combinations as homogeneous catalysts," *Journal of the Chemical Society, Chemical Communications*, no. 9, pp. 536–538, 1983.

[71] B. J. Fisher and R. Eisenberg, "Electrocatalytic reduction of carbon dioxide by using macrocycles of nickel and cobalt," *Journal of the American Chemical Society*, vol. 102, no. 24, pp. 7361–7363, 1980.

[72] J. D. Froehlich and C. P. Kubiak, "The homogeneous reduction of CO_2 by $Ni(cyclam)^+$: increased catalytic rates with the addition of a CO scavenger," *Journal of the American Chemical Society*, vol. 137, no. 10, pp. 3565–3573, 2015.

[73] R. Angamuthu, P. Byers, M. Lutz, A. L. Spek, and E. Bouwman, "Electrocatalytic CO_2 conversion to oxalate by a copper complex," *Science*, vol. 327, no. 5963, pp. 313–315, 2010.

[74] J. P. Collin, A. Jouaiti, and J. P. Sauvage, "Electrocatalytic properties of (tetraazacyclotetradecane)nickel(2^+) and $Ni_2(biscyclam)^{4+}$ with respect to carbon dioxide and water reduction," *Inorganic Chemistry*, vol. 27, no. 11, pp. 1986–1990, 1988.

[75] M. Beley, J. P. Collin, R. Ruppert, and J. P. Sauvage, "Electrocatalytic reduction of carbon dioxide by nickel $cyclam^{2+}$ in water: study of the factors affecting the efficiency and the selectivity of the process," *Journal of the American Chemical Society*, vol. 108, no. 24, pp. 7461–7467, 1986.

[76] M. Beley, J.-P. Collin, R. Ruppert, and J.-P. Sauvage, "Nickel (II)-cyclam: an extremely selective electrocatalyst for reduction of CO_2 in water," *Journal of the Chemical Society, Chemical Communications*, no. 19, pp. 1315-1316, 1984.

[77] C. Costentin, S. Drouet, M. Robert, and J.-M. Savéant, "A local proton source enhances CO_2 electroreduction to CO by a molecular Fe catalyst," *Science*, vol. 338, no. 6103, pp. 90–94, 2012.

[78] E. A. Mohamed, Z. N. Zahran, and Y. Naruta, "Efficient electrocatalytic CO_2 reduction with a molecular cofacial iron porphyrin dimer," *Chemical Communications*, vol. 51, no. 95, pp. 16900–16903, 2015.
[79] L. Chen, Z. Guo, X.-G. Wei et al., "Molecular catalysis of the electrochemical and photochemical reduction of CO_2 with earth-abundant metal complexes. selective production of CO vs HCOOH by switching of the metal center," *Journal of the American Chemical Society*, vol. 137, no. 34, pp. 10918–10921, 2015.
[80] D. J. Darensbourg, A. Rokicki, and M. Y. Darensbourg, "Facile reduction of carbon dioxide by anionic Group 6b metal hydrides. Chemistry relevant to catalysis of the water-gas shift reaction," *Journal of the American Chemical Society*, vol. 103, no. 11, pp. 3223-3224, 1981.
[81] D. J. Darensbourg, M. Y. Darensbourg, L. Y. Goh, M. Ludvig, and P. Wiegreffe, "Reaction of $(Cy3P)_2Ni(H)(CH_3)$ with carbon dioxide. Formation of an hydridonickel formate complex, $HNi(O_2CH)(Cy_3P)_2$," *Journal of the American Chemical Society*, vol. 109, no. 24, pp. 7539-7540, 1987.
[82] Y. Musashi and S. Sakaki, "Theoretical study of ruthenium-catalyzed hydrogenation of carbon dioxide into formic acid. Reaction mechanism involving a new type of σ-bond metathesis," *Journal of the American Chemical Society*, vol. 122, no. 16, pp. 3867–3877, 2000.
[83] A. Urakawa, M. Iannuzzi, J. Hutter, and A. Baiker, "Towards a rational design of ruthenium CO_2 hydrogenation catalysts by Ab initio metadynamics," *Chemistry–A European Journal*, vol. 13, no. 24, pp. 6828–6840, 2007.
[84] C. Bo and A. Dedieu, "Carbon dioxide interaction with pentacarbonylhydridochromate(1-): theoretical study of the thermodynamic aspects," *Inorganic Chemistry*, vol. 28, no. 2, pp. 304–309, 1989.
[85] S. Sakaki and K. Ohkubo, "Ab initio MO study of carbon dioxide insertion into a copper(I)-hydrogen bond. Semi-quantitative understanding of changes in geometry, bonding, and electron distribution during the reaction," *Inorganic Chemistry*, vol. 28, no. 13, pp. 2583–2590, 1989.

12

Biogenic Synthesis of Copper and Silver Nanoparticles using Green Alga *Botryococcus braunii* and its Antimicrobial Activity

Anju Arya,[1] Khushbu Gupta,[1] Tejpal Singh Chundawat,[1] and Dipti Vaya[2]

[1]*Department of Applied Sciences, The North Cap University, Sector 23-A, Gurugram 122017, Haryana, India*
[2]*Department of Chemistry, Amity University, Gurugram, Haryana, India*

Correspondence should be addressed to Tejpal Singh Chundawat; chundawatchem@yahoo.co.in

Academic Editor: Zhe-Sheng Chen

The spread of infectious diseases and the increase in the drug resistance among microbes has forced the researchers to synthesize biologically active nanoparticles. Improvement of the ecofriendly procedure for the synthesis of nanoparticles is growing day-by-day in the field of nanobiotechnology. In the present study, we use the extract of green alga *Botryococcus braunii* for the synthesis of copper and silver nanoparticles. The characterization of copper and silver nanoparticles was carried out by using UV-Vis spectroscopy, Fourier transform infrared spectroscopy (FTIR), X-ray diffraction (XRD), and scanning electron spectroscopy (SEM). FTIR measurements showed all functional groups having control over reduction and stabilization of the nanoparticles. The X-ray diffraction pattern revealed that the particles were crystalline in nature with a face-centred cubic (FCC) geometry. SEM micrographs have shown the morphology of biogenically synthesized metal nanoparticles. Furthermore, these biosynthesized nanoparticles were found to be highly toxic against two Gram-negative bacterial strains *Pseudomonas aeruginosa* (MTCC 441) *and Escherichia coli* (MTCC 442), two Gram-positive bacterial strains *Klebsiella pneumoniae* (MTCC 109) and *Staphylococcus aureus* (MTCC 96), and a fungal strain *Fusarium oxysporum* (MTCC 2087). The zone of inhibition was measured by the agar well plate method, and furthermore, minimum inhibitory concentration was determined by the broth dilution assay.

1. Introduction

Nowadays, nanotechnology is a notable efficacious field for research work [1]. Nanotechnology is the branch of science and technology, which deals with the production of substances in size less than 100 nm scale as nanoparticles [2]. Among other nanoparticles, metal nanoparticles have raised attention over the last few decades because they have larger surface area per weight or to volume and many characteristics; biological, thermal, chemical, dielectric, electrical, physical, mechanical, electronic, magnetic, and optical properties make them attractive tools for research work [3]. Therefore, nanoparticles are studied as the building blocks of the next generation of technology with applications in different industrial areas. In particular, metal and metal oxide nanoparticles are receiving optimum attention in a large variety of applications [4]. Nanoparticles of transition metals are an important class of semiconductors, which have applications in magnetic storage media, solar energy transformation electronics, gas sensors, and catalysis [5–7]. The nanoparticles play up the most important role in different disciplines such as health care, screening and medicines, drug delivery system, antisense, tissue biotechnology, cosmetics, applications of gene engineering, and in many other fields [8, 9].

Currently biological species like algae are in great use for the nanoparticles synthesis. Both live and dead biomass of algae is used for the biogenic synthesis of nanoparticles that is why they are called as bionanofactories. Algae have good metal uptake capacity, and thus, the biological method with the use of algae is cost-effective and ecofriendly [10]. Algae contain large amount of the reducing agent which reduces metal salts to their respective metal nanoparticles without any hazardous by-products. Aqueous extract of algae contains secondary metabolites such as polysaccharide, proteins, tannins, and steroids as bioactive molecules [11–13].

Synthesized silver and copper nanoparticles also showed antimicrobial activity [14–16]. The nanoparticles have effective antibacterial and antifungal property due to their large surface area, which allow them a better contact with microbes [17]. The nanoparticles can disorder the helical structure of DNA by cross-linking within the nucleic acids stands, and they also disrupt the biochemical processes [18].

Nanoparticles with antimicrobial activity are very advantageous in reducing acute toxicity, lowering cost, and overcoming resistance as compared with other prevalent antibiotics [19, 20]. Antibiotics can sustain for long run in the form of nanoparticles than any other large or small molecules [21].

The previous findings of research on biogenic synthesis of silver and copper nanoparticles using green algae have been unexplored; there are few literatures found on silver nanoparticles [22–26] but on copper nanoparticles less and none using green alga *Botryococcus braunii.* The present study has shown that the use of green alga *Botryococcus braunii* as biofactory for synthesis of silver and copper nanoparticles and their antimicrobial activities. To best of our knowledge and understanding, there is no report on synthesis of silver and copper nanoparticles using green alga *Botryococcus braunii.* The synthesized nanoparticles were characterized by different techniques used such as UV-visible spectroscopy, scanning electron microscopy, X-ray diffraction, and Fourier infrared spectroscopy. Furthermore, antimicrobial activity against bacterial and fungal species was also exhibited by biogenically synthesized silver and copper nanoparticles.

2. Materials and Methods

2.1. Materials. Green alga *Botryococcus braunii* was collected from Udaisagar Lake Udaipur (Rajasthan, India). The reagents agar-agar, copper acetate, and silver nitrate are of analytical grade and were purchased from Sigma-Aldrich. Bacterial strains like *Pseudomonas aeruginosa* (MTCC 441), *Escherichia coli* (MTCC 442), *Klebsiella pneumoniae* (MTCC 109), and *Staphylococcus aureus* (MTCC 96) and a fungal strain *Fusarium oxysporum* (MTCC 2087) were purchased from the microbial type collection, Chandigarh, India.

2.2. Methods

2.2.1. Isolation and Culturing of Green Alga and Algal Extract Preparation. Green alga was isolated by the serial dilution method and grown on the Chu-13 nutrient medium solidified by 1.5% agar-agar. The algal colonies appearing after three weeks of incubation were isolated and inoculated into the liquid medium. For experiments, alga was grown for algal biomass in a incubator at 27 ± 1°C, 1.2 ± 0.2 Klux light intensity, and 16 : 8 hrs light: dark cycle in the nutrient medium. Grown algal biomass was centrifuged, shade dried, and taken 5 g of algal biomass in a 250 ml Erlenmeyer flask along with 100 ml of distilled water. Mixture was autoclaved for 15 min and filtered hot through the Whatman No. 1 filter paper. The filtered extract was centrifuged, and the supernatant was used as the reducing agent for preparing metal nanoparticles. Prepared algal extract was kept at 4°C in a refrigerator for future use.

2.2.2. Synthesis of Metal Nanoparticles

(1) Biogenic Synthesis of Copper Nanoparticles. 5 ml algal extract was added dropwise into 50 ml of 1 mM aqueous copper acetate in a 100 ml Erlenmeyer flask with vigorous stirring at 100°C for 24 h. Simultaneously, a positive control of copper acetate aqueous solution and algal extract and a negative control containing only copper acetate aqueous solution were maintained under the same conditions. In the positive control within three hours, the light sky blue solution changed to dark brown. It was the indication of formation of copper nanoparticles, but in the negative control, the colour remains unchanged. The progress of the process was regularly monitored by observing colour change and recording UV-visible spectrum. After the completion of the process, the above reaction mixture was centrifuged for 15 min and the obtained material was subsequently redispersed and washed with deionised water to remove debris and any uncoordinated biomolecules. This process of separation and washing was carried out thrice to certain severance of copper nanoparticles. These copper nanoparticles were dried in a oven for further characterization.

(2) Biogenic Synthesis of Silver Nanoparticles. 5 ml algal extract mixed with 45 ml and 1 mM silver nitrate aqueous solution in a 100 ml Erlenmeyer flask were put on magnetic stirrer at room temperature for 3 h. Simultaneously, a positive control of silver nitrate aqueous solution and algal extract and a negative control containing only silver nitrate aqueous solution were maintained under the same conditions. The progress of the process was regularly monitored by observing colour change and recording UV-visible spectrum. In the positive control, the initial light pale yellow solution turned to reddish brown, indicating formation of silver nanoparticles, but in the negative control, no change in colour was found. After the reaction reached saturation, the solution was centrifuged for 20 min, and the obtained pellet was washed with deionised water to remove impurities. This process of centrifugation and washing was carried out thrice to get a better separation of nanoparticles. The obtained silver nanoparticles were oven-dried at 55°C for 5 h.

2.2.3. Characterization of Metal Nanoparticles. The bioreduction of metal ions in solution was monitored using a UV-Vis NIR spectrophotometer (Model no. Cary series) taking a spectra from 200 to 800 nm for each sample against distilled water as blank. The dried powders of copper and silver nanoparticles were used for characterizations. FTIR analyses were carried out on FTIR (PerkinElmer) in the range of 4000–450 cm^{-1} using dried powder of metal nanoparticles. Samples for analysis were prepared at ambient conditions and mixed with KBr, and X-ray diffraction measurements were carried out on the Philips Xpert pro XRD system (DY 1650). The shape and size of copper and silver nanoparticles were analyzed by using scanning electron microscopy (SEM) images obtained with the help of scanning electron microscope (Model-FEI Quanta 200 SEM).

2.2.4. Antimicrobial Activity of Synthesized Metal Nanoparticles. Antimicrobial activity of green-synthesized nanoparticles was evaluated by the agar well plate method, and minimum inhibitory concentration was determined by microbroth dilution assay. Two Gram-negative bacterial strains *Pseudomonas aeruginosa* (MTCC 441) *and Escherichia coli* (MTCC 442), two Gram-positive bacterial strains *Klebsiella pneumoniae* (MTCC 109) and *Staphylococcus aureus* (MTCC 96), and a fungal strain *Fusarium oxysporum* (MTCC 2087) were used in the investigation. Bacterial cultures were maintained in Petri plates containing the nutrient agar (NA) medium at 37°C, and the fungus *Fusarium oxysporum* was maintained in potato dextrose agar at 25°C. All the cultures were subcultured on regular basis and stored at 4°C.

2.2.5. Agar Well Plate Assay. The agar well diffusion method was used to evaluate the antimicrobial activity of synthesized nanoparticles. The nutrient agar containing the microbial inoculum was spread all over the Petri plate. Allow it to cool for some time, and then a well of diameter 8–10 mm was punched aseptically with a sterile cock borer or a tip. 20–40 ml of 1000 μg/ml concentration of nanoparticles was added into the well, and then the agar plates were incubated under sterile conditions depending upon the test microorganism. The nanoparticles diffuse into the agar media and inhibit the growth of microbes. Then, the antimicrobial activity of the nanoparticles was detected by the appearance of the inhibition zone around the agar well. Zone was measured with the help of a transparent ruler from one edge to the other edge of the clear area.

2.2.6. Broth Dilution Method. For this method, all synthesized nanoparticles were diluted to obtain a stock solution having 2000 μg/ml concentration. For the primary screening, these synthesized nanoparticles were diluted obtaining 1000 μg/ml, 500 μg/ml, and 250 μg/ml concentrations. The nanoparticles found active against microbes in the primary screening were further diluted in the secondary screening to obtain 200 μg/ml, 100 μg/ml, 50 μg/ml, 25 μg/ml, 12.5 μg/ml, and 6.250 μg/ml concentrations. Each synthesized nanoparticles were diluted using DMSO as a diluent. The highest dilution showing at least 99% inhibition zone is taken as MIC. The MIC value is very much affected by the size of the inoculums. The test mixture should contain approximately 10^8 number of organisms/ml.

3. Results and Discussion

In this research, we have used green alga *Botryococcus braunii* for biogenic synthesis of copper and silver nanoparticles. Synthesized copper and silver nanoparticles were examined for antimicrobial activity against bacterial and fungal species. The alga-mediated biogenic synthesis of both nanoparticles as copper and silver is given in Figure 1. Synthesized metal nanoparticles were initially confirmed by visual observation as change in colour. Furthermore, UV-visible spectra have been regularly used for characterization of biogenically synthesized nanoparticles. Surface plasmon resonance phenomenon provides a convenient signature to indicate the formation of copper and silver nanoparticles by change in colour in the reaction mixture [27]. Aqueous extract of green alga *B. braunii* was added into aqueous solution of copper acetate and silver nitrate for copper and silver nanoparticles, respectively. The formation of copper and silver nanoparticles was characterized by colour change from light sky blue to dark brown and light pale yellow to reddish brown through visual observation means, indicating the reduction of copper and silver ions into their respective nanoparticles [27, 28]. In this reduction process, metal nanoparticles scatter and absorb light at a certain wavelength due to the resonant excitations of charge density at the interface between a conductor and an insulator, the phenomena called surface plasmon resonance [29].

3.1. UV-Vis Spectroscopy. UV-visible spectrum of silver and copper nanoparticles synthesized by *B. braunii* exhibited the absorption peaks around 490 nm and 258 nm, respectively (Figure 2). The synthesized silver nanoparticles show absorbance at 490 nm according to previous findings [22]. In case of copper nanoparticles, two peaks were found: first one at 258 nm is attributed to formation of copper nanoparticles [14, 30, 31], and second one at 460 nm ascribed to the presence of the oxide shell around nanoparticles [32]. These peaks have already been recorded for various metal nanoparticles which ranged from 2 to 100 nm in size [33]. The synthesized silver and copper nanoparticles covered with biomolecules are well dispersed in solutions and fairly stable up to 3 months as indicated by retention of brown colour of the solution. In case of copper nanoparticles, these results clearly indicate the formation of mixture of CuO and Cu_2O nanoparticles as previously reported in literature [32, 34].

3.2. Fourier Transform Infrared (FTIR). Fourier transform infrared (FTIR) spectrum of the experimental samples revealed two types of vibrations as stretching and bending in the wavelength range of 4000–450 cm^{-1}. FTIR spectrum measurements were demonstrated to identify the major functional groups present in green alga *B. braunii* to examine their possible involvement in the synthesis of copper and silver nanoparticles. Different peaks positioned at 3435.88, 2923.49, 2852.33, 2092.30, 1637.82, 1559.61, 1414.42, 1384.79, 1069.01, 1056.17, 837.53, 781.32, 714.25, 695.06, 657, 618.16, and 532.74 cm^{-1} in FTIR spectrum of algal extract of green alga *B. braunii*. The peak at 3435.88 cm^{-1} is due to N-H and O-H stretching vibrations [35]. 2923.49 and 2852.33 cm^{-1} bands arose due to asymmetrical C-H stretching vibrations of $-CH_2$ and $-CH_3$ [36]. The IR peak at 2092.30 cm^{-1} is due to the alkynes groups present in the lipids and nitrile group (-CN) present in proteins of algae [37]. The 1637.82 cm^{-1} peak is a characteristic of N-H bending vibrations in amide of

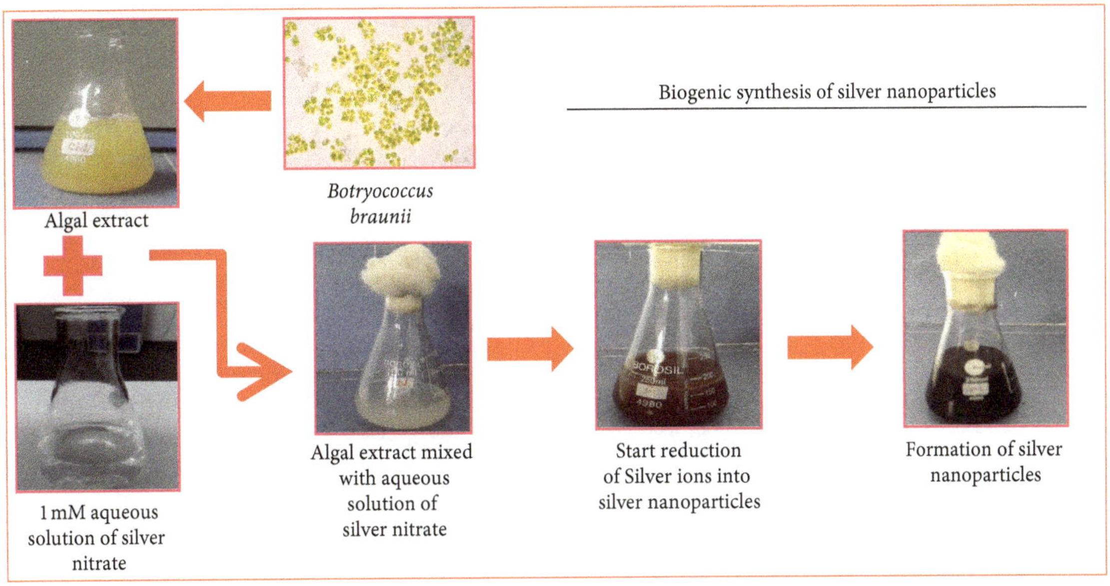

(a)

(b)

FIGURE 1: (a) Biogenic synthesis of silver nanoparticles; (b) biogenic synthesis of copper nanoparticles.

protein as a capping agent [38]. The peak at 1559.61 cm^{-1} showed the presence of the carboxyl group and the weak band at 1414.42 and 618.16 cm^{-1} due to COO^- in the amino acid residue of protein [36] [39]. The peak observed around 1384.79 cm^{-1} can be assigned to C-N stretching vibrations of amine. C-H bending vibrations by carbohydrates (glucose residue by the C-OH bond) showed the peak at 1037.17 cm^{-1}. 873.53, 781.32, 695.06, and 532.74 cm^{-1} bands were demonstrated due to O-C=O bending vibrations of CO_3^{-2}, C-H rocking of lipids, and N-H wagging of amine and alkyl halide, respectively. The results of the present study have shown that the hydroxyl groups have strong ability to interact with nanoparticles. The main peaks existing in the spectrum of alga are also present in the spectra of synthesized silver and copper nanoparticles with lower intensities and slight shift. Therefore, it may be evidenced that proteins, polysaccharides, amides, and long-chain fatty acids are responsible biomolecules for bioreduction, capping, and stabilizing agents.

3.3. Scanning Electron Microscopy (SEM). The shape and size of both biogenically synthesized nanoparticles were elucidated with the help of scanning electron microscopy (SEM) (Figures 3(a) and 3(b)). The cubical, spherical, and truncated

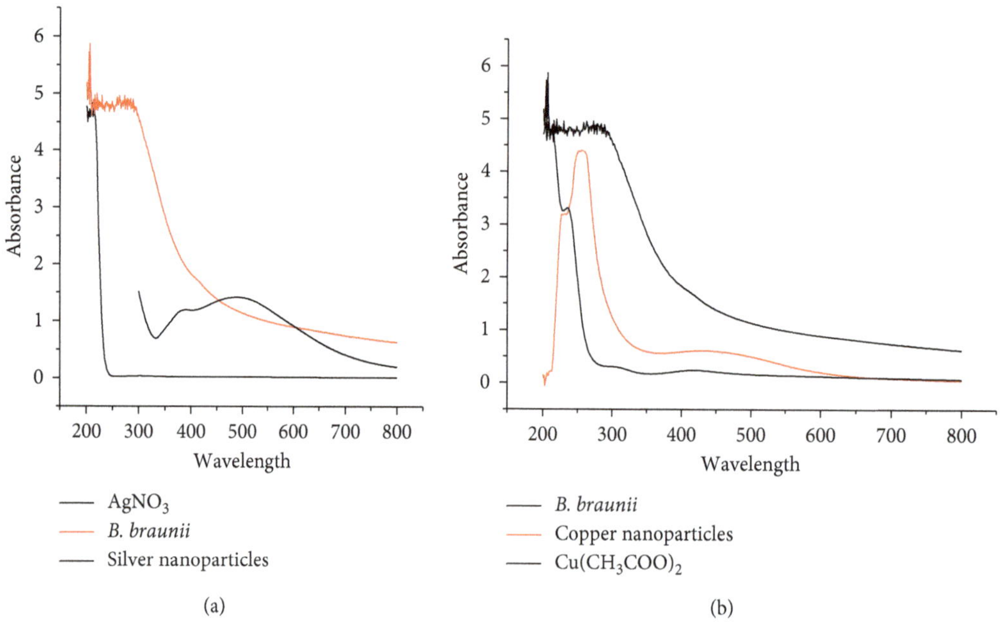

FIGURE 2: UV-Vis spectra of (a) silver nanoparticles and (b) copper nanoparticles.

FIGURE 3: Scanning electron microscopy (SEM) images of green-synthesized (a) silver nanoparticles and (b) copper nanoparticles.

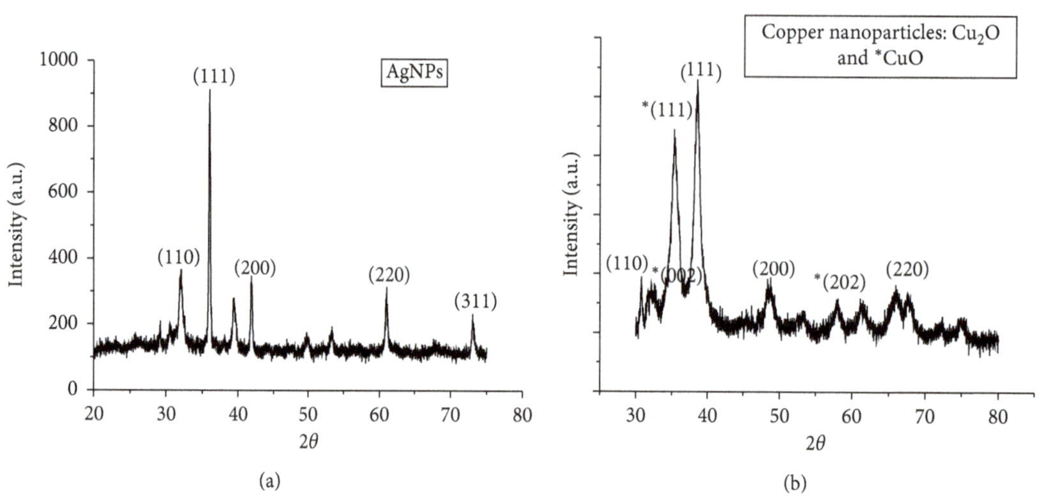

FIGURE 4: XRD patterns of biogenically synthesized (a) silver nanoparticles and (b) copper nanoparticles.

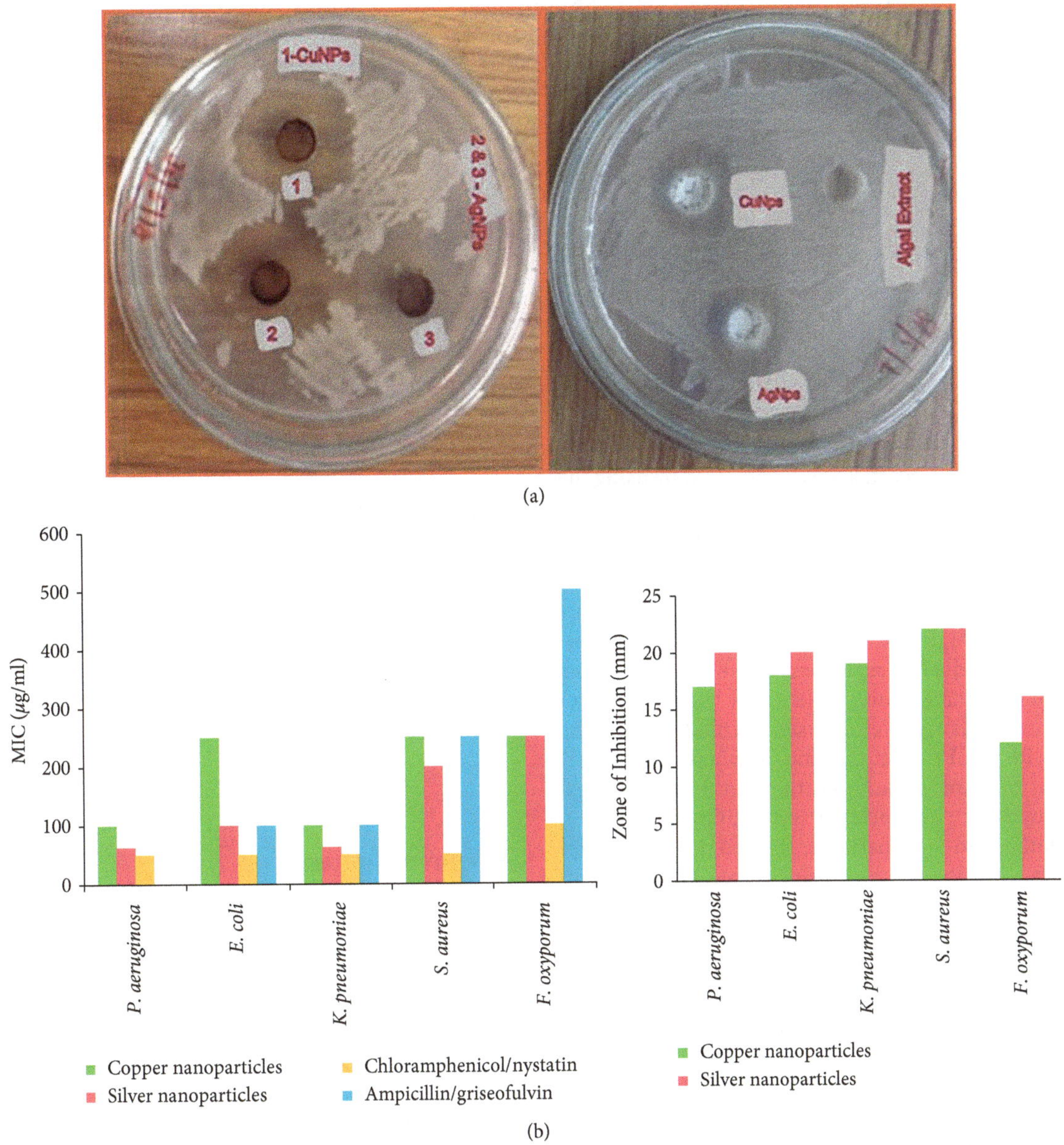

FIGURE 5: (a) Antibacterial assay: zone of inhibition against *Staphylococcus aureus* and *Pseudomonas aeruginosa*. (b) MIC value and zone of inhibition against microbes.

triangular shapes of silver nanoparticles were observed from scanning electron microscopy images. Copper nanoparticles observed from SEM are cubical and spherical with an elongated shape. The size of algal synthesized silver and copper nanoparticles was found to be in range of 40–100 nm and 10–70 nm, respectively. These results of SEM were confirmed by previously reported literature [14, 25].

3.4. X-Ray Diffraction. Furthermore, evidence for biosynthesis of silver and copper nanoparticles was confirmed by the X-ray diffraction (XRD) analysis (Figure 4). The X-ray diffraction (XRD) analysis explained the structure of silver and copper nanoparticles. In Figure 5(a), XRD patterns for silver nanoparticles between 2θ of 20–80° are shown. It is indexed by JCPDS card no (04–0783). XRD patterns show the diffraction peaks of silver nanoparticles at 2θ values for 33°, 36°, 42°, 62°, and 73° which matched with 110, 111, 200, 220, and 311 lattice planes of face-centred cubic structure, respectively [35, 40]. XRD confirmed that silver nanoparticles are crystalline in nature with the face-centred cubic structure. Average crystalline size calculated by the Debye Scherer equation is 89 nm. Furthermore, algal-synthesized copper nanoparticles in Figure 5(b) demonstrate the major XRD peaks at 30.8°, 38.66°, 48.66°, and 68.26° diffracted from the (110), (111), (200), and (220) planes of the face-centred cubic structure, respectively. It is indexed by JCPDS card no. (71–4610). In addition, three other peaks were also found at 2θ of 32.3°, 35.4°, and 58.1 °can be assigned to (002), (111), and (202) indicating the presence of the oxide shell around the copper nanoparticles. These peaks of the CuO phase were reported in previous findings [14, 41]. The average size of

Table 1: Antibacterial and antifungal activity of biogenically synthesized copper and silver nanoparticles.

Nanoparticles (500 μg/ml)	Zone of inhibition diameter (mm)				
	Gram-negative bacterial strains		Gram-positive bacterial strains		Fungal strain
	Pseudomonas aeruginosa	*Escherichia coli*	*Klebsiella pneumoniae*	*Staphylococcus aureus*	*Fusarium oxysporum*
Copper	17 ± 1.56	18 ± 0.1	19 ± 0.0	22 ± 0.88	12 ± 0.58
Silver	20 ± 1.78	20 ± 0.8	21 ± 0.60	22 ± 0.75	16 ± 1.5
	All values are mean ± SD				
	Minimum inhibitory concentration (μg/ml)				
Copper	100	250	100	250	250
Silver	62.5	100	62.5	200	250
Ampicillin	—	100	100	250	—
Chloramphenicol	50	50	50	50	—
Nystatin	—	—	—	—	100
Griseofulvin	—	—	—	—	500

synthesized copper nanoparticles was calculated by using the Debye Scherer equation found to be around 58 nm which is correlated with that obtained from SEM images.

3.5. Antimicrobial Activity of Synthesized Silver and Copper Nanoparticles. Biological activity of algal extract and synthesized copper and silver nanoparticles [42, 43] was done against both bacteria (Gram positive and Gram negative) and fungus using the agar well diffusion method [44, 45] The well filled with algal extract did not show any zone of inhibition, but the nanoparticles synthesized from that algal culture showed both antibacterial and antifungal activity.

The synthesized nanoparticles showed the inhibition zone against all of the test organisms (Table 1). Maximum zone of inhibition was found against *Staphylococcus aureus* (22 mm). The next maximum zone of inhibition was recorded in *Klebsiella pneumoniae* (20 mm) followed by *Escherichia coli* (18 mm) and *Pseudomonas aeruginosa* (17 mm), and the minimum zone of inhibition was recorded against *Fusarium oxysporum* (12 mm). Furthermore, the nanoparticles synthesized by the green route are found to be highly effective against Gram-negative bacteria. However, the minimum inhibitory concentration [46] required to inhibit the growth of the microbes is less in the case of silver as compared with copper (Figure 5). These synthesized nanoparticles show least activity towards tested fungus *Fusarium oxysporum*. Even biogenically synthesized silver nanoparticles showed a lower MIC value than the positive control drug chloramphenicol against both Gram-positive and Gram-negative bacteria. Nystatin and griseofulvin were used as the positive control for fungus *Fusarium oxysporum*.

4. Conclusion

This present work demonstrated an ecofriendly and convenient method used for synthesis of silver and copper nanoparticles using green alga *Botryococcus braunii*. Aqueous extract of green alga can reduce silver and copper ions into silver and copper nanoparticles and has the potential to stabilise them. Biogenically synthesized nanoparticles were characterised by different techniques such as UV-visible spectroscopy, Fourier transform infrared spectroscopy, scanning electron microscopy, and X-ray diffraction and exhibited antibacterial and antifungal activity. The biogenic synthesis of metal nanoparticles can be a promising process for production of other metal and metal oxide nanoparticles which can have environmental, valuable, pharmaceutical, medical, and biotechnological applications.

Acknowledgments

The authors are thankful to The NorthCap University for providing facilities for research work.

References

[1] M. Amin, F. Anwar, M. R. S. A. Janju, M. A. Iqbal, and U. Rashid, "Green synthesis of silver nanoparticles through reduction with *Solanum xanthocarpum* L. Berry extract: characterization, antimicrobial and urease inhibitory activities against *Helicobacter pylori*," *International Journal of Molecular Sciences*, vol. 13, no. 8, pp. 9923–9941, 2012.

[2] P. Mohanpuria, N. K. Rana, and S. K. Yadav, "Biosynthesis of nanoparticles: technological concepts and future applications," *Journal of Nanoparticle Research*, vol. 10, no. 3, pp. 507–517, 2008.

[3] I. Khan, K. Saeed, and I. Khan, "Nanoparticles: properties, applications and toxicities," *Arabian Journal of Chemistry*, 2017.

[4] A. Dashora and K. Sharma, "Green synthesis of nanoparticles and their applications," *Advanced Science, Engineering and Medicine*, vol. 10, no. 6, pp. 523–541, 2018.

[5] S. J. Ahmadi, M. Outokesh, M. Hosseinpour, and T. Mousavand, "A simple granulation technique for preparing high-porosity nano copper oxide (II) catalyst beads," *Particuology*, vol. 9, pp. 480–485, 2011.

[6] N. Ramgir, N. Datta, M. Kaur et al., "Metal oxide nanowires of chemoresistive gas sensors: issue, challenges and prospects,"

Colloids and Surfaces A: Physicochemical and Engineering Aspects, 2013.

[7] A. E. Shalana, M. M. Rashada, Y. Yu, M. L. Cantub, and M. S. A. Abdel-Mottaleb, "Controlling the microstructure and properties of the titania nanopowders for high efficiency dye sensitized solar cells," *Electrochimica Acta*, vol. 89, pp. 469–478, 2013.

[8] N. Sharma and P. Sharma, "Industrial and biotechnological applications of algae: a review," *Journal of Advances in Plant Biology*, vol. 1, no. 1, pp. 1-25, 2017.

[9] S. Sinha, P. Pan, P. Chanda, and S. K. Sen, "Nanoparticles fabrications using ambient biological resource," *Journal of Applied Biosciences*, vol. 19, pp. 1113–1130, 2009.

[10] M. Bilal, T. Rasheed, J. E. Sosa-Hernández, A. Raza, F. Nabeel, and H. M. N. Iqbal, "Biosorption: an interplay between marine algae and potentially toxic elements—a review," *Marine Drugs*, vol. 16, no. 2, p. 65, 2018.

[11] N. Salvador, A. G. Garreta, L. Lavelli, and M. A. Ribera, "Antimicrobial activity of Iberian microalgae," *Scientia Marina*, vol. 71, no. 1, pp. 101–113, 2007.

[12] S. Sangeetha, N. B. Dhayanithi, and N. Sivakumar, "Antibacterial activity of *Sargassum longifolium* and *Gracilaria corticata* from gulf of Manar against selected common shrimp pathogens," *International Journal of Pharma and Bio Sciences*, vol. 5, pp. 76–82, 2014.

[13] J. Jin, C. Dupré, J. Legrand, and D. Grizeau, "Extracellular hydrocarbon and intracellular lipid accumulation are related to nutrient-sufficient conditions in pH-controlled chemostat cultures of the microalga *Botryococcus braunii SAG* 30.81," *Algal Research*, vol. 17, pp. 244–252, 2016.

[14] Y. Abboud, T. Saffaj, A. Chagraoui, A. E. I. Bouari, K. Brouzi, and O. Tanane, "Biosynthesis, characterization and antimicrobial activity of copper oxide nanoparticles (CONPs) produced using brown alga extract (*Bifurcaria bifurcate*)," *Applied Nanoscience*, vol. 4, no. 5, pp. 571–576, 2013.

[15] J. S. Kim, E. Kuk, K. N. Yu, J. H. Kim, S. J. Park et al., "Antimicrobial effects of silver nanoparticles," *Nanomedicine: Nanotechnology, Biology and Medicine*, vol. 3, no. 1, pp. 95–101, 2007.

[16] S. Thakur, S. Sharma, S. Thakur, and R. Rai, "Green synthesis of copper nano-particles using Asparagus adscendens roxb. Root and leaf extract and their antimicrobial activities," *International Journal of Current Microbiology and Applied Sciences*, vol. 7, no. 4, pp. 683–694, 2018.

[17] P. Prema and S. Thangapandiyan, "In vitro antibacterial activity of gold nanoparticles capped with polysaccharide stabilizing agents," *International Journal of Pharmacy and Pharmaceutical Sciences*, vol. 5, pp. 310–314, 2013.

[18] G. Speranza, G. Gottardi, C. Pederzolli et al., "Role of chemical interactions in bacterial adhesion to polymer surfaces," *Biomaterials*, vol. 25, pp. 2029–2037, 2004.

[19] S. Pal, Y. K. Tak, and J. M. Song, "Dose the antibacterial activity of silver nanoparticles depend on the shape of the nanoparticle? A study of the gram-negative bacterium *Escherichia coli*," *Applied and Environmental Microbiology*, vol. 73, no. 6, pp. 1712–1720, 2007.

[20] E. Weir, A. Lawlor, A. Whelan, and F. Regan, "The use of nanoparticles in anti-microbial materials and their characterization," *Analyst*, vol. 133, no. 7, pp. 835–845, 2008.

[21] M. Kumar, A. Curtis, and C. Hoskins, "Application of nanoparticles technologies in the combat against antimicrobial resistance," *Pharmaceutics*, vol. 10, no. 1, p. 11, 2018.

[22] F. S. Ahmadi, A. Tanhacin, and M. H. Pirkohi, "Biosynthesis of silver nanoparticles using *Chlamydomonas reinhardtii* and its inhibitory effect on growth and virulence of *Listeria monocytogenes*," *Iranian Journal of Biotechnology*, vol. 14, no. 3, pp. 163–168, 2015.

[23] J. Jena, N. Pradhan, R. R. Nayak et al., "Microalgae *Scendesmus sp*: a potential low cost green machine for silver nanoparticles synthesis," *Journal of Microbiology and Biotechnology*, vol. 24, no. 4, pp. 522–533, 2014.

[24] T. Kathirawan, A. Sunderamannicka, N. Shamgam, and T. Balasubramanan, "Green Synthesis of silver nanoparticles using marine alga *Caulrepa racemosa* and their antibacterial activity against some human pathogens," *Applied Nanoscience*, vol. 5, no. 4, pp. 499–544, 2014.

[25] S. N. Sinha, D. Paul, N. Halder, D. Sengupta, and S. K. Patra, "Green synthesis of silver noanoparticles using fresh water green algae *Pithophora oedogonia* (Mont.) wittrock and evaluation of their antibacterial activity," *Applied Nanoscience*, vol. 5, no. 6, pp. 703–709, 2014.

[26] M. Soleimani and M. Habibi-Pirkoohi, "Biosynthesis of silver nanoparticles using *Chlorella vulgaris* and evaluation of the antibacterial efficacy against *Staphylococcus aureus Avicenna*," *Journal of Medical Biotechnology*, vol. 9, no. 3, pp. 120–125, 2016.

[27] N. Krithiga, A. Jayachitra, and A. Rajalakshmi, "Synthesis, characterization and analysis of the effect of copper oxide nanoparticles in biological system," *Indian Journal of Nanoscience*, vol. 1, pp. 6–15, 2013.

[28] M. K. Swamy, K. M. Sudipta, K. Jayanta, and S. Balasubramayana, "The green synthesis, characterization, and evalution of the biological activities, characterization, of silver nanoparticles synthesized from Leptadenia reticulate leaf extract," *Applied Nanoscience*, vol. 5, no. 1, 2014.

[29] C. Noguez, "Surface plasmons on metal nanoparticles: the influence of shape and physical environmental," *Journal of Physical Chemistry C*, vol. 111, no. 10, pp. 3806–3819, 2007.

[30] K. Borgohain, N. Murase, and S. Mahamuni, "Synthesis and properties of Cu_2O quantum particles," *Journal of Applied Physics*, vol. 92, no. 3, pp. 1292–1297, 2002.

[31] R. L. Kimber, E. A. Lewis, F. Parmeggiani et al., "Biosynthesis and characterization of copper nanoparticles using shewanella oneidensis: application for click chemistry small," *Small*, vol. 14, no. 10, 2018.

[32] M. Yin, C. K. Wu, Y. Lou et al., "Copper oxide nanocrystals," *Journal of the American Chemical Society*, vol. 127, no. 26, pp. 9506–9511, 2005.

[33] B. K. Ravindra and A. H. Rajsab, "A comparative study on biosynthesis of silver nanoparticles using four different fungal species," *International Journal of Pharmacy and Pharmaceutical Sciences*, vol. 6, pp. 372–376, 2014.

[34] A. Rahman, A. Ismail, D. Jumbianti, S. Magdalena, and H. Sudrajat, "Synthesis of copper oxide nanoparticles by using *Phormidium cyanobacterium*," *Indonesian Journal of Chemistry*, vol. 9, pp. 355–360, 2009.

[35] E. F. Aboelfetoh, R. A. EI-Shenody, and M. M. Ghobara, "Eco-friendly synthesis of silver nanoparticles using green alga (*Caulerpa serrulata*): reaction optimization, catalytic and antibacterial activities," *Environmental Monitoring and Assessment*, vol. 189, no. 7, p. 349, 2017.

[36] B. Sharma, D. D. Purkayastha, S. Hazra et al., "Biosynthesis of gold nanoparticles using fresh water green alga *Prasiola crispa*," *Materials Letters*, vol. 116, pp. 94–97, 2013.

[37] A. Gaebler, R. Milan, L. Straub, D. Hoelper, L. Kuerschner, and C. Thiele, "Alkyne lipids as substrates for click-chemistry based in vitro enzymatic assays," *Journal of Lipid Research*, vol. 54, no. 8, pp. 2282–2290, 2013.

[38] S. Shende, A. Gade, and M. Rai, "Large Scale synthesis and antibacterial activity of fungal derived silver nanoparticles," *Environmental Chemistry Letters*, vol. 15, no. 3, pp. 427–434, 2016.
[39] M. Ramakrishana, D. R. Babu, R. M. Gengan, S. Chandra, and G. N. Rao, "Green synthesis of gold nanoparticles using marine algae and evaluation of their catalytic activity," *Journal of Nanostructure in Chemistry*, vol. 6, no. 1, pp. 1–13, 2015.
[40] S. Ashokkumar, S. Ravi, V. Kathiravan, and S. Velmurugan, "Synthesis of silver nanoparticles using *Indicum* leaf extract and their antibacterial activity," *Spectrochimica Acta Part A: Molecular and Biomolecular Spectroscopy*, vol. 134, pp. 34–39, 2015.
[41] D. P. Volanti, D. Keyson, L. S. Cavalcante et al., "Synthesis and characterization of CuO flower-nanostructure processing by a domestic hydrothermal microwave," *Journal of Alloys and Compounds*, vol. 459, no. 1-2, pp. 537–542, 2008.
[42] J. Annamalai and T. Nallamuthu, "Green synthesis of silver nanoparticles: characterization and determination of antibacterial potency," *Applied Nanoscience*, vol. 6, no. 2, pp. 259–265, 2016.
[43] V. P. P. N. Kumar, U. Shameem, K. Pratap, R. L. Kalyani, and S. V. N. Pammi, "Green synthesis of copper oxide nanoparticles using *Aloe vera* leaf extract and its antibacterial activity against fish bacterial pathogens," *BioNanoScience*, vol. 5, no. 3, pp. 135–139, 2015.
[44] S. Magaldi, S. Mata-Essayag, and C. Hartungde Capriles, "Well diffusion for antifungal susceptibility testing," *International Journal of Infectious Diseases*, vol. 8, no. 1, pp. 39–45, 2004.
[45] C. Valgas, S. M. DeSouza, and E. F. A. Smânia, "Screening methods to determine antibacterial activity of natural products," *Brazilian Journal of Microbiology*, vol. 38, no. 2, pp. 369–380, 2007.
[46] M. Sarkar, D. V. Reneer, and J. A. Carlyon, "Sialyl-Lewis x-independent infection of human myeloid cells by anaplasma phagocytophilum strains HZ and HGE1," *Infection and Immunity*, vol. 75, no. 12, pp. 5720–5725, 2007.

13

Hydrolysis of Methionine- and Histidine-Containing Peptides Promoted by Dinuclear Platinum(II) Complexes with Benzodiazines as Bridging Ligands: Influence of Ligand Structure on the Catalytic Ability of Platinum(II) Complexes

Snežana Rajković,[1] Beata Warżajtis,[2] Marija D. Živković,[3] Biljana Đ. Glišić,[1] Urszula Rychlewska,[2] and Miloš I. Djuran[4]

[1]*Department of Chemistry, Faculty of Science, University of Kragujevac, R. Domanovića 12, 34000 Kragujevac, Serbia*
[2]*Faculty of Chemistry, Adam Mickiewicz University, Umultowska 89B, 61-614 Poznań, Poland*
[3]*Department of Pharmacy, Faculty of Medical Sciences, University of Kragujevac, Svetozara Markovića 69, 34000 Kragujevac, Serbia*
[4]*Serbian Academy of Sciences and Arts, Knez Mihailova 35, 11000 Belgrade, Serbia*

Correspondence should be addressed to Snežana Rajković; snezana@kg.ac.rs

Academic Editor: Giovanni Natile

Dinuclear platinum(II) complexes, $[\{Pt(en)Cl\}_2(\mu\text{-qx})]Cl_2\cdot 2H_2O$ (**1**), $[\{Pt(en)Cl\}_2(\mu\text{-qz})](ClO_4)_2$ (**2**), and $[\{Pt(en)Cl\}_2(\mu\text{-phtz})]Cl_2\cdot 4H_2O$ (**3**), were synthesized and characterized by different spectroscopic techniques. The crystal structure of **1** was determined by single-crystal X-ray diffraction analysis, while the DFT M06-2X method was applied in order to optimize the structures of **1–3** The chlorido Pt(II) complexes **1–3** were converted into the corresponding aqua species **1a–3a**, and their reactions with an equimolar amount of Ac–L–Met–Gly and Ac–L–His–Gly dipeptides were studied by ^1H NMR spectroscopy in the pH range $2.0 < pH < 2.5$ at 37°C. It was found that, in all investigated reactions with the Ac–L–Met–Gly dipeptide, the cleavage of the Met–Gly amide bond had occurred, but complexes **2a** and **3a** showed lower catalytic activity than **1a**. However, in the reactions with Ac–L–His–Gly dipeptide, the hydrolysis of the amide bond involving the carboxylic group of histidine was observed only with complex **1a**. The observed disparity in the catalytic activity of these complexes is thought to be due to different relative positioning of nitrogen atoms in the bridging qx, qz, and phtz ligands and consequent variation in the intramolecular separation of the two platinum(II) metal centers.

1. Introduction

Selective cleavage of peptides and proteins is an important procedure in biochemistry and molecular biology. However, the extreme kinetic inertness of the amide bond, with an estimated half-life between 250 and 600 years under physiological conditions [1], makes this procedure particularly interesting from the chemical point of view. This remarkable kinetic stability of the amide bond is required for its function but presents a challenge when there is a physiological need to break it. Typical proteolytic enzymes such as carboxypeptidase A contain a zinc(II) ion in their active sites, thus suggesting that small metal coordination complexes may play the role of a protease [2]. In recent years, a number of metal complexes including zinc(II) [3, 4], cobalt(III) [5], iron(II/III) [6–9], copper(II) [10–12], nickel(II) [13, 14], cerium(IV) [15], and zirconium(IV) [16, 17] have been found to be effective at promoting the hydrolysis of unactivated amide bonds in peptides and proteins. Besides the basic requirement that the complex should be able to promote amide bond cleavage, the selectivity of the cleavage remains a big challenge. In recent years, a great deal of interest has focused on the study of the interactions of mononuclear Pt(II) [18–20] and Pd(II) [19–33] complexes with methionine- and histidine-containing peptides. These interactions are of cardinal importance after the discovery that aqua derivatives of the investigated Pt(II) and Pd(II) complexes can be promising reagents for the hydrolytic cleavage of the abovementioned peptides. In general, it was shown that these complexes bind to the heteroatom in the side chain of methionine [18–25] or histidine [19, 20, 26–34] and promote cleavage of the amide bond involving the carboxylic group of the anchoring amino acid. Recent studies in one of our laboratories showed that dinuclear diazine-bridged Pt(II) complexes are very efficient in promoting the hydrolysis of the amide bond in methionine-containing peptides [35]. It was found that two Pt(II) ions bridged with one aromatic pyrazine (pz) ligand are more efficient in the hydrolysis of the methionine-containing peptides than a single Pt(II) ion in the corresponding mononuclear complex. The better catalytic activity of pyrazine-bridged dinuclear Pt(II) complexes in comparison with the corresponding mononuclear complexes has been explained by possible cooperation between two pyrazine-bridged Pt(II) ions. The cooperation between two metal ions as an important advantage of their dimerization was also demonstrated in the hydrolysis of DNA, RNA, and their models catalyzed by polynuclear metal complexes and metalloenzymes [36–43]. More recently, we have compared catalytic properties of two aqua dinuclear Pt(II) complexes with those of pyrazine and pyridazine (pydz) ligands, $[\{Pt(en)(H_2O)\}_2(\mu\text{-pz})]^{4+}$ and $[\{Pt(en)(H_2O)\}_2(\mu\text{-pydz})]^{4+}$, with peptides containing amino acids methionine and/or histidine in the side chain. The results obtained from that study showed that the $[\{Pt(en)(H_2O)\}_2(\mu\text{-pydz})]^{4+}$ complex reacts only with methionine sulfur atom and promotes the sole cleavage of the amide bond involving the carboxylic group of this anchoring amino acid, while the analogous pyrazine Pt(II) aqua dimer reacts with both residues, promoting cleavage of the amide bonds involving the carboxylic groups of both amino acids, methionine and histidine [35, 44]. The X-ray data for the corresponding chloride complexes, $[\{Pt(en)Cl\}_2(\mu\text{-pydz})]Cl_2$ and $[\{Pt(en)Cl\}_2(\mu\text{-pz})]Cl_2$, confirmed the hidden position of the Pt(II) centers in the pydz-bridged Pt(II) complex caused by their close proximity due to the *ortho*-positioning of the two diazine nitrogen atoms (the two Pt(II) centers were only 3.2535(4) Å apart), compared to the separation of 6.7890(3) Å in the analogous pz-bridged Pt(II) dimer.

As a continuation of our interest in the reactions between methionine- and histidine-containing peptides and dinuclear Pt(II) complexes with six-membered aromatic diazines as bridging ligands, in the present study, we describe synthesis and characterization of three new dinuclear platinum(II) complexes, $[\{Pt(en)Cl\}_2(\mu\text{-qx})]Cl_2\cdot 2H_2O$ (**1**), $[\{Pt(en)Cl\}_2(\mu\text{-qz})](ClO_4)_2$ (**2**), and $[\{Pt(en)Cl\}_2(\mu\text{-phtz})]Cl_2\cdot 4H_2O$ (**3**) (qx = quinoxaline, qz = quinazoline, phtz = phthalazine, and en = ethylenediamine). Complexes were characterized by NMR (^1H and ^{13}C), UV-Vis, and IR spectroscopic techniques. The crystal structure of **1** was determined by single-crystal X-ray diffraction analysis, while the DFT M06-2X method was applied in order to optimize the structures of **1–3**. The complexes **1–3** were converted into the corresponding aqua derivatives, and their catalytic activities in the hydrolysis of the *N*-acetylated Ac–L–His–Gly and Ac–L–Met–Gly dipeptides were investigated.

2. Materials and Methods

2.1. Materials. Distilled water was demineralized and purified to a resistance greater than 10 MΩ·cm. The compounds D_2O, DNO_3, NaOD, ethylenediamine (en), quinoxaline (1,4-benzodiazine, qx), quinazoline (1,3-benzodiazine, qz), phthalazine (2,3-benzodiazine, phtz), and $K_2[PtCl_4]$ were obtained from Sigma-Aldrich Chemical Co. All common chemicals were of reagent grade. The dipeptides L-histidyl-glycine (L-His–Gly) and L-methionyl-glycine (L-Met–Gly) were obtained from Bachem A.G. The terminal amino group in these dipeptides was acetylated by a standard method [21]. The $[Pt(en)Cl_2]$ complex was synthesized according to a procedure published in the literature [28, 45, 46]. The purity of the complex was checked by elemental microanalyses and NMR (^1H and ^{13}C) spectroscopy.

2.2. Synthesis of Dinuclear Platinum(II) Complexes **1–3**. The complexes $[\{Pt(en)Cl\}_2(\mu\text{-qx})]Cl_2\cdot 2H_2O$ (**1**), $[\{Pt(en)Cl\}_2(\mu\text{-qz})](ClO_4)_2$ (**2**), and $[\{Pt(en)Cl\}_2(\mu\text{-phtz})]Cl_2\cdot 4H_2O$ (**3**) were synthesized by modification of the procedure published in the literature [35, 44, 47, 48].

The mononuclear $[Pt(en)Cl_2]$ complex was converted into the corresponding monodimethylformamide complex $[Pt(en)(dmf)Cl]NO_3$ by treatment with 0.98 equivalent of $AgNO_3$. The suspension of 0.1487 g (0.456 mmol) of $[Pt(en)Cl_2]$ in 10.0 mL of dimethylformamide (dmf) was added to the solution of 0.0759 g (0.447 mmol) of $AgNO_3$ in 5 mL of dmf. The mixture was stirred overnight at room temperature in the dark. The precipitated AgCl was removed by filtration, and resulting pale yellow dmf solution of $[Pt(en)(dmf)Cl]NO_3$ was used as the starting material for the preparation of the required quinoxaline-, quinazoline-, or phthalazine-bridged platinum(II) complexes.

The dmf solution of the ligand L (0.0297 g; 0.228 mmol) (L is quinoxaline, qx; quinazoline, qz; and phthalazine, phtz) was added dropwise to the solution of $[Pt(en)(dmf)Cl]NO_3$ complex. The mixture was stirred at room temperature in the dark for 24 h. The solvent was then rotary evaporated, and the residue was washed with ether. The crude product was dissolved in a minimal amount of aqueous solution of LiCl (0.5 M) for **1** and **3** or $LiClO_4$ (0.5 M) for **2**. The obtained solutions were left overnight in the dark. The pale yellow precipitate of the dinuclear platinum(II) complex was removed by filtration, washed with methanol and then ether, and air-dried. Depending on the type of the bridging ligand L, the yield of complexes **1–3** was between 35 and 40%. Elemental analysis for **1** ($Pt_2C_{12}H_{26}N_6Cl_4O_2$): found: C, 17.87; H, 3.20; and N, 10.15%; calc.: C, 17.61; H, 3.20; and N, 10.27%. ^1H NMR (200 MHz, D_2O): $\delta = 2.76$ (m, 8H, en), 8.41 (dd, 2H, C6H and C7H, qx), 9.59 (s, 2H, C2H and C3H, qx), and 9.73 (dd, 2H, C5H and C8H, qx) ppm. ^{13}C NMR (50 MHz, D_2O): $\delta = 49.42$ and 49.77 (en), 130.09 (C6, C7), 135.93 (C5, C8), 145.26 (C4a, C8a), and 151.77 (C2, C3) ppm. IR (KBr, ν, cm^{-1}): ~3415 (O-H stretch); 3275-3027

(N-H stretch); and 1631, 1597, 1587, and 1500 (C=N/C=C, quinoxaline group stretch). UV-Vis (H_2O, λ_{max}, nm): 247 ($\varepsilon = 2.5 \cdot 10^4\ M^{-1} \cdot cm^{-1}$) and 339 ($\varepsilon = 9.8 \cdot 10^3\ M^{-1} \cdot cm^{-1}$). Elemental analysis for **2** ($Pt_2C_{12}H_{22}N_6Cl_4O_8$): found: C, 15.63; H, 2.60; and N, 8.94%; calc.: C, 15.83; H, 2.44; and N, 9.23%. ^1H NMR (200 MHz, D_2O): $\delta = 2.81$ (m, 8H, en), 8.16 (m, 2H, C6H and C7H, qz), 8.51 (m, 2H, C5H and C8H, qz), 9.52 (d, H, C4H, qz), and 10.12 (d, H, C2H, qz) ppm. ^{13}C NMR (50 MHz, D_2O): $\delta = 50.24$ (en), 128.58 (C4a), 130.14 (C5), 132.81 (C8), 135.11 (C6), 143.10 (C7), 152.39 (C8a), 168.56 (C4), and 178.11 (C2) ppm. IR (KBr, ν, cm^{-1}): 3233 and 3138 (N-H stretch); 1620 and 1587 (C=N/C=C quinazoline group stretch); and 1096, 1062, and 623 (perchlorate counterion). UV-Vis (H_2O, λ_{max}, nm): 237 ($\varepsilon = 4.7 \cdot 10^4\ M^{-1} \cdot cm^{-1}$) and 328 ($\varepsilon = 8.1 \cdot 10^3\ M^{-1} \cdot cm^{-1}$). Elemental analysis for **3** ($Pt_2C_{12}H_{30}N_6Cl_4O_4$): found: C, 16.45; H, 3.58; and N, 9.57%; calc.: C, 16.87; H, 3.54; and N, 9.84%. ^1H NMR (200 MHz, D_2O): $\delta = 2.83$ (m, 8H, en), 8.40 (m, 4H, C5H, C6H, C7H, and C8H, phtz), and 10.05 (s, 2H, C1H and C4H, phtz) ppm. ^{13}C NMR (50 MHz, D_2O): $\delta = 51.27$ (en), 130.96 (C4a, C8a), 131.04 (C5, C8), 140.50 (C6, C7), and 164.32 (C1, C4) ppm. IR (KBr, ν, cm^{-1}): ~3489 (O-H stretch); 3156 and 3046 (N-H stretch); and 1625 (C=N/C=C, phthalazine group stretch). UV-Vis (H_2O, λ_{max}, nm): 236 ($\varepsilon = 5.1 \cdot 10^4\ M^{-1} \cdot cm^{-1}$) and 329 ($\varepsilon = 8.7 \cdot 10^3\ M^{-1} \cdot cm^{-1}$).

2.3. Preparation of Aqua Platinum(II) Complexes **1a–3a**. The dinuclear chlorido complexes **1–3** were converted into the corresponding aqua derivatives, $[\{Pt(en)(H_2O)\}_2(\mu\text{-}L)]^{4+}$, where L = qx (**1a**), qz (**2a**), and phtz (**3a**), according to a previously published method [49] by treatment with 3.98 (complexes **1** and **3**) and 1.98 (complex **2**) equivalents of $AgNO_3$. In each case, the precipitated white solid product was removed by filtration in the dark, and the fresh solutions of the aqua complexes **1a–3a** were kept in a refrigerator and used in the further experiments.

2.4. Measurements. Elemental microanalyses for carbon, hydrogen, and nitrogen parameters were performed by the Microanalytical Laboratory, Faculty of Chemistry, University of Belgrade. All pH measurements were realized at ambient temperature using a Mettler Toledo SevenCompact S220-U pH meter calibrated with Mettler Toledo-certified buffer solutions of pH 4.00 and 7.00. The results were not corrected for the deuterium isotope effect. The UV-Vis spectra were recorded on a Shimadzu double-beam spectrophotometer equipped with thermostated 1.00 cm quartz Suprasil cells after dissolving the corresponding platinum(II) complex in water over the wavelength range of 200–500 nm. The concentration of the dinuclear platinum(II) complexes was 5×10^{-5} M. The IR spectra were recorded on a PerkinElmer Spectrum One FT-IR spectrometer using the KBr pellet technique, over the range of 4000–450 cm^{-1}. The NMR spectra of platinum(II) complexes **1–3** and aromatic *N*-heterocyclic ligands were recorded at 25°C in D_2O-containing TSP (sodium 3-(trimethylsilyl)propionate) as the internal reference on a Varian Gemini 2000 spectrometer (^1H at 200 MHz; ^{13}C at 50 MHz). Chemical shifts are reported in parts per million (ppm), and scalar couplings are reported in hertz (Hz). 0.005 g of each compound was dissolved in 0.6 mL of D_2O, and this solution was transferred into a 5 mm NMR tube. All the NMR spectra were processed using the Varian VNMR software (version 6.1, revision C). Fresh solutions of the aqua derivatives of dinuclear platinum(II) complexes **1a–3a** and Ac–L–Met–Gly and Ac–L–His–Gly dipeptides were prepared separately and then mixed in 1 : 1 molar ratio. The initial concentrations of dipeptide and aqua complex solutions were 20 mM. All reactions were performed in the pH range $2.0 < pH < 2.5$ at 37 °C.

2.5. Crystallographic Data Collection and Refinement of the Structure of **1**. Diffraction data for $[\{Pt(en)Cl\}_2(\mu\text{-qx})]Cl_2 \cdot 2H_2O$ complex (**1**) were measured with an Xcalibur kappa geometry diffractometer using CrysAlisPro software [50] and monochromated Mo Kα radiation ($\lambda = 0.71073$ Å). Crystal data and experimental details are summarized in Table S1. The structure was solved by direct methods using SHELXS-86 [51] and refined by full-matrix least-squares calculations on F^2 with SHELXL [52]. The intensity data were corrected for absorption effects [50]. Anisotropic displacement parameters were refined for all nonhydrogen atomic positions. Hydrogen atoms attached to the carbon and nitrogen atoms were placed in calculated positions (methylene C–H = 0.97, aromatic C–H = 0.93, and amine N–H = 0.89 Å). Water hydrogens have been located on subsequent difference Fourier maps, and their bond lengths were standardized to a value of 0.85 Å. During the refinement, isotropic displacement parameters for H atoms were assigned to be 20% higher than the isotropic equivalent of the atom to which the H atom was bonded. All H atoms were refined as riding. MERCURY [53] was used to prepare drawings. Selected bond distances and angles are reported in Table S2. Geometrical parameters describing intermolecular hydrogen bonds and stacking interactions are listed in Tables S3 and S4, respectively.

2.6. Quantum-Mechanical Methods. The M06-2X functional [54] in combination with the cc-pVTZ basis set [55, 56] for the H, C, N, and Cl atoms and the LanL2TZ(f) basis set [57] for the platinum atoms was used to optimize the geometries of the studied systems. M06-2X is a hybrid metadensity functional and has been recommended for the main group and transition metals thermochemistry and kinetics [58]. To take into account the effect of the solvent, the polarizable continuum model (PCM) [59] was used with water chosen as the solvent. All structures were fully optimized without any geometric constraints. The optimized structures were confirmed to be potential energy minima by vibrational frequency calculations at the same level of theory because no imaginary frequencies were found.

All DFT calculations were performed using the Gaussian 09 program package [60]. The M06-2X method and cc-PVTZ basis set were employed as implemented in the software package, while the LanL2TZ(f) basis set for platinum was obtained from EMSL Basis Set Exchange (https://bse.pnl.gov/bse/portal).

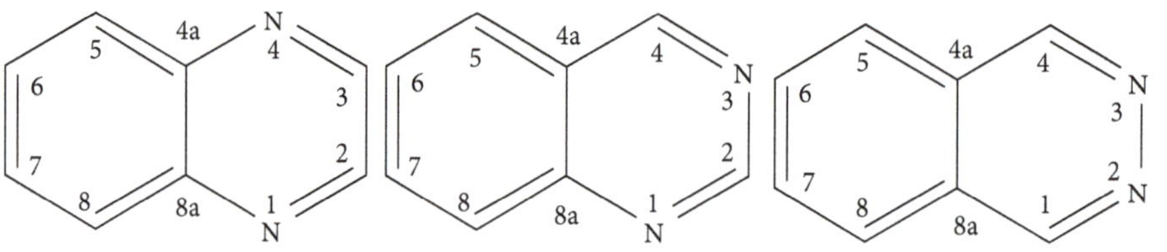

Quinoxaline (qx) Quinazoline (qz) Phthalazine (phtz)

(a)

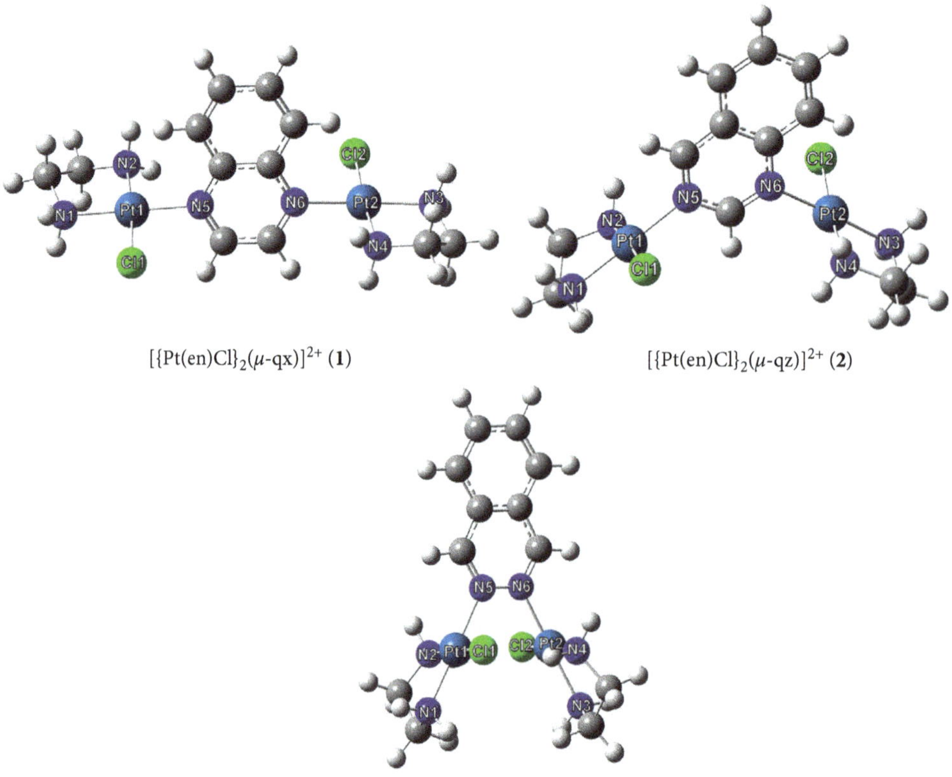

$[\{Pt(en)Cl\}_2(\mu\text{-qx})]^{2+}$ (**1**) $[\{Pt(en)Cl\}_2(\mu\text{-qz})]^{2+}$ (**2**)

$[\{Pt(en)Cl\}_2(\mu\text{-phtz})]^{2+}$ (**3**)

(b)

Figure 1: Schematic drawing of the *N*-heterocycles used in this study (a) and the structures of dinuclear platinum(II) complexes **1–3** calculated at the M06-2X(PCM)/cc-pVTZ+LanL2TZ(f) level of theory. The numbering scheme of carbon atoms in *N*-heterocycles is in agreement with IUPAC recommendations for the fused ring system and does not match the one applied in the X-ray study of **1**.

3. Results and Discussion

*3.1. Synthesis and Structural Features of the Dinuclear Platinum(II) Complexes **1–3**.* Three aromatic *N*-heterocycles, quinoxaline (qx), quinazoline (qz), and phthalazine (phtz) (Figure 1(a)), were used as the bridging ligands between two {Pt(en)Cl} units. All these *N*-heterocycles contain two nitrogen atoms within one ring but at different positions, that is, 1,4 for qx, 1,3 for qz, and 2,3 for phtz, resulting in their different steric and electronic properties. However, despite these differences, they all reacted with Pt(II) ion to form dinuclear species $[\{Pt(en)Cl\}_2(\mu\text{-qx})]Cl_2{\cdot}2H_2O$ (**1**), $[\{Pt(en)Cl\}_2(\mu\text{-qz})](ClO_4)_2$ (**2**), and $[\{Pt(en)Cl\}_2(\mu\text{-phtz})]Cl_2{\cdot}4H_2O$ (**3**) (en is a bidentate-coordinated ethylenediamine). The stoichiometries of **1–3** were confirmed by elemental microanalysis, and the structures emerged from NMR (^1H and ^{13}C), IR, and UV-Vis spectroscopic methods. Crystals of **1** were obtained after the crude product, resulting from the reaction of mononuclear $[Pt(en)Cl_2]$ and quinoxaline, was dissolved in a minimal amount of water saturated with LiCl; the crystal structure of this complex was determined by single-crystal X-ray analysis. Simultaneous attempts to crystallize complexes **2** and **3** from their amorphous powders using different solvents (water, methanol, acetone, chloroform, and dimethylformamide) were unsuccessful. Accordingly, the structures of the complexes **1–3** have been optimized by means of the DFT M06-2X method.

3.1.1. NMR Characterization. Ambient temperature NMR spectra of platinum(II) complexes **1–3** and the corresponding *N*-heterocyclic ligands were measured in D_2O. The ^1H and ^{13}C NMR chemical shifts as well as $\Delta(^1H)_{coord}$ and $\Delta(^{13}C)_{coord}$ coordination shifts for **1–3** determined in respect

TABLE 1: NMR (^{1}H and ^{13}C) chemical and coordination shifts (δ, ppm), alongside multiplicities and coupling constants (J_{H-H}, Hz), for the *N*-heterocyclic ligands and the corresponding dinuclear platinum(II) complexes **1–3** in D_2O as a solvent with TSP as the internal standard.

Atom position	^{1}H		^{13}C	
	qx	**1**	qx	**1**
2,3	8.44, s	9.59, s (+1.15)	145.67	151.77 (+6.10)
5,8	7.51, m	9.73, dd, $J = 6.7$; 3.3 Hz (+2.22)	131.94	135.93 (+3.99)
6,7	7.51, m	8.41, dd, $J = 6.7$; 3.3 Hz (+0.90)	128.83	130.09 (+1.26)
4a,8a	—	—	141.85	145.26 (+3.41)
	qz	**2**	qz	**2**
2	9.09, s	10.12, d, $J = 11.0$ Hz (+1.03)	163.39	178.11 (+14.72)
4	8.84, s	9.52, d, $J = 8.8$ Hz (+0.68)	156.05	168.56 (+12.51)
5	7.80, m	8.51, m	129.07	130.14 (+1.07)
8		(+0.71)	131.53	132.81 (+1.28)
6	7.79, m	8.16, m	130.60	135.11 (+4.51)
7		(+0.37)	138.42	143.10 (+4.68)
4a	—	—	126.98	128.58 (+1.60)
8a	—	—	150.76	152.39 (+1.63)
	phtz	**3**	phtz	**3**
1,4	9.18, s	10.05, s (+0.87)	154.27	164.32 (+10.05)
6,7	7.88, m	8.40, m (+0.52)	136.61	140.50 (+3.89)
5,8	7.88, m	8.40, m (+0.52)	129.38	131.04 (+1.66)
4a,8a	—	—	129.04	130.96 (+1.92)

The aliphatic methylene protons of the bidentatedly coordinated en ligand in **1–3** give a singlet in the region 2.78–2.84 ppm, while the resonance for the corresponding carbon atoms is in the region 49.42–51.27 ppm. s = singlet; d = doublet; dd = doublet of doublets; m = multiplet.

to those for the uncoordinated *N*-heterocycles are listed in Table 1.

The chemical shifts of the used *N*-heterocycles in D_2O are almost identical with those reported in literature for the NMR spectra of these compounds measured in other solvents [61, 62]. In the aromatic region, the ^{1}H NMR spectra of **1** and **2** consist of two multiplets corresponding to the protons of the condensed benzene ring (C5H and C8H, and C6H and C7H, resp.), with the chemical shifts significantly differing from those of the uncoordinated qx and qz ligands. Contrary to this, all benzene protons C5H–C8H in phtz-containing complex **3** give rise to one multiplet. Besides the resonances corresponding to the benzene ring, those due to the protons of the diazine ring can be detectable in the aromatic region of the ^{1}H NMR spectra. Thus, a singlet due to the C2H and C3H, and C1H and C4H for **1** and **3**, respectively, is observed in their spectra. On the other hand, diazine protons C2H and C4H for qz-containing complex **2** are nonequivalent, which results in the appearance of the two doublets. In addition to the characteristic ^{1}H NMR resonances for the aromatic protons of the complexes **1–3**, the aliphatic CH_2 group of the bidentatedly coordinated en ligand gives a singlet in the region 2.78–2.84 ppm. This singlet is on the same chemical shift as that for the mononuclear $[Pt(en)Cl_2]$ complex.

As can be seen from Table 1, the resonances for the aromatic protons of the complexes **1–3** are downfield shifted with respect to those for the uncoordinated *N*-heterocycles. The downfield shifting for the protons in the used *N*-heterocycles after their platination can be ascribed to a delocalization of the charge deficiency (cation formation by Pt(II) coordination) throughout all the rings in the molecules as anticipated [61, 63]. It is important to note that the ^{1}H chemical shifts for the complexes **2** and **3** which contain quinazoline- and phthalazine-bridging ligands, respectively, are in agreement with those for the structurally similar platinum(II) complexes reported previously [64, 65].

The ^{13}C NMR spectra of **1–3** in D_2O due to the aromatic carbons display four (**1** and **3**) and eight (**2**) distinct signals and are noticeably different from those of the free *N*-heterocyclic ligands; their addition to the D_2O solution of the complexes results in the appearance of another set of ^{13}C signals (Table 1). As a consequence of the Pt(II) complexation of the investigated *N*-heterocyclic ligands, all ring carbons are deshielded (up to +14.72 ppm for C2 in complex **2**). The chemical shift of the methylene carbons of ethylenediamine in **1–3** is identical to that of these carbons of $[Pt(en)Cl_2]$ complex ($\delta = 49.42$–51.27 ppm).

3.1.2. IR and UV-Vis Characterization. The IR and UV-Vis spectral data for the dinuclear platinum(II) complexes **1–3** are listed in the Materials and Methods (*vide infra*). The IR spectra of the complexes measured in the wavenumber range of 4000–450 cm^{-1} show the bands attributable to the vibration of the coordinated *N*-heterocyclic ligand, as well as those due to the bidentatedly coordinated en ligand, crystalline water molecules (**1** and **3**), and perchlorate counteranion (**2**). Thus, a broad absorption at ~3400 cm^{-1} is assigned to the stretching vibration of the OH group and confirms the presence of crystalline water molecules in **1** and **3** [66]. Besides, the complexes **1–3** exhibit two very strong and sharp bands at ~3200 and 3100 cm^{-1}, which were assigned, respectively, to the asymmetric and symmetric stretching vibrations of the coordinated amino group of en ligand [67]. Complex **2** exhibits a very strong band with two submaxima at 1096 and 1062 cm^{-1} and a strong one at 623 cm^{-1} which can be attributed to the ν(ClO) and δ(OClO)

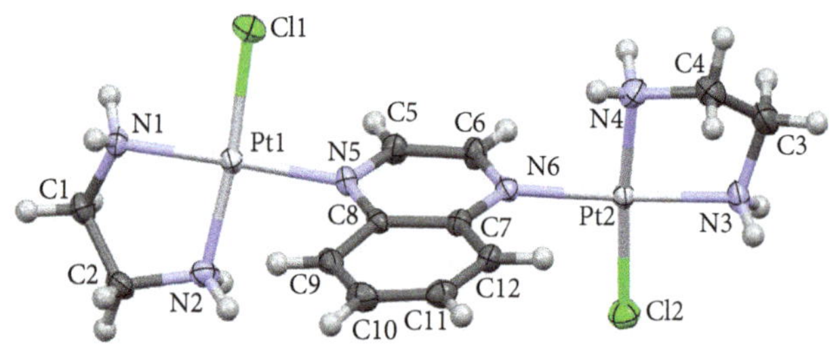

FIGURE 2: Perspective view of the dinuclear complex cation $[\{Pt(en)Cl\}_2(\mu\text{-qx})]^{2+}$ in the crystal of **1** at 295 K. Atomic displacement ellipsoids are drawn at the 40% probability level.

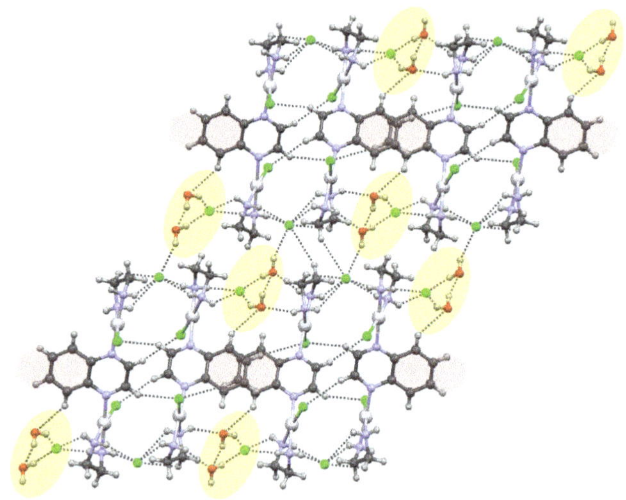

FIGURE 3: Arrangement of molecules constituting the multicomponent crystal of $[\{Pt(en)Cl\}_2(\mu\text{-qx})]Cl_2{\cdot}2H_2O$ (**1**). Hydrogen bonds are marked by dotted lines, and $\pi\cdots\pi$ interactions operate between partially overlapping molecules that stack along the *b*-direction. Structural voids filled by water molecules and chloride ions are shaded in yellow, while benzene rings involved in stacking interactions are shaded in purple.

modes, respectively, of the uncoordinated perchlorates [68]. The presence of the two bands attributed to the asymmetric stretching vibration of ClO_4^- ion can be the consequence of its participation in hydrogen bonding interactions, which results in the lowering of the point group symmetry from T_d to C_{2v}, leading to "pseudomonodentate" spectroscopic behaviour of ClO_4^- [68, 69].

The shape of UV-Vis spectra and the values of λ_{max} are similar for **1–3**, indicating the same bidentate-bridging coordination mode of the corresponding *N*-heterocycle to the Pt(II) ion (Figure S1 in the Supplementary Material). In all complexes, the absorbance peaks at higher energy are due to $\pi \to \pi^*$ transitions in the aromatic *N*-heterocycles [70], and they show significant red shifts compared to those in the free ligands. The second absorbance peak at ~330 nm observed for the complexes corresponds to LMCT (ligand-to-metal charge transfer) transition.

3.1.3. Description of the Crystal Structure of ***1***. The molecular structure and labeling scheme of **1** are shown in Figure 2. X-ray analysis has confirmed that the $[\{Pt(en)Cl\}_2(\mu\text{-qx})]Cl_2{\cdot}2H_2O$ complex is a dinuclear complex of Pt(II) bridged by the quinoxaline ligand. Each Pt(II) ion exhibits an approximately square planar coordination, with one Pt–Cl bond, one Pt–N bond to the qx ligand, and two Pt–N bonds to the same chelating diamine (en) ligand. The Pt···Pt distance is 6.8217(7) Å, comparable with the mean value of 6.82(6) Å obtained from 47 observations (24 hits) for crystal structures containing pyrazine-bridged Pt(II) fragments, deposited in the CSD [71]. Two Cl ligands and two en chelate rings are mutually *trans* oriented. The two Pt–N(en) distances within each en chelate slightly differ; the bond that is *trans* to the Pt–Cl bond is consistently longer than the one that is *trans* to the Pt–N(qx) bond (average values 2.035(1) Å versus 2.013(3) Å). The coordinated diazine ring of qx is significantly inclined with respect to the Pt(II) coordination planes. The dihedral angles between the heterocyclic ring plane and the square plane around each of the two Pt(II) ions amount to 72.76(16)° and 89.79(16)°. Noticeably, in the centrosymmetric $[\{Pt(en)Cl\}_2(\mu\text{-pz})]Cl_2$ complex, the dihedral angle between the pyrazine ring plane and the square plane around Pt(II) amounts to only 58.4(1) Å [35]. The diamine rings adopt the usual twist conformation of the same helicity within one complex molecule. The quinoxaline part is not strictly planar; the two rings are inclined at 4.0(2)°.

There are relatively short intramolecular Pt···H–$C_{benzene}$ contacts of 2.81 and 2.75 Å to Pt1 and Pt2 ions, respectively, which might be connected with a noticeable twisting of the benzene moiety with respect to the Pt–N···N–Pt line. The quoted H···Pt distances belong to the shortest within the platinum(II) complexes with pyrazine and related ligands [71].

The crystal packing of **1** is driven by numerous hydrogen bonds of the NH···Cl, NH···O, OH···Cl, and OH···O type (for geometrical parameters describing these interactions, see Table S3 in the Supplementary Material), supplemented by CH···Cl and CH···O intermolecular interactions. The complex cations form double-molecular columns, extending along the *b*-direction, within which the aromatic rings are involved in weak $\pi\cdots\pi$ interactions (geometrical parameters describing these interactions are included in Table S4). The neighbouring Pt(II) ions along the *b* direction are at a distance of 6.4872(3) Å, while the shortest intermolecular Pt···Pt distance is only 4.6100(5) Å (Pt1···Pt2 at $0.5 - x, 0.5 + y$, and $0.5 - z$) and operates between molecules that are doubly bridged by an uncoordinated Cl3 ion acting as an acceptor of two relatively strong N-H···Cl hydrogen bonds (Table S3; Figure 3). The uncoordinated chloride anions and water molecules are located in crystal voids (Figure 3). The presence of water molecules in this crystal structure can be contrasted with the absence of any solvent molecules in the crystal structures of related Pt(II) complexes, namely, $[\{Pt(en)Cl\}_2(\mu\text{-pz})]Cl_2$ [35] and $[\{Pt(en)Cl\}_2(\mu\text{-pydz})]Cl_2$ [44]. Clearly, the incorporation of an extra benzene ring to the ligand causes noticeable difficulties in close packing of the complex cationic species and consequent inclusion of water molecules in order to fill structural voids. It also causes replacement of anion···π interactions present in the crystal structure of $[\{Pt(en)Cl\}_2(\mu\text{-pz})]Cl_2$ by parallel-displaced slightly attractive $\pi\cdots\pi$ stacking interactions (Figure 3).

3.1.4. Computational Studies. The structures of the dinuclear platinum(II) complexes **1–3** were optimized in water at the M06-2X(PCM)/cc-pVTZ + LanL2TZ(f) level of theory. The optimized structures of **1–3** are shown in Figure 1(b), and the values of the calculated bond lengths and angles are presented in Table S2. This table also compares the DFT-calculated parameters for complex **1** to those inferred from the corresponding X-ray structure. As can be seen, the calculated bond lengths and angles for this complex show very good agreement with the corresponding X-ray data. This supports our recently obtained results showing that the M06-2X method can reproduce to a reasonable good extent the experimental structural parameters for platinum(II) complexes with aromatic *N*-heterocycles [72]. In addition, this method has been shown as appropriate for evaluation of the electron-donating properties of aromatic *N*-heterocycles [73]. The error of the calculated Pt–N and Pt–Cl bond lengths was found to be ~0.03 and 0.04 Å, respectively, whereas calculations reproduced very well nonequivalency of the two Pt–N(en) bonds, that is, the Pt–N(en) bond which is in the *trans* position to the coordinated nitrogen atom of qx and the Pt–N(en) bond which is slightly shorter than the remaining Pt–N(en) bonds (Table S2). Moreover, the calculated N1–Pt1–N2 and N3–Pt2–N4 angles in the five-membered chelate rings show good agreement with the experimentally found ones; these angles (approximately 83°) significantly deviate from the ideal angle of 90°.

From the DFT study, we found that the complexes **2** and **3** have the same square planar geometry as crystallographically characterized complex **1** (Figure 1(b)). These complexes contain two Pt(II) ions, which are coordinated to two nitrogen atoms of the chelating en ligand, nitrogen of the bridging qz and phtz ligands, and chlorine in the fourth coordination site. The Pt–N and Pt–Cl bond lengths in these complexes are in the expected range and compare well with those in **1** and the other platinum(II) complexes with the same [N_3Cl] coordination sphere, such as $[\{Pt(en)Cl\}_2(\mu\text{-pz})]^{2+}$ and $[\{Pt(en)Cl\}_2(\mu\text{-pydz})]^{2+}$ [35, 44]. Furthermore, the calculations for **2** and **3** provide the same difference between the Pt–N(en) bond lengths as in **1**, as well as significant deviation of the N1–Pt1–N2 and N3–Pt2–N4 angles within the five-membered chelate rings from the angle of 90°.

Similarly to $[\{Pt(en)Cl\}_2(\mu\text{-pydz})]^{2+}$ complex, in which the intramolecular distance between two Pt(II) ions was found to be 3.2535(4) Å [44], in $[\{Pt(en)Cl\}_2(\mu\text{-phtz})]^{2+}$ (**3**), two Pt(II) ions are only 3.2392 Å apart. It is important to note that the aromatic *N*-heterocycles, pyridazine and phthalazine, which act as bridging ligands in these two complexes, contain two nitrogen atoms in the *ortho*-position. The intramolecular distance between two Pt(II) ions in **3** is shorter than the sum of van der Waals radii for two Pt(II) ions which amounts to 3.50 Å [74]. On the other hand, this distance in **1** and **2** is calculated to be 6.8596 and 5.7914 Å, respectively, and is significantly longer than the sum of van der Waals radii for Pt(II) ions. This is a direct consequence of the *para*- and *meta*-positions of the two nitrogen donors in qx and qz ligands, respectively.

*3.2. Reactions of Aqua Derivatives of Dinuclear Platinum(II) Complexes **1–3** with Ac–L–Met–Gly and Ac–L–His–Gly Dipeptides.* The chlorido Pt(II) complexes **1–3** were converted into the corresponding aqua species **1a–3a**, $[\{Pt(en)(H_2O)\}_2(\mu\text{-qx})]^{4+}$ (**1a**), $[\{Pt(en)(H_2O)\}_2(\mu\text{-qz})]^{4+}$ (**2a**), and $[\{Pt(en)(H_2O)\}_2(\mu\text{-phtz})]^{4+}$ (**3a**), and their reactions with Ac–L–Met–Gly and Ac–L–His–Gly were studied by ^1H NMR spectroscopy. Complexes **1a–3a** and the corresponding dipeptide were reacted in 1 : 1 molar ratio, and all reactions were performed in the pH range $2.0 < pH < 2.5$ at 37°C. In the reaction between **1a–3a** and Ac–L–Met–Gly, only one Pt(II)–peptide product was obtained. In this product, complexes **1a–3a** are monodentatedly coordinated to the dipeptide through its methionine sulfur atom (Figure 4(a)). The binding of **1a–3a** to the methionine side chain was observed by the simultaneous decline of the resonance at 2.11 ppm due to the S-methyl protons of the free dipeptide and the growth of a resonance at 2.35–2.55 ppm, corresponding to the S-methyl protons of the dipeptide coordinated to platinum (II) [18–20]. The platinum(II)-dipeptide products formed in these reactions are intermediate species, and they promote hydrolysis of the Met–Gly amide bond in Ac–L–Met–Gly dipeptide.

The amount of the hydrolyzed Met–Gly amide bond in these reactions was determined by integration of the resonance for the glycine protons of the dipeptide attached to platinum(II) (4.02 ppm) and that for these protons of the free glycine (3.76 ppm). The changes in concentrations of the free glycine and nonhydrolyzed Ac–L–Met–Gly bound to platinum(II) were determined every 30 min during 24 h. During this time, the total amount of the platinum(II)-peptide product and free glycine was always equal to the initial concentration of Ac–L–Met–Gly dipeptide. The time dependence of the hydrolytic cleavage of the Met–Gly amide bond in the reaction between **1a–3a** and Ac–L–Met–Gly dipeptide is given in Figure 5. From this Figure, it can be concluded that the rate of the hydrolysis of the Ac–L–Met–Gly dipeptide resulting from the reaction with complexes **1a–3a** decreases in the following order: **1a** > **2a** > **3a**.

When an equimolar amount of complexes **1a–3a** was reacted with Ac–L–His–Gly dipeptide, under the above-mentioned experimental conditions, no reaction was observed by proton NMR spectroscopy for the complexes **2a** and **3a** during 48 h. However, in the reaction between **1a** and this dipeptide, two platinum(II)-dipeptide products were observed in the reaction mixture after 30 min of reaction time (Figure 4(b)). These products were distinguished by observing the changes in the chemical shifts of two imidazole C2H and C5H protons with respect to those of free Ac–L–His–Gly dipeptide ($\delta_{C2H} = 8.61$ and $\delta_{C5H} = 7.33$ ppm). Additionally, the chemical shifts of these two protons for the investigated products were compared with those previously reported for the platinum(II)-peptide complexes obtained from the reactions of Ac–L–His–Gly with various dinuclear platinum(II) complexes [44, 75]. Based on this, we have found that the two platinum(II)-dipeptide products were constitutional isomers with a monodentate coordination of **1a** *via* either N3 ($\delta_{C2H} = 8.12$ ppm and $\delta_{C5H} = 7.21$ ppm) or N1 ($\delta_{C2H} = 7.95$ ppm and $\delta_{C5H} = 6.89$ ppm) atom to the

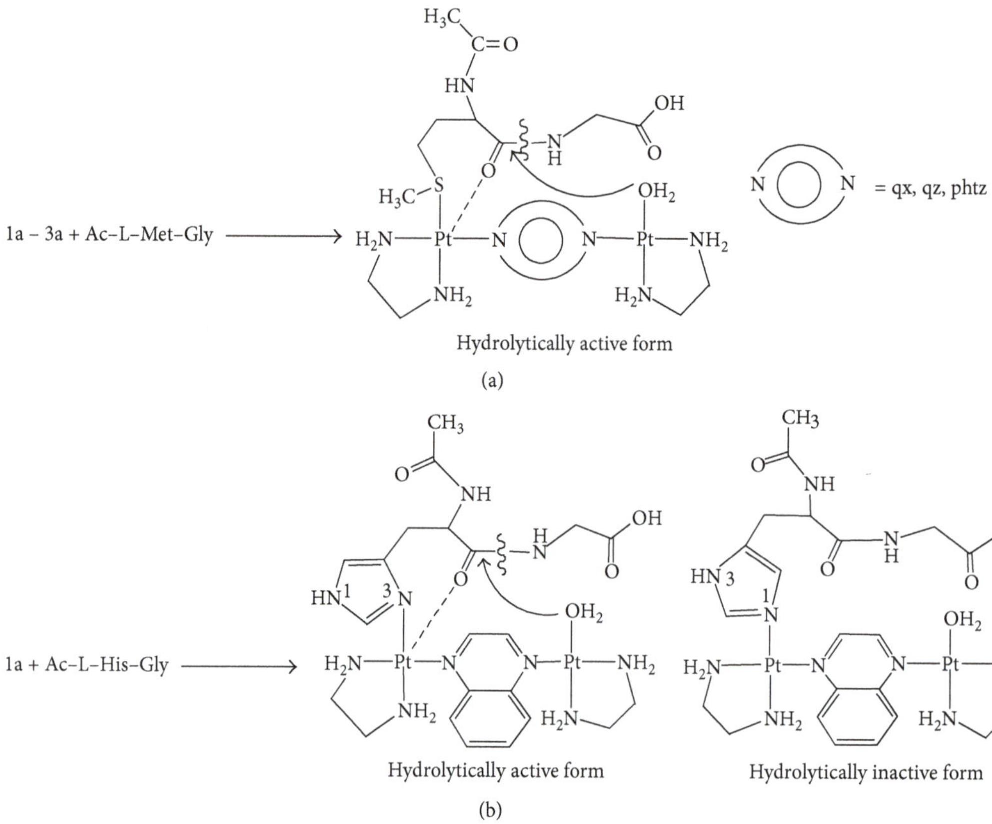

FIGURE 4: Schematic presentation of the hydrolytic reaction of Ac-L-Met-Gly (a) and Ac-L-His-Gly (b) in the presence of complexes **1a–3a** in the pH range 2.0 < pH < 2.5 in D_2O as a solvent and at 37°C. In the reaction with Ac-L-His-Gly, only 1a was catalytically active complex.

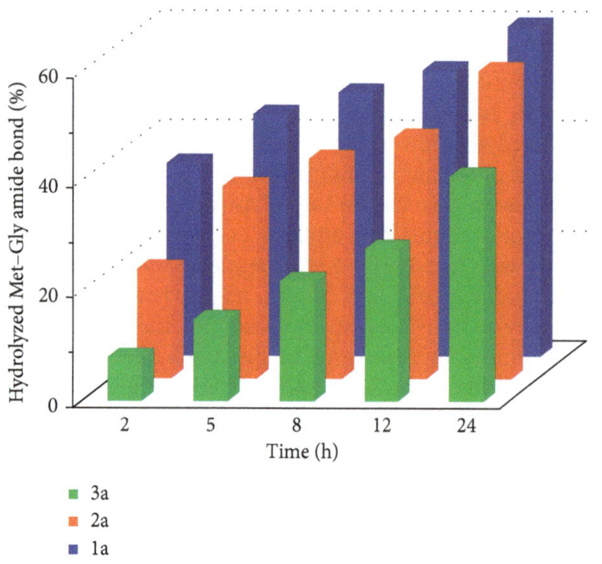

FIGURE 5: Time dependence of the hydrolytic cleavage of the Met-Gly amide bond in Ac-L-Met-Gly dipeptide in the presence of equimolar amount of **1a–3a** complexes in the pH range 2.0 < pH < 2.5 and at 37°C.

imidazole ring. When the reaction between **1a** and Ac-L-His-Gly was monitored with time, a new signal at 3.64 ppm appeared in the ^1H NMR spectrum which showed a tendency to increase during the passage of time. We have found that this signal belonged to the methylene protons of the free amino acid glycine. Upon addition of glycine to the reaction mixture, this resonance was enhanced. Additionally, the resonance at 3.83 ppm due to glycine protons of Ac-L-His-Gly dipeptide was decreasing gradually with the passing time. We assumed that these changes in the ^1H NMR spectrum resulted from the hydrolysis of the His-Gly amide bond in Ac-L-His-Gly dipeptide. This assumption was based on our previous findings that only monodentate coordination of the dinuclear platinum(II) complex to the N3 imidazole nitrogen atom of histidine-containing peptides promotes cleavage of the amide bond involving the carboxylic group of the anchoring amino acid histidine [44, 75]. The concentrations of Ac-L-His-Gly and the hydrolysis products were determined from the known initial concentration of the dipeptide and from the integrated resonance of the free glycine. The cleavage of Ac-L-His-Gly was regioselective, and about 35% of the His-Gly bond in this dipeptide has cleaved after 48 h.

Complexes **1a–3a** showed different catalytic activities in the hydrolytic cleavage of Ac-L-Met-Gly and Ac-L-His-Gly dipeptides which resulted from the presence of different *N*-heterocycles as the bridging ligands in these complexes. In the presence of Ac-L-Met-Gly dipeptide, all complexes reacted with the methionine sulfur atom and cleaved the Met-Gly amide bond in this dipeptide, although with

TABLE 2: Comparison of the catalytic activities between **1a** and $[\{Pt(en)(H_2O)\}_2(\mu\text{-pz})]^{4+}$ [35] and **3a** and $[\{Pt(en)(H_2O)\}_2(\mu\text{-pydz})]^{4+}$ [44] complexes in the hydrolysis of Ac–L–Met–Gly dipeptide.

Platinum(II) complex	Hydrolyzed Met-Gly amide bond (%)		
	2 h	12 h	24 h
$[\{Pt(en)(H_2O)\}_2(\mu\text{-qx})]^{4+}$ (**1a**)	35	52	60
$[\{Pt(en)(H_2O)\}_2(\mu\text{-pz})]^{4+}$	60	85	88
$[\{Pt(en)(H_2O)\}_2(\mu\text{-phtz})]^{4+}$(**3a**)	8	28	41
$[\{Pt(en)(H_2O)\}_2(\mu\text{-pydz})]^{4+}$	12	36	54

a different reaction rate. In a given time interval, the percentage of the hydrolyzed dipeptide varied depending on the type of the complex used for the reaction and decreased in the following order: **1a** > **2a** > **3a**. We compared catalytic activities of **1a** with $[\{Pt(en)(H_2O)\}_2(\mu\text{-pz})]^{4+}$ [35] and **3a** with $[\{Pt(en)(H_2O)\}_2(\mu\text{-pydz})]^{4+}$ [44] in the hydrolysis of Ac–L–Met–Gly dipeptide (Table 2). All reactions were performed under the same experimental conditions, in the pH range 2.0–2.5 and at 37°C. Complexes **1a** and $[\{Pt(en)(H_2O)\}_2(\mu\text{-pz})]^{4+}$, with quinoxaline- and pyrazine-bridging ligands, respectively, have *para*-positioned nitrogen atoms. By contrast, complexes **3a** and $[\{Pt(en)(H_2O)\}_2(\mu\text{-pydz})]^{4+}$, with phthalazine- and pyridazine-bridging ligands, respectively, have nitrogen atoms in an *ortho*-position. As can be seen from Table 2, complexes **1a** and **3a** showed lower catalytic activity than $[\{Pt(en)(H_2O)\}_2(\mu\text{-pz})]^{4+}$ and $[\{Pt(en)(H_2O)\}_2(\mu\text{-pydz})]^{4+}$ complexes, respectively. This can be attributed to the hindering role of the additional benzene ring in quinoxaline- and phthalazine-bridging ligands.

In the presence of Ac–L–His–Gly dipeptide, only complex **1a** reacted with the histidine side chain, promoting cleavage of the His–Gly amide bond, while the other two complexes **2a** and **3a** remained inactive. Obviously, this had to be connected with different positioning of the nitrogen atoms in the bridging qx, qz, and phtz ligands and consequent variation in intramolecular distances between the two platinum(II) ions in these dimetallic complexes. As mentioned above, the DFT calculated values for the intramolecular distances between two platinum(II) ions in the corresponding chlorido complexes **1–3** were 6.8596, 5.7914, and 3.2392 Å, respectively. One would expect these values to be nearly the same in complexes **1a–3a**, which are formed upon displacement of a chlorido ligand by a water molecule. The significantly shorter Pt···Pt distances in complexes **2a** and **3a** in comparison with **1a** indicate an increased steric crowding in the former two complexes, which is a direct cause of a total inhibition of the reaction with Ac–L–His–Gly and slower cleavage of the amide bond in Ac–L–Met–Gly dipeptides. Similarly to **1a**, the $[\{Pt(en)(H_2O)\}_2(\mu\text{-pz})]^{4+}$ complex cation, with the pz-bridging ligand having two nitrogen atoms in *para*-position, reacts with the histidine side chain and promotes the cleavage of the His–Gly amide bond [35]. Like for **2a** and **3a**, no reaction with the histidine side chain was observed for $[\{Pt(en)(H_2O)\}_2(\mu\text{-pydz})]^{4+}$ complex, with nitrogen atoms in the bridging ligands positioned in *ortho*-position of the aromatic ring [44].

4. Conclusions

Three isomeric six-membered aromatic diazines with an additional fused benzene ring, quinoxaline, quinazoline, and phthalazine, were shown as good bridging ligands between two {Pt(en)Cl} units forming dinuclear complexes **1–3**. From the DFT study, it was found that complexes **2** and **3** have the same square planar geometry as crystallographically characterized complex **1**. Compared to the $[\{Pt(en)Cl\}_2(\mu\text{-pz})]Cl_2$ parent compound [35], complex **1** contains an additional benzene ring and its hindering role is noticeable in the crystal structure on both molecular and supramolecular levels. The relative position of the two nitrogen atoms in the bridging aromatic diazine ring has a substantial effect on the catalytic activity of the aqua derivatives of the corresponding complexes **1–3** with methionine- and histidine-containing dipeptides. All complexes bind to the methionine side chain of Ac–L–Met–Gly and promote cleavage of the amide bond involving the methionine carboxylic group. However, only aqua platinum(II) complex with the quinoxaline-bridging ligand shows the catalytic ability in the reaction with Ac–L–His–Gly dipeptide. Our present results concerning dinuclear platinum(II) complexes with benzene-fused aromatic diazines as bridging ligands along with those previously reported for analogous platinum(II) complexes with pyrazine and pyridazine [35, 44, 48, 75] showed that selective cleavage of a peptide molecule containing both methionine and histidine side chains can be achieved by dinuclear platinum(II) complexes containing *ortho*- and *meta*-positioned nitrogen atoms in the aromatic diazine-bridging ligand.

Acknowledgments

This work was financially supported by the Ministry of Education, Science and Technological Development of the Republic of Serbia under Grant no. 172036, the Supra-MedChem@Balkans.Net SCOPES Institutional Partnership (Project no. IZ74Z0_160515), and the Serbian Academy of Sciences and Arts (Project no. F128).

Supplementary Materials

Figure S1: the UV-Vis spectra for dinuclear platinum(II) complexes **1–3**, which indicated the same bidentate-bridging coordination mode of the corresponding *N*-heterocycle to the Pt(II) ion. Table S1: crystal data and experimental details for $[\{Pt(en)Cl\}_2(\mu\text{-qx})]Cl_2{\cdot}2H_2O$ complex (**1**). Table S2: the selected bond distances and angles for $[\{Pt(en)Cl\}_2(\mu\text{-qx})]Cl_2{\cdot}2H_2O$ complex (**1**). Table S3: Geometrical parameters describing intermolecular hydrogen bonds. Table S4: Geometrical parameters describing stacking interactions. (*Supplementary Materials*)

References

[1] A. Radzicka and R. Wolfenden, "Rates of uncatalyzed peptide bond hydrolysis in neutral solution and the transition state affinities of proteases," *Journal of the American Chemical Society*, vol. 118, no. 26, pp. 6105–6109, 1996.

[2] D. W. Christianson and W. N. Lipscomb, "Carboxypeptidase A," *Accounts of Chemical Research*, vol. 22, no. 2, pp. 62–69, 1989.

[3] E. Bamann, J. G. Hass, and H. Trapmann, "Metallionenkatalytische zerlegung der (-CO–NH–)-bindung in peptiden und N-acetyl-aminosäuren," *Archiv der Pharmazie*, vol. 294, no. 9, pp. 569–580, 1961.

[4] M. Yashiro, Y. Sonobe, A. Yamamura, T. Takarada, M. Komiyama, and Y. Fujii, "Metal-ion-assisted hydrolysis of dipeptides involving a serine residue in a neutral aqueous solution," *Organic and Biomolecular Chemistry*, vol. 1, no. 4, pp. 629–632, 2003.

[5] K. W. Bentley and E. H. Creaser, "Qualitative determination of *N*-terminal amino acids of peptides and proteins with cobalt(III) chelates," *Biochemical Journal*, vol. 135, no. 3, pp. 507–511, 1973.

[6] R. Miyake, K. Murakami, J. T. Owens et al., "Dimeric association of *Escherichia coli* RNA polymerase α subunits, studied by cleavage of single-cysteine α subunits conjugated to iron-(S)-1-[p-(bromoacetamido)benzyl]ethylenediaminetetraacetate," *Biochemistry*, vol. 37, no. 5, pp. 1344–1349, 1998.

[7] N. Ettner, J. W. Metzger, T. Lederer et al., "Proximity mapping of the tet repressor-tetracycline-Fe^{2+} complex by hydrogen peroxide mediated protein cleavage," *Biochemistry*, vol. 34, no. 1, pp. 22–31, 1995.

[8] E. Heyduk and T. Heyduk, "Mapping protein domains involved in macromolecular interactions: a novel protein footprinting approach," *Biochemistry*, vol. 33, no. 32, pp. 9643–9650, 1994.

[9] A. Mustaev, M. Kozlov, V. Markovtsov, E. Zaychikov, L. Denissova, and A. Goldfarb, "Modular organization of the catalytic center of RNA polymerase," *Proceedings of the National Academy of Sciences*, vol. 94, no. 13, pp. 6641–6645, 1997.

[10] Y. Fujii, T. Kiss, T. Gajda et al., "Copper(II)-*cis,cis*-1,3,5-triaminocyclohexane complex-promoted hydrolysis of dipeptides: kinetic, speciation and structural studies," *Journal of Biological Inorganic Chemistry*, vol. 7, no. 7-8, pp. 843–851, 2002.

[11] E. L. Hegg and J. N. J. Burstyn, "Hydrolysis of unactivated peptide bonds by a macrocyclic copper(II) complex: Cu([9]aneN3)Cl2 hydrolyzes both dipeptides and proteins," *Journal of the American Chemical Society*, vol. 117, no. 26, pp. 7015-7016, 1995.

[12] L. Zhang, Y. Mei, Y. Zhang, S. Li, X. Sun, and L. Zhu, "Regioselective cleavage of myoglobin with copper(II) compounds at neutral pH," *Inorganic Chemistry*, vol. 42, no. 2, pp. 492–498, 2003.

[13] A. Krezel, E. Kopera, A. M. Protas, J. Poznanski, A. Wysłouch-Cieszynska, and W. Bal, "Sequence-specific Ni(II)-dependent peptide bond hydrolysis for protein engineering. Combinatorial library determination of optimal sequences," *Journal of the American Chemical Society*, vol. 132, no. 10, pp. 3355–3366, 2010.

[14] E. Kopera, A. Krezel, A. M. Protas et al., "Sequence-specific Ni(II)-dependent peptide bond hydrolysis for protein engineering: reaction conditions and molecular mechanism," *Inorganic Chemistry*, vol. 49, no. 14, pp. 6636–6645, 2010.

[15] T. Takarada, M. Yashiro, and M. Komiyama, "Catalytic hydrolysis of peptides by cerium(IV)," *Chemistry*, vol. 6, no. 21, pp. 3906–3913, 2000.

[16] M. Kassai and K. B. Grant, "Tuning Zr(IV)-assisted peptide hydrolysis at near-neutral pH," *Inorganic Chemistry Communications*, vol. 11, no. 5, pp. 521–525, 2008.

[17] M. Kassai, R. G. Ravi, S. J. Shealy, and K. B. Grant, "Unprecedented acceleration of zirconium(IV)-assisted peptide hydrolysis at neutral pH," *Inorganic Chemistry*, vol. 43, no. 20, pp. 6130–6132, 2004.

[18] I. E. Burgeson and N. M. Kostić, "Selective hydrolysis of unactivated peptide bonds, promoted by platinum(II) complexes anchored to amino acid side chains," *Inorganic Chemistry*, vol. 30, no. 23, pp. 4299–4305, 1991.

[19] N. M. Milović and N. M. Kostić, "Palladium(II) and platinum(II) complexes as synthetic peptidases," in *Metal Ions in Biological Systems*, A. Sigel and H. Sigel, Eds., vol. XXXVIII, pp. 145–186, Marcel Dekker Inc., New York, NY, USA, 2001.

[20] S. Rajković, M. D. Živković, C. Kállay, I. Sóvágó, and M. I. Djuran, "A study of the reactions of a methionine- and histidine-containing tetrapeptide with different Pd(II) and Pt(II) complexes: selective cleavage of the amide bond by platination of the peptide and steric modification of the catalyst," *Dalton Transactions*, no. 39, pp. 8370–8377, 2009.

[21] L. Zhu and N. M. Kostić, "Toward artificial metallopeptidases: mechanisms by which platinum(II) and palladium(II) complexes promote selective, fast hydrolysis of unactivated amide bonds in peptides," *Inorganic Chemistry*, vol. 31, no. 19, pp. 3994–4001, 1992.

[22] L. Zhu and N. M. Kostić, "Selective hydrolysis of peptides, promoted by palladium aqua complexes: kinetic effects of the leaving group, pH, and inhibitors," *Journal of the American Chemical Society*, vol. 115, no. 11, pp. 4566–4570, 1993.

[23] L. Zhu and N. M. Kostić, "Hydrolytic cleavage of peptides by palladium(II) complexes is enhanced as coordination of peptide nitrogen to palladium(II) is suppressed," *Inorganica Chimica Acta*, vol. 217, no. 1-2, pp. 21–28, 1994.

[24] E. N. Korneeva, M. V. Ovchinnikov, and N. M. Kostić, "Peptide hydrolysis promoted by polynuclear and organometallic complexes of palladium(II) and platinum(II)," *Inorganica Chimica Acta*, vol. 243, no. 1-2, pp. 9–13, 1996.

[25] D. P. Ašanin, S. Rajković, D. Molnar-Gabor, and M. I. Djuran, "Hydrolysis of the peptide bond in *N*-Acetylated L-methionylglycine catalyzed by various palladium (II) complexes: dependence of the hydrolytic reactions on the nature of the chelate ligand in *cis*-$[Pd(L)(H_2O)_2]^{2+}$ complexes," *Monatshefte für Chemie/Chemical Monthly*, vol. 135, no. 12, pp. 1445–1453, 2004.

[26] T. N. Parac and N. M. Kostić, "New selectivity and turnover in peptide hydrolysis by metal complexes. A palladium(II) aqua complex catalyzes cleavage of peptides next to the histidine residue," *Journal of the American Chemical Society*, vol. 118, no. 1, pp. 51–58, 1996.

[27] T. N. Parac and N. M. Kostić, "Effects of linkage isomerism and of acid–base equilibria on reactivity and catalytic turnover in hydrolytic cleavage of histidyl peptides coordinated to palladium(II). Identification of the active complex between palladium(II) and the histidyl residue," *Journal of the American Chemical Society*, vol. 118, no. 25, pp. 5946–5951, 1996.

[28] S. U. Milinković, T. N. Parac, M. I. Djuran, and N. M. Kostić, "Dependence of hydrolytic cleavage of histidine-containing peptides by palladium(II) aqua complexes on the coordination modes of the peptides," *Journal of the Chemical Society, Dalton Transactions*, no. 16, pp. 2771–2776, 1997.

[29] X. Chen, L. Zhu, H. Yan, X. You, and N. M. Kostić, "Kinetic study of stereochemical and other factors governing hydrolytic cleavage of a peptide ligand in binuclear palladium(II) complexes," *Journal of the Chemical Society, Dalton Transactions*, no. 13, pp. 2653–2658, 1996.

[30] M. I. Djuran and S. U. Milinković, "NMR study of the interaction of palladium(II) complexes with some histidine-containing peptides: effects of the mode of coordination on hydrolytic reactions," *Monatshefte für Chemie*, vol. 130, no. 5, pp. 613–622, 1999.

[31] M. I. Djuran and S. U. Milinković, "Hydrolysis of amide bond in histidine-containing peptides promoted by chelated amino acid palladium(II) complexes: dependence of hydrolytic pathway on the coordination modes of the peptides," *Polyhedron*, vol. 18, no. 27, pp. 3611–3616, 1999.

[32] M. I. Djuran and S. U. Milinković, "Selective hydrolysis of the unactivated peptide bond in *N*-acetylated L-histidylglycine catalyzed by various palladium(II) complexes: dependence of the hydrolysis rate on the steric bulk of the catalyst," *Polyhedron*, vol. 19, no. 8, pp. 959–963, 2000.

[33] Z. D. Petrović, M. I. Djuran, F. W. Heinemann, S. Rajković, and S. R. Trifunović, "Synthesis, structure, and hydrolytic reaction of trans-dichlorobis(diethanolamine-N)palladium(II) with *N*-acetylated L-histidylglycine dipeptide," *Bioorganic Chemistry*, vol. 34, no. 5, pp. 225–234, 2006.

[34] M. D. Živković, S. Rajković, U. Rychlewska, B. Warzajtis, and M. I. Djuran, "A study of the reactions of methionine- and histidine-containing peptides with palladium(II) complexes: the key role of steric crowding on palladium(II) in the selective cleavage of the peptide bond," *Polyhedron*, vol. 26, no. 7, pp. 1541–1549, 2007.

[35] D. P. Ašanin, M. D. Živković, S. Rajković, B. Warzajtis, U. Rychlewska, and M. I. Djuran, "Crystallographic evidence of anion···π interactions in the pyrazine bridged {[Pt(en)Cl]$_2$(μ-pz)}Cl$_2$ complex and a comparative study of the catalytic ability of mononuclear and binuclear platinum(II) complexes in the hydrolysis of *N*-acetylated L-methionylglycine," *Polyhedron*, vol. 51, pp. 255–262, 2013.

[36] N. Strater, W. N. Lipscomb, T. Kabunde, and B Krebs, "Two-metal ion catalysis in enzymatic acyl- and phosphoryl-transfer reactions," *Angewandte Chemie International Edition in English*, vol. 35, no. 18, pp. 2024–2055, 1996.

[37] N. Strater, T. Kabunde, P. Tucker, H. Witzel, and B. Krebs, "Crystal structure of a purple acid phosphatase containing a dinuclear Fe(III)-Zn(II) active site," *Science*, vol. 268, no. 5216, pp. 1489–1492, 1995.

[38] M. Deitrich, D. Munstermann, H. Suerbaum, and H. Witzel, "Purple acid phosphatase from bovine spleen. Interactions at the active site in relation to the reaction mechanism," *European Journal of Biochemistry*, vol. 199, no. 1, pp. 105–113, 1991.

[39] D. E. Wilcox, "Binuclear metallohydrolases," *Chemical Reviews*, vol. 96, no. 7, pp. 2435–2458, 1996.

[40] M. A. Aquino, J. Lim, and A. G. Sykes, "Mechanism of the reaction of phosphate with purple acid phosphatase," *Journal of the Chemical Society, Dalton Transactions*, no. 13, pp. 2135-2136, 1992.

[41] E. Hough, L. K. Hansen, B. Birknes et al., "High-resolution (1.5 Å) crystal structure of phospholipase C from *Bacillus cereus*," *Nature*, vol. 338, pp. 357–360, 1989.

[42] A. Volbeda, A. Lahm, F. Sakiyama, and D. Suck, "Crystal structure of *Penicillium citrinum* P1 nuclease at 2.8 A resolution," *EMBO Journal*, vol. 10, no. 7, pp. 1607–1618, 1991.

[43] E. E. Kim and H. W. Wyckoff, "Reaction mechanism of alkaline phosphatase based on crystal structures: two-metal ion catalysis," *Journal of Molecular Biology*, vol. 218, no. 2, pp. 449–464, 1991.

[44] S. Rajković, U. Rychlewska, B. Warżajtis, D. P. Ašanin, M. D Živković, and M. I. Djuran, "Disparate behavior of pyrazine and pyridazine platinum(II) dimers in the hydrolysis of histidine- and methionine-containing peptides and unique crystal structure of {[Pt(en)Cl]$_2$(μ-pydz)}Cl$_2$ with a pair of NH···Cl–···HN hydrogen bonds supporting the pyridazine bridge," *Polyhedron*, vol. 67, pp. 279–285, 2014.

[45] H. Hohmann and R. van Eldik, "Rate and equilibrium data for substitution reactions of diaqua(ethylenediamine)palladium(II) with chloride in aqueous solution," *Inorganica Chimica Acta*, vol. 174, no. 1, pp. 87–92, 1990.

[46] M. D. Živković, D. P. Ašanin, S. Rajković, and M. I. Djuran, "Hydrolysis of the amide bond in *N*-acetylated L-methionylglycine catalyzed by various platinum(II) complexes under physiologically relevant conditions," *Polyhedron*, vol. 30, no. 6, pp. 947–952, 2011.

[47] S. Komeda, G. V. Kalayda, M. Lutz et al., "New isomeric azine-bridged dinuclear platinum(II) complexes circumvent cross-resistance to cisplatin," *Journal of Medicinal Chemistry*, vol. 46, no. 7, pp. 1210–1219, 2003.

[48] S. Rajković, D. P. Ašanin, M. D Živković, and M. I. Djuran, "Synthesis of different pyrazine-bridged platinum(II) complexes and ^{1}H NMR study of their catalytic abilities in the hydrolysis of the *N*-acetylated L-methionylglycine," *Polyhedron*, vol. 65, pp. 42–47, 2013.

[49] G. Mahal and R. van Eldik, "Kinetics and mechanism of the formation, aquation, and base hydrolysis reactions of a series of monodentate carbonato complexes of palladium(II)," *Inorganic Chemistry*, vol. 24, no. 24, pp. 4165–4170, 1985.

[50] *CrysAlisPro 1.171.38.46, Program for Data Collection, Cell Refinement and Data Reduction, Rigaku Oxford Diffraction*, The Woodlands, TX, USA, 2015.

[51] G. M. Sheldrick, "A short history of SHELX," *Acta Crystallographica Section A Foundations of Crystallography*, vol. 64, no. 1, pp. 112–122, 2008.

[52] G. M. Sheldrick, "Crystal structure refinement with *SHELXL*," *Acta Crystallographica Section C Structural Chemistry*, vol. 71, no. 1, pp. 3–8, 2015.

[53] I. J. Bruno, J. C. Cole, P. R. Edgington et al., "New software for searching the Cambridge Structural Database and visualizing crystal structures," *Acta Crystallographica Section B Structural Science*, vol. 58, no. 3, pp. 389–397, 2002.

[54] Y. Zhao and D.G. Truhlar, "A new local density functional for main-group thermochemistry, transition metal bonding, thermochemical kinetics, and noncovalent interactions," *Journal of Chemical Physics*, vol. 125, no. 19, pp. 194101–194118, 2006.

[55] T. H. Dunning Jr., "Gaussian basis sets for use in correlated molecular calculations. I. The atoms boron through neon and hydrogen," *Journal of Chemical Physics*, vol. 90, no. 2, pp. 1007–1023, 1989.

[56] D. E. Woon and T. H. Dunning Jr., "Gaussian basis sets for use in correlated molecular calculations. III. The atoms aluminum through argon," *Journal of Chemical Physics*, vol. 98, no. 2, pp. 1358–1371, 1993.

[57] P. J. Hay and W. R. Wadt, "Ab initio effective core potentials for molecular calculations. Potentials for the transition metal atoms Sc to Hg," *Journal of Chemical Physics*, vol. 82, no. 1, pp. 270–283, 1985.

[58] Y. Zhao and D. G. Truhlar, "The M06 suite of density functionals for main group thermochemistry, thermochem-

ical kinetics, noncovalent interactions, excited states, and transition elements: two new functionals and systematic testing of four M06-class functionals and 12 other functionals," *Theoretical Chemistry Accounts*, vol. 120, no. 1–3, pp. 215–241, 2008.

[59] J. Tomasi, B. Mennucci, and R. Cammi, "Quantum mechanical continuum solvation models," *Chemical Reviews*, vol. 105, no. 8, pp. 2999–3094, 2005.

[60] M. J. Frisch, G. W. Trucks, H. B. Schlegel et al., *Gaussian 09, Revision D.01*, Gaussian Inc., Wallingford, CT, USA, 2013.

[61] B. Đ. Glišić, B. Warżajtis, N. S. Radulović, U. Rychlewska, and M. I. Djuran, "Gold(III) complexes with phenazine and quinoxaline: the role of molecular symmetry in intra- and intermolecular interactions," *Polyhedron*, vol. 87, pp. 208–214, 2015.

[62] B. Đ. Glišić, L. Senerovic, P. Comba et al., "Silver(I) complexes with phthalazine and quinazoline as effective agents against pathogenic *Pseudomonas aeruginosa* strains," *Journal of Inorganic Biochemistry*, vol. 155, pp. 115–128, 2006.

[63] B. Warżajtis, B. Đ. Glišić, N. S. Radulović, U. Rychlewska, and M. I. Djuran, "Gold(III) complexes with monodentate coordinated diazines: an evidence for strong electron-withdrawing effect of Au(III) ion," *Polyhedron*, vol. 79, pp. 221–228, 2014.

[64] G. V. Kalayda, S. Komeda, K. Ikeda, T. Sato, M. Chikuma, and J. Reedijk, "Synthesis, structure, and biological activity of new azine-bridged dinuclear platinum(II) complexes–a new class of anticancer compounds," *European Journal of Inorganic Chemistry*, vol. 2003, no. 24, pp. 4347–4355, 2003.

[65] P. O. Ongoma and D. Jaganyi, "Substitution reactions in dinuclear platinum(II) complexes: an evaluation of the influence of the diazine-bridging ligand on reactivity," *Transition Metal Chemistry*, vol. 38, no. 5, pp. 587–601, 2013.

[66] B. Đ. Glišić, I. Aleksić, P. Comba et al., "Copper(II) complexes with aromatic nitrogen-containing heterocycles as effective inhibitors of quorum sensing activity in *Pseudomonas aeruginosa*," *RSC Advances*, vol. 6, no. 89, pp. 86695–86709, 2016.

[67] S. Chattopadhyay, P. Chakraborty, M. G. B. Drew, and A. Ghosh, "Nickel(II) complexes of terdentate or symmetrical tetradentate Schiff bases: evidence of the influence of the counter anions in the hydrolysis of the imine bond in Schiff base complexes," *Inorganica Chimica Acta*, vol. 362, no. 2, pp. 502–508, 2009.

[68] K. N. Lazarou, I. Chadjistamatis, A. Terzis, S. P. Perlepes, and C. P. Raptopoulou, "Complexes derived from the copper(II)/succinamic acid/*N,N′,N″*-chelate tertiary reaction systems: synthesis, structural and spectroscopic studies," *Polyhedron*, vol. 29, no. 7, pp. 1870–1879, 2009.

[69] K. N. Lazarou, C. P. Raptopoulou, S. P. Perlepes, and V. Psycharis, "Complexes derived from the general copper(II)/maleamic acid/*N,N′,N″*-chelate reaction systems: synthetic, reactivity, structural and spectroscopic studies," *Polyhedron*, vol. 28, no. 15, pp. 3185–3192, 2009.

[70] J.-A. Zhang, M. Pan, J.-Y. Zhang et al., "Syntheses, structures and bioactivities of silver(I) complexes with a tridentate heterocyclic N- and S-ligand," *Polyhedron*, vol. 28, no. 1, pp. 145–149, 2009.

[71] F. H. Allen, "The Cambridge Structural Database: a quarter of a million crystal structures and rising," *Acta Crystallographica Section B Structural Science*, vol. 58, no. 3, pp. 380–388, 2002.

[72] M. D. Živković, S. Rajković, B. Đ. Glišić, N. S. Drašković, and M. I. Djuran, "Hydrolysis of the amide bond in histidine- and methionine-containing dipeptides promoted by pyrazine and pyridazine palladium(II)-aqua dimers: comparative study with platinum(II) analogues," *Bioorganic Chemistry*, vol. 72, pp. 190–198, 2017.

[73] R. G. A. R. Maclagan, S. Gronert, and M. Meot-Ner, "Protonated polycyclic aromatic nitrogen heterocyclics: proton affinities, polarizabilities, and atomic and ring charges of 1–5-ring ions," *Journal of Physical Chemistry A*, vol. 119, no. 1, pp. 127–139, 2015.

[74] A. Bondi, "van der Waals volumes and radii," *Journal of Physical Chemistry*, vol. 68, no. 3, pp. 441–451, 1964.

[75] S. Rajković, M. D. Živković, B. Warȧzajtis, U. Rychlewska, and M. I. Djuran, "Synthesis, spectroscopic and X-ray characterization of various pyrazine-bridged platinum(II) complexes: ^{1}H NMR comparative study of their catalytic abilities in the hydrolysis of methionine- and histidine-containing dipeptides," *Polyhedron*, vol. 117, pp. 367–376, 2016.

14

Synthesis of Pyrazolone Derivatives and their Nanometer Ag(I) Complexes and Physicochemical, DNA Binding, Antitumor, and Theoretical Implementations

Ismail Althagafi,[1] Nashwa M. El-Metwaly,[1,2] Marwa G. Elghalban,[1,2] Thoraya A. Farghaly,[1,3] and Abdalla M. Khedr[1,4]

[1]*Chemistry Department, College of Applied Sciences, Umm Al-Qura University, Mecca, Saudi Arabia*
[2]*Chemistry Department, Faculty of Science, Mansoura University, Mansoura, Egypt*
[3]*Department of Chemistry, Faculty of Science, University of Cairo, Giza 12613, Egypt*
[4]*Chemistry Department, Faculty of Science, Tanta University, Tanta, Egypt*

Correspondence should be addressed to Ismail Althagafi; iithagafi@uqu.edu.sa

Academic Editor: Claudio Pettinari

Four pyrazolone derivatives and their corresponding silver complexes were synthesized and characterized. Based on elemental analysis, 1 : 2 (M : L) molar ratio was suggested for all inspected complexes. ^{1}H, ^{13}C NMR, mass, UV-Vis, TGA, and IR were the spectral tools used for describing the formulae. Moreover, XRD patterns and SEM pictures were used to evaluate the particle sizes which appeared strongly in nanometer range. CT-DNA study is the major consideration in this study, to test the interacting ability of all synthesized cationic complexes towards cell DNA. Each binding constant was computed and correlated with the Hammett sigma constant. Antitumor activity was examined upon three carcinoma cell lines (MCF-7, HepG2, and HCT116). The high efficiency was recorded towards MCF-7 (breast carcinoma) cell line. Kinetic studies yield essential parameters to assert on the rule of metal atom on thermal feature of organic compounds. Molecular modeling was implemented to optimize the structures of compounds. Also, molecular docking was achieved to obtain a clear view about proposed drug behavior within the affected cells. This was achieved through comparing the calculated internal energy values of all docking complexes. All the tested compounds displayed a significant interaction with breast cancer protein (strong matching with practical result) followed by DNA polymerase protein.

1. Introduction

After discovering the cisplatin as an effective antitumor drug, the interest of researchers in the coordination chemistry increased to design highly effective complexes as antiantitumor operators [1–3]. But, the limitations of the clinical success of cisplatin due to critical side effects and resistance that cause retrogression [4] have stimulated searches for other metal complexes. The synthesis for series silver(I) complexes as antitumor agents conquers a considerable place in cancer chemotherapy, as evidenced by their remarkable therapeutic potential reflected in many reported research works [5]. The silver ion is of considerable solicitude in the design of coordination complexes due to the resilience of its coordination sphere and the ease with which it varies its coordination number (from 2 to 4, rarely 5 or 6). Coordination complexes of silver are known to be convertible through changes in ligand geometry, rigidity or functionality, and/or by modifications to the counter-ion or solvent system [6]. But, silver often tends to prepare coordination polymers [7, 8]. The main factor in the effectiveness of anti-cancer compounds is their ability to interact with the DNA of the cancer cell which prevents its division until the cell will die [9]. Ag^{+} is uniquely interesting because it binds exclusively to the bases rather than the backbone of DNA [10, 11]. On the other hand, the wide range of biological activities of pyrazolone as antitumor [12], anti-inflammatory [13] and antimicrobial

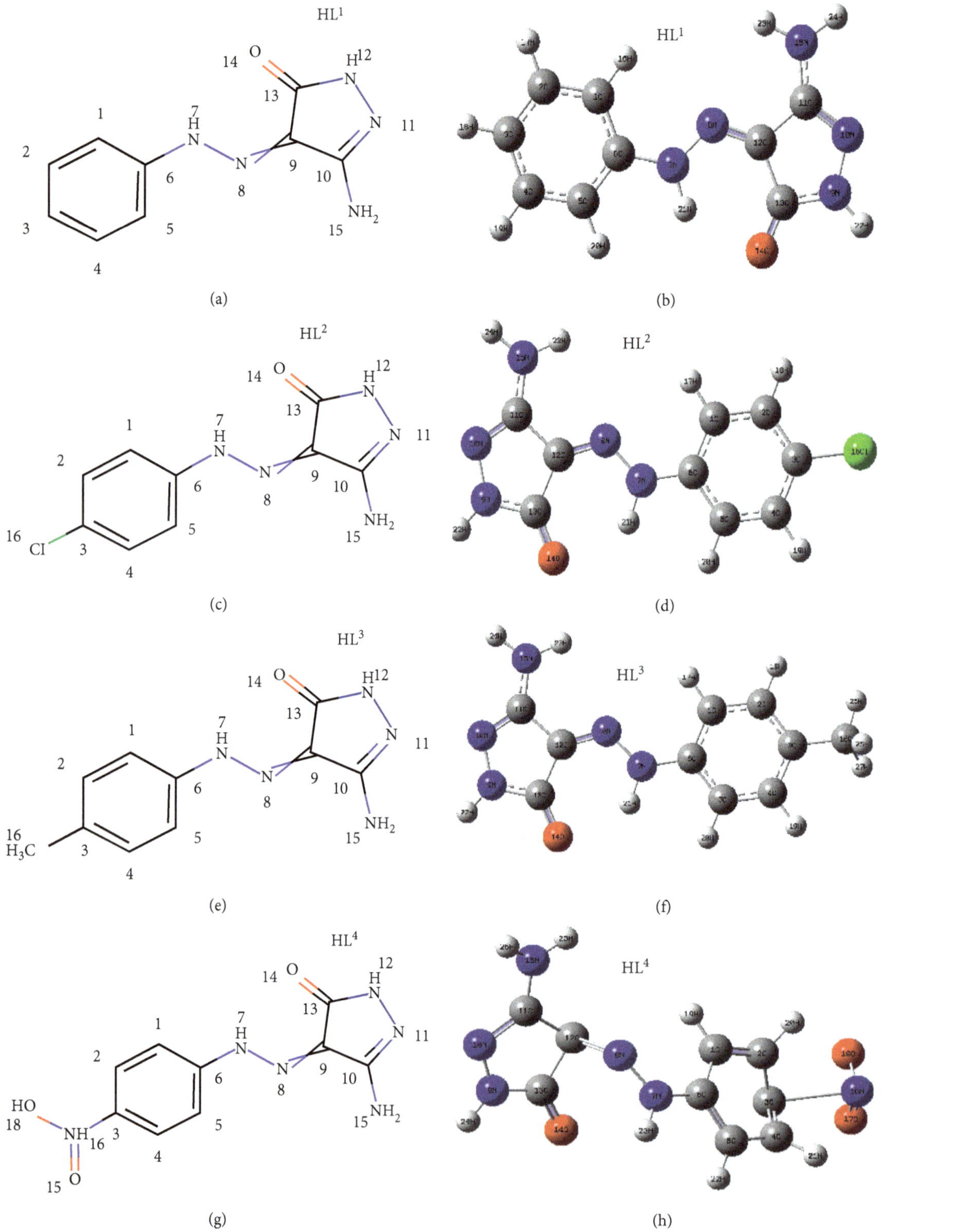

FIGURE 1: The best structural forms of pyrazolone ligands (HL^1–HL^4).

[14] makes it a basic objective to synthesize biologically active transition metal complexes as it is known that the formation of the complexes enhances the biological efficiency of the ligand [15]. Based on all of the afore-mention, we focused in this work on the synthesis of nonpolymeric silver complexes with pyrazolone to improve its efficiency as anticancer drug for three carcinoma cell lines and to study their interaction with DNA. The structure of novel complexes will be investigated through all possible spectral and analytical tools. Computational studies (i.e., Gaussian 09 molecular modeling, QSAR, and

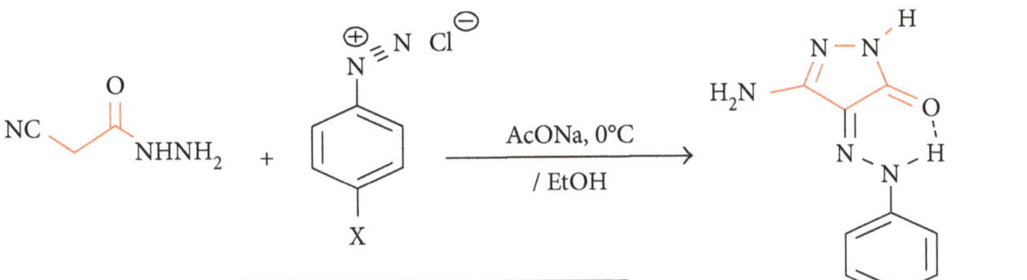

Ligand Number	X
HL_1	H
HL_2	Cl
HL_3	CH_3
HL_4	NO_2

Scheme 1: Synthesis of pyrazolone ligands (HL^1–HL^4).

Autodock tools) will be applied to prove and interpret the efficiency of novel complexes.

2. Experimental Work

2.1. Chemical Reagents. All chemicals used to prepare pyrazolone derivatives (ethyl cyanoacetate, aniline, 4-chloroaniline, 4-toluidine, 4-nitroaniline, hydrazine hydrate, ethanol, sodium acetate trihydrate, sodium nitrite, and HCl) were used as received without any pretreatment. They were purchased from Fluka and Sigma-Aldrich. Also, $AgNO_3$ used for preparing the complexes was PDH.

2.2. Preparations

2.2.1. Synthesis of 3-Amino-4-(2-substituted phenylhydrazono)-1-H-pyrazol-5(4H)-one. Derivatives of pyrazolone (HL^{1-4}) were synthesized and fully characterized in accordance with the previous studies [16, 17]. A fresh prepared solution of cyanoacetic hydrazide (0.99 g, 0.01 mol) in ethanol (50 ml) was coupled in presence of sodium acetate trihydrate (1.36 g, 0.01 mol). The coupling was carried out on 0.01 mol arenediazonium chloride (prepared by diazotizing corresponding aniline derivative) in 6 M hydrochloric acid (6 ml) with cold sodium nitrite solution (10 ml, 1 M; as known) in ice bath (at 0–5°C). The reaction mixture was set in a refrigerator for 3 hrs. The precipitating compound was filtered off and dried. Refluxing ethanolic arylhydrazone solution for 2 hr gives the aimed ligands, 3-amino-4-(2-substituted phenylhydrazono)-4,5-dihydropyrazol-5-ones (HL^{1-4}) as appeared in Scheme 1. Whereas, their scientific names are HL^1, 3-amino-4-(2-phenylhydrazono)-1*H*-pyrazol-5(4*H*)-one; HL^2, 3-amino-4-(2-(4-chlorophenyl) hydrazono)-1*H*-pyrazol-5(4*H*)-one; HL^3, 3-amino-4-(2-(4-mehylphenyl)hydrazono)-1*H*-pyrazol-5(4*H*)-one and HL^4, 3-amino-4-(2-(4-nitrophenyl)hydrazono)-1*H*-pyrazol-5 (4*H*)-one. The prepared pyrazolone compounds were structurally optimized to clarify the orientation of active groups (Figure 1) using Gaussian 09 software. ^1H NMR (DMSO-d_6) spectra of four ligands (Figure 1S) were assigned as follows: HL^1, δ: 5.80 (s, 2H, NH_2), 7.10–7.54 (m, 6H, Ar–H, and amidic-NH), 10.53 (s, 1H, NH). HL^2, δ: 5.84 (s, 2H, NH_2), 7.42–7.59 (m, 5H, Ar–H and amidic-NH), 10.53 (s, 1H, NH). HL^3, δ: 2.23 (s, 3H, CH_3), 5.71 (s, 2H, NH_2), 7.01–7.60 (m, 5H, Ar–H and amidic-NH), 11.41 (s, 1H, NH), and HL^4, δ: 6.02 (s, 2H, NH_2), 7.73–8.25 (m, 5H, Ar–H and amidic-NH), 10.66 (s, 1H, NH). Also, mass spectral analysis of HL^2 and HL^3 ligands (Figure 2S) displays the following fragmentation peaks starting with molecular ion; m/z = 239 ($M^+ + 2$, 21.76); 238 ($M^+ + 1$, 8.61); 237 (M^+, 64.69); 126 (100); 111 (26.80); 98 (36.92); 91 (9.09) and 218 ($M^+ + 1$, 13); 217 (M^+, 100); 126 (39.26); 111 (1.32); 106 (38.82); 77 (46.96), respectively.

2.2.2. Synthesis of Silver(I) Complexes. 2 mmol $AgNO_3$ was dissolved in a little amount of ethanol. Then, the solution was added gradually to 2 mmol from each ligand solution (in DMF). The mixture was stirred and refluxed at 50–60°C for 15–30 min. The complexes appeared in the reaction medium without external additions. The main reaction product was precipitated, then filtered off and washed with ethanol followed by diethyl ether, and after that, dried over $CaCl_2$ desiccators.

2.3. DNA Binding Methodology. The binding efficiency of calf thymus-DNA (CT-DNA) has been deliberated by spectrophotometric titration method. 50 mg from CT-DNA were dissolved by continuous stirring overnight in deionized water (pH = 7.0) and preserved at 4°C. Tris(hydroxymethyl)-aminomethane was used to prepare 5.0 mM tris-HCl buffer (pH = 7.2) after adding 50 mM NaCl in deionized water. The prepared buffering solution is perfectly free from protein, based on the value measured for absorbance ratio (1.8–1.9; A260/A280) at 260 and 280 nm [18]. The concentration of prepared DNA (5.25×10^{-4} M) was computed by knowing molar absorptivity coefficient which is 6600 $M^{-1} \cdot cm^{-1}$ at 260 nm. 2.5×10^{-5} M was the concentration used from each complex. At room temperature, the UV-Vis scanning process was carried out at the fixed range (200–900 nm) by 1 cm quartz cuvette. A gradual increment for CT-DNA concentration from 0.00 till 1.58×10^{-4} mol·L^{-1} was done, towards a fixed concentration from each tested compound. The same CT-DNA amount was added to a reference buffer solution with each measurement, to omit the absorbance of free DNA. Intrinsic binding constant (K_b) of the interaction between compound and CT-DNA was computed by; $[DNA]/(\epsilon a - \epsilon f) = [DNA]/(\epsilon b - \epsilon f) + 1/K_b\ (\epsilon a - \epsilon f)$, equation [19]. Where, [DNA] is the molar concentration from DNA, ϵa is the extinction coefficient observed at each DNA concentration by; A_{obs}/[compound], ϵf is the extinction coefficient for each free tested compound in solution. Moreover, ϵb is the extinction coefficient for the fully bonded compound towards DNA. By drawing the relation of $[DNA]/(\epsilon a - \epsilon f)$ versus [DNA], K_b is given by the following ratio: slope/intercept.

2.4. In Vitro Antitumor Study. The biological feature of synthesized pyrazolone derivatives (HL^{1-4}) and their Ag(I) complexes were screened towards three carcinoma cell lines (MCF-7, HEPG-2, and HCT-116) by sulforhodamine B (SRB). Cells were kept in RPMI medium enriched with 100 μg/ml streptomycin, 100 units/ml penicillin, and 10% heat-inactivated genetic bovine serum inhibit, 5% (v/v) CO_2 atmosphere at 37°C, and cells were subcultured two times/week. The growing cells were gathered using trypsin–EDTA (0.25%) and plated in 96-well plates at 1000 cells/well. Cells were treated to extraction for 72 h and subsequently fixed with TCA (10%) for 1 h at 4°C. After washing times, cells were exposed to SRB solution (0.4%) for 10 min in a dark place and subsequently washed with glacial acetic acid (1%). After keeping overnight, Tris–HCl was used to dissolve SRB-stained cells, and the intensity of color was measured to be 540 nm. The viability of cell was measured by Trypan blue (0.4) stain, which permits us to differentiate viable (nonstained) from dead (stained) cells. Viable and dead cells were counted with Nikon microscope implementing hemocytometer. Surviving cells percentage = (1.00 – (number of blue cells/number of total cells)) × 100. Statistical analysis was executed by using Sigma Plot version 12.0.

2.5. Equipments and Physical Analysis. Carbon, hydrogen, and nitrogen contents were performed at Perkin-Elmer 2400 CHN elemental analyzer. The silver content was evaluated gravimetrically using standard method [20]. The molar conductivities of metal complexes were obtained by JENWAY model 4070 conductance bridge (in DMSO solvent). KBr-IR and ^{13}C, ^{1}H NMR spectra were obtained by JASCO FT-IR-4100 spectrophotometer (400–4000 cm^{-1}) and Burker 500 MHz, respectively. The purity of most compounds was checked from mass spectra at 70 eV by using AEIMS 30 mass spectrometer with heating rate 40°C/min and mass range (50–1000). Electronic spectra were recorded by UV_2 Unicam UV/Vis spectrophotometer at room temperature using 1 cm quartz cell (in DMSO solvent). Thermal analysis was accomplished using Shimadzu thermogravimetric analyzer by heating rate of 10°C min^{-1} under nitrogen (20–900°C). X-ray diffraction patterns (XRD) were accomplished by Rigaku diffractometer using Cu/Kα radiation. Scanning electron microscopy (SEM) images were extracted by the Joel JSM-6390 equipment. Antitumor activity was tested in microbiology center.

2.6. Computational Implementation

2.6.1. Kinetics. Essential thermodynamic parameters were extracted over two degradation stages from TGA curves. All degradation processes were carried out at a satisfactory rate which introduces a good computation for kinetic parameters along most degradation stages. The kinetic parameters were calculated for the first and second degradation stages. The chosen steps exhibit the decomposition for great masses from the coordination sphere, and order (n) and energy of activation (E) were computed. Many researches were dealing with kinetic calculations [21–29] and proposed equations stated for such purpose. Among these methods, Coats and Redfern [23] and Horowitz and Metzger [28]were used in this study.

2.6.2. Structural Optimization. Utilizing Gaussian 09 software [30], the structural optimization for pyrazolone derivatives (HL^{1-4}) and their Ag(I) complexes were executed. DFT/B3LYP is a proper method achieving best configurations and computational files. The files extracted from the optimization process (log and chk files) were visualized by Gauss-View screen [31] to produce essential parameters. Also, HOMO and LUMO images were extracted from chk file visualization.

2.6.3. QSAR Calculation. QSAR parameters calculations were carried out over all synthesized compounds after optimization process by applying HyperChem software program (8.1). Firstly, preoptimization was accomplished by soft adjustment procedures, through molecular mechanics force field (MM^+) followed by semiempirical (AM1). The program implemented over structural forms without fixing any parameter to execute the optimization at equilibrium. The minimization of energy was done by Polak–Ribière conjugated gradient algorithm. QSAR parameters were computed for all derivates and their Ag(I) complexes, for comparison.

2.6.4. Molecular Docking Methodology. The docking process was executed between protein receptors (3s7s, 3e1r, and 4dk7 assign for breast, colon, and liver cancers, respectively) and pyrazolone derivatives (HL^{1-4}) (proposed inhibitors). Moreover, the docking process was done over CT-DNA polymerase (1bpb). The selected protein receptors were the same for the pathogen of tumor cells investigated in the experimental part to assert on the results. Autodock tools 4.2 is a software used for docking methodology. Gasteiger partial charges were added over elements of tested compounds (pyrazolones). The rotatable bonds were omitted and nonpolar hydrogen atoms were ruling out. The energy levels of docked complexes were computed after addition of fundamental hydrogen atoms, Kollman united atom type charges, and salvation parameters [32]. Affinity (grid) maps of ××Å grid points and 0.375 Å spacing were generated

TABLE 1: Analytical and physical data for Ag(I)-pyrazolone (HL^{1-4}) complexes.

Compounds (empirical formula, calcd./found)	Λ_m ($\Omega^{-1}\cdot cm^2\cdot mol^{-1}$)	Color	Elemental analysis (%) calcd. (found)			
			C	H	N	M
(1) $HL^1(C_9H_9N_5O)$ (203.20)	—	Red	53.20 (53.18)	4.46 (4.46)	34.47 (34.44)	—
(2) $[Ag(HL^1)_2](NO_3)$ (576.27)	57.31	Brown	37.52 (37.51)	3.15 (3.16)	26.74 (26.72)	18.72 (18.72)
(3) $HL^2(C_9H_8N_5OCl)$ 237.64/239	—	Red	45.49 (45.50)	3.39 (3.41)	29.47 (29.45)	—
(4) $[Ag(HL^2)_2](NO_3)$ 645.15/647.66	55.21	Reddish brown	33.51 (33.52)	2.50 (2.50)	23.88 (23.89)	6.72 (16.71)
(5) $HL^3(C_{10}H_{11}N_5O)$ 217.23/218	—	Faint brown	55.29 (55.31)	5.10 (5.11)	32.24 (32.25)	—
(6) $[Ag(HL^3)_2](NO_3)$ $2H_2O$ 640.36/605	51.06	Brown	37.51 (37.50)	4.09 (4.11)	24.06 (24.10)	6.84 (16.86)
(7) $HL^4(C_9H_8N_6O_3)$ (248.20)	—	Faint brown	43.55 (43.55)	32.49 (32.47)	33.86 (33.85)	—
(8) $[Ag(HL^4)_2](NO_3)$ 666.27/665.22	56.12	Deep brown	32.45 (32.43)	2.42 (2.41)	27.33 (27.35)	6.19 (16.20)

utilizing Autogrid program [33]. Also, Vander Waal forces and electrostatic terms were acquired, through the use of autodock parameter set-dependent and distance-dependent dielectric functions, respectively. The docking was accomplished using Solis and Wets local search method and Lamarckian Genetic algorithm (LGA) [34]. Initial position, orientation, and torsions of tested molecule (inhibitor) were set indiscriminately. All rotatable torsions were expelled during the docking process. Each experiment is the average value for 10 different runs that were set to close after maximum of 250000 energy assessments. 150 is the used population size. During the process, translational step of 0.2 Å, quaternion, and torsion steps of 5 were used.

3. Results and Discussion

3.1. Physicochemical Properties. Common specific features are aggregated in Table 1. All synthesized complexes have high melting points (›300°C), while that of free organic derivatives (HL^{1-4}) were 246, 300, 252, and 300°C. The analytical results are proper with 1 : 2 molar ratio between silver ion and chelating ligands. The molar conductivity (in DMSO) which measured for all complexes (1 mmol) is coinciding with conductivity feature. The conductivity values (Λ_m) are confined in the range of conducting feature for monoanionic conjugated nitrate with cationic coordination sphere [35]. This is proper feature with high acidic anions as nitrate.

3.2. IR and ^1H, ^{13}C NMR Assignments. IR spectral analysis of pyrazolone (HL^{1-4}) derivatives (Table 1S) displays significant bands assigned for the following vibrations; $\nu_{as}(NH_2)$, $\nu_s(NH_2)$, ν(NHs), $\delta(NH_2)$, δ(NH), ν(C=N), and ν(C=O). Observable lower shifted appearance for $\nu(NH_2)$, $\delta(NH_2)$, and ν(C=N) bands suggests the bidentate contribution mode of pyrazolones towards silver atom. However, more or less shifted appearance of NHs and C=O donors introduces their ruling out from coordination [36]. Each coordination sphere includes two coordinating ligands towards one central atom, in which the orientation of donors according to central atom is the cause for chelating mode (by NH_2 and C=N)and also has the major impact for ruling out of other donors (NHs and C=O). The appearance of new band at ≈1383 cm^{-1} refers to $\nu(NO_3)$ about the ionic attachment of nitrate. Moreover, new band appeared at the range 509–560 cm^{-1} in all complexes, refers to ν(M–N) vibration [37–39]. ^1H and ^{13}C NMR (DMSO-d_6) spectral data of investigated complexes are tabulated (Table 2). ^1H NMR spectra (Figure 3S) display a considerable down field shift for NH_2 group by an observable ppm value, whereas ^{13}C NMR spectral data represent the absence of significant alteration in hybridization attitude upon carbon atoms inside the forms. This is omitting hydrogen relocation through tautomerism followed by the enolization process.

3.3. Electronic Spectra. The preparation of AgNPs has been assured through UV-Vis spectral analysis, which carried out for all complexes in the DMSO solvent (10^{-3} M). The measurements process was accomplished at 510, 440, 420, and 480 nm for four Ag(I)-HL (1–4) complexes, respectively. The excessive presence for AgNPs was expected from the deep color of Ag(I) complexes. The dark color of complex solution points to the formation of AgNPs. The electronic spectra of four complexes reflect a well-defined Plasmon band between 420 and 510 nm in the nanoscale range, according to silver nanoparticles association. According to theoretical and experimental studies [40], the optical features of AgNPs were reported based on electronic spectral band position. The band at 420 nm for Ag(I)-HL^3 complex solution is referring to the presence of small spherical silver nanoparticles. The spherical particles display λ_{max} band at shorter wave length, while the nonspherical shape of particles shifts the λ_{max} to longer wave length [41]. The higher the intensity of the complex color, the higher the concentration of AgNPs [42]. The tetrahedral configuration of Ag(I) complexes (Figure 2) was proposed based on analytical, IR, ^1H, ^{13}C NMR, and mass spectral analysis, which is acceptable for d^{10} systems.

3.4. Microanalysis Mass Spectra. Two organic ligands were chosen for this study (HL^2 and HL^3), the molecular ion peaks recorded were $m/z = 239$ ($M^+ + 2$, calcd. 237.63) and 218 ($M^+ + 1$, calcd. 217.23), respectively. A high agreement between each recorded M^+ and the calculated one supports high purity. Three silver complexes were studied to assert on the formulae proposed (Figure 4S). In the $[Ag(HL^2)_2](NO_3)$ complex spectrum, a well-defined parent peak was appeared at $m/z = 647.66$ ($M^+ + 2$ (1.2), calcd. 645.15) is followed by a successive fragmentation corresponding to peaks at 344 (3.81), 243 (11.78), 238 (7.55), 220 (14.50), 197 (81.93), 140

Table 2: DMSO-d_6, ^1H NMR, and ^{13}C NMR data (ppm) for Ag(I)-pyrazolone complexes.

Compound	δ(s, NH_2)	δ(m, Ar-CHs, amidic-NH); (s, CH_3)	δ(s, NH_h)	δ(CH_3)	δ(2C, Ar), (2C, Ar)	δ(C, Ar), (p-C, Ar)	δ(C=O)	δ2(C=N)
Ag(I) + HL^1	5.86	7.13–7.52	10.55	—	115.48, 128.89	141.32, 118.35	161.21	150.47
Ag(I) + HL^2	5.90	7.43–7.60	10.56	—	117.39, 129.69	141.34, 125.61	161.30	150.37
Ag(I) + HL^3	5.86	7.21–7.44; 2.3	10.54	20.92	115.60, 130.33	139.90, 129.30	161.89	153.81
Ag(I) + HL^4	6.09	7.78–8.10	10.69	—	117.36, 128.38	148.33, 143.09	158.04	150.25

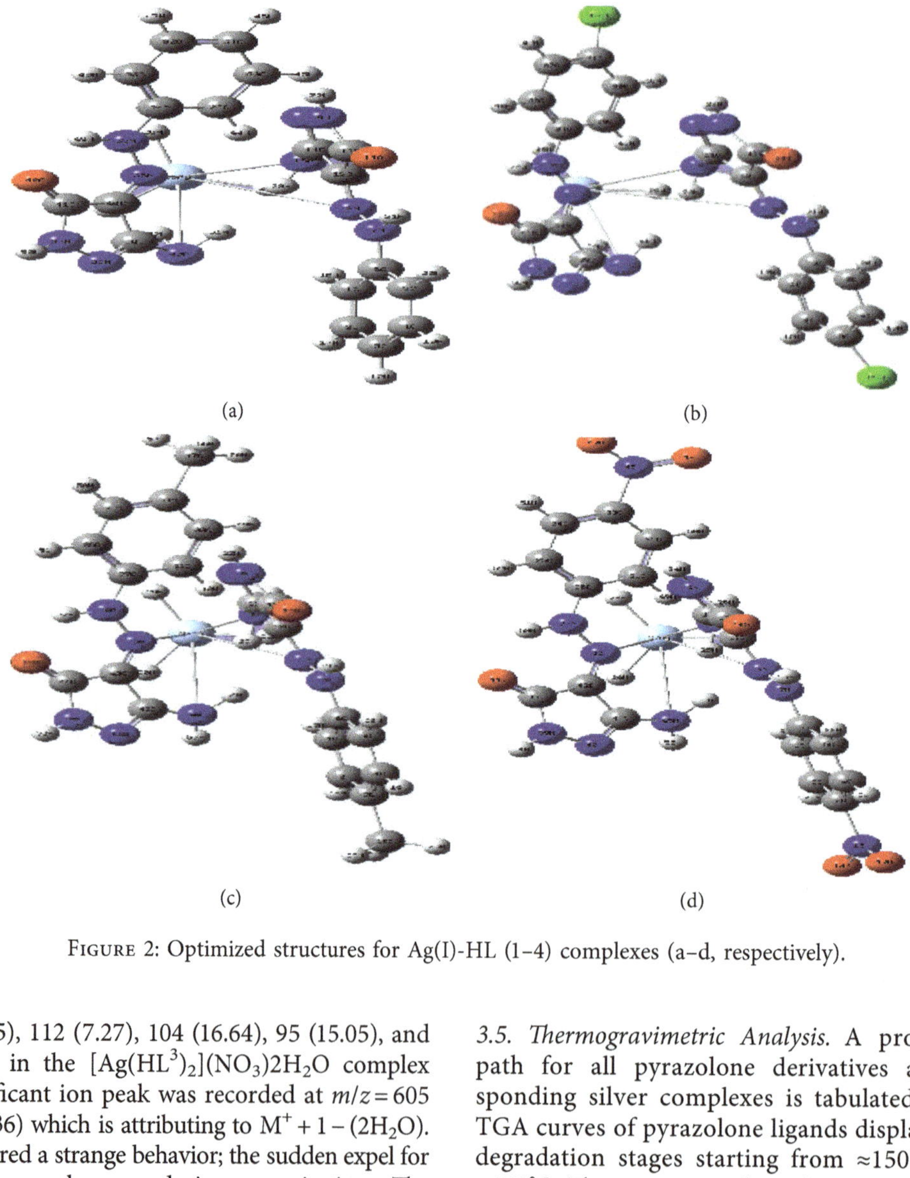

Figure 2: Optimized structures for Ag(I)-HL (1–4) complexes (a–d, respectively).

(15.61), 127 (39.25), 112 (7.27), 104 (16.64), 95 (15.05), and 95 (65.29). Also, in the $[Ag(HL^3)_2](NO_3)2H_2O$ complex spectrum, a significant ion peak was recorded at $m/z = 605$ (25.94, calcd. 640.36) which is attributing to $M^+ + 1 - (2H_2O)$. This is not considered a strange behavior; the sudden expel for water molecules may happen during vaporization. The vaporization is a prefragmentation process before electron bombardment [43]. Essential fragmentation peaks recorded are 546 (54.26), 386 (5.28), 370 (100), 268 (12.45), 243 (19.73), and 202 (1.06). Moreover, in the $[Ag(HL^4)_2](NO_3)$ complex spectrum, the recorded molecular ion peak attributes to $M^+ - 1$ ($m/z = 665.22$) (calcd. 666.27). The more proper fragmentation perception was displayed in Scheme 2.

3.5. Thermogravimetric Analysis. A proper degradation path for all pyrazolone derivatives and their corresponding silver complexes is tabulated (Table 2S). The TGA curves of pyrazolone ligands display two significant degradation stages starting from ≈150°C and ended to ≈620°C. The successive degradation process had three to four stages (ended at ≈800°C). All the coordination spheres around the silver atom were completely destroyed along the temperature range applied. Moreover, the residue includes AgN_3 or silver atom meanwhile polluted by carbons. A satisfactory agreement between calculated and found weight loss corresponding to the expelled particles reflects the exact assignments for step borders,

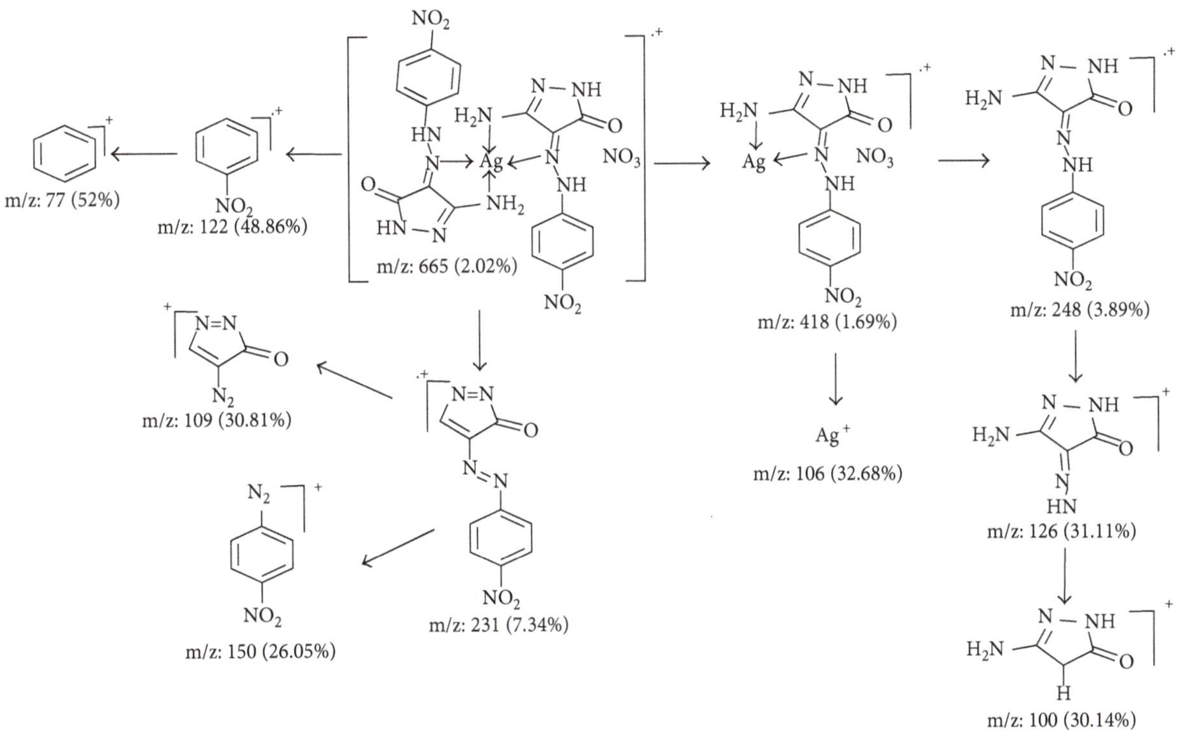

SCHEME 2: The fragmentation perception for $[Ag(HL^4)_2](NO_3)$ complex.

which suggests that well computed kinetics may be obtained.

3.6. XRD and SEM Analysis. XRD patterns of pyrazolone ligands and their silver complexes (solid samples) were recorded (Figures 3 and 5S). This analysis gives a good insight about the crystal lattice for tested compounds. Applying referenced methods, the purity of each compound can be asserted through comparative study with the reactants patterns [44]. Moreover, essential lattice parameters were deducted from high intense peak in patterns (FWHM). All tested compounds show nanocrystalline feature, which clearly impacted on the sharpness of patterns peaks. This may be due to the fixation of compounds' nucleus, as the ligands family and silver central atom. Particle sizes, 2θ, relative intensity (%), d spacing, and FWHM are represented in Table 3. The sizes of crystals were estimated from the Deby–Scherrer equation: $\beta = 0.94\lambda/(S\cos\theta)$, where S is the crystallite size, θ is the angle of diffraction, and β is FWHM, Cu/Kα $(\lambda) = 1.5406$ Å.

The d-spacing upon two following layers was extracted from the Bragg equation: $n\lambda = 2d\sin(\theta)$ at $n = 1$. The crystallite sizes calculated for all tested compounds appeared excellently in nanometer range. This feature devotes the application towards antitumor activity as well as DNA binding study. Also, ε index (crystal strain) and δ index (dislocation density) were calculated by known relations [45] equation. The two indexes are reflecting the dislocation inside the compound network. The lower their values, the higher the quality of compounds. Also, Miller indexes (hkl) were calculated on the bases of cubic feature for investigated

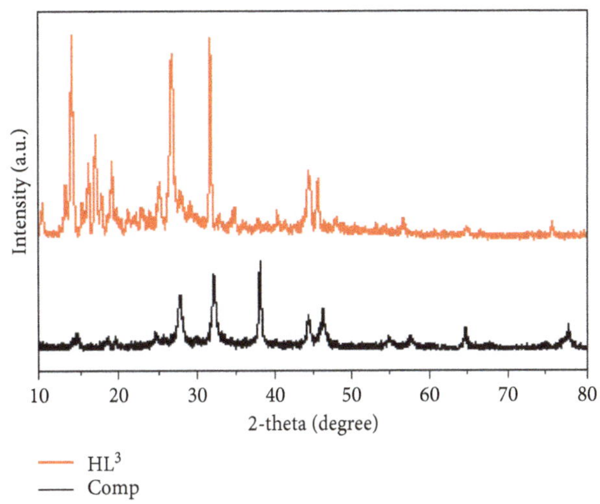

FIGURE 3: X-ray diffraction pattern of pyrazolone, HL^3, and its AgNPs complex.

crystals. The calculated values display mainly the parallelism of planes towards y- and z-axes while, intercepts with x-axis, except for Ag(I)-HL^2 and Ag(I)-HL^4 complexes. This reflects the relatively perfect crystal building towards the axes with most compounds. SEM is another technique used to obtain a clear view about the surface topography for tested compounds (Figure 6S). The images of four organic ligands were found as needles or sticks by a regular way. Also, the high aggregation observed with HL^4 ligand may reflect its distinguished nanometer size of separate crystals which coincide with XRD data. However, the images of complexes are completely different from that of corresponding ligands,

TABLE 3: XRD computed data for Ag(I)-pyrazolone (HL^{1-4}) complexes.

Compounds	Size (Å)	2θ	Intensity	d-spacing (Å)	ε	$\delta(Å^{-2})$	FWHM	1000 $\sin^2\theta$	a^2	hkl
(1) HL^1	6.872	12.96	1572	6.8255	0.1191	0.0214	0.2121	12.737	46.586	100
(2) $[Ag(HL^1)_2](NO_3)$	5.585	10.92	222	8.0956	0.1737	0.0153	0.2605	9.054	65.536	100
(3) HL^2	6.460	14.60	298	6.0623	0.1470	0.0272	0.2216	16.145	36.752	100
(4) $[Ag(HL^2)_2](NO_3)$	5.880	26.68	384	3.3385	0.0682	0.0897	0.2531	53.236	11.146	331
(5) HL^3	2.842	14.12	570	6.2673	0.2650	0.0255	0.5134	15.107	39.277	100
(6) $[Ag(HL^3)_2](NO_3)2H_2O$	2.561	38.02	254	2.3648	0.1105	0.1788	0.5982	106.102	5.593	100
(7) HL^4	1.980	27.65	2086	3.2236	0.1954	0.0962	0.7532	57.101	10.391	100
(8) $[Ag(HL^4)_2](NO_3)$	4.604	28.08	172	3.1752	0.0829	0.0992	0.3242	58.85	10.083	311

which initially reflects the purity. Moreover, the spherical shape is the feature observed, which indicates the impact of metal central atom in the aggregation process of particles.

3.7. DNA Binding Studies. The degree of interaction of Ag(I) complexes against CT-DNA was examined by the spectrophotometric titration method. The freshly prepared solutions (as reported in experimental part) were scanned along the applied wave length range (200–900 nm) against referenced solutions (blank). The intrinsic binding constants (K_b) were calculated at λ_{max} for each charge transfer band; 364.7, 366.4, 365.1, and 458 nm attributing to the binding yields of silver complexes (1–4) and DNA. The absorption relationships for all tested compounds are displayed in Figure 7S. An observable bathochromic effect for new CT band is conjugated with progressive increase DNA amount [46]. The red shift observed is found in the range 1–4 nm, with all investigated compounds. This indicates the formation of new compounds yielded from the binding with DNA helix. Also, a gradual altitude in the absorption is the other effective indicator. The higher the binding constant, the higher the stability of coupled DNA helix. The coupling process is mainly proper with occluding inside major and minor grooves of DNA, whereas the coupling process attributing to silver complexes may prolonged by electrostatic attraction for cationic coordination sphere as well as occlusion inside grooves. The coupling stabilization may be facilitated by the broadness of occluded compound surface as well as a charged coordination sphere (with silver complex). The occlusion process causes reorganization inside CT-DNA, which required partial disassembly or deterioration for the double helix at the exterior phosphate leading to form cavity suitable for entering [47]. The bathochromic shift is corelated with the substituent's effect by a direct relation with electron withdrawing character (σR). Hammett's equation established the significant effect of p-substituent's on the binding constant value (σR and K_b) [48]. The constant of binding was arranged by 1.29×10^4, 1.82×10^4, 3.57×10^3, and 2.4×10^4, attributing to complexes for HL^{1-4}, respectively, in agreement with Hammett's relation. The binding constants reflect an acceptable interaction between the complexes and CT-DNA. The higher the inductive effect, the higher the interacting ability with DNA helix. Other essential plots against σR (Figure 4) were extracted to clarify the relationships as; ΔE, PK_a, (docking parameter), and log P. A direct relation was shinned with ΔE (energy gaps) and PK_a plots, while a reverse with log P. The effective impact for log P value is appearing with the minimization of its value.

3.8. Antitumor Study. All synthesized complexes were screened against three cell lines (HEPG2, MCF-7, and HCT116) to establish a view about their antitumor activity. The obtained results propose the cytotoxic feature of complexes towards the tested cell lines in comparison with congruent ligands. The IC_{50} values are represented in Table 4 and graphed in Figure 8S. The cytotoxic indicator IC_{50} is considered a good anticancer parameter on behalf of other significant indexes such as DNA fragmentation or morphology of the cell. Based on this vision, experiments were conducted on all compounds to identify their impact on reducing cells or not through morphological index. To achieve that, cells were treated with IC_{50} for selected high impacted compounds (HL^{3-4} and their complexes). MCF-7 breast cancer was the chosen cell line. After treatment with IC_{50} of selected compounds, MCF-7 cells were incubated for 24 hours. Cells were fixed using Carnoy's fixative and stained by fluorescent stain [49]. The estimation was based on the nucleus shape, nucleus density, foci presence, and the number of cells. The outcome data clearly showed that nucleus remained in its regular shape and that there were no condensation and no degradation for detected nucleus (Figure 8S). Lack of micronucleus inside cytoplasm revealed that ligands and their Ag(I) complexes are not clastogenic agents, which may consider a very good feature. Also, cells preserved their capability to produce foci which means that the compounds may have no impact towards penetrating layers and have no ability to solve cell-to-cell connections. This may be considered useful for them because they do not help cells to spread, where they can be killed in situ (Figure 5). In terms of cell numbers, picture has assured the compatibility and regularity with the datum proposed by SRB assay. In the light of former results and discussion, ligands and complexes may be treated as in situ cytotoxic but not genototoxic.

3.9. Computational Utilizations

3.9.1. Kinetics. The kinetics calculations are a significant study that gives clear view about the impact of metal atom occluded with organic molecules upon the thermal degradation attitude over whole compound (Table 5). E, ΔH, ΔS,

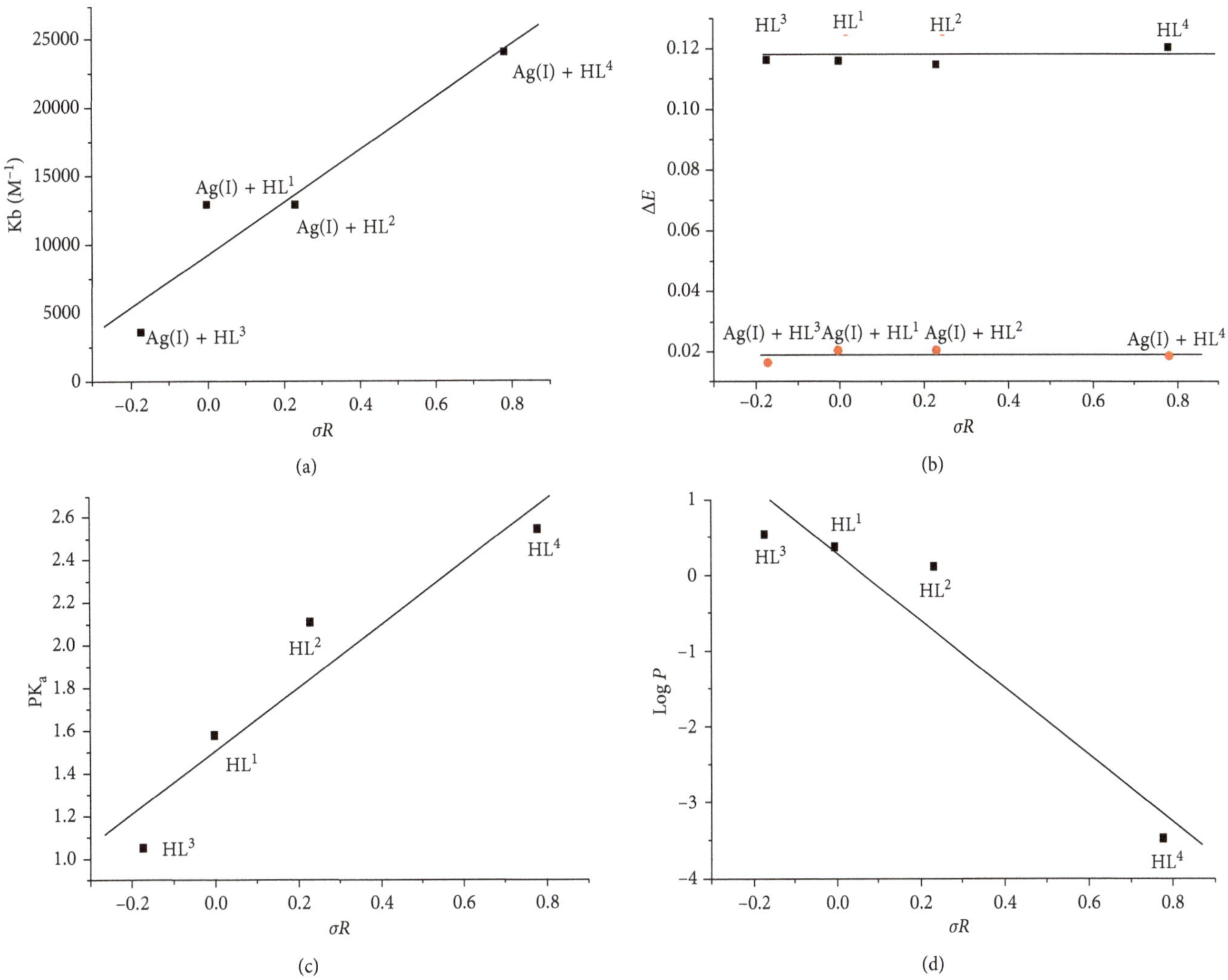

Figure 4: Hammett's relation between effect of *p*-substituent (σR) versus ΔE, PK_a (docking parameter), log *P*, and intrinsic binding constants (K_b) of Ag(I)-HL^{1-4} complexes.

and ΔG are the parameters computed over two degradation stages in each investigating compound. These parameters were calculated based on evaluating fraction decompose (α) at variable temperatures inside step borders. Coats and Redfern [23] and Horowitz and Metzger equations [28] are the two comparative methods used. The relationship graphs are exhibited in Figures 9AS and 9BS. $\Delta H = E - RT$ and $\Delta G = \Delta H - T\Delta S$ were the equations used. Entropy of activation (ΔS) was extracted from the equation $\Delta S^* = R\ln(Ah/K_bT_s)$, where h, K_b, and T_s are Plank's constant, Boltzmann constant, and DTG peak midpoint temperature, respectively. The following notes are grouped into calculated values: (i) The values of activation energy are suffering reduction with complexes in comparison with free ligands. This is a good evidence for the weakness appearing on the bonds of organic compound as a result of metal coordination. (ii) There is an observable increase in the activation energy from one step to another due to the reduction of decomposition rate (endothermic steps) from the original molecule. This will lead to increase of ΔG values in the following stages. (iii) The altitude of activation entropies (ΔS, by negative sign) for complexes upon their corresponding ligands reflects the high-ordered complexes than the free ligands and/or slow reaction rate.

3.9.2. Structural Optimization Parameters. Applying Gaussian 09 software, the best orientation for atoms inside the structures was extracted. DFT/B3LYP is a proper method for this purpose. The optimization procedure exerted essential output files (chk and log). The visualization for chk file, HOMO, and LUMO images and their energy values were obtained. The frontier, HOMO, and LUMO images belong to free ligands and their complexes are displayed in Figure 6 and Figures 10AS and 10BS. The HOMO level for pyrazolone ligands were centered on pyrazol-5(4*H*)-one, moiety which includes the atoms interested in bonding. While, the LUMO level were spread over a whole molecule. However, the attitude of HOMO and LUMO levels in silver complexes were centered on the metal atom. This is normal with such electron-dense atom as silver. All effective

TABLE 4: Cytotoxic activity of tested complexes against human cancer cell lines.

Cell type	IC_{50} (μg/ml)								
	HL^1	Ag(I)-HL^1	HL^2	Ag(I)-HL^2	HL^3	Ag(I)-HL^3	HL^4	Ag(I)-HL^4	Doxorubicin
MCF-7	60.4	60.8	57.7	78.8	14.5	12.81	1.976	16.66	0.168
HepG2	71.3	62.32	63.52	88.5	17.8	11.88	15.86	12.52	0.777
HCT116	36.5	26.8.5	52.6	45.8	20.4	12.8	14.8	8.42	0.9255

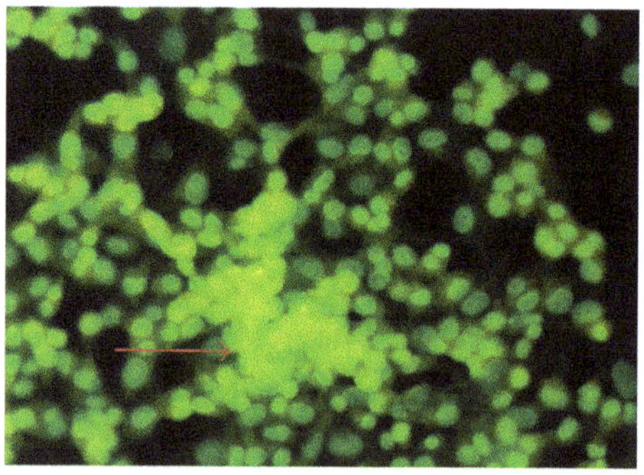

FIGURE 5: The arrow appeared is attributing to the foci of MCF-7 despite IC_{50} of variable compounds.

parameters tabulated in Table 3S were computed concerning to frontier energy gaps ($E_{LUMO} - E_{HOMO}$, ΔE). The indexes χ (electronegativity), μ (chemical potential), η (global hardness), S (global softness), ω (electrophilicity), and σ (absolute softness) were calculated using referenced equations [50, 51]. Using electrophilicity index (ω), the toxicity and reactivity of tested compounds can be evaluated. This will give an acceptable expectation for the biological feature of compounds by which, the excellence for complexes upon their corresponding ligands will be expected with the progression for Ag (I)-HL^4 complex. As well as, the constancy and electron affinity can be evaluated from two opposing indexes; η and σ [52, 53]. Also, the superiority was recorded with the complexes in comparison with their congruous ligands. Other two opposing indexes (S and η) attributing to the degree of softness displayed the soft character of all compounds with the priority of complexes. This indicates an expected distinct biological behavior for complexes. Furthermore, frontier energy gaps in complexes were reduced from original ligands. This is due to the effective impact of metal atom, which is favored in biological activity [54].

3.9.3. Considerable Log File Parameters and QSAR. Other additive parameters were deducted from log file data, from which important data are drawn (Table 4S) that give accurate specifications for structural form and confirmed it. Significant charges of coordinating atoms (N^8, N^{15}, N^{36}, and N^{43}) before (free ligand) and after complexation suffer reduction overall. Moreover, the net charge of silver ion is highly reduced due to charge transfer accompanied with coordination ($L \rightarrow M$). Two bond lengths contributing with donor atoms (C^{11}–N^{15} and C^{12}–N^8) were estimated for ligands only, while they could not be technically computed for complexes. So, effective comparison was absent. Oscillator strength (f) was extracted for all studied compounds and compared. An overall lessening was recorded with the complexes values. This is a good indicator for facilitating electronic transition inside the whole complexes molecules [55]. The dipole moment values were extracted and display variable features in comparative view. Two of them suffer enhancement in complexes and the others significantly reduced. The enhancement may reflect increase in polarity over a molecule, whilst, the reduction purpose a reverse feature. QSAR parameters (Table 6) were obtained after optimizing the structures of all tested compounds using HyperChem (8.1) program. Surface area and the volume of tested compounds showed a significant increase in complexes compared to ligands. Also, reactivity and polarizability were generally enhanced in complexes with the priority of Ag(I)-HL^4 complex in agreement with the previously displayed. Moreover, log P values were estimated and displayed excellence with Ag(I)-HL^4 complex and its ligand. Partition coefficient (log P) points to the biological feature of tested compound by a reverse relation.

3.9.4. Docking Interpretation. In the last few decades, a great revolution is seen in drug design industry, which extended to preemptive theoretical studies to show the expected biological efficiency of proposed drug. Autodock 4.2. tools software is one of these tools used for this purpose. The program designed for simulation about the interaction between proposed drug (inhibitor) and protein receptors attributes to infected cells. 3s7s, 3e1r, 4dk7, and 1bpb were the receptors used for breast, colon, and liver cancers besides CT-DNA polymerase. The docking feature of all organic compounds was studied to evaluate the inhibition activity towards the selected proteins. The output internal energies computed over PDB files are tabulated (Table 5S). pK_a (dissociation constant) value was calculated for each compound, which is considered a biopharmaceutical measure for drug-likeness feature. This constant is reflecting the degree of ionization for compound. The high pK_a value means high degree of polarity around hydrogen atoms of function groups. This will lead to increase in the number of H-bonding between the proposed drug (ligand) and the protein receptor. Also, a direct relation between pK_a value and σR (Figure 4) goes parallel with the former aspect. Computed different energies showed distinct correlations for docked drugs (ligands) with breast cancer followed by DNA polymerase proteins [56]. The best docking complexes (Figure 7), (3s7s-HL^{1-4}) display high occlusion for drugs (ligands) which centered inside protein helix. Also, the binding towards DNA (1bpb) is highly satisfactory for the inhibition feature while the other docking complexes with 3e1r and 4dk7 proteins display binding in moderate level

TABLE 5: Kinetic parameters by using Coats–Redfern (CR) and Horowitz–Metzger (HM) methods.

Comp.	Step	Method	Kinetic parameters					
			E (J·mol^{-1})	A (S^{-1})	ΔS (J·mol^{-1}·K^{-1})	ΔH (J·mol^{-1})	ΔG (Jmol^{-1})	r
(1) HL^1	1st	CR	5.67E+04	9.77E+03	−1.72E+02	5.27E+04	1.34E+05	0.99993
		HM	6.43E+04	1.07E+05	−1.52E+02	6.04E+04	1.33E+05	
	2nd	CR	3.58E+04	1.30E+06	−1.35E+02	3.02E+04	1.22E+05	
		HM	4.55E+04	9.47E+00	−2.33E+02	3.99E+04	1.98E+05	
(2) $[Ag(HL^1)_2](NO_3)$	1st	CR	4.49E+04	9.46E+03	−1.73E+02	4.09E+04	1.26E+05	0.99993
		HM	5.25E+04	2.48E+03	−1.84E+02	4.84E+04	1.39E+05	
	2nd	CR	3.43E+04	2.20E+06	−1.31E+02	2.85E+04	1.20E+05	
		HM	4.61E+04	7.76E+00	−2.35E+02	4.03E+04	2.05E+05	
(3) HL^2	1st	CR	3.62E+04	4.94E+04	−1.59E+02	3.21E+04	1.11E+05	0.99993
		HM	4.42E+04	2.28E+02	−2.04E+02	4.00E+04	1.42E+05	
	2nd	CR	4.16E+04	1.09E+06	−1.37E+02	3.54E+04	1.37E+05	
		HM	5.55E+04	2.57E+01	−2.25E+02	4.94E+04	2.16E+05	
(4) $[Ag(HL^2)_2](NO_3)$	1st	CR	4.32E+04	2.34E+04	−1.66E+02	3.88E+04	1.27E+05	
		HM	5.14E+04	6.58E+02	−1.96E+02	4.70E+04	1.51E+05	
	2nd	CR	5.85E+04	1.04E+05	−1.57E+02	5.22E+04	1.70E+05	
		HM	7.38E+04	5.11E+02	−2.01E+02	6.75E+04	2.19E+05	
(5) HL^3	1st	CR	3.45E+04	6.69E+04	−1.57E+02	3.04E+04	1.09E+05	0.99993
		HM	4.46E+04	2.29E+02	−2.04E+02	4.04E+04	1.43E+05	
	2nd	CR	2.18E+04	1.79E+07	−1.14E+02	1.56E+04	9.97E+04	
		HM	3.37E+04	4.47E−01	−2.59E+02	2.76E+04	2.19E+05	
(6) $[Ag(HL^3)_2](NO_3)2H_2O$	1st	CR	4.73E+04	2.24E+04	−1.64E+02	4.41E+04	1.07E+05	0.99993
		HM	5.47E+04	2.76E+05	−1.43E+02	5.15E+04	1.07E+05	
	2nd	CR	4.30E+04	2.08E+04	−1.67E+02	3.88E+04	1.24E+05	
		HM	5.16E+04	1.05E+03	−1.92E+02	4.74E+04	1.46E+05	
(7) HL^4	1st	CR	3.97E+04	2.89E+04	−1.64E+02	3.56E+04	1.17E+05	0.99993
		HM	4.79E+04	6.18E+02	−1.96E+02	4.37E+04	1.41E+05	
	2nd	CR	3.72E+04	1.94E+06	−1.32E+02	3.10E+04	1.29E+05	
		HM	4.98E+04	8.93E+00	−2.34E+02	4.36E+04	2.17E+05	
(8) $[Ag(HL^4)_2](NO_3)$	1st	CR	3.59E+04	1.41E+05	−1.51E+02	3.15E+04	1.12E+05	0.99993
		HM	4.46E+04	1.15E+02	−2.10E+02	4.02E+04	1.52E+05	
	2nd	CR	5.76E+04	1.22E+05	−1.55E+02	5.14E+04	1.67E+05	
		HM	6.96E+04	2.96E+02	−2.05E+02	6.35E+04	2.16E+05	

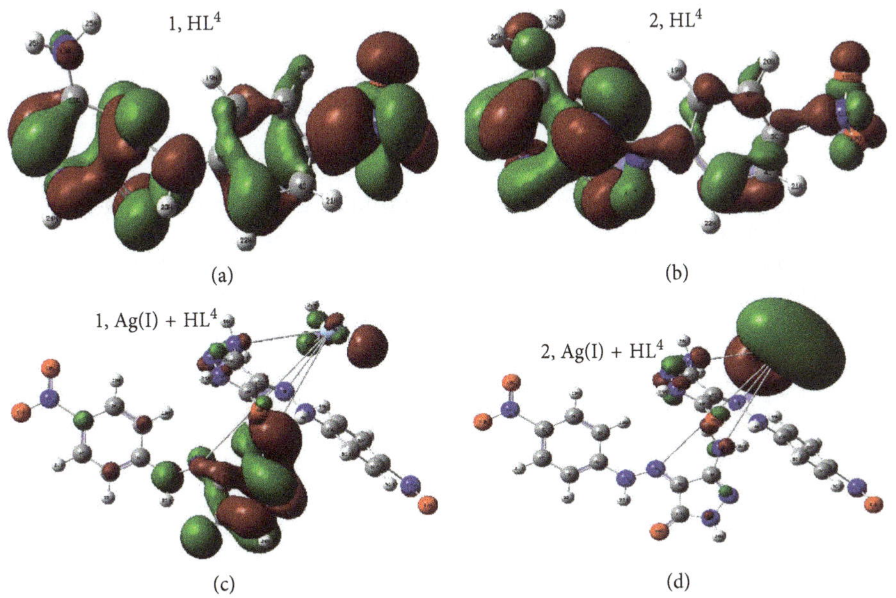

FIGURE 6: Images of frontier molecular orbital's (HOMO,1 and LUMO,2) for HL^4 ligand and its Ag(I) complex.

Table 6: QSAR computation for pyrazolone ligands and their Ag(I) complexes.

Function	HL^1	Ag(I)-HL^1	HL^2	Ag(I)-HL^2	HL^3	Ag(I)-HL^3	HL^4	Ag(I)-HL^4
Surface area (approx) (Å^2)	303.45	570.44	338.10	648.39	345.78	685.84	351.56	677.32
Surface area (grid) (Å^2)	396.20	703.28	421.12	755.88	424.55	773.68	427.04	648.34
Volume (Å^3)	611.36	1234.02	655.15	1322.62	664.82	1364.65	728.68	1410.40
Hydration energy (kcal/mol)	16.61	−20.40	−16.23	−20.24	−15.36	−18.13	−29.03	−41.97
Log P	0.36	−0.89	0.14	−1.33	0.52	−0.58	−3.48	−8.57
Reactivity (Å^3)	58.85	114.45	63.57	123.89	63.13	123.02	64.10	124.95
Polarizability (Å^3)	21.17	42.00	23.10	45.86	23.01	45.67	23.08	45.82

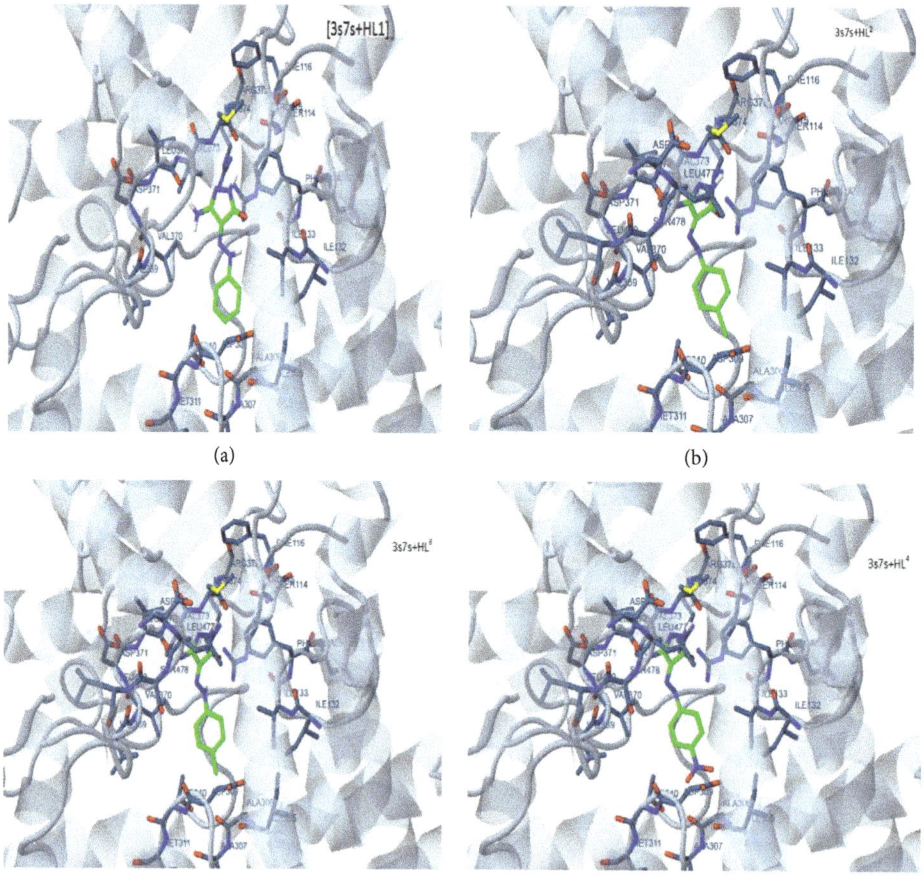

(a) (b) (c) (d)

Figure 7: High interacting complexes for 3s7s protein with HL^{1-4} ligands.

(Figure 11S). 2D plots (Figure 12S) verify the intense H-bonding with 3s7s protein in comparison with others. This study shows the promising inhibition activity towards breast cancer as well as CT-DNA which strongly and strangely agreeable with practical data that reported in our study.

4. Conclusion

A series of tetrahedral Ag(I)-pyrazolone complexes were synthesized and characterized. All compounds investigated were found in the nanometer range based on UV-Vis and XRD spectra. CT-DNA study yields intrinsic binding constants by the arrangement attributing to Hammett's postulation. Also, antitumor activity offered a promising feature for the Ag(I)-HL^4 complex. Structural formulae of all synthesized compounds were optimized theoretically. Also, simulation for the interaction between tested drugs and types from infected cell proteins was achieved.

Acknowledgments

Dr. Ismail Althagafi and coauthors would like to thank Deanship of Scientific Research at the Umm Al-Qura University (project ID: 43405078) for the financial support to this research extracted from the project.

Supplementary Materials

Table 1S: characteristic IR spectral bands (cm^{-1}) for Ag(I)-pyrazolone (HL^{1-4}) complexes. Table 2S: plausible TGA data for Ag(I)-pyrazolone (HL^{1-4}) complexes. Table 3S: energy parameters (eV) applying DFT method. Table 4S: significant log file data for pyrazolone ligands and their Ag(I) complexes. Table 5S: energy values (kcal/mol) for pyrazolone-protein docked complexes. Figure 1S: ^1H NMR spectra of pyrazolone derivatives, $HL^{(1-4)}$. Figure 2S: mass spectra of two elected ligands. Figure 3S: ^1H NMR spectra of three Ag(I)-pyrazolone derivatives. Figure 4S: mass spectra of selected ligands and two Ag(I) complexes. Figure 5S: X-ray patterns of three pyrazolone derivatives and their Ag(I) complexes. Figure 6S: SEM images of pyrazolone derivatives, $HL^{(1-4)}$ and their Ag(I) complexes. Figure 7S: spectrophotometric plots for DNA binding towards Ag(I) complexes (1–4). Figure 8S: dose response curves of HL (1–4) ligands and their Ag(I) complexes against MCF-7, HEPG-2 and HCT-116 cancer cells, cells were treated with various concentrations of ligands and complexes incubated for 72 hrs. %5 CO_2. Figure 9AS: kinetic plots using Coats–Redfern and Horowitz–Metzger for pyrazolone ligands, HL (1–4). Figure 9BS: kinetic plots using Coats–Redfern and Horowitz–Metzger for pyrazolone-Ag(I) complexes. Figure 10AS: images of frontier molecular orbital's (HOMO,1 and LUMO,2) for pyrazolone ligands (HL^{1-3}). Figure 10BS: images of frontier molecular orbital's (HOMO,1 and LUMO,2) for Ag(I)-HL^{1-3} complexes (A–C, respectively). Figure 11S: interacting complexes appeared with 3e1r, 4dk7, and 1bpb proteins and HL^{1-4} ligands. Figure 12S: 2D plots for interacting complexes appeared with 3e1r, 4dk7, 1bpb, and 3s7s proteins proteins and HL^{1-4} ligands. (*Supplementary Materials*)

References

[1] F. Marchetti, C. Pettinari, and R. Pettinari, "Acylpyrazolone ligands: synthesis, structures, metal coordination chemistry and applications," *Coordination Chemistry Reviews*, vol. 249, no. 24, pp. 2909–2945, 2005.

[2] I. Yousuf, F. Arjmand, S. Tabassum, L. Toupet, R. A. Khan, and M. A. Siddiqui, "Mechanistic insights into a novel chromone-appended Cu(II) anticancer drug entity: *in vitro* binding profile with DNA/RNA substrates and cytotoxic activity against MCF-7 and HepG2 cancer cells," *Dalton Transactions*, vol. 44, no. 22, pp. 10330–10342, 2015.

[3] F. Arjmand, M. Muddassir, Y. Zaidi, and D. Ray, "Design, synthesis and crystal structure determination of dinuclear copper-based potential chemotherapeutic drug entities; *in vitro* DNA binding, cleavage studies and an evaluation of genotoxicity by micronucleus test and comet assay," *Medicinal Chemistry Communications*, vol. 4, no. 2, pp. 394–405, 2013.

[4] J. Sastry and S. J. Kellie, "Severe neurotoxicity, ototoxicity and nephrotoxicity following high-dose cisplatin and amifostine," *Pediatric Hematology and Oncology*, vol. 22, no. 5, pp. 441–445, 2005.

[5] T. Siciliano, M. Deblock, H. Hindi et al., "Synthesis and anticancer properties of gold(I) and silver(I) N-heterocyclic carbene complexes," *Journal of Organometallic Chemistry*, vol. 696, no. 5, pp. 1066–1071, 2011.

[6] R. P. Feazell, C. E. Carson, and K. K. Klausmeyer, "Variability in the structures of luminescent [2-(aminomethyl)pyridine] silver(I) complexes: effect of ligand ratio, anion, hydrogen bonding, and π-stacking," *European Journal of Inorganic Chemistry*, vol. 2005, no. 16, pp. 3287–3297, 2005.

[7] S. Roy, T. K. Mondal, A. Layek, R. Saha, and C. Sinha, "Structure, spectra and electrical conductivity of copper(I) and silver(I) phosphino bridging mixed ligand complexes with coumarinyl Schiff base," *Inorganica Chimica Acta*, vol. 469, pp. 523–535, 2018.

[8] M. Nath, H. Singh, G. Eng, and X. Q. Song, "New di- and tri-organotin(IV) derivatives of tyrosinylphenylalanine as models for metaleprotein interactions: synthesis and structural characterization. Crystal structure of Me_2Sn(Tyr-Phe)$ MeOH," *Journal of Organometallic Chemistry*, vol. 693, no. 15, pp. 2541–2550, 2008.

[9] J. Wei, J. Xiao, J. Yu, X. Yi, S. Liu, and G. Liu, "Synthesis and structural characterization of silver(I) and gold(I) complexes of N, N′-diisobutyloxycarbonyl-N′,N″-(1,3-propylene)-bisthiourea," *Polyhedron*, vol. 137, pp. 176–181, 2017.

[10] S. M. Swasey, L. E. Leal, O. Lopez-Acevedo, J. Pavlovich, and E. G. Gwinn, "Silver (I) as DNA glue: Ag^+-mediated guanine pairing revealed by removing Watson–Crick constraints," *Scientific Reports*, vol. 5, no. 1, pp. 10163–10171, 2015.

[11] P. R. O'Neill, L. R. Velazquez, D. G. Dunn, E. G. Gwinn, and D. K. Fygenson, "Hairpins with poly-C loops stabilize four types of fluorescent Agn:DNA," *Journal of Physical Chemistry C*, vol. 113, no. 11, pp. 4229–4233, 2009.

[12] P. S. Cankara, B. Çaliskan, I. Durmaz, R. Atalay, and E. Banoglu, "Synthesis and preliminary mechanistic evaluation of 5-(p-tolyl)-1-(quinolin-2-yl) pyrazole-3-carboxylic acid amides with potent antiproliferative activity on human cancer cell lines," *European Journal of Medicinal Chemistry*, vol. 87, pp. 140–149, 2014.

[13] B. V. Kendre, M. G. Landge, and S. R. Bhusare, "Synthesis and biological evaluation of some novel pyrazole, isoxazole, benzoxazepine, benzothiazepine and benzodiazepine derivatives bearing an aryl sulfonate moiety as antimicrobial and anti-inflammatory agents," *Arabian Journal of Chemistry*, 2015.

[14] R. V. Ragavan, V. Vijayakumar, and N. S. Kumari, "Synthesis and antimicrobial activities of novel 1,5-diaryl pyrazoles," *European Journal of Medicinal Chemistry*, vol. 45, no. 3, pp. 1173–1180, 2010.

[15] F. Saad, N. El-Metwaly, T. A. Farghaly et al., "Illustration for series of new metal ion complexes extracted from pyrazolone derivative, spectral, thermal, QSAR, DFT/B3LYP, docking and antitumor investigations," *Journal of Molecular Liquids*, vol. 229, pp. 614–627, 2017.

[16] R. Prajuli, J. Banerjee, and H. Khanal, "Synthesis of some pyrazolone derivatives and evaluation of its antibacterial and cytotoxic activity," *Oriental Journal of Chemistry*, vol. 31, no. 4, pp. 2099–2106, 2015.

[17] M. K. A. Ibrahim and M. R. H. El-Moghayar, "Synthesis of azoles and fused azoles from a-arylhydrazononitriles," *Indian Journal of Chemistry*, vol. 26, pp. 832–835, 1987.

[18] M. S. Nair, D. Arish, and R. S. Joseyphus, "Synthesis, characterization, antifungal, antibacterial and DNA cleavage studies

of some heterocyclic Schiff base metal complexes," *Journal of Saudi Chemical Society*, vol. 16, no. 1, pp. 83–88, 2012.

[19] F. H. Stootman, D. M. Fisher, A. Rodger, and J. R. Aldrich-Wright, "Improved curve fitting procedures to determine equilibrium binding constants," *Analyst*, vol. 131, no. 10, pp. 1145–1151, 2006.

[20] A. I. Vogel, *Text Book of Quantitative Inorganic Analysis*, Longman, London, UK, 1986.

[21] G. A. Bain and J. F. Berry, "Diamagnetic corrections and Pascal's constants," *Journal of Chemical Education*, vol. 85, no. 4, pp. 532–536, 2008.

[22] E. Moukhina, "Determination of kinetic mechanisms for reactions measured with thermoanalytical instruments," *Journal of Thermal Analysis and Calorimetry*, vol. 109, no. 3, pp. 1203–1214, 2012.

[23] W. Coats and J. P. Redfern, "Kinetic parameters from thermogravimetric data," *Nature*, vol. 201, no. 4914, pp. 68–73, 1964.

[24] T. Ozawa, "A new method of analyzing thermogravimetric," *Bulletin of the Chemical Society of Japan*, vol. 38, no. 11, pp. 1881–1886, 1965.

[25] L. C. D. Cides, A. A. S. Araújo, M. Santos-Filho, and J. R. Matos, "Thermal behaviour, compatibility study and decomposition kinetics of glimepiride under isothermal and non-isotermal conditions," *Journal of Thermal Analysis and Calorimetry*, vol. 84, no. 2, pp. 441–445, 2006.

[26] S. Hochreuther, R. Puchta, and R. Eldik, "Thermodynamic kinetic studies on novel dinuclear platinum(ii) complexes containing bidentate *N,N*-donor ligands," *Inorganic Chemistry*, vol. 50, no. 18, pp. 8984–8996, 2011.

[27] M. J. Starink, "The determination of activation energy from linear heating rate experiments: a comparison of the accuracy of isoconversion methods," *Thermochimica Acta*, vol. 404, no. 1-2, pp. 163–176, 2003.

[28] H. W. Horowitz and G. A. Metzger, "A new analysis of thermogravimetric traces," *Analytical Chemistry*, vol. 35, no. 10, pp. 1464–1468, 1963.

[29] X. Wu and A. K. Ray, "Density-functional study of water adsorption on the $PuO_2(110)$ surface," *Physical Review B*, vol. 65, no. 8, pp. 85403–85408, 2002.

[30] M. J. Frisch, G. W. Trucks, H. B. Schlegel et al., *Gaussian 09, Revision D*, Gaussian Inc., Wallingford, CT, USA, 2010.

[31] R. Dennington, T. Keith, and J. Millam, *Gauss View, Version 4.1.2*, Semichem Inc., Shawnee Mission, KS, USA, 2007.

[32] H. Zhang, Y. Wang, H. Zhu, Z. Fei, and J. Cao, "Binding mechanism of triclocarban with human serum albumin: effect on the conformation and activity of the model transport protein," *Journal of Molecular Liquids*, vol. 247, pp. 281–288, 2017.

[33] G. M. Morris, D. S. Goodsell, R. S. Halliday et al., "Automated docking using a Lamarckian genetic algorithm and an empirical binding free energy function," *Journal of Computational Chemistry*, vol. 19, no. 14, pp. 1639–1662, 1998.

[34] C. Hamzaçebi and F. Kutay, "Continuous functions minimization by dynamic random search technique," *Applied Mathematical Modelling*, vol. 31, no. 10, pp. 2189–2198, 2007.

[35] İ. Gönül, A. Burak, S. Karaca, O. Şahin, and S. Serin, "Novel copper(II) complexes of two tridentate ONN type ligands: synthesis, characterization, electrical conductivity and luminescence properties," *Inorganica Chimica Acta*, vol. 477, pp. 75–83, 2018.

[36] A. B. P. Lever, *Inorganic Electronic Spectroscopy*, Elsevier, Amsterdam, Netherlands, 1986.

[37] N. M. El-Metwaly, R. M. El-shazly, I. M. Gabr, and A. A. El-Asmy, "Physical and spectroscopic studies on novel vanadyl complexes of some substituted thiosemicarbazides," *Spectrochimica Acta Part A*, vol. 61, no. 6, pp. 1113–1119, 2005.

[38] K. S. Abu-Melhaa and N. M. El-Metwally, "Spectral and thermal studies for some transition metal complexes of bis (benzylthiocarbohydrazone) focusing on EPR study for Cu (II) and VO $^{2+}$," *Spectrochimica Acta Part A*, vol. 70, no. 2, pp. 277–283, 2008.

[39] A. A. Abou-Hussen, N. M. El-Metwaly, E. M. Saad, and A. A. El-Asmy, "Spectral, magnetic, thermal and electrochemical studies on phthaloyl bis (thiosemicarbazide) complexes," *Journal of Coordination Chemistry*, vol. 58, no. 18, pp. 1735–1749, 2005.

[40] K. P. Bankura, D. Maity, M. M. R. Mollick et al., "Synthesis, characterization and antimicrobial activity of dextran stabilized silver nanoparticles in aqueous medium," *Carbohydrate Polymers*, vol. 89, no. 4, pp. 1159–1165, 2012.

[41] M. Baia, L. Baia, S. Astilean, and J. Popp, "Surface-enhanced Raman scattering efficiency of truncated tetrahedral Ag nanoparticle arrays mediated by electromagnetic couplings," *Applied Physical Letters*, vol. 88, no. 14, pp. 143121–143123, 2006.

[42] E. Garuglieri, E. Meroni, C. Cattò, F. Villa, F. Cappitelli, and D. Erba, "Effects of sub-lethal concentrations of silver nanoparticles on a simulated intestinal prokaryotic–eukaryotic interface," *Frontiers in Microbiology*, vol. 8, p. 2698, 2017.

[43] M. S. Refat and N. M. El-Metwaly, "Spectral, thermal and biological studies of Mn(II) and Cu(II) complexes with two thiosemicarbazide derivatives," *Spectrochimica Acta Part A*, vol. 92, pp. 336–346, 2012.

[44] B. D. Cullity, *Elements of X-ray Diffraction*, Addison–Wesley Inc., Boston, MA, USA, 1993.

[45] S. Velumani, X. Mathew, P. J. Sebastian, S. K. Narayandass, and D. Mangalaraj, "Structural and optical properties of hot wall deposited CdSe thin films," *Solar Energy Materials and Solar Cells*, vol. 76, no. 3, pp. 347–358, 2003.

[46] J. S. Ritch, T. Chivers, K. Ahmad, M. Afzaal, and P. O'Brien, "Synthesis, structures, and multinuclear NMR spectra of tin (II) and lead(II) complexes of tellurium-containing imidodiphosphinate ligands: preparation of two morphologies of phase-pure PbTe from a single-source precursor," *Inorganic Chemistry*, vol. 49, no. 3, pp. 1198–1205, 2010.

[47] J.-M. Chen, W. Wei, X.-L. Feng, and T.-B. Lu, "CO_2 fixation and transformation by a dinuclear copper cryptate under acidic conditions," *Chemistry–An Asian Journal*, vol. 2, no. 6, pp. 710–714, 2007.

[48] A. A. El-Bindary, A. Z. El-Sonbati, A. F. Shoair, and A. S. Mohamed, "Spectroscopic, thermal analysis and antimicrobial activities of supramolecular Cu(II) complexes," *Journal of Molecular Liquids*, vol. 220, pp. 409–425, 2016.

[49] C. Fosset, B. A. McGaw, and M. D. Reid, "A non-radioactive method for measuring Cu uptake in HepG2 cells," *Journal of Inorganic Biochemistry*, vol. 99, no. 5, pp. 1018–1022, 2005.

[50] U. El-Ayaan, N. M. El-Metwally, M. M. Youssef, and S. A. A. El Bialy, "Perchlorate mixed–ligand copper(II) complexes of β-diketone and ethylene diamine derivatives: thermal, spectroscopic and biochemical studies," *Spectrochimica Acta A*, vol. 68, no. 5, pp. 1278–1286, 2007.

[51] R. K. Ray and G. R. Kauffman, "EPR spectra and covalency of bis(amidinourea/O-alkyl-1-amidinourea)copper(II) complexes part II. Properties of the CuN_4^{2-} chromophore," *Inorganica Chimica Acta*, vol. 173, no. 2, pp. 207–214, 1990.

[52] R. C. Chikate and S. B. Padhye, "Transition metal quinone–thiosemicarbazone complexes 2: magnetism, ESR and redox behavior of iron(II), iron(III), cobalt(II) and copper

(II) complexes of 2-thiosemicarbazido-1,4-naphthoquinone," *Polyhedron*, vol. 24, no. 13, pp. 1689–1700, 2005.

[53] S. Sagdinc, B. Köksoy, K. F. Andemirli, and S. H. Bayari, "Theoretical and spectroscopic studies of 5-fluoro-isatin-3-(N-benzylthiosemicarbazone) and its zinc(II) complex," *Journal of Molecular Structure*, vol. 917, no. 2-3, pp. 63–70, 2009.

[54] I. Fleming, *Frontier Orbital's and Organic Chemical Reactions*, Wiley, London, UK, 1976.

[55] S. K. Tripathi, R. Muttineni, and S. K. Singh, "Extra precision docking, free energy calculation and molecular dynamics simulation studies of CDK2 inhibitors," *Journal of Theoretical Biology*, vol. 334, pp. 87–100, 2013.

[56] N. Terakado, S. Shintani, Y. Nakahara et al., "Expression of Cu,Zn-SOD, Mn-SOD and GST-pi in oral cancer treated with preoperative radiation therapy," *Oncology Reports*, vol. 7, pp. 1113–1117, 2000.

15

Enhanced Adsorption of Selenium Ions from Aqueous Solution using Iron Oxide Impregnated Carbon Nanotubes

Omer Y. Bakather,[1,2] Ahmad Kayvani Fard,[3,4] Ihsanullah,[5] Majeda Khraisheh,[6] Mustafa S. Nasser,[7] and Muataz Ali Atieh[3,4]

[1]*Department of Chemical Engineering, Jazan University, P.O. Box 706, Jazan 45142, Saudi Arabia*
[2]*Department of Chemical Engineering, Hadhramout University, Hadhramout, Yemen*
[3]*Qatar Environment and Energy Research Institute, Hamad Bin Khalifa University, Qatar Foundation, P.O. Box 5825, Doha, Qatar*
[4]*College of Science and Engineering, Hamad Bin Khalifa University, Qatar Foundation, P.O. Box 5825, Doha, Qatar*
[5]*Department of Chemical Engineering, King Fahd University of Petroleum & Minerals, Dhahran 31261, Saudi Arabia*
[6]*Chemical Engineering Department, Qatar University, P.O. Box 2713, Doha, Qatar*
[7]*Gas Processing Centre, College of Engineering, Qatar University, P.O. Box 2713, Doha, Qatar*

Correspondence should be addressed to Muataz Ali Atieh; mhussien@hbku.edu.qa

Academic Editor: Spyros P. Perlepes

The aim of this research was to investigate the potential of raw and iron oxide impregnated carbon nanotubes (CNTs) as adsorbents for the removal of selenium (Se) ions from wastewater. The original and modified CNTs with different loadings of Fe_2O_3 nanoparticles were characterized using high resolution transmission electron microscopy (HRTEM), scanning electron microscopy (SEM), X-ray diffractometer (XRD), Brunauer, Emmett, and Teller (BET) surface area analyzer, thermogravimetric analysis (TGA), zeta potential, and energy dispersive X-ray spectroscopy (EDS). The adsorption parameters of the selenium ions from water using raw CNTs and iron oxide impregnated carbon nanotubes (CNT-Fe_2O_3) were optimized. Total removal of 1 ppm Se ions from water was achieved when 25 mg of CNTs impregnated with 20 wt.% of iron oxide nanoparticles is used. Freundlich and Langmuir isotherm models were used to study the nature of the adsorption process. Pseudo-first and pseudo-second-order models were employed to study the kinetics of selenium ions adsorption onto the surface of iron oxide impregnated CNTs. Maximum adsorption capacity of the Fe_2O_3 impregnated CNTs, predicted by Langmuir isotherm model, was found to be 111 mg/g. This new finding might revolutionize the adsorption treatment process and application by introducing a new type of nanoadsorbent that has super adsorption capacity towards Se ions.

1. Introduction

Selenium (Se) is unique nonmetal chemical element with five known oxidative states under the form of elemental selenium Se^0, namely, 0, $^-1$, $^-2$, $^+4$, and $^+6$ [1]. Although it is an important trace element for many organisms, exceeding homeostatic levels is considered toxic. Se is typically found on the earth's crust, rocks, and sedimentary soils [2]. Recent reports state that a significant amount of Se emissions into the atmosphere and aquatic environment is related to industrial and mining activities in addition to agricultural drainage run-off [3–5].

Tan et al. [6], in a recent extensive review, reported the different types of organic and inorganic Se species available. The inorganic form of Se is typically found in surface and ground water in a number of reported forms such as Se^{-2} (selinide), $SeO_3{}^{-2}$ (selenite), $SeO_4{}^{-2}$ (selenate), and Se^0, which is the species nonsoluble form. If found in high concentration levels in waste or surface water, Se can cause serious environmental problems [7, 8]. The maximum level for Se in surface water was set at 5.0 μg/L by the United States Environmental Protection Agency (USEPA) in 1999. Therefore, due to the increased interest in Se treatment and the more stringent environmental consent levels for its concentration in surface water, the value set in 1999 is being updated by USEPA in 2014 [1, 5, 9]. Furthermore, there is variation in the maximum allowable levels of Se in drinking water worldwide. According to the World Health Organization [10] the value is set at 40 μg/L, while the European Union (EU) sets the Se levels at a much lower value of 10 μg/L [11]. In order to meet the drinking water standards for Se in addition to treating industrial, mining wastewaters and agricultural run-offs, a varied number of treatment processes are reported in the open literature with various degrees of complexity and advancement. Comprehensive reviews conducted by Robberecht and Van Grieken [4] and Tan et al. [6] offer an excellent overview of all current known treatments used for Se. It can be seen from the two recent reviews that although a great variety of biological [12–14], chemical [15–20], and physical treatments technologies [9, 21, 22] were developed in recent years, no single treatment offers a complete and cost effective scheme for treating Se from waste or drinking water. While biological treatments stand as one of the best options for treating Se, challenges related to long-term stability of the biogenic selenium and the predicting of the fate of the bioreduced Se in the environment are two major concerns. On the other hand, adsorption has been reported as an effective lower cost treatment technology for Se removal [23]. Varied adsorbents have been tested including low cost alternative materials [4], activated carbons, and other metal-oxide-enhanced materials. Of the latter, metal-oxide-enhanced materials with ferrihydrite have been reported as the most promising materials for the treatment of water containing SeO_3^{-2} ions [16]. Reports showed that the strong affinity of SeO_3^{-2} to iron hydroxide surfaces resulted in around 99% removal rate for a wide pH range (3–8). In contrast, the process was not effective for SO_4^{-2} and its presence acted as a competitor for the desorption sites causing suppressed removal rates [20]. The inadequate SO_4^{-2} removal was also related to interacting between the various Se species and the surface of the adsorbent. While SeO_3^{-2} was reported to form inner-sphere adsorption on iron hydroxide adsorbents, SO_4^{-2}, on the other hand, was adsorbed in an outer sphere manner resulting in an ineffective removal [16]. Accordingly, the development of a process to remove effectively most of the possible inorganic forms of Se is still required. With this in mind and due to their high surface areas, recent studies reported the use of various nanomaterials for the removal of the Se oxyanions. Example such as hydrophilic magnetic nanoparticle-graphene oxide composite was reported recently by Fu et al. [16]. A two-step reaction was used in the preparation of the adsorbents in addition to deposition of iron oxide nanoparticles at high temperatures. The authors reported a maximum removal of 95% of Se (VI) ions at pH ~ 2. While it is not easy to compare the maximum capacity of an adsorbent for the Se removal given the varied preparation methods, costs, and experimental condition, a recent study using a new iron oxide nanoparticles reported a higher removal rates compared with literature reported values [18]. Having said that, the values are still small and the search for other adsorbents is still required for cost effective Se removal from water.

Accordingly, this study is focused on developing multi-walled carbon nanotubes based novel materials for effective removal of Se from aqueous solutions. In recent years, multiwalled carbon nanotubes (CNTs) have been explored widely due to their unique properties [24–26] and have been extensively used in nanotechnology, optics, electronics, material science, and water treatment [27]. CNTs were also reported for the adsorption of heavy metal ions [28–32] and organic compounds [33–36]. Despite the most recent comprehensive review articles and to the best of our knowledge, no studies were reported on the use of multiwalled carbon nanotubes for Se removal. In this work, iron oxide impregnated CNTs were employed for the adsorption of the selenium ions from water. The effect of iron oxide loading, pH, dosage of CNTs, contact time, and initial concentration were studied on the removal of selenium. The raw CNTs and impregnated CNTs were characterized using field emission scanning electron microscopy (FE-SEM), high transmission electron microscopy (HR-TEM), thermogravimetric analysis (TGA), X-ray diffractometer (XRD), energy dispersive X-ray spectroscopy (EDS), Brunauer, Emmett, and Teller (BET) nitrogen adsorption technique, and zeta potential. The adsorption equilibrium data were correlated by the Langmuir and Freundlich isotherms and the kinetic data were analyzed using two kinetic models.

2. Experimental

2.1. Materials. All solvents used in this study were of analytical grade and purchased from Sigma-Aldrich Co. Ltd. Ethanol liquid (98%, purity) was used as a solvent and iron (III) nitrate as a precursor of iron nanoparticles and selenium dioxide (SeO_2) was used as source of selenium ions in the water.

2.2. Production of Carbon Nanotubes. Floating catalyst chemical vapor deposition reactor was used for the production of CNTs. The experimental setup used and reaction conditions are reported previously by Fard et al. [26]. Briefly, injected vertical chemical vapor deposition (FC-CVD) with quartz tube 100 mm in diameter and 1200 mm in length with flanges fixed at both ends was used to synthesis CNTs. Xylene was used as source of hydrocarbon and the argon gas was used to flush the air from the system, while the hydrogen gas was used as a carrier and reacting gas. Purity of CNTs produced was >96%.

2.3. Impregnation of CNTs. The iron oxide nanoparticles were impregnated on the surface of CNTs by a wet impregnation method. For 5% iron oxide loadings, 361 mg of Iron (III) nitrate nonahydrate and 1 g of CNTs were dissolved separately in ethanol solution and sonicated for 30 minutes to ensure uniform mixing. Upon further mixing of the two solutions and further sonication, the solution was kept in a furnace at 80°C overnight to evaporate the ethanol. Finally, the product was calcined at 350°C for 3 hours in the convection oven to ensure effective attachment of the iron oxide particles onto the surface of CNTs. After cooling, the composite of CNTs with 5% iron oxide NP is synthesized. To produce CNTs with 10% and 20% iron oxide loading, 722 mg and 1.44 g Iron (III) nitrate nonahydrate are mixed with 1 g of CNTs, respectively. Details of the preparation are found elsewhere [37].

2.4. Characterization of Carbon Nanotubes

2.4.1. Crystal Structure. Powder X-ray diffraction (XRD) patterns were recorded using a Rigaku MiniFlex-600. The X-ray diffractometer with Cu Kα radiation $\lambda = 1.54$ Å at a rate of 0.4% over Bragg angles ranging from 10 to 90° was used for the analysis. The operating voltage and current were maintained at 40 kV and 15 mA, respectively.

2.4.2. Surface Structure. Scanning electron microscopy (SEM) was performed using FEI Quanta 200 Environmental Scanning Electron Microscope (ESEM) with a resolution of 5 nm and magnification 200K to observe the morphology and structure of the material. Also, the morphological and structural analysis of CNTs was conducted using transmission electron microscopy (TEM) (CM12, Philips).

2.4.3. Point of Zero Charge. To measure the surface charge and zeta potential Zetasizer (Nano ZS 90, Malvern Instruments Ltd., Malvern, UK) equipped with a 4.0 mW internal laser was used. The Zetasizer works on the principle of dynamic light scattering (DLS). The measurements were performed at room temperature (25°C) with a scattering angle of 90°.

2.4.4. Other Chemical Properties and Surface Area. To analyze physical and chemical properties of CNTs with respect to temperature, the thermogravimetric analyses were performed using a TGA analyzer (SDT, Q600) at a heating rate of 10°C/min in air. The surface areas of CNTs were measured by N_2 adsorption at 77 K using BET surface area analyzer (Micromeritics ASAP 2020).

2.5. Preparation of the Selenium Stock Solution. Specific amount of SeO_2 was dissolved in deionized water to prepare the stock solution. SeO_2 dissolves in water to form selenous acid (SeO_2^{-3}). The pH of the stock solution was adjusted by using 0.1 M NaOH or 0.1 M HNO_3 and maintained by the addition of buffer solutions.

2.6. Se Sorption Experiments. In order to assess the effectiveness of the new adsorbent, batch adsorption experiments were conducted at room temperature in 1 L glass beakers. 50 mL of selenium solution of required concentration was placed in the flasks, covered, and mounted on the mechanical rotary shaker (MPI Lab Shaker) to ensure adequate mixing. Different experimental runs were conducted to study the effect of solution pH, contact time, CNTs dosage, and initial Se (IV) ions concentration on the adsorption of Se ions. Inductively coupled plasma mass spectroscopy (ICP-MS) was employed to analyze the concentrations of the samples. The adsorption capacity (Q) and removal efficiency (RE) were calculated as follows [37]:

$$Q = \frac{(C_i - C_f) \times V}{W_g}$$
$$\text{RE}\,(\%) = \frac{(C_i - C_f)}{C_i} \times 100, \quad (1)$$

where C_i (mg/L) is the initial concentration of selenium in the water, C_f (mg/L) is the final concentration of the selenium in the water, V (L) is the volume of the water, and W_g is the mass of CNTs.

2.7. Adsorption Isotherms Models. The Langmuir and Freundlich isotherms were used to study the adsorption performance and to determine the adsorption capacity for the adsorbent [37]. The Langmuir adsorption isotherm is expressed as follows:

$$Q_e = \frac{q_m K_L C_e}{1 + K_L C_e}, \quad (2)$$

where Q_e (mg/g) and q_m (mg/g) are the amount adsorbed at equilibrium and the maximum adsorption capacity, respectively, while C_e (mg/L) is the equilibrium adsorbate concentration and K_L is Langmuir constant.

Equation (2) can be linearized as follows:

$$\frac{C_e}{Q_e} = \frac{C_e}{q_m} + \frac{1}{K_L q_m}. \quad (3)$$

Freundlich isotherm is expressed as follows:

$$Q_e = K_f C_e^{1/n}. \quad (4)$$

Equation (4) can be linearized as follows:

$$\log Q_e = \frac{1}{n} \log C_e + \log K_f, \quad (5)$$

where n and K_f are the empirical constants.

2.8. Adsorption Kinetics. In order to find the maximum selenium removal by CNTs and to model the experimental data, two well-known kinetic models, pseudo-first-order and pseudo-second-order models, were used in this study.

The Lagergren pseudo-first-order model proposes that the rate of sorption is proportional to the number of sites unoccupied by the adsorbate [37]. The linear form of the pseudo-first-order equation can be expressed as follows:

$$\ln(Q_e - Q_t) = \ln Q_e - k_1 t, \quad (6)$$

where Q_t is the sorption capacity (mg/g) at any preset time interval (t) and k_1 is the pseudo-first-order rate constant (min^{-1}). A graph of $\ln(Q_e - Q_t)$ versus time is plotted and the constant is found. Additionally, the adsorption data

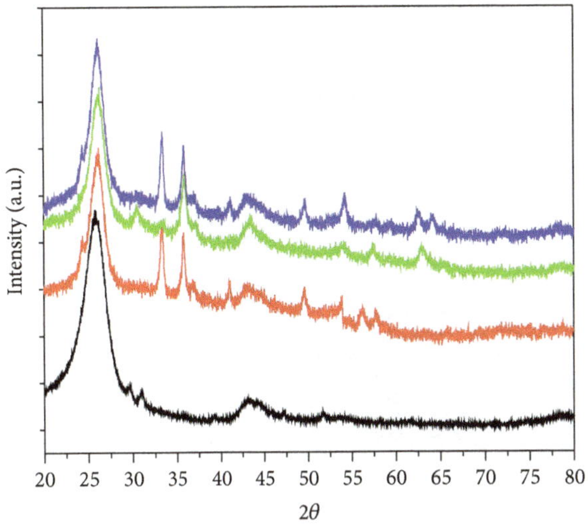

FIGURE 1: XRD analysis of raw CNTs, and CNT-5% iron oxide, CNT-10% iron oxide, and CNT-20% iron oxide.

were modeled to the pseudo-second-order kinetic model [37] where its linear form is expressed as follows:

$$\frac{t}{Q_t} = \frac{1}{k_2 Q_e^2} + \frac{t}{Q_e}, \quad (7)$$

where k_2 is the pseudo-second-order rate constant (g/mg·min). By plotting t/Q_t versus time, straight lines were obtained and the constants, k_2 and Q_e, were found.

3. Results and Discussion

3.1. Characterization of Carbon Nanotubes. Figure 1 shows the X-ray diffraction patterns of CNTs and impregnated CNTs with Fe_2O_3 nanoparticles. The XRD diffraction pattern of pure α-Fe_2O_3 is similar to that of CNTs impregnated Fe_2O_3 nanoparticles confirming the presence of α-Fe_2O_3 crystal nanoparticles on the surfaces of CNTs. The distinct peaks of α-Fe_2O_3 crystalline structure are found at 2θ of 34, 36, 42, 50, 54, 63, 65, 72, and 75. The characteristic peak of CNTs was observed at 2θ of 27 which corresponds to C (002) and indicative of proper and undamaged graphite structure of the CNTs. The other characteristic diffraction peaks of graphite are at 2θ of 43°, 45°, and 77° and associated with C (100), C (101), and C (110) diffractions of graphite, respectively. Peaks indexed at C (002), C (100), and C (101) are indication of hexagonal structure of CNTs and the presence of C (002) peak in the XRD data confirms the multiwalled nature of the carbon nanotubes [42].

The sharp peak at the 2θ of 25.5° of the raw and impregnated CNTs is indicative of undamaged (no impurities in lattice) and developed graphite structure. Defects would have been observed in the CNTs if the 2θ peak of 25.5° was broader and a shift of the peak diffraction towards lower angles was detected. Peaks at 31, 44, and 52.5 are indicative of the multiwalled nature of the carbon nanotubes. The results observed by the peaks in Figure 1 are a clear testament to the highly crystalline, uniform, highly ordered, and pure CNTs. The results here are further confirmed by SEM and TEM images.

In addition to XRD, the raw and impregnated CNTs were characterized further using field-scanning electron microscopy (FE-SEM), high resolution transmission electron microscopy (HR-TEM), thermogravimetry (TGA) techniques, XRD, BET surface area, and zeta potential.

The morphologies of these samples were obtained by SEM as shown in Figure 2. The diameter of the CNTs, with sponge-like structure, varied from 20 to 50 nm with an average diameter of 25 nm. The surface of CNTs after impregnation with Fe_2O_3 showed no surface changes and appeared to be agglomerated and untangled.

High resolution transmission electron microscopy (HR-TEM) was carried out to characterize the size, structure, and purity of the iron oxide nanoparticles doped and virgin CNTs. The raw CNTs TEM image presented in Figure 3 clearly indicates a highly ordered CNTs crystalline structure with diameter ranging from 10 to 30 nm and length from 10 to 30 μm. In addition, it can be noted that the clear fringes of the graphitic sheets are well separated by 0.35 nm and are aligned with a tilted angle of about 2° toward the tube axis. The TEM images of CNTs doped with Fe_2O_3 nanoparticles are presented in Figures 3(b)–3(d) in order to substantiate the presence of iron oxide nanoparticle on the surfaces of CNTs. The white iron oxide nanoparticles with some spherical and irregular shapes are shown in the TEM images. The size of Fe_2O_3 nanoparticles is estimated to be around 1–5 nm which are somehow distributed evenly and on some locations agglomerated slightly causing increase in the nanoparticle size.

To confirm the presence of iron oxide and also to experimentally find out the percentage of iron oxide nanoparticles on surface of CNTs, EDS analysis was conducted and the results are shown in Table 1.

The Fe_2O_3 content of the impregnated CNTs was also investigated using TGA. The experiment was carried out using air at a heating rate of 10°C/min. The thermograms are shown in Figure 4. Under this operating condition, the raw CNTs decomposed and oxidized completely as verified by the TGA curve in Figure 4. As Fe_2O_3 content on surface of CNTs increased, higher residual yield was found which corresponded to the presence of iron oxide NPs on the surface of CNT [37]. TGA provides an accurate estimate of the loading of iron oxide NPs doped on surface of CNTs by comparing the residues for the complete oxidation of the raw and impregnated CNTs. Therefore, CNTs with 1%, 10%, and 20% Fe_2O_3 loading had residual yields of 6 wt.%, 8 wt.%, and 17 wt.%, respectively. Moreover it can be noted that the increase in Fe_2O_3 NP loading resulted in a decrease of the decomposition temperature. It can also be inferred, from the TGA analysis shown in Figure 4, that raw CNTs start to decompose at 540°C while the 5%, 10%, and 20% loaded CNTs with Fe_2O_3 decomposed at lower temperatures (450, 430, and 410°C, resp.). According to Chiang et al. [43] the earlier decomposition of CNTs impregnated with metallic NPs can be explained by reduction of the thermal stability of CNTs by

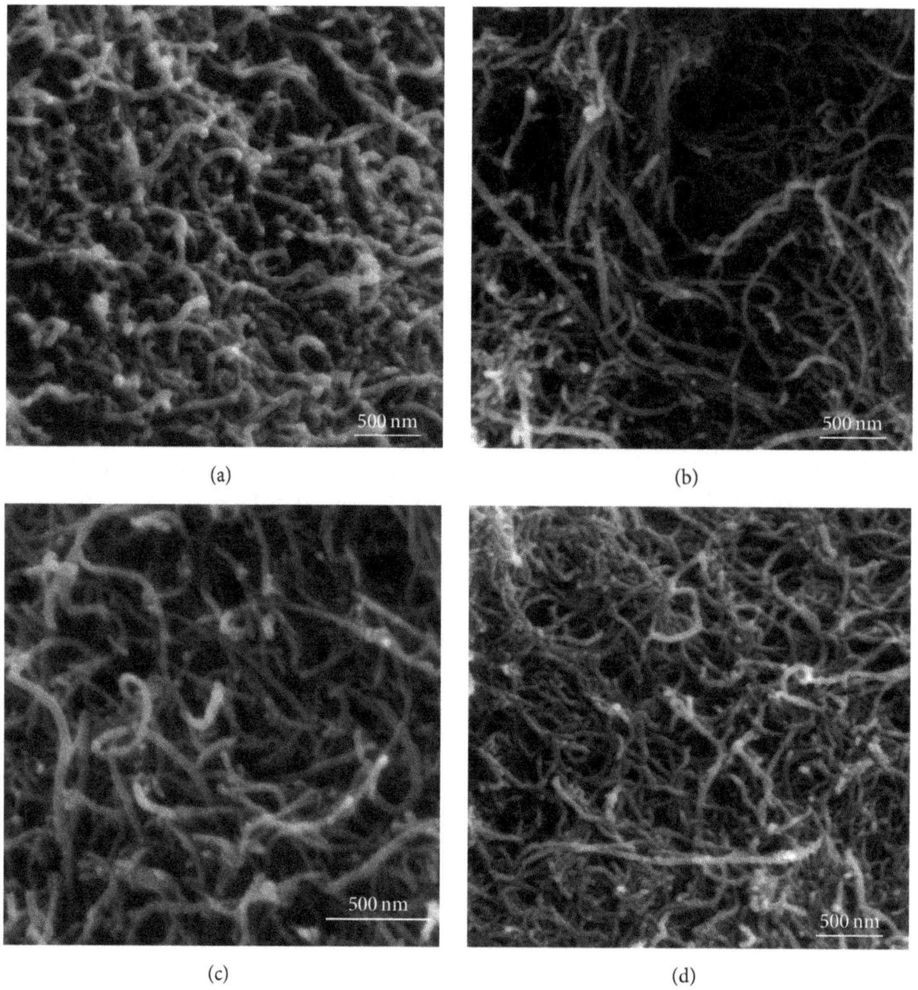

FIGURE 2: SEM images of (a) raw CNTs, (b) CNT-5% iron oxide, (c) CNT-10% iron oxide, and (d) CNT-20% iron oxide.

catalyzing the low-temperature oxidation of CNTs. Therefore, it can be inferred that the presence of nanosized iron oxide particles with high surface area altered the thermal stability of the CNTs and catalyzed oxidation of impregnated CNTs under the air compared to the pristine CNTs. The nanosized metallic particles might act as a heating accelerant that progresses the heat transfer to the surface of the CNTs and enhance the oxidation process.

The surface area of raw CNTs and impregnated CNTs was measured using BET surface area analyzer. As shown in Figure 5, the surface area of raw CNTs and CNTs impregnated with 5%, 10%, and 20% iron oxide nanoparticles was found to be 137.7, 226.6, 295.4, and 360 m^2/g, respectively. The presence of the nanosized iron oxide particles doubled the surface area upon using 20 wt.% iron oxide nanoparticles. Increasing the surface area inherently increases the number of adsorption sites available for the removal of selenium and therefore results in higher adsorption capacity.

Finally, the surface charge of the CNTs and impregnated CNTs with iron oxide nanoparticles was measured and the results are illustrated in Figure 6. When the Fe_2O_3 nanoparticles were loaded onto the surface of CNTs the negative sign of zeta potential on the surface of carbon nanotubes decreased due to neutralizing the repulsive effects of the electrical double layers. However, selenium (SeO_3^{2-}) has very large negative zeta potential (−0.37 V). As the negative sign of zeta potential decreases, the electrostatic attraction would make and attachment between CNTs and selenium ions more likely [38]. Thus, it is important to decrease the electrostatic repulsion barrier between the selenium ions and CNTs to further improve the adsorption process. It is clear from the zeta potential measurements in Figure 6 that increasing the percentage of Fe_2O_3 nanoparticles onto the surface of CNTs caused a reduction in the negative sign of zeta potential. The trends obtained here are in good agreement with the adsorption measurements carried out in this work, where increasing the loading of Fe_2O_3 nanoparticle onto the surface of CNTs enhanced the removal of the selenium ions from the water. On the other hand, loading Fe_2O_3 NPs onto the surface of CNTs alters the point of zero charge (PZC) of the composite. PZC is a pH value at which material has zero zeta potential. Raw CNTs have PZC at pH of 4.6 while CNTs with 5%, 10%, and 20% Fe_2O_3 loading have PZC at pH of 5.2, 5.6, and 5.9, respectively.

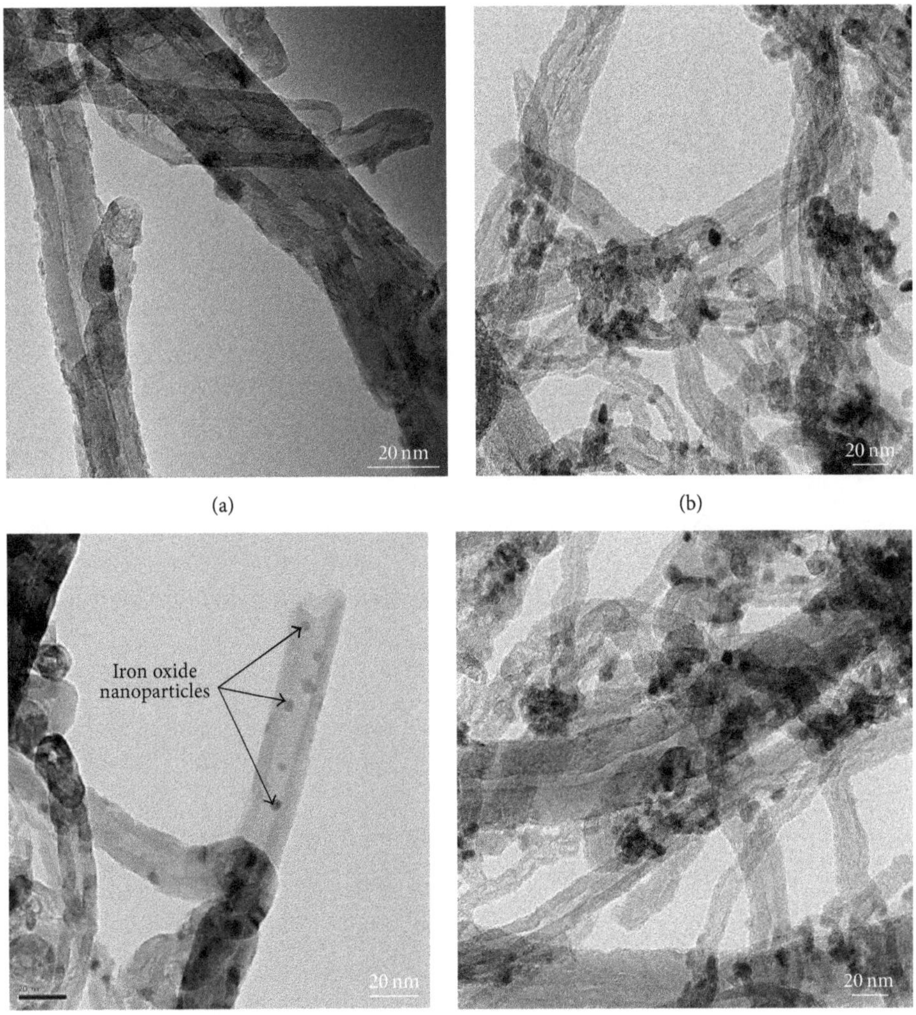

FIGURE 3: HRTEM images of (a) raw CNTs, (b) CNT-5% iron oxide, (c) CNT-10% iron oxide, and (d) CNT-20% iron oxide.

TABLE 1: EDS analysis of raw CNTs and CNTs with different percentage of Fe_2O_3.

CNTs sample	Raw CNTs	$CNT\text{-}Fe_2O_3$ (5%)	$CNT\text{-}Fe_2O_3$ (10%)	$CNT\text{-}Fe_2O_3$ (20%)
Element	Weight %	Weight %	Weight %	Weight %
C	98.50	91.80	84.53	56.4
O	1.50	3.65	4.44	25.52
Fe	0	4.55	11.03	18.08
Total %	*100*	*100*	*100*	*100*

3.2. Effect of pH. The pH of the solution plays an important role in the adsorption of selenium ions on the adsorbent surface as it is dependent on the surface properties of the impregnated CNTs and distribution of selenium ions in water.

The effect of pH on the removal of selenium ions is presented in Figure 7. The adsorption of selenium species was higher at lower pH and the removal was observed to decrease with increase in pH for the impregnated CNTs. The maximum removal was observed at pH 1. This higher removal at lower pH was due to the higher positive surface charges as indicated by zeta potential of the CNTs that favors the adsorption of anion (e.g., SeO_3^{2-}) [44]. This can be explained by release of OH ions when anion or weak acid is adsorbed onto hydroxide, which has favor toward adsorption of SeO_3^{2-} at low pH. Zhang et al. [38] used activated carbon doped with iron oxide for the removal of selenium from water. The authors reported that selenium removal sharply dropped after pH 8 and the maximum removal occurred in acidic solutions with pH between 1 and 3. A similar trend was previously observed in case of selenium ions adsorption on iron oxide nanoparticle [8], iron-coated GAC [38], hematite [45], soil [46], hydroxyapatite [39, 47], goethite and hydrous ferric oxide [48, 49], aluminum oxide coated sand [40], metal oxide nanoparticles [50], goethite [51], nanoscale zero-valent iron [52], and cellulose microcolumn [53].

It is anticipated that at lower pH, the adsorbent surface carries a additional positive charge and the anionic species are preferably adsorbed on the surface due to electrostatic

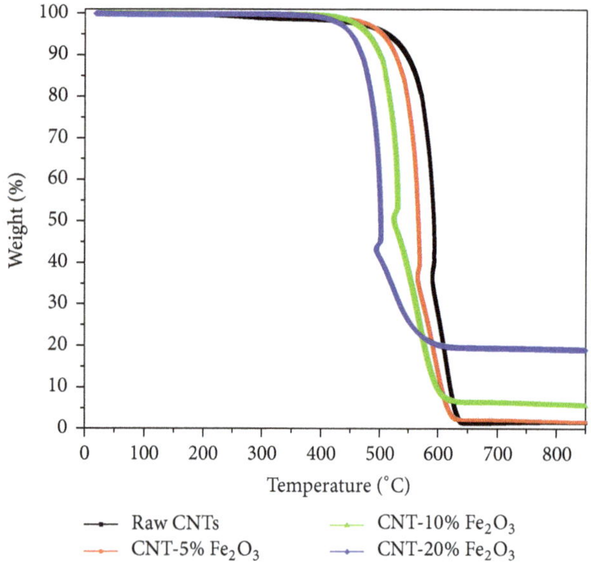

FIGURE 4: TGA analysis of raw CNTs and CNT-5% iron oxide, CNT-10% iron oxide, and CNT-20% iron oxide.

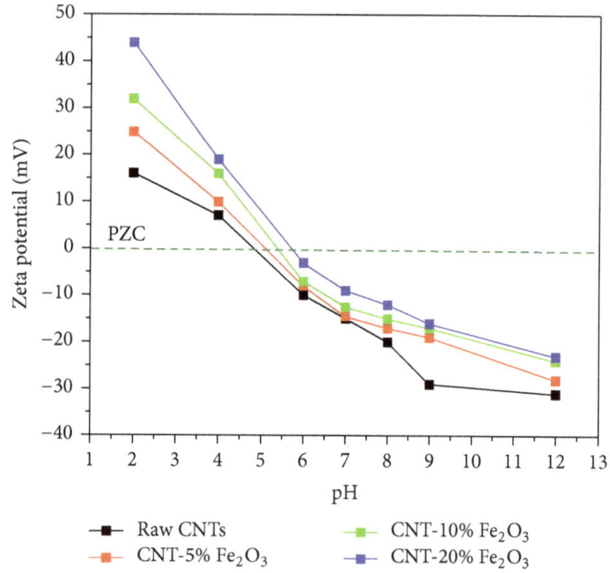

FIGURE 6: Zeta potential of raw CNTs and CNTs impregnated with 1%, 10%, and 20% Fe_2O_3.

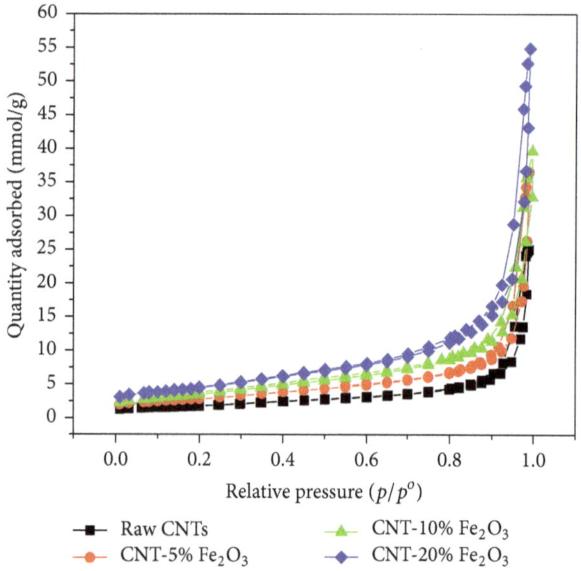

FIGURE 5: BET adsorption of N_2 at 77 K for raw CNTs and CNT-5% iron oxide, CNT-10% iron oxide, and CNT-20% iron oxide.

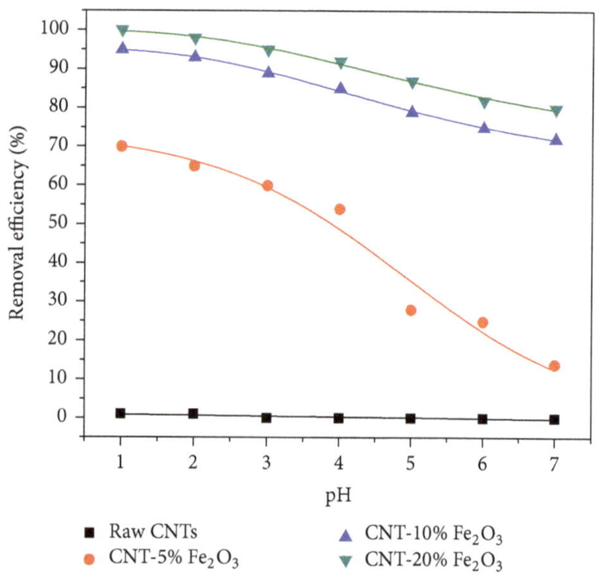

FIGURE 7: Effect of pH on the removal of selenium ions by raw and impregnated CNTs with different loading of iron oxide (agitation speed = 50 rpm, CNTs dosage = 10 mg, time = 6 hr, and initial concentration = 1 ppm).

attractions [8, 46]. Monteil-Rivera et al. [47], however, reported that Se removal by hydroxyapatite increases with pH from pH 7 to 8.5, and it decreases above pH 8.5. In general, the adsorption capacity of Se was decreased slightly upon increasing the pH of solution. This might be due to the increase in the negative charge of the adsorbent surface and the consequent competition between the OH^- ions and selenium ions for the available adsorption sites [44, 45]. Figure 7 also reveals that raw CNTs were not efficient in removing Se ions from solution as shown in the flat trend that it depicts in Figure 6. The adsorption of selenium was less than 1% at pH between 1 and 2. Negligible (close to zero) was recorded at pH of 2. In contrast, it can be seen from the trends in Figure 7 that vast improvement on the removal of Se from solution was achieved upon impregnating the surface of raw CNTs with iron oxide.

The percentage of iron oxide loading on the CNTs surface influenced the removal rate of Se. In general, it was observed at 20% loading rate that near 100% removal of Se was achieved compared to 93% and 65% for a loading rate of 10 and 5% iron oxide, respectively, at a pH of 1 for example. A number of factors may be attributed to this marked change. For example, the attachment of iron oxide particles onto the

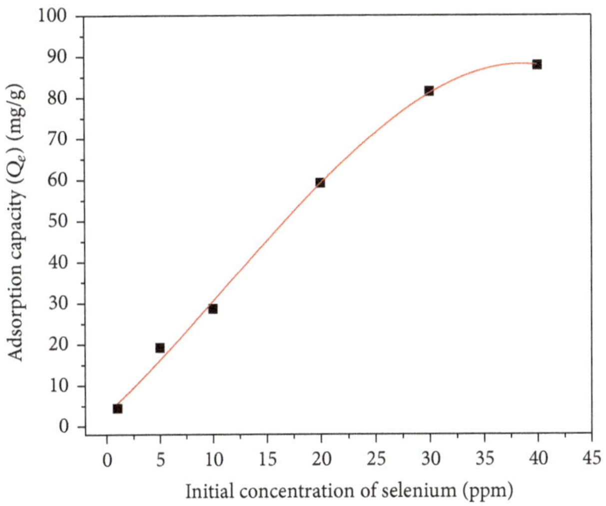

FIGURE 8: Effect of initial concentration of selenium on adsorption capacity of CNTs-20% iron oxide (agitation speed = 150 rpm, pH = 6, CNTs dosage = 10 mg, and time = 6 hr).

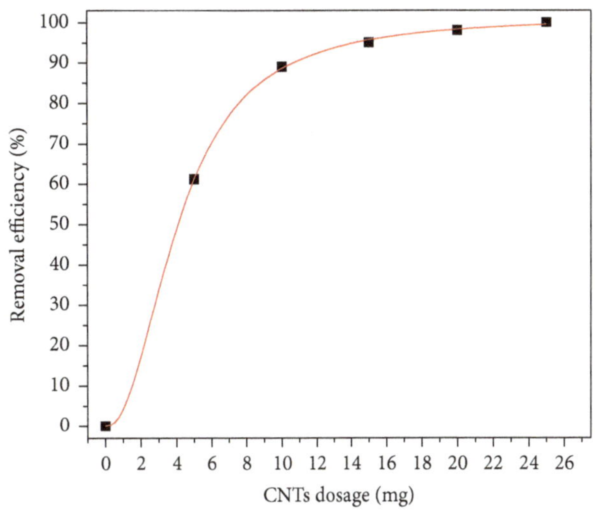

FIGURE 9: Effect of CNTs-20% iron oxide dosage on selenium removal (agitation speed = 150 rpm, pH = 6, time = 6 hr, and initial concentration = 1 ppm).

surface of CNTs may provide ample adsorption sites for the selenium ions to interact with. Also, it is well known that surfaces of metal oxides nanoparticles in aqueous solution are covered with hydroxyl groups [18]. Therefore, anion adsorption occurs by positive adsorbent surface charge (less negative sign compared to anion). Generally, increasing the pH causes decrease in the adsorbent surface charge and, accordingly, decreases in the adsorbent capacity [18, 39, 54]. As a result, when pH is increased the adsorbent surface is negatively charged (more negative sign) and leads to repulsion between negatively charged adsorbent particles and selenium anions. This repulsion causes termination of the adsorption process and also leads to the release of adsorbed selenium anions on surface of CNTs at higher pH to the water (desorption process). Based on the results reported in Figure 7, CNTs impregnated with 20 wt.% of iron oxide were selected in the later experiments to study the effects of other variables such as the initial Se concentration, CNTs dosage, contact time, and kinetics and isotherms models.

3.3. Effect of Initial Concentration. Figure 8 depicts the impact of initial Se concentration on the adsorption capacity of iron oxide impregnated CNTs. In general, the trend showed a marked increase of adsorption capacity with the increase in Se initial concentration. Figure 8 shows that at Se initial concentration between 5 and 20 ppm, the adsorption rate was steep and the adsorption was fast as can be seen from the slope of the first trend line. Above 20 pmm the adsorption slowed down and a plateau (second trend line) can be observed. The decline in the rate of adsorption at initial concentrations higher (as can be seen from the slope of the trend line) than 30 ppm (Figure 8) may be attributed to the saturation of all adsorption sites on the surface of CNTs. The higher adsorption capacity at higher Se concentration may be due to increase in the mass transfer (driving force) of selenium ions towards the iron oxide impregnated CNTs surfaces [18]. The highest adsorption capacity was about 88 mg/g at an initial Se concentration of 40 ppm as shown in Figure 8.

3.4. Effect of CNTs Dosage. The amount of CNTs added to the solution was varied between 5 and 25 mg in order to study the required optimum amount of adsorbent required to carry out the adsorption duty. In the experiments, the contact time, agitation speed, and pH were kept constant at 6 hr, 150 rpm, and 6, respectively. The experimental results are shown in Figure 9. Adsorption percentage (%) of selenium was plotted as a function of adsorbent dosage. Selenium ions adsorption was increased with increasing CNTs dosage due to the increase in the adsorption sites on CNTs surfaces resulting in the increase amount of adsorbed selenite ions. Selenium ions were completely removed from the solution using only 25 mg of CNTs. Therefore, the results show that the impregnated CNTs were suitable to adsorb selenium ions completely when there was sufficient CNTs surface area in the solution. Although the data was not shown here, the adsorption capacity was high at low dosages and keeps reducing at higher dosages. It is clearly agreed that as adsorbent dosage increases, the number of available adsorption sites increases, too. Alternatively, the decrease in adsorption capacity with increase in the adsorbent dosage is mostly related to the unsaturated nature of the adsorption sites through the adsorption process.

3.5. Effect of Contact Time. The effect of contact time on the adsorption of selenium ions is presented in Figure 10. It can be clearly seen that the rate of adsorption of selenium increased at initial period of contact time and then it decreased gradually with time until the adsorption reached an equilibrium point. The adsorption of selenium has increased rapidly during the first 30 minutes to reach about 65% removal. After that a slight increase was observed in the adsorption to

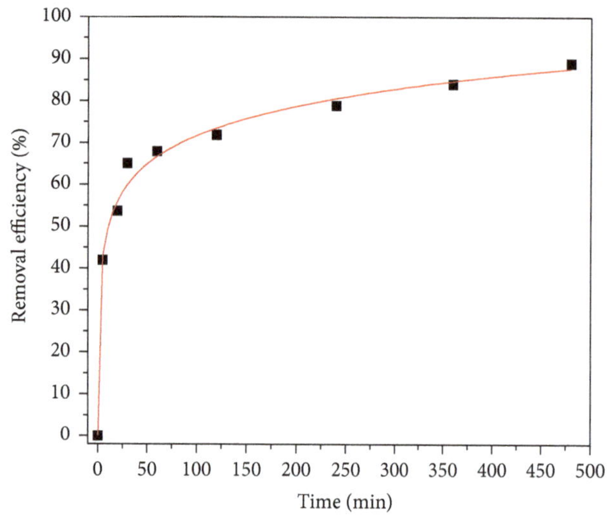

FIGURE 10: Effect of contact time on selenium removal (agitation speed = 150 rpm, pH = 6, CNTs dosage = 10 mg, and initial concentration = 1 ppm).

reach the maximum removal of selenium within four hours. The fast rate and then slow rate of adsorption suggest that selenium ions were first adsorbed on exterior surfaces of the CNTs during the initial time of contact. When the exterior surface gets saturated, the selenium ions diffused into the pores of the CNTs and were adsorbed at the interior surface of the CNTs.

3.6. Adsorption Kinetic Models. The modeling of the kinetics for selenium adsorption on CNTs was investigated using pseudo-first-order and pseudo-second-order kinetic models. Linearized plots of the two models are shown in Figure 11.

From Figure 11, the selenium adsorption on Fe_2O_3 impregnated CNTs does not fit very well to the pseudo-first-order model, as the R^2 values are less than 0.5. On the other hand, it does fit perfectly the pseudo-second-order model with R^2 value being almost 1. The rate constant for the 20% Fe_2O_3 loaded CNTs is found to be 0.016 g/mg·h.

The result obtained in this study is in good agreement with those reported in literature as most of the solid-liquid adsorption processes tend to conform to the pseudo-second-order model. Moreover, the pseudo-second-order model in this study can be defended by two-step linear relationship supporting the chemisorption nature of the process which is considered as rate-controlling mechanism as well [55]. The two-step linear mechanism between the composite CNTs (indicated as C and XC as activate adsorption site on CNTs) and Se ions is as follows:

$$\begin{aligned} 2C + Se^{2+} &\longrightarrow Se \\ 2XC + Se^{2+} &\longrightarrow SeC_2 + 2X^{+}. \end{aligned} \tag{8}$$

As it was found in Figure 11, the adsorption is pseudo-second-order and the rate-limiting step is the chemical adsorption between Se ions and surface of CNTs through sharing or the exchange of electrons. Similar trends are reported for selenium ion adsorption in literature where the adsorption mechanism fit to pseudo-second-order model [38, 39].

TABLE 2: Parameters of Langmuir and Freundlich adsorption isotherm models of selenium.

Langmuir			Freundlich		
Q_m (mg/g)	K_L (Lmg^{-1})	R^2	n	K_f (mg$^{(1-1/n)}$L$^{1/n}$g^{-1})	R^2
111	0.158	0.879	1.74	16	0.98

3.7. Adsorption Isotherms Models. The adsorption data is modeled using the Freundlich and Langmuir isotherm models. As shown in Figure 12, the ability of the Freundlich model to fit the experimental data was studied by generating a plot of $\ln Q_e$ versus $\ln C_e$ with the intercept value of K_f and the slope of n. It can be observed from Table 3 and Figure 12 that Freundlich isotherm model best fit the data ($R^2 = 0.98$). From Figure 12(a) the Freundlich constants K_f and n for the CNTs were found to be 16 and 1.74, respectively. The value of n or slope of the fit is an indication of sorption intensity or surface heterogeneity. When the slope gets closer to zero, the system tends to become more heterogeneous and a value of the slope greater than unity implies a favorable process and indicative of cooperative sorption. Therefore, as value of n increases, the sorption process between selenium ion and CNTs is more favorable and this means there is better bonding between sorbent and selenium. However, the Langmuir isotherm model (Figure 12(a)) fit reasonably R^2 value of 0.879. The maximum adsorption capacity of the iron oxide impregnated CNTs, as predicted by isotherm model, was found to be 111 mg/g.

Interestingly, composite made of CNTs and iron oxide nanoparticle showed very superior adsorption capacity compared to each one of the mentioned adsorbent in Table 2, where the capacity could reach maximum of 111 mg/g for selenium removal. There are few factors which contributed to the very high adsorption capacity compared to other materials in Table 3. Superior surface area, high zeta potential, and special surface structure of CNTs suggest that CNTs have great potential for use as contaminant adsorbents in wastewater treatment. The only limiting factor which might affect usage of CNTs in the application such as environmental protection or water treatment is their production cost. It is not very far if we predict that the cost will further reduce by advancement in technology and CNTs then could be one of the potential for application such as water treatment.

4. Conclusions

The study revealed that CNTs impregnated with 20 wt.% of iron oxide showed 100% removal of selenium ions in 6 hours with an initial concentration of 1 ppm at pH 6, adsorbent dosage of 25 mg, and agitation speed of 150 rpm. The adsorption data very well fitted to the Freundlich model with maximum adsorption capacity of the iron oxide impregnated CNTs predicted by Langmuir isotherm model

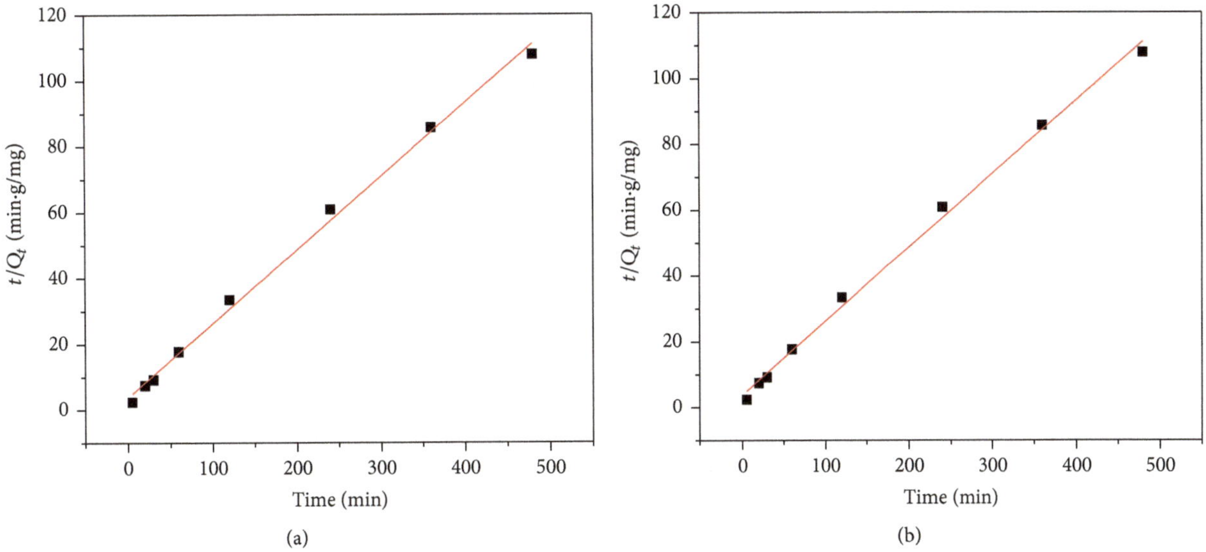

FIGURE 11: Adsorption kinetic for selenium removal: (a) pseudo-first-order and (b) pseudo-second-order model (agitation speed = 150 rpm, pH = 6, CNTs dosage = 10 mg, and initial concentration = 1 ppm).

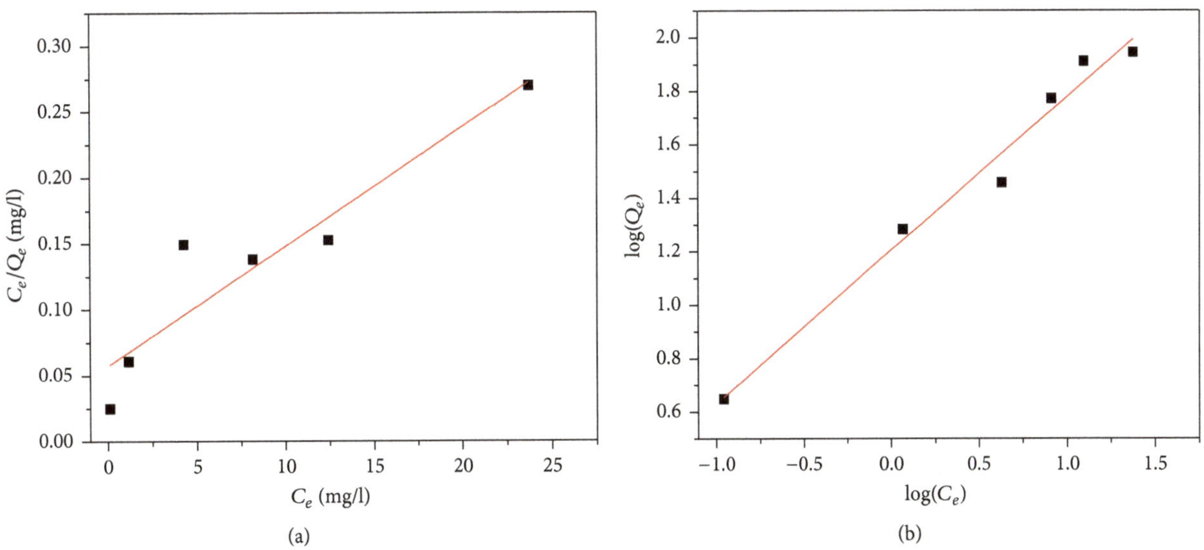

FIGURE 12: Adsorption isotherm models of selenium: (a) Langmuir and (b) Freundlich.

TABLE 3: Adsorption capacity of different materials for adsorption of selenium from water.

Type of adsorbent	Experimental conditions	Max. adsorption capacity (mg/g)	Reference
Iron oxide nanoparticle	pH = 4, initial concentration = 0.01 mg/L	15.1	[8]
Chitosan–clay composite	pH = 4, initial concentration = 0.1 mg/L, $T = 30^{\circ}C$	18.4	[15]
Iron-coated GAC	pH = 2–8, initial concentration = 2 mg/L, $T = 45^{\circ}C$	2.89	[38]
Nanocrystalline hydroxyapatite	pH = 5, initial concentration = 0.01 mg/L, $T = 30^{\circ}C$	1.94	[39]
Aluminum oxide coated sand	pH = 4.80, initial concentration = 1.2 mM	1.08	[40]
Sulfuric acid treated rice husk	pH = 1.5, initial concentration = 100 mg/L $T = 45^{\circ}C$	40.92	[41]
Iron oxide impregnated CNTs	pH = 6, CNTs dosage = 10 mg and initial concentration = 1 ppm	111	This study

to be 111 mg/g. The date was correlated very well to pseudo-second-order kinetic model and rate of constant was found to be 0.016 g/mg·h. The highest adsorption capacity of iron oxide impregnated CNTs suggested that it can be employed effectively for the removal of selenium ions from water.

Acknowledgments

The authors would like to acknowledge the support provided by King Fahd University of Petroleum & Minerals (KFUPM) for sponsoring and funding this work.

References

[1] J. Das, D. Das, G. P. Dash, and K. M. Parida, "Studies on Mg/Fe hydrotalcite-like-compound (HTlc): I. Removal of inorganic selenite (SeO_3^{2-}) from aqueous medium," *Journal of Colloid and Interface Science*, vol. 251, no. 1, pp. 26–32, 2002.

[2] K. S. Dhillon and S. K. Dhillon, "Selenium in groundwater and its contribution towards daily dietary Se intake under different hydrogeological zones of Punjab, India," *Journal of Hydrology*, vol. 533, pp. 615–626, 2016.

[3] J. Lessa, A. Araujo, G. Silva, L. Guilherme, and G. Lopes, "Adsorption-desorption reactions of selenium (VI) in tropical cultivated and uncultivated soils under Cerrado biome," *Chemosphere*, vol. 164, pp. 271–277, 2016.

[4] H. Robberecht and R. Van Grieken, "Selenium in environmental waters: determination, speciation and concentration levels," *Talanta*, vol. 29, no. 10, pp. 823–844, 1982.

[5] S. Santos, G. Ungureanu, R. Boaventura, and C. Botelho, "Selenium contaminated waters: an overview of analytical methods, treatment options and recent advances in sorption methods," *Science of the Total Environment*, vol. 521-522, no. 1, pp. 246–260, 2015.

[6] L. C. Tan, Y. V. Nancharaiah, E. D. van Hullebusch, and P. N. Lens, "Selenium: environmental significance, pollution, and biological treatment technologies," *Biotechnology Advances*, vol. 34, no. 5, pp. 886–907, 2016.

[7] C. Hu, Q. Chen, G. Chen, H. Liu, and J. Qu, "Removal of Se(IV) and Se(VI) from drinking water by coagulation," *Separation and Purification Technology*, vol. 142, pp. 65–70, 2015.

[8] G. Zelmanov and R. Semiat, "Selenium removal from water and its recovery using iron (Fe^{3+}) oxide/hydroxide-based nanoparticles sol (NanoFe) as an adsorbent," *Separation and Purification Technology*, vol. 103, pp. 167–172, 2013.

[9] N. Rajamohan and M. Rajasimman, "Biosorption of Selenium using activated plant based sorbent—effect of variables, isotherm and kinetic modeling," *Biocatalysis and Agricultural Biotechnology*, vol. 4, no. 4, pp. 795–800, 2015.

[10] H. G. Gorchev and G. Ozolins, "WHO guidelines for drinking-water quality," *WHO Chronicle*, vol. 38, no. 3, pp. 104–108, 2011.

[11] G. Kallis and D. Butler, "The EU water framework directive: measures and implications," *Water Policy*, vol. 3, no. 2, pp. 125–142, 2001.

[12] G. Gonzalez-Gil, P. N. L. Lens, and P. E. Saikaly, "Selenite reduction by anaerobic microbial aggregates: Microbial community structure, and proteins associated to the produced selenium spheres," *Frontiers in Microbiology*, vol. 7, article 571, 2016.

[13] T. Kagami, T. Narita, M. Kuroda et al., "Effective selenium volatilization under aerobic conditions and recovery from the aqueous phase by Pseudomonas stutzeri NT-I," *Water Research*, vol. 47, no. 3, pp. 1361–1368, 2013.

[14] L. C. Staicu, E. D. Van Hullebusch, M. A. Oturan, C. J. Ackerson, and P. N. L. Lens, "Removal of colloidal biogenic selenium from wastewater," *Chemosphere*, vol. 125, pp. 130–138, 2015.

[15] N. Bleiman and Y. G. Mishael, "Selenium removal from drinking water by adsorption to chitosan-clay composites and oxides: batch and columns tests," *Journal of Hazardous Materials*, vol. 183, no. 1–3, pp. 590–595, 2010.

[16] Y. Fu, J. Wang, Q. Liu, and H. Zeng, "Water-dispersible magnetic nanoparticle-graphene oxide composites for selenium removal," *Carbon*, vol. 77, pp. 710–721, 2014.

[17] J. H. Kwon, L. D. Wilson, and R. Sammynaiken, "Sorptive uptake of selenium with magnetite and its supported materials onto activated carbon," *Journal of Colloid and Interface Science*, vol. 457, pp. 388–397, 2015.

[18] A. W. Lounsbury, J. S. Yamani, C. P. Johnston, P. Larese-Casanova, and J. B. Zimmerman, "The role of counter ions in nano-hematite synthesis: implications for surface area and selenium adsorption capacity," *Journal of Hazardous Materials*, vol. 310, pp. 117–124, 2016.

[19] L. D. Mafu, B. B. Mamba, and T. A. M. Msagati, "Synthesis and characterization of ion imprinted polymeric adsorbents for the selective recognition and removal of arsenic and selenium in wastewater samples," *Journal of Saudi Chemical Society*, vol. 20, no. 5, pp. 594–605, 2013.

[20] V. C. Schwindt, J. S. Ardenghi, P. Bechthold et al., "Selenium adsorption at different coverages on Fe(100) and Fe(111): a DFT study," *Applied Surface Science*, vol. 315, no. 1, pp. 252–260, 2014.

[21] M. Lenz and P. N. L. Lens, "The essential toxin: the changing perception of selenium in environmental sciences," *Science of the Total Environment*, vol. 407, no. 12, pp. 3620–3633, 2009.

[22] M. O. M. Sharrad, H. Liu, and M. Fan, "Evaluation of FeOOH performance on selenium reduction," *Separation and Purification Technology*, vol. 84, pp. 29–34, 2012.

[23] S. Klas and D. W. Kirk, "Understanding the positive effects of low pH and limited aeration on selenate removal from water by elemental iron," *Separation and Purification Technology*, vol. 116, pp. 222–229, 2013.

[24] P. J. F. Harris, *Carbon Nanotube Science: Synthesis, Properties and Applications*, Cambridge University Press, Cambridge, UK, 2009.

[25] D. Kang, X. Yu, M. Ge, F. Xiao, and H. Xu, "Novel Al-doped carbon nanotubes with adsorption and coagulation promotion for organic pollutant removal," *Journal of Environmental Sciences*, vol. 54, pp. 1–12, 2017.

[26] A. K. Fard, G. Mckay, Y. Manawi, Z. Malaibari, and M. A. Hussien, "Outstanding adsorption performance of high aspect ratio and super-hydrophobic carbon nanotubes for oil removal," *Chemosphere*, vol. 164, pp. 142–155, 2016.

[27] X. Liu, M. Wang, S. Zhang, and B. Pan, "Application potential of carbon nanotubes in water treatment: a review," *Journal of Environmental Sciences*, vol. 25, no. 7, pp. 1263–1280, 2013.

[28] Ihsanullah, F. A. Al-Khaldi, B. Abu-Sharkh et al., "Effect of acid modification on adsorption of hexavalent chromium

(Cr(VI)) from aqueous solution by activated carbon and carbon nanotubes," *Desalination and Water Treatment*, vol. 57, no. 16, pp. 7232–7244, 2016.

[29] Y.-H. Li, J. Ding, Z. Luan et al., "Competitive adsorption of Pb^{2+}, Cu^{2+} and Cd^{2+} ions from aqueous solutions by multiwalled carbon nanotubes," *Carbon*, vol. 41, no. 14, pp. 2787–2792, 2003.

[30] G. P. Rao, C. Lu, and F. Su, "Sorption of divalent metal ions from aqueous solution by carbon nanotubes: a review," *Separation and Purification Technology*, vol. 58, no. 1, pp. 224–231, 2007.

[31] F. Yu, Y. Wu, J. Ma, and C. Zhang, "Adsorption of lead on multi-walled carbon nanotubes with different outer diameters and oxygen contents: kinetics, isotherms and thermodynamics," *Journal of Environmental Sciences*, vol. 25, no. 1, pp. 195–203, 2013.

[32] N. M. M. Fourier, J. N. Sahu, E. C. Abdullah, and N. S. Jayakumar, "Rapid adsorption of toxic Pb(II) ions from aqueous solution using multiwall carbon nanotubes synthesized by microwave chemical vapor deposition technique," *Journal of Environmental Sciences*, vol. 45, pp. 143–155, 2016.

[33] A. Abbas, B. A. Abussaud, Ihsanullah, N. A. H. Al-Baghli, M. Khraisheh, and M. A. Atieh, "Benzene removal by iron oxide nanoparticles decorated carbon nanotubes," *Journal of Nanomaterials*, vol. 2016, Article ID 5654129, 10 pages, 2016.

[34] H. A. Asmaly, B. Abussaud, Ihsanullah et al., "Evaluation of micro- and nano-carbon-based adsorbents for the removal of phenol from aqueous solutions," *Toxicological and Environmental Chemistry*, vol. 97, no. 9, pp. 1164–1179, 2015.

[35] Ihsanullah, H. A. Asmaly, T. A. Saleh, T. Laoui, V. K. Gupta, and M. A. Atieh, "Enhanced adsorption of phenols from liquids by aluminum oxide/carbon nanotubes: comprehensive study from synthesis to surface properties," *Journal of Molecular Liquids*, vol. 206, pp. 176–182, 2015.

[36] H. Ding, X. Li, J. Wang, X. Zhang, and C. Chen, "Adsorption of chlorophenols from aqueous solutions by pristine and surface functionalized single-walled carbon nanotubes," *Journal of Environmental Sciences*, vol. 43, pp. 187–198, 2016.

[37] A. Kayvani Fard, T. Rhadfi, G. Mckay et al., "Enhancing oil removal from water using ferric oxide nanoparticles doped carbon nanotubes adsorbents," *Chemical Engineering Journal*, vol. 293, pp. 90–101, 2016.

[38] N. Zhang, L.-S. Lin, and D. Gang, "Adsorptive selenite removal from water using iron-coated GAC adsorbents," *Water Research*, vol. 42, no. 14, pp. 3809–3816, 2008.

[39] S. Kongsri, K. Janpradit, K. Buapa, S. Techawongstien, and S. Chanthai, "Nanocrystalline hydroxyapatite from fish scale waste: preparation, characterization and application for selenium adsorption in aqueous solution," *Chemical Engineering Journal*, vol. 215-216, pp. 522–532, 2013.

[40] W.-H. Kuan, S.-L. Lo, M. K. Wang, and C.-F. Lin, "Removal of Se(IV) and Se(VI) from water by aluminum-oxide-coated sand," *Water Research*, vol. 32, no. 3, pp. 915–923, 1998.

[41] E. I. El-Shafey, "Sorption of Cd(II) and Se(IV) from aqueous solution using modified rice husk," *Journal of Hazardous Materials*, vol. 147, no. 1-2, pp. 546–555, 2007.

[42] J. H. Lehman, M. Terrones, E. Mansfield, K. E. Hurst, and V. Meunier, "Evaluating the characteristics of multiwall carbon nanotubes," *Carbon*, vol. 49, no. 8, pp. 2581–2602, 2011.

[43] I. W. Chiang, B. E. Brinson, A. Y. Huang et al., "Purification and characterization of single-wall carbon nanotubes (SWNTs) obtained from the gas-phase decomposition of CO (HiPco process)," *Journal of Physical Chemistry B*, vol. 105, no. 35, pp. 8297–8301, 2001.

[44] R. Dobrowolski and M. Otto, "Preparation and evaluation of Fe-loaded activated carbon for enrichment of selenium for analytical and environmental purposes," *Chemosphere*, vol. 90, no. 2, pp. 683–690, 2013.

[45] M. Duc, G. Lefèvre, and M. Fédoroff, "Sorption of selenite ions on hematite," *Journal of Colloid and Interface Science*, vol. 298, no. 2, pp. 556–563, 2006.

[46] K.-H. Goh and T.-T. Lim, "Geochemistry of inorganic arsenic and selenium in a tropical soil: effect of reaction time, pH, and competitive anions on arsenic and selenium adsorption," *Chemosphere*, vol. 55, no. 6, pp. 849–859, 2004.

[47] F. Monteil-Rivera, M. Fedoroff, J. Jeanjean, L. Minel, M.-G. Barthes, and J. Dumonceau, "Sorption of selenite (SeO_3^{2-}) on hydroxyapatite: an exchange process," *Journal of Colloid and Interface Science*, vol. 221, no. 2, pp. 291–300, 2000.

[48] A. Manceau and L. Charlet, "The mechanism of selenate adsorption on goethite and hydrous ferric oxide," *Journal of Colloid And Interface Science*, vol. 168, no. 1, pp. 87–93, 1994.

[49] K. F. Hayes, C. Papelis, and J. O. Leckie, "Modeling ionic strength effects on anion adsorption at hydrous oxide/solution interfaces," *Journal of Colloid And Interface Science*, vol. 125, no. 2, pp. 717–726, 1988.

[50] Y. Tianmeng, B. Liu, and J. Liu, "Adsorption of selenite and selenate by metal oxides studied with fluorescent DNA probes for analytical application," *Journal of Analysis and Testing*, vol. 1, article 2, 2017.

[51] Z. Nie, N. Finck, F. Heberling, T. Prüessmann, C. Liu, and J. Lutzenkirchen, "Adsorption of selenium and strontium on Goethite: an EXAFS study and surface complexation modeling of the ternary systems," *Environmental Science and Technology*, vol. 51, no. 7, pp. 3751–3758, 2017.

[52] S. O. Adio, M. H. Omar, M. Asif, and T. A. Saleh, "Arsenic and Selenium removal from water using Biosynthesized Nanoscale zero-valent iron: a factorial design analysis," *Process Safety and Environmental Protection*, vol. 107, pp. 518–527, 2017.

[53] M. Min, C. Shen, L. Fang et al., "Design of a selective regenerable cellulose microcolumn for selenium efficient recovery and economic determination," *Chemical Engineering Research and Design*, vol. 117, pp. 7773–783, 2017.

[54] D. S. Han, B. Batchelor, and A. Abdel-Wahab, "Sorption of selenium(IV) and selenium(VI) to mackinawite (FeS): effect of contact time, extent of removal, sorption envelopes," *Journal of Hazardous Materials*, vol. 186, no. 1, pp. 451–457, 2011.

[55] Y. S. Ho and G. McKay, "Pseudo-second order model for sorption processes," *Process Biochemistry*, vol. 34, no. 5, pp. 451–465, 1999.

16

Chemical Composition and Porosity Characteristics of Various Calcium Silicate-Based Endodontic Cements

Seok Woo Chang

Department of Conservative Dentistry, School of Dentistry, Kyung Hee University, Seoul, Republic of Korea

Correspondence should be addressed to Seok Woo Chang; swc2007smc@gmail.com

Academic Editor: Claudio Pettinari

Chemical composition and porosity characteristics of calcium silicate-based endodontic cements are important determinants of their clinical performance. Therefore, the aim of this study was to investigate the chemical composition and porosity characteristics of various calcium silicate-based endodontic cements: MTA-angelus, Bioaggregate, Biodentine, Micromega MTA, Ortho MTA, and ProRoot MTA. The specific surface area, pore volume, and pore diameter were measured by the porosimetry analysis of N2 adsorption/desorption isotherms. Chemical composition and powder analysis by scanning electron microscope (SEM) and energy dispersive spectroscopy (EDS) were also carried out on these endodontic cements. Biodentine and MTA-angelus showed the smallest pore volume and pore diameter, respectively. Specific surface area was the largest in MTA-angelus. SEM and EDS analysis showed that Bioaggregate and Biodentine contained homogenous, round and small particles, which did not contain bismuth oxide.

1. Introduction

Mineral trioxide aggregate (MTA) was introduced in endodontic field as root end filling material and perforation repair material in early 1990s [1]. Due to its superior biocompatibility [2] and sealing ability [3], MTA has been widely used for perforation repair [4], root end filling [5], pulp capping [6], one-visit apexification [7], and pulpal revascularization [8]. However, MTA has been described to have drawbacks such as long setting time [9], tooth discoloration potential [10], and handling difficulty [11]. To overcome these drawbacks, many calcium silicate-based cements such as MTA-angelus [12], Bioaggregate [13], Biodentine [12], Micromega MTA (MM-MTA) [6], and Ortho MTA [14] have been introduced in market and showed good clinical and experimental results.

There are many reports that proved superior sealing ability of MTA in the MTA-tooth interface [15, 16]. However, the porosity existing in MTA itself has not been studied extensively [17–19]. Considering that the porosity of MTA is related to its ability to resist microbial penetration and leakage [20], there is relative lack of knowledge on this issue currently.

Thus, the aim of this study was to investigate the pore volume, pore diameter, and the specific surface area of various commercial calcium silicate-based endodontic cements. The surface morphology and chemical compositions of these cements were also investigated.

2. Materials and Methods

2.1. Materials Used. The materials used in this study were MTA-angelus, Bioaggregate, Biodentine, MM-MTA, Ortho MTA, and ProRoot MTA. The compositions of these materials are listed in Table 1.

2.2. BET Surface Area and Porosimetry Analyzer. Surface area and pore structure were measured by N2 adsorption/desorption isotherms (ASAP 2020 series) at 77 and 273 K for nitrogen and carbon dioxide within relative pressures from 0 to 1.0 and from 0 to 0.03, respectively. Before analysis, the samples were degassed in the degas port of the adsorption analyzer at 423 K for 10 hours. The surface area, pore volume, and pore diameter were analyzed using ASAP 2020 v3.00 software (Micromeritics, Norcross, GA, USA).

TABLE 1: Names and compositions of calcium silicate-based endodontic cements which were used in this study.

Products	Compositions
MTA-angelus (Londrina, PR, Brazil)	Tricalcium silicate, dicalcium silicate, tricalcium aluminate, tetracalcium aluminoferrite, bismuth oxide (MSDS)
Bioaggregate (Diadent, Burnaby, Canada)	Tricalcium silicate, dicalcium silicate, tantalum pentoxide, calcium phosphate monobasic, amorphous silicon oxide (MSDS)
Biodentine (Septodont, St. Maur-des-Fossés, France)	Tricalcium silicate, dicalcium silicate, calcium carbonate and oxide, iron oxide, zirconium oxide [21]
MM-MTA (MicroMega, Besançon, France)	Mixture of several mineral oxides and bismuth oxides (MSDS)
Ortho MTA (BioMTA, Seoul, Korea)	Calcium carbonate, silicon dioxide, aluminum oxide, dibismuth trioxide (MSDS)
ProRoot MTA (Dentsply, Tulsa, OK, USA)	Portland cement, bismuth oxide (MSDS)

TABLE 2: Porosity in tested calcium silicate-based endodontic cements.

	MTA-angelus	Bioaggregate	Biodentine	MM-MTA	Ortho MTA	ProRoot MTA
S_A	6.2	5.5	4.0	3.5	4.5	3.2
V_{pore}	0.016	0.014	0.0080	0.0086	0.014	0.0097
d_{pore}	9.3	11.9	13.5	21.5	11.1	17.3

S_A: specific surface area (m^2/g) calculated by BET equation; V_{pore}: pore volume (cm^3/g) calculated by BJH equation; d_{pore}: pore diameter (nm) calculated by BJH equation.

2.3. Scanning Electron Microscope (SEM) and Energy Dispersive Spectroscopy (EDS) Analysis. The morphology of the powders and chemical constitutions was measured on JEOL JSM-6700 scanning electron microscope. Prior to SEM measurement, the samples were coated with platinum using sputter for 45 seconds.

3. Results

3.1. BET Surface Area and Porosimetry Analysis. The specific surface area (m^2/g), pore volume (cm^3/g), and pore diameter (nm) values of all the samples are listed in Table 2. Specific surface area was the largest in MTA-angelus and the smallest in ProRoot MTA. Pore volume was the largest in MTA-angelus and the smallest in Biodentine. Pore diameter was the largest in MM-MTA and the smallest in MTA-angelus.

3.2. Scanning Electron Microscope (SEM) and Energy Dispersive Spectroscopy (EDS) Analysis. MTA-angelus (Figure 1) showed multiple aggregates of round particles. EDS analysis showed that these round particles are mainly composed of calcium and silica. Among these round particles, long spindle-shaped particles were shown. EDS analysis showed that these long spindle-shaped particles were mainly composed of bismuth.

Bioaggregate (Figure 2) showed relatively homogenous aggregates of small round particles. EDS analysis showed that these particles were mainly composed of calcium, silicon, and tantalum. Bioaggregate did not contain bismuth.

Biodentine (Figure 3) showed that relatively large particles were covered with small particles. EDS analysis showed that these particles were mainly composed of calcium and silicon.

MM-MTA (Figure 4) also showed the mixtures of relatively larger particles and smaller particles. EDS analysis showed that these particles were mainly composed of calcium and silicon.

Ortho MTA (Figure 5) showed large particles, small particles, and long spindle-shaped particles at the same time. All these particles were shown to be mainly composed of calcium and silicon.

ProRoot MTA (Figure 6) showed relatively homogenous particles which are mainly composed of calcium and silicon.

4. Discussion

Porosity of mineral trioxide aggregate is important in that it is related to bacterial leakage [20]. However, there are few studies which investigated the porosity of MTA [17–19]. Regarding the porosity characteristics, one previous study [17] reported that the apparent porosity of ProRoot MTA was 29.36% while that of Dycal was 9.04%. However, this study used Archimedes' principle to calculate the porosity of MTA samples. In this reason, this study had a limitation that it could not give information regarding the characteristics such as pore diameter and specific surface area of MTA.

Porosity-related properties of a certain material are specific surface area (m^2/g), pore volume (cm^3/g), and pore diameter [22]. Most previous studies which investigated MTA porosity used mercury intrusion porosimetry [18, 19]. It was reported that the detection range of mercury intrusion porosimetry is from 3 nm to 200 μm, whereas that of N2 adsorption/desorption isotherms is from 0.3 nm to 300 nm [22]. According to this report [22], N2 adsorption/desorption isotherms can detect the small pores which could not be detected by mercury intrusion porosimetry. In this reason, the study of porosity of MTA using N2 adsorption/desorption isotherms as well as mercury intrusion porosimetry could be regarded as ideal methods.

Element	Weight %	Atomic %
O K	14.92	63.46
Ca K	6.43	10.92
Bi M	78.65	25.62
Totals	100.00	

Element	Weight %	Atomic %
O K	40.82	62.50
Si K	5.09	4.44
Ca K	54.09	33.06
Totals	100.00	

FIGURE 1: SEM and EDS analysis results of MTA-angelus.

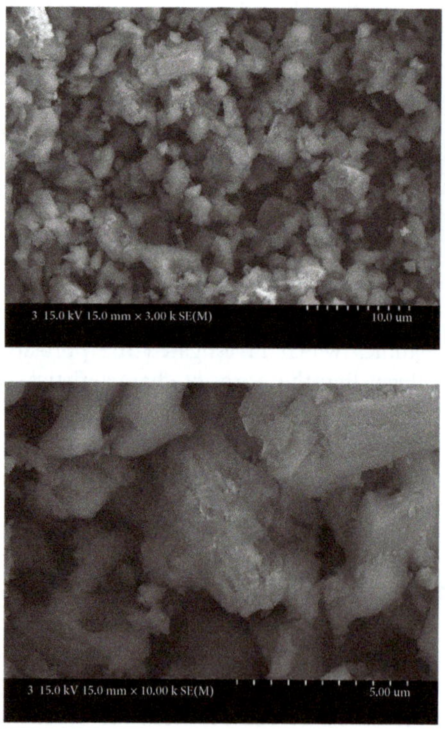

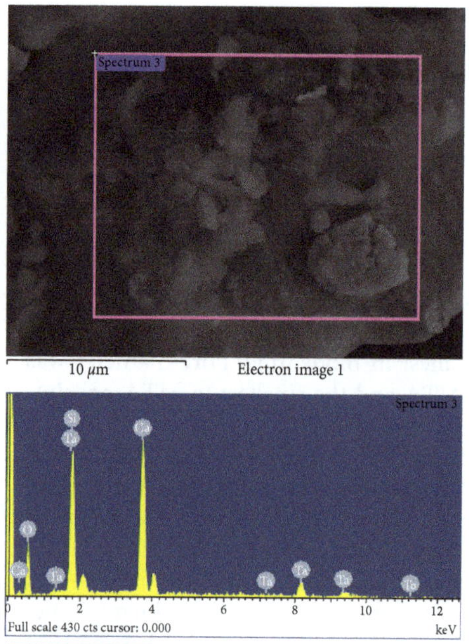

Element	Weight %	Atomic %
O K	31.92	59.06
Si K	9.53	10.05
Ca K	37.05	27.37
Ta M	21.50	3.52
Totals	100.00	

FIGURE 2: SEM and EDS analysis results of Bioaggregate.

The previous studies reported that the pore volume for ProRoot MTA was 0.1025 cm^3/g at pH 7.4 [19]. The pore volume for ProRoot MTA was 0.0097 cm^3/g in this study. This difference could be attributed to the experimental conditions such as time elapsed for MTA setting and the environment around the MTA setting.

The pore volume inside the specimen was the largest in MTA-angelus group (0.016 cm^3/g). The pore volume inside

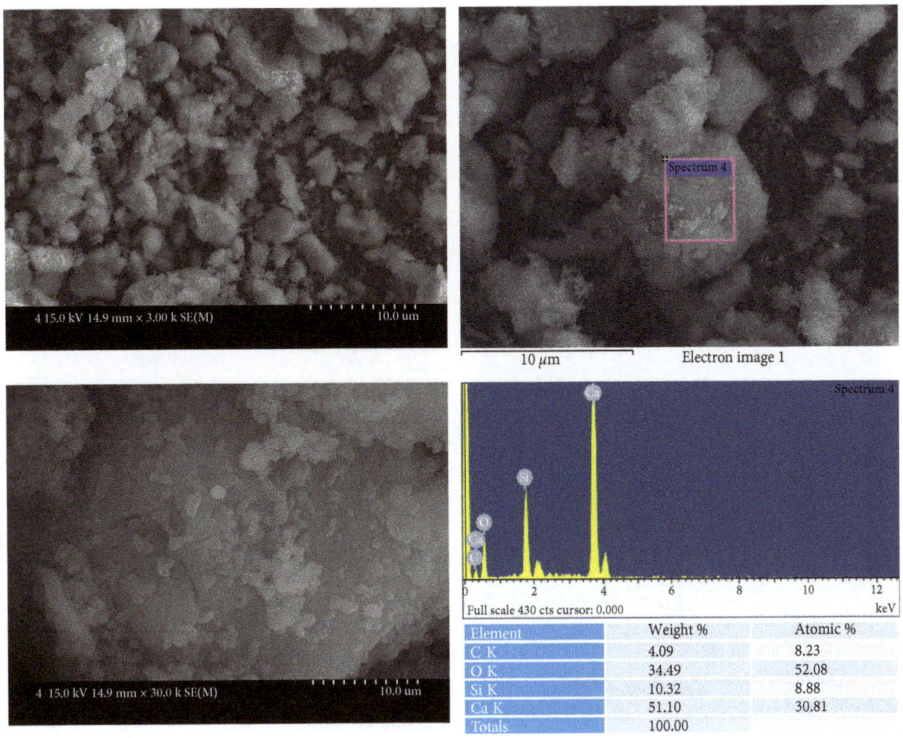

Figure 3: SEM and EDS analysis results of Biodentine.

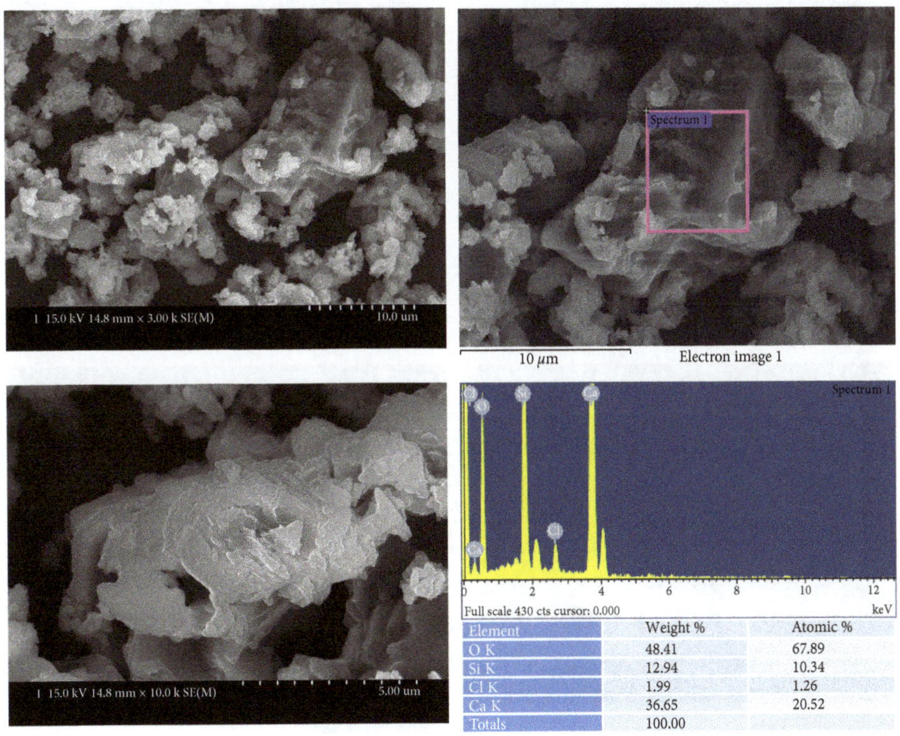

Figure 4: SEM and EDS analysis results of MM-MTA.

Bioaggregate and Ortho MTA was the same and was 0.014 cm^3/g. The pore volume of MM-MTA was 0.0086 cm^3/g. The pore volume of Biodentine was the smallest of all the tested groups. (0.0080 cm^3/g).

In addition to the total pore volume, the size of pore diameter is important [19]. Unfortunately, there has been no study which evaluated pore diameters of mineral trioxide aggregate. In the present study, pore diameter was the largest

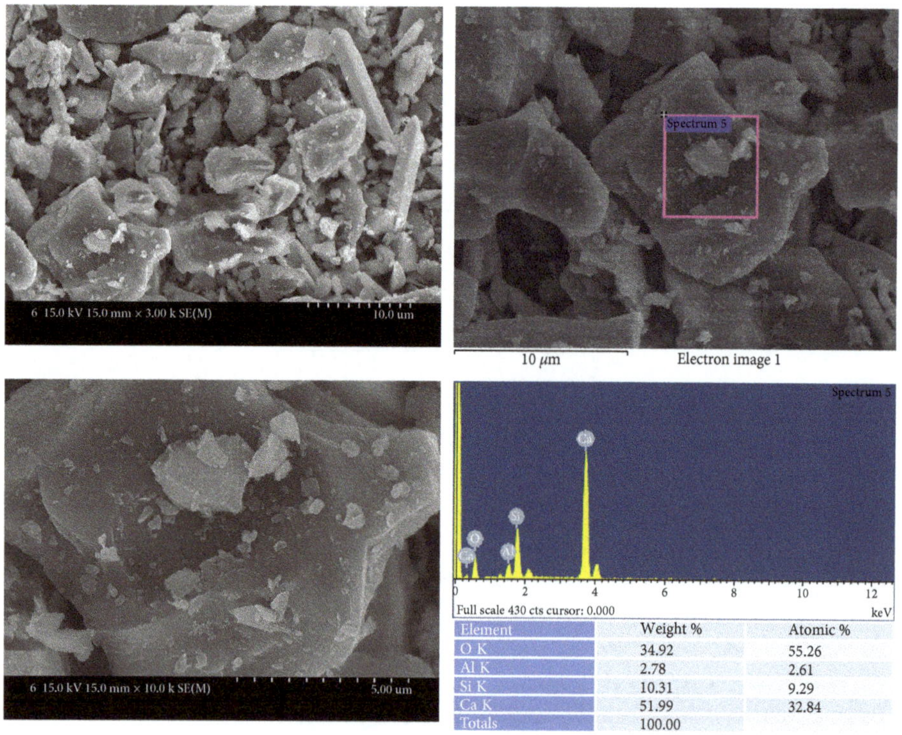

FIGURE 5: SEM and EDS analysis results of Ortho MTA.

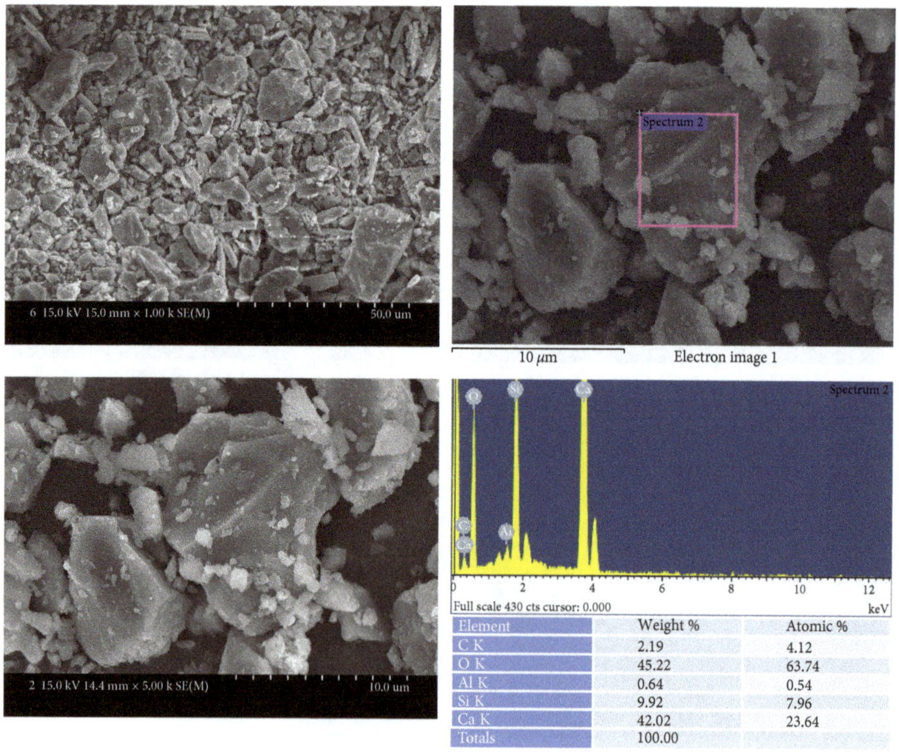

FIGURE 6: SEM and EDS analysis results of ProRoot MTA.

in MM-MTA (21.5 nm) and decreased in the order of ProRoot MTA, Biodentine, Bioaggregate, Ortho MTA, and MTA-angelus. MTA-angelus has the smallest pore diameter, and it was 9.3 nm. Considering that the average size of *Enterococcus faecalis* (representative endodontic bacterium) is 0.6–2.5 μm [23], it is quite unlikely that bacteria could penetrate well-condensed and hydrated MTA. Another characteristic investigated in this study was specific surface area. Specific surface area could affect the adhesion of contacting cells [24]. The larger surface area is considered to be the more favorable condition to cellular adhesion [24]. In the present study, the specific surface area was the largest in MTA-angelus and decreased in the order of Bioaggregate, Ortho MTA, Biodentine, MM-MTA, and ProRoot MTA. ProRoot MTA has the smallest specific surface area, and it was 3.2 m^2/g. The effect of these different specific surface areas should be investigated further in future study.

5. Conclusion

In conclusion, this study showed that Biodentine and MTA-angelus showed the smallest pore volume and pore diameter, respectively, which could be regarded as superior physicochemical properties from the perspective of clinical endodontics.

References

[1] M. Torabinejad, A. F. Rastegar, J. D. Kettering, and T. R. Pitt Ford, "Bacterial leakage of mineral trioxide aggregate as a root-end filling material," *Journal of Endodontics*, vol. 21, no. 3, pp. 109–112, 1995.

[2] M. Torabinejad, C. U. Hong, T. R. Pitt Ford, and J. D. Kettering, "Cytotoxicity of four root end filling materials," *Journal of Endodontics*, vol. 21, no. 10, pp. 489–492, 1995.

[3] M. Torabinejad, P. W. Smith, J. D. Kettering, and T. R. Pitt Ford, "Comparative investigation of marginal adaptation of mineral trioxide aggregate and other commonly used root-end filling materials," *Journal of Endodontics*, vol. 21, no. 6, pp. 295–299, 1995.

[4] C. Main, N. Mirzayan, S. Shabahang, and M. Torabinejad, "Repair of root perforations using mineral trioxide aggregate: a long-term study," *Journal of Endodontics*, vol. 30, no. 2, pp. 80–83, 2004.

[5] M. Torabinejad, T. R. Pitt Ford, D. J. McKendry, H. R. Abedi, D. A. Miller, and S. P. Kariyawasam, "Histologic assessment of mineral trioxide aggregate as a root-end filling in monkeys," *International Endodontic Journal*, vol. 42, no. 5, pp. 408–411, 1997.

[6] S. W. Chang, W. J. Bae, J. K. Yi et al., "Odontoblastic differentiation, inflammatory response, and angiogenic potential of 4 calcium silicate-based cements: Micromega MTA, ProRoot MTA, RetroMTA, and experimental calcium silicate cement," *Journal of Endodontics*, vol. 41, no. 9, pp. 1524–1529, 2015.

[7] S. Simon, F. Rilliard, A. Berdal, and P. Machtou, "The use of mineral trioxide aggregate in one-visit apexification treatment: a prospective study," *International Endodontic Journal*, vol. 40, no. 3, pp. 186–197, 2007.

[8] S. Corbella, G. Ferrara, A. El Kabbaney, and S. Taschieri, "Apexification, apexogenesis and regenerative endodontic procedures: a review of the literature," *Minerva Stomatologica*, vol. 63, no. 11-12, pp. 375–389, 2014.

[9] C. J. Chung, E. Kim, M. Song, J. W. Park, and S. J. Shin, "Effects of two fast-setting calcium-silicate cements on cell viability and angiogenic factor release in human pulp-derived cells," *Odontology*, vol. 104, no. 2, pp. 143–151, 2016.

[10] J. C. Ramos, P. J. Palma, R. Nascimento et al., "1-year in vitro evaluation of tooth discoloration induced by 2 calcium silicate-based cements," *Journal of Endodontics*, vol. 42, no. 9, pp. 1403–1407, 2016.

[11] L. Cianconi, P. Palopoli, V. Campanella, and M. Mancini, "Composition and microstructure of MTA and Aureoseal Plus: XRF, EDS, XRD and FESEM evaluation," *European Journal of Paediatric Dentistry*, vol. 17, no. 4, pp. 281–285, 2016.

[12] S. W. Chang, S. Y. Lee, H. J. Ann, K. Y. Kum, and E. C. Kim, "Effects of calcium silicate endodontic cements on biocompatibility and mineralization-inducing potentials in human dental pulp cells," *Journal of Endodontics*, vol. 40, no. 8, pp. 1194–1200, 2014.

[13] S. W. Chang, S. Y. Lee, K. Y. Kum, and E. C. Kim, "Effects of ProRoot MTA, Bioaggregate, and Micromega MTA on odontoblastic differentiation in human dental pulp cells," *Journal of Endodontics*, vol. 40, no. 1, pp. 113–118, 2014.

[14] S. W. Chang, S. H. Baek, H. C. Yang et al., "Heavy metal analysis of ortho MTA and ProRoot MTA," *Journal of Endodontics*, vol. 37, no. 12, pp. 1673–1676, 2011.

[15] D. C. Bird, T. Komabayashi, L. Guo, L. A. Opperman, and R. Spears, "In vitro evaluation of dentinal tubule penetration and biomineralization ability of a new root-end filling material," *Journal of Endodontics*, vol. 38, no. 8, pp. 1093–1096, 2012.

[16] M. Parirokh and M. Torabinejad, "Mineral trioxide aggregate: a comprehensive literature review–Part III: clinical applications, drawbacks, and mechanism of action," *Journal of Endodontics*, vol. 36, no. 3, pp. 400–413, 2010.

[17] M. G. Gandolfi, F. Siboni, C. M. Primus, and C. Prati, "Ion release, porosity, solubility, and bioactivity of MTA Plus tricalcium silicate," *Journal of Endodontics*, vol. 40, no. 10, pp. 1632–1637, 2014.

[18] I. R. Oliveira, V. C. Pandolfelli, and M. Jacobovitz, "Chemical, physical and mechanical properties of a novel calcium aluminate endodontic cement," *International Endodontic Journal*, vol. 43, no. 12, pp. 1069–1076, 2010.

[19] M. A. Saghiri, K. Asgar, M. Lotfi, K. Karamifar, P. Neelakantan, and J. L. Ricci, "Application of mercury intrusion porosimetry for studying the porosity of mineral trioxide aggregate at two different pH," *Acta Odontologica Scandinavica*, vol. 70, no. 1, pp. 78–82, 2012.

[20] M. A. Saghiri, M. Lotfi, A. M. Saghiri et al., "Effect of pH on sealing ability of white mineral trioxide aggregate as a root-end filling material," *Journal of Endodontics*, vol. 34, no. 10, pp. 1226–1229, 2008.

[21] I. About, "Biodentine: from biochemical and bioactive properties to clinical applications," *Giornale Italiano di Endodonzia*, vol. 30, no. 2, pp. 81–88, 2016.

[22] S. Westermarck, *Use of Mercury Porosimetry and Nitrogen Adsorption in Characterisation of the Pore Structure of Mannitol and Microcrystalline Cellulose Powders, Granules, and Tablets*, Academic dissertation, Department of Pharmacy, University of Helsinki, Helsinki, Finland, 2000.

[23] Available from: https://catalog.hardydiagnostics.com/cp_prod/Content/hugo/Enterococcus.htm, 2017.

[24] M. Chen, P. K. Patra, M. L. Lovett, D. L. Kaplan, and S. Bhowmick, "Role of electrospun fibre diameter and corresponding specific surface area (SSA) on cell attachment," *Journal of Tissue Engineering and Regenerative Medicine*, vol. 3, no. 4, pp. 269–279, 2009.

17

Ir-6: A Novel Iridium (III) Organometallic Derivative for Inhibition of Human Platelet Activation

Ren-Shi Shyu,[1,2] Themmila Khamrang,[2,3] Joen-Rong Sheu,[2] Chih-Wei Hsia,[2] Marappan Velusamy,[3] Chih-Hsuan Hsia,[2] Duen-Suey Chou,[4] and Chao-Chien Chang[4,5]

[1]*Division of Nephrology, Department of Internal Medicine, Min-Sheng General Hospital, Taoyuan 330, Taiwan*
[2]*Graduate Institute of Medical Sciences, College of Medicine, Taipei Medical University, Taipei 110, Taiwan*
[3]*Department of Chemistry, North Eastern Hill University, Shillong 793022, India*
[4]*Department of Pharmacology, School of Medicine, College of Medicine, Taipei Medical University, Taipei 110, Taiwan*
[5]*Department of Cardiology, Cathay General Hospital, Taipei 106, Taiwan*

Correspondence should be addressed to Chao-Chien Chang; change@seed.net.tw

Academic Editor: Konstantinos Tsipis

Platelet activation has been reported to play a major role in arterial thrombosis, cancer metastasis, and progression. Recently, we developed a novel Ir(III)-based compound, [Ir(Cp*)1-(2-pyridyl)-3-(4-dimethylaminophenyl)imidazo[1,5-a]pyridine Cl]BF_4 or Ir-6 and assessed its effectiveness as an antiplatelet drug. Ir-6 exhibited higher potency against human platelet aggregation stimulated by collagen. Ir-6 also inhibited ATP-release, intracellular Ca^{2+} mobilization, P-selectin expression, and the phosphorylation of phospholipase Cγ2 (PLCγ2), protein kinase C (PKC), v-Akt murine thymoma viral oncogene (Akt)/protein kinase B, and mitogen-activated protein kinases (MAPKs), in collagen-activated platelets. Neither the adenylate cyclase inhibitor SQ22536 nor the guanylate cyclase inhibitor 1H-[1,2,4]oxadiazolo[4,3-a]quinoxalin-1-one significantly reversed the Ir-6-mediated inhibition of collagen-induced platelet aggregation. Moreover, Ir-6 did not considerably diminish OH radical signals in collagen-activated platelets or Fenton reaction solution. At 2 mg/kg, Ir-6 markedly prolonged the bleeding time in experimental mice. In conclusion, Ir-6 plays a crucial role by inhibiting platelet activation through the inhibition of signaling pathways, such as the PLCγ2–PKC cascade and the subsequent suppression of Akt and MAPK activation, thereby ultimately inhibiting platelet aggregation. Therefore, Ir-6 is a potential therapeutic agent for preventing or treating thromboembolic disorders or disrupting the interplay between platelets and tumor cells, which contributes to tumor cell growth and progression.

1. Introduction

Platelets are anucleate blood cells that play a crucial role in thrombosis under both physiological and pathological conditions. They are required for maintaining the integrity of the vascular system and are the first-line of defense against hemorrhage. On encountering a subendothelial matrix exposed by an injury to a blood vessel, platelets adhere to the matrix and become activated and adhesive to other platelets, leading to further aggregation [1]. During platelet activation, the release of several mediators (e.g., ATP and thromboxane A_2) occurs simultaneously with relative intracellular Ca^{2+} ($[Ca^{2+}]_i$) mobilization, attracting additional platelets to the site of the injured endothelium and consequently thickening the initial platelet monolayer. Finally, fibrinogen binds to its specific platelet receptor, completing the final common pathway for platelet aggregation.

Platelets and their activation are the key events, which play critical role in cancer progression [2, 3]. The effect of platelets on malignancy development has been proposed to be a controlled process that triggers cancer growth. Therefore, inhibition of platelet aggregation is expected to be a novel therapeutic target for reducing the formation of platelet–tumor complexes [4]. Transition metal complexes, including those of iridium (Ir), have been explored for at least a decade as platforms for producing innovative molecules with anticancer properties. Although several biological studies have shown that Ir-based compounds exhibit potent anticancer activity with relatively few side effects, no study has investigated the effects of Ir compounds on platelet aggregation to date.

In recent years, researchers have paid more attention on Ir(III) compounds since they found powerful antitumor activity with low cytotoxicity toward normal tissues [5, 6]. Likewise, Ir complexes show excellent antiangiogenic effects through activating various antiangiogenic signaling pathways [5]. In our earlier studies, we have shown that phenol [7] and anisole [8] substituted imidazo[1,5-a]pyridine ligand-based iridium(III) complexes and their antiplatelet and antithrombotic activities. In continuation of these studies on the effect of electron donors on imidazo[1,5-a]pyridine-based ligand, we introduced dimethyl aniline as strong electron donor group and studied its antiplatelet activities. In addition, we developed a new biologically active Ir(III) derivative Ir-6 as shown in Figure 1(a). Moreover, a previous study had reported the photophysical and photochemical properties of Ir(III)-cyclometalated complexes that contain the luminescent ligands [9]. Hence, we used the same ligand to synthesise the Ir-6 complex [Ir(Cp*)1-(2-pyridyl)-3-(4-dimethylaminophenyl) imidazo[1,5-a] pyridine Cl]BF4 to perform this study. Although several in vitro and in vivo anticancer activities have demonstrated in Ir-based compounds, to date, no study has investigated their effects on platelet aggregation. Our preliminary findings revealed the potent antiplatelet activity of Ir-6 in human platelets; hence, we further examined the molecular mechanism and activity of Ir-6 against platelet activation. The present study is a primary step for investigating whether the potent activity of Ir-6 against cancer progression is due to its ability to inhibit platelet aggregation effectively.

2. Materials and Methods

2.1. Chemicals. Thrombin, collagen, arachidonic acid (AA), luciferin-luciferase, U46619, phorbol 12,13-dibutyrate (PDBu), nitroglycerin (NTG), heparin, prostaglandin E_1 (PGE_1), 5,5-dimethyl-1-pyrroline N-oxide (DMPO), SQ22536, 1H-[1, 2, 4]oxadiazolo[4,3-a]quinoxalin-1-one (ODQ), LY294002, SB203580, PD98059, SP600125, and bovine serum albumin (BSA) were purchased from Sigma (St. Louis, MO, USA). Fura-2AM was purchased from Molecular Probes (Eugene, OR, USA). An anti-phospho-p38 mitogen-activated protein kinase (MAPK) Ser^{182} monoclonal antibody (mAb) was purchased from Santa Cruz Biotechnology (Santa Cruz, CA, USA). Anti-p38 MAPK, anti-phospho-c-Jun N-terminal kinase (JNK) (Thr^{183}/Tyr^{185}), and anti-p44/42 extracellular signal-regulated kinase (ERK) mAbs as well as anti-phospholipase Cγ2 (PLCγ2), anti-phospho (Tyr^{759}) PLCγ2, anti-phospho-(Ser) protein kinase C (PKC) substrate (pleckstrin; p-p47), anti-JNK, and anti-phospho-p44/p42 ERK (Thr^{202}/Tyr^{204}) polyclonal antibodies (pAbs) were purchased from Cell Signaling (Beverly, MA, USA). Anti-phospho-protein kinase B (Akt) (Ser^{473}) and anti-Akt mAbs were purchased from Biovision (Mountain View, CA, USA). An anti-pleckstrin (p47) pAb was purchased from GeneTex (Irvine, CA, USA). Hybond-P polyvinylidene fluoride (PVDF) membranes horseradish peroxidase- (HRP-) conjugated donkey anti-rabbit immunoglobulin G (IgG) and sheep anti-mouse IgG were purchased from Amersham (Buckinghamshire, UK). A fluorescein isothiocyanate (FITC) anti-human CD42P (P-selectin) mAb was purchased from BioLegend (San Diego, CA, USA).

2.2. Synthesis of [Ir(Cp)(L)Cl]BF$_4$(Ir-6).* To 10 mL of a methanolic solution of 1-(2-pyridyl)-3-(4-dimethylaminophenyl) imidazo[1,5-a]pyridine (L) (0.12 g, 0.4 mM) [9], a solution of $[Ir(Cp^*)(Cl)_2]_2$ (0.16 g, 0.2 mM) in 10 mL methanol was added dropwise, and the solution was stirred at room temperature for 3 h. Subsequently, NH_4BF_4 (200 mg, 0.6 mM) was added to the solution, and the solution slowly changed color from pale yellow to orange. After 24 h, the solution was evaporated, and the solid obtained was filtered. The residue was washed with diethyl ether (40 mL) and dried under vacuum. The desired products were recrystallized from a mixture of dichloromethane and hexane as orange microcrystals. ^1H NMR (400 MHz, dimethyl sulfoxide [DMSO]-d6) δ 8.86–8.84 (*d*, 1H, $J = 8$ Hz), 8.50–8.44 (*m*, 2H), 8.21–8.18 (*t*, 1H, $J = 6$ Hz), 8.01–7.99 (*d*, 2H, $J = 8$ Hz), 7.59–7.49 (*m*, 2H), 7.19–7.16 (*t*, 1H, $J = 6$ Hz), 7.03–7.01 (*d*, 2H, $J = 8$ Hz), 3.6 (*s*, 15H), 1.36 (*s*, 6H); UV–Vis (λ_{abs}, nm) (ε, $M^{-1}\cdot cm^{-1}$): 400 (1447), 304 (2576), 242 (1468); ESI-MS (m/z): 677.17 $[M\text{-}BF_4]^+$ (Figure 1(a)).

2.3. Platelet Aggregation. The institutional review board of Taipei Medical University (TMU-JIRB-N201612050), Taiwan, approved this study and conformed to the directives of the Declaration of Helsinki. All human volunteers involved in this study provided informed consent. Human platelet suspensions were prepared as described previously [10]. Human blood samples were collected from adult volunteers who had not taken any drugs or substances that could affect with the tests for at least 14 days before collecting the samples; the collected blood samples were mixed with an acid-citrate-dextrose solution. After centrifugation, the platelet-rich plasma (PRP) was supplemented with 0.5 μM PGE_1 and 6.4 IU/mL heparin. Tyrode's solution containing 3.5 mg/mL BSA was used to prepare the final suspension of washed human platelets. The final Ca^{2+} concentration in the Tyrode's solution was 1 mM. Platelet aggregation was assessed using a lumiaggregometer (Payton Associates, Scarborough, ON, Canada) as described previously [10]. The platelet suspensions (3.6×10^8 cells/mL) preincubated with various concentrations of Ir-6 or a solvent control (0.1% DMSO) for 3 min before the addition of various agonists (i.e., collagen). The extent of platelet aggregation was calculated and expressed as a percentage relative to the control (without Ir-6) in light transmission units. In the ATP-release assay, a 20 μL of luciferin–luciferase was added 1 min before the addition of agonist; the amount of ATP release was compared with that released by the control platelets.

2.4. Measurement of $[Ca^{2+}]_i$ Mobilization. The $[Ca^{2+}]_i$ concentration was determined using Fura-2AM as described

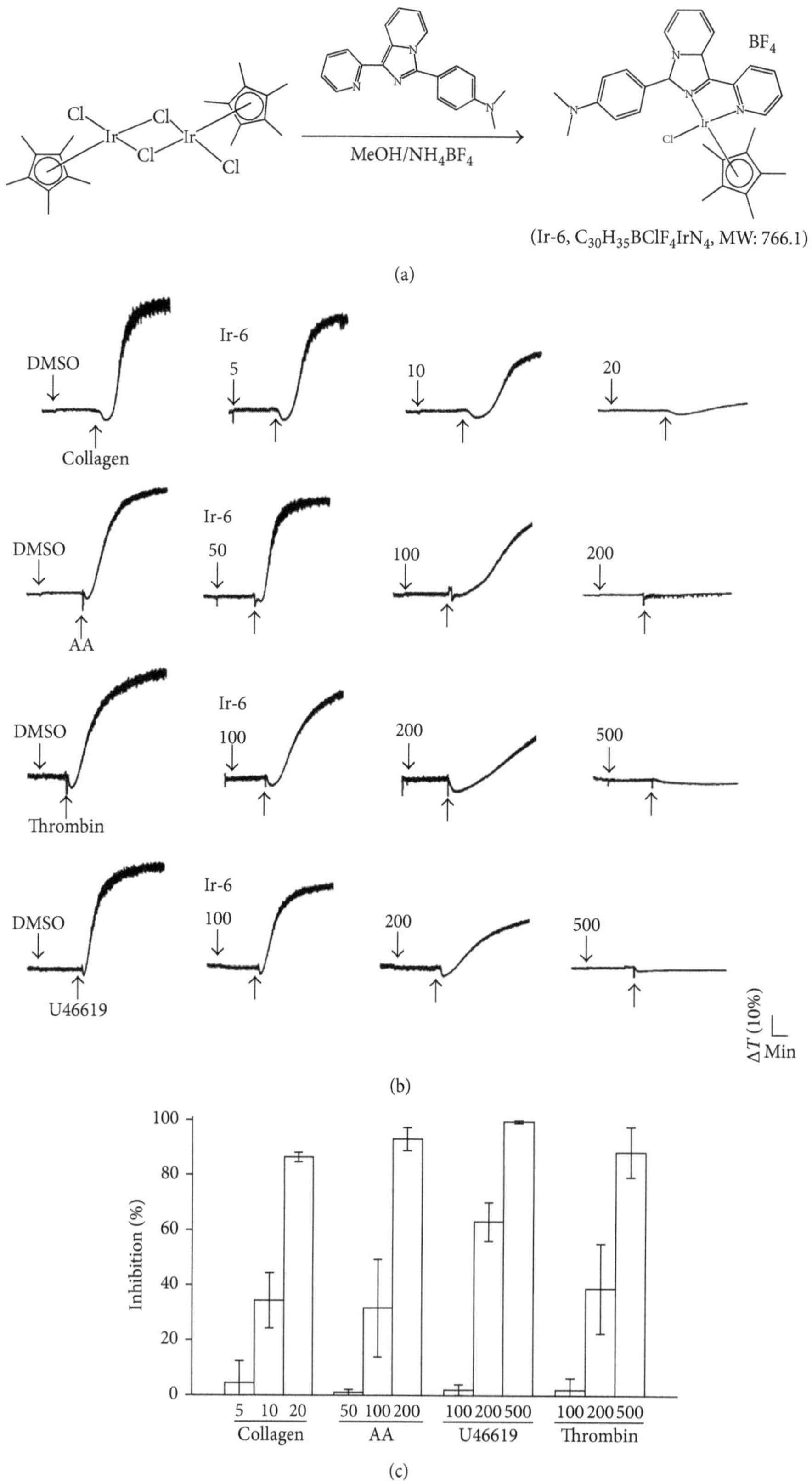

FIGURE 1: Comparison among the levels of Ir-6-mediated inhibition of platelet aggregation induced by agonists in washed human platelets. (a) Synthesis of complex (Ir-6) $[Ir(Cp^*)(L)Cl]BF_4$. (b) Platelets (3.6×10^8 cells/mL) were preincubated with the solvent control (0.1% DMSO) or Ir-6 (5–500 μM) and subsequently treated with collagen (1 μg/mL), arachidonic acid (AA; 120 μM), thrombin (0.01 U/mL), or U46619 (1 μM) to stimulate platelet aggregation. (c) Concentration–response histograms of Ir-6 representing the levels of inhibition of platelet aggregation stimulated by various agonists. Data are presented as the mean ± standard the error of the mean (n = 4).

previously [10]. In brief, citrated whole blood was centrifuged at 120 ×g for 10 min, and the supernatant was collected and incubated with 5 μM Fura-2AM for 1 h. Human platelet suspensions were prepared as described in the previous section. The Fura-2AM-loaded platelets were washed and preincubated with Ir-6 in the presence of 1 mM $CaCl_2$ and stimulated by collagen. Fura-2 fluorescence was measured using a spectrofluorometer (Hitachi FL Spectrophotometer F-4500, Tokyo, Japan) at excitation wavelengths of 340 and 380 nm and an emission wavelength of 510 nm.

2.5. Detection of Lactate Dehydrogenase (LDH). To detect the LDH activity, washed human platelets (3.6×10^8 cells/mL) were preincubated with 20–100 μM Ir-6 or the solvent control (0.1% DMSO) for 20 min at 37°C. An aliquot of the supernatant (10 μL) was placed on a Fuji Dri-Chem slide LDH-PIII (Fuji, Tokyo, Japan), and the absorbance was measured at 540 nm by using a UV-Vis spectrophotometer (UV-160; Shimadzu, Japan). A maximal value (MAX) of LDH was recorded in sonicated platelets.

2.6. Platelet Surface P-Selectin Expression Analysis by Flow Cytometric. To this analysis, washed platelet suspensions were prepared as described previously [8]. Aliquots of the platelet suspensions (3.6×10^8 cells/mL) were preincubated with Ir-6 (10 and 20 μM) or the solvent control (0.1% DMSO) and FITC-P-selectin (2 μg/mL) for 3 min. Collagen (1 μg/mL) was subsequently added to trigger platelet activation. The suspensions were then assayed for fluorescein-labeled platelets by using a flow cytometer (FAC Scan System, Becton Dickinson, San Jose, CA, USA). Data were collected from 50,000 platelets per experimental group, and the platelets were distinguished according to their characteristic forward and orthogonal light-scattering profiles. All experiments were repeated at least four times to ensure reproducibility.

2.7. Immunoblotting. For the immunoblotting analysis, washed platelets (1.2×10^9 cells/mL) were preincubated with Ir-6 (10 and 20 μM) or the solvent control (0.1% DMSO) for 3 min, and then collagen was added to trigger platelet activation. After 10 min, the reaction was stopped by adding EDTA, and the resulting platelets were resuspended in 200 μL lysis buffer. Proteins (approximately 80 μg) were separated through sodium dodecyl sulfate polyacrylamide gel electrophoresis (SDS-PAGE) on a 12% gel. The separated proteins were transferred to PVDF membranes by using a Bio-Rad semidry transfer unit (Bio-Rad, Hercules, CA, USA). The membranes were then blocked with Tris-buffered saline in Tween 20 (TBST; 10 mM Tris-base, 100 mM NaCl, and 0.01% Tween 20) containing 5% BSA for 1 h and were then probed using various primary antibodies. The membranes were subsequently incubated with HRP-conjugated anti-mouse IgG or anti-rabbit IgG (diluted 1 : 3000 in TBST) for 1 h. An enhanced chemiluminescence system was used to detect immunoreactive bands, and the optical densities (OD) of the bands were quantified using Bio-profil Biolight (version V2000.01; Vilber Lourmat, Marne-la-Vallée, France).

2.8. Detection of OH· Radical Formation in Both Platelets and Fenton Reaction Solution by Electron Spin Resonance Spectrometry. Electron spin resonance (ESR) spectrometry (Bruker EMX ESR, Billerica, MA, USA) analysis was performed as described previously [11]. Either platelet suspension (3.6×10^8 cells/mL) or Fenton reaction solution (50 μM $FeSO_4$ + 2 mM H_2O_2) was preincubated with 0.1% DMSO or Ir-6 (10 and 20 μM) for 3 min, with or without the addition of 1 μg/mL collagen. After 5 min, 100 μM DMPO was added to the suspensions before ESR spectrometry. The ESR spectral signals were recorded using a quartz flat cell designed for aqueous solutions. The spectrometer was operated at 20 mW and 9.78 GHz, with a scan range of 100 G and a receiver gain of 5×10^4. The modulation amplitude was 1 G, and the time constant was 164 ms. Each sample was scanned for 42 s, and each spectrum was the sum of three scans.

2.9. Tail Bleeding Time Assay in Mice. To investigate whether Ir-6 has bleeding risk, the bleeding time assay was performed through transection of the tails in male ICR mice. To this, after 30 min of intraperitoneal administration of Ir-6 (1 or 2 mg/kg), the tails of mice were cut at a 3 mm distance from the tip. The tails were directly placed in tubes filled with normal saline at 37°C to measure the bleeding time, which was recorded until the bleeding completely stopped. The animal experiments were conformed to the Guide for the Care and Use of Laboratory Animals (8th edition, 2011) and obtained an affidavit of approval for animal use from Taipei Medical University (LAC-2016-0395), Taiwan.

2.10. Statistical Analysis. The experimental results are expressed as the mean ± standard error of the means (S.E.M.) and are accompanied by the number of observations (n). The values of n refer to the number of experiments, and each experiment was conducted using different blood donors. The unpaired Student's t-test was used to determine the significance of differences between the control and experimental mice. The differences between multiple groups in other experiments were assessed through analysis of variance (ANOVA). When ANOVA indicated significant differences among the group means, the groups were compared using the Student–Newman–Keuls method. In the analysis, P values < 0.05 were considered statistically significant. Statistical analyses were performed using SAS (version 9.2; SAS Inc., Cary, NC, USA).

3. Results

3.1. Effects of Ir-6 on Platelet Aggregation in Washed Human Platelets. As shown in Figures 1(b) and 1(c), pretreatment of Ir-6 (5–20 μM) intensely and concentration-dependently inhibited collagen-induced aggregation in washed human platelets. In response to 120 μM AA stimulation, Ir-6 gradually

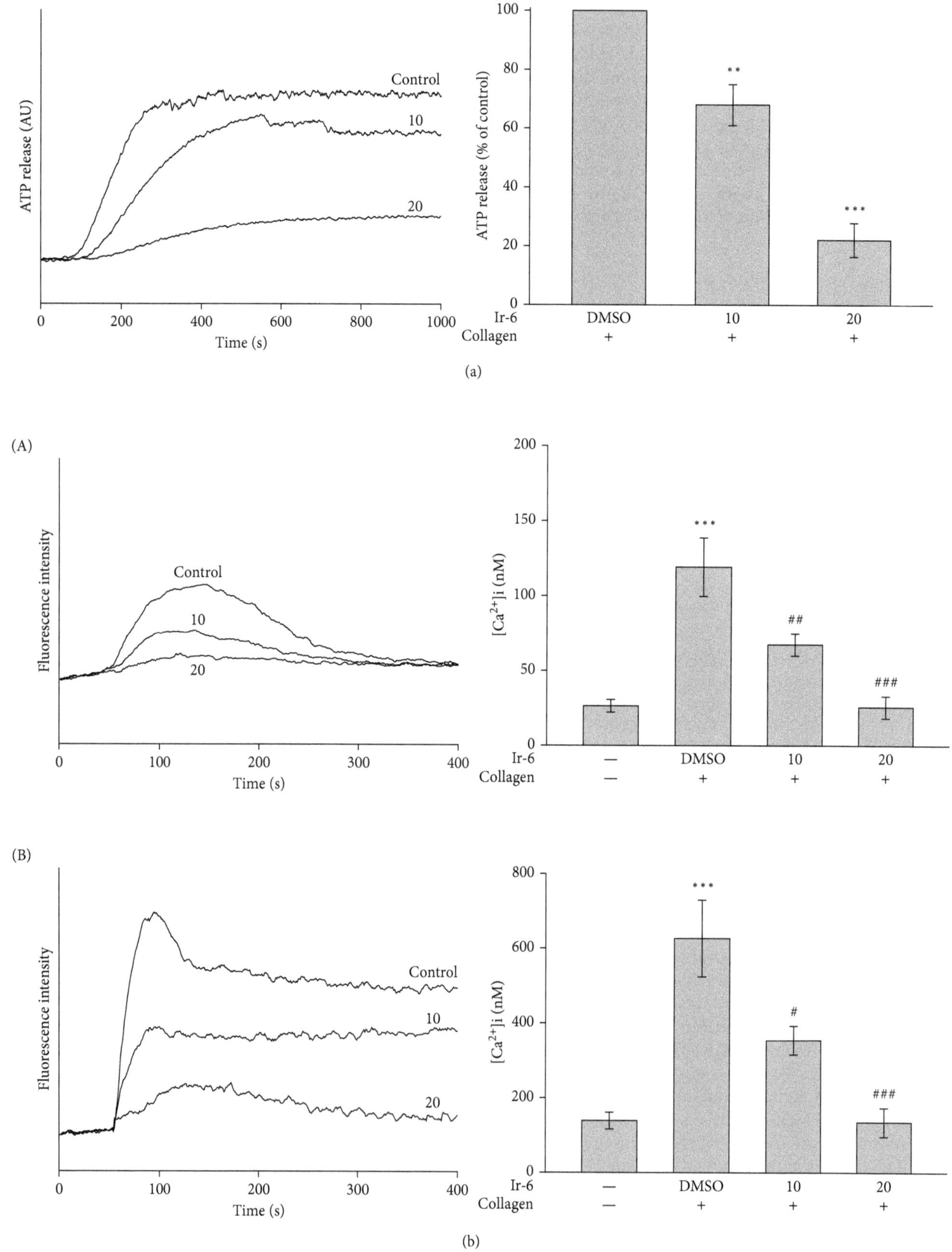

FIGURE 2: Continued.

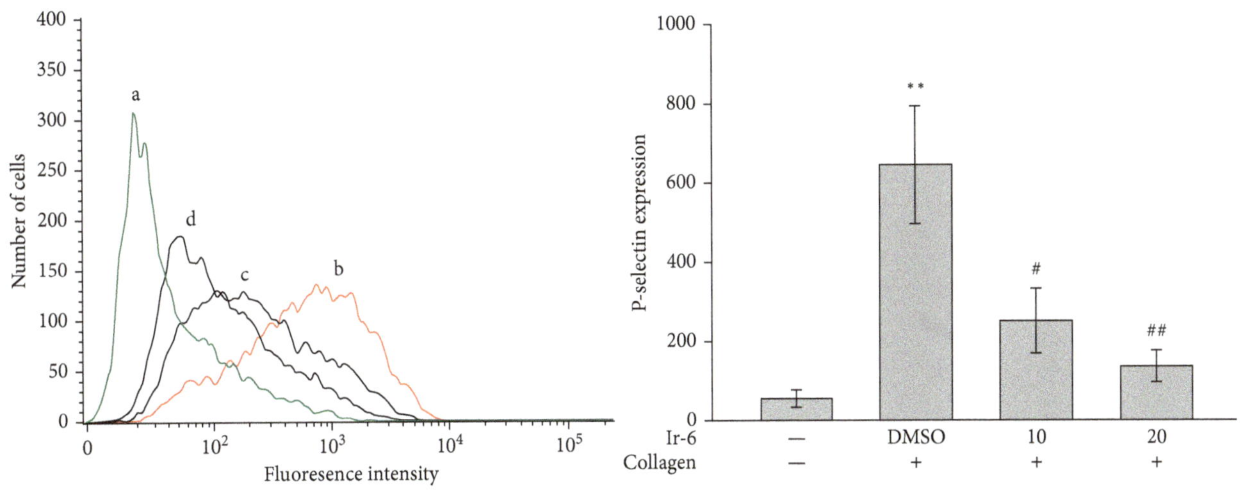

(c)

FIGURE 2: Effect of Ir-6 on ATP-release reaction, relative $[Ca^{2+}]_i$ mobilization, and surface FITC-P-selectin expression in washed human platelets. Platelets (3.6×10^8 cells/mL) were preincubated with the solvent control (0.1% DMSO) or Ir-6 (10 and 20 μM). Subsequently, collagen (1 μg/mL) was added to trigger either (a) the ATP-release reaction (AU; arbitrary unit) or (b) relative $[Ca^{2+}]_i$ mobilization (nM) in calcium-free Tyrode's solution (A) or Tyrode's solution (B). (c) Washed platelets (3.6×10^8/mL) were preincubated with the solvent control (0.1% DMSO) or Ir-6 (10 and 20 μM) and FITC-P-selectin (2 μg/mL) for 3 min and then stimulated using collagen (1 μg/mL). The corresponding statistical data are shown in the right panel of each Figures 2(a)–2(c). Data are presented as the mean ± standard error of the mean ($n = 4$). $^{*}P < 0.01$ and $^{***}P < 0.001$, compared with the DMSO-treated group (a) or resting control (b and c); $^{\#}P < 0.05$, $^{\#\#}P < 0.01$, and $^{\#\#\#}P < 0.001$, compared with the DMSO-treated group (b and c).

inhibited aggregation of platelets even at 50–200 μM. Furthermore, at 100–500 μM, Ir-6 exhibited relatively weak activity against platelet aggregation induced by 0.01 U/mL thrombin or 1 μM U46619, a prostaglandin endoperoxide analogue; this result indicated that Ir-6 had more potent activity against collagen-induced aggregation than t that induced by other agonists AA, thrombin, and U46619 (Figure 1(c)). The used solvent control (0.1% DMSO) did not ominously affect platelet aggregation. In the subsequent experiments, 1 μg/mL collagen was used as an agonist for investigating the possible mechanisms through which Ir-6 inhibits human platelet activation.

3.2. Ir-6 Inhibits ATP Release, Relative $[Ca^{2+}]_i$ Mobilization, and Surface P-Selectin Expression. Platelet activation is connected with the release of granular contents such as ATP and Ca^{2+} and surface P-selectin expression, which leads to strong platelet aggregation. In the present study, Ir-6 (10 and 20 μM) inhibited the ATP-release reaction stimulated by 1 μg/mL collagen (Figure 2(a)). Collagen stimulates $[Ca^{2+}]_i$ by promoting the entry of Ca^{2+} into the cytosol from two resources; Ca^{2+} is released from intracellular stores and enters to the platelet across the cell membrane. Figure 2(b), (A) shows responses to collagen without 1 mM $CaCl_2$. Under this condition, influx of calcium is expected to be markedly diminished, and therefore any change in $[Ca^{2+}]_i$ may be attributed to discharge from intracellular stores into the cytoplasma. Pretreatment with IL-6 (10 and 20 μM) obviously reduced relative $[Ca^{2+}]_i$ mobilization both in calcium-free Tyrode's solution (resting control, 26.5 ± 4.3 nM; collagen-stimulated, 119.1 ± 19.5 nM; 10 μM Ir-6, 67.3 ± 7.3 nM; and 20 μM Ir-6, 25.8 ± 7.4 nM; $n = 4$), Figure 2(b), (A) and in Tyrode's solution (resting control, 139.4 ± 22.4 nM; collagen-stimulated, 626.8 ± 102.6 nM; 10 μM Ir-6, 354.3 ± 38.7 nM; and 20 μM Ir-6, 135.6 ± 38.7 nM; $n = 4$), Figure 2(b), (B) in platelets stimulated by 1 μg/mL collagen. In addition, in quiescent (resting) platelets, P-selectin is located on the inner wall of the α-granules. Platelet activation exposes the inner walls of the granules to the outside of the cell [12]. Ir-6 treatment markedly reduced collagen-induced surface P-selectin expression, as demonstrated by the statistical data in the right panel of Figure 2(c) (resting control, 55.7 ± 21.7; collagen-activated, 645.0 ± 148.9; 10 μM Ir-6, 250.7 ± 82.0; 20 μM Ir-6, 136.0 ± 39.7; $n = 4$).

3.3. Effect of Ir-6 on Cyclic Nucleotide Formation and LDH Release in Washed Human Platelets. Both 100 μM SQ22536, an adenylate cyclase inhibitor and 10 μM ODQ, a guanylate cyclase inhibitor, significantly reversed the inhibition of collagen-induced platelet aggregation mediated by 1 μM PGE_1 or 10 μM NTG (Figure 3(a)). However, neither SQ22536 nor ODQ significantly reversed the inhibition of collagen-induced platelet aggregation mediated by 20 μM Ir-6 (Figure 3(a)), indicating that the mechanisms of Ir-6-mediated inhibition of platelet aggregation do not involve the enhancement of cyclic nucleotide synthesis. Moreover, the aggregation curves of platelets preincubated with 100 μM Ir-6 for 10 min and subsequently washed two times with Tyrode's solution were not significantly different from those of platelets preincubated with the solvent control (0.1% DMSO) under equivalent conditions (Figure 3(c)). This finding indicates that the effects of Ir-6 on platelet

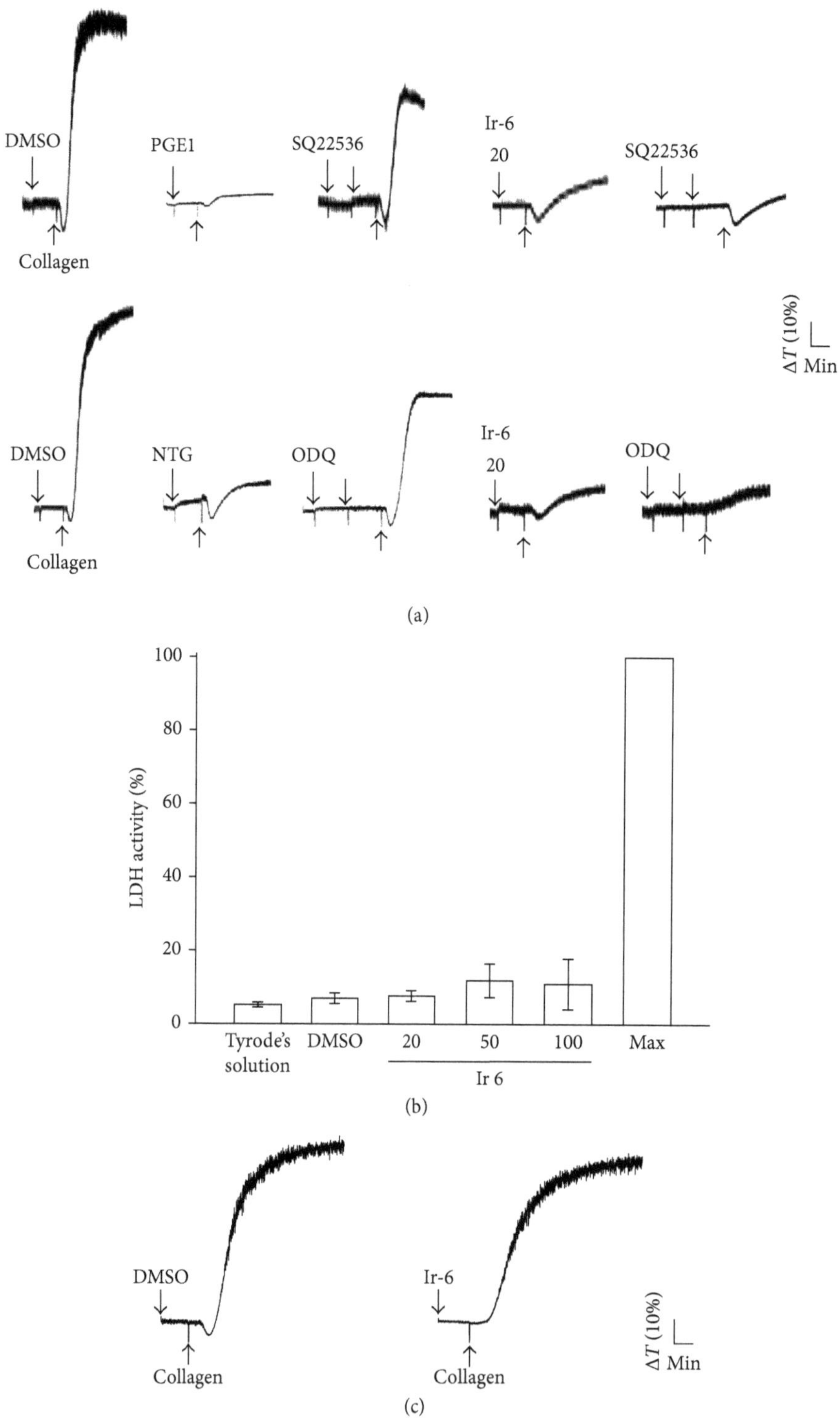

FIGURE 3: Effect of Ir-6 on cyclic nucleotide formation, LDH release, and cytotoxicity in human platelets. (a) Washed platelets (3.6×10^8 cells/mL) were preincubated with 1 μM PGE1, 10 μM NTG, or 20 μM Ir-6 with or without 100 μM SQ22536 or 10 μM ODQ and were subsequently treated with collagen (1 μg/mL) to induce platelet aggregation. (b) Washed platelets (3.6×10^8 cells/mL) were preincubated with the solvent control (0.1% DMSO) or with various concentrations of Ir-6 (20, 50, and 100 μM) for 20 min, and a 10 μL aliquot of the supernatant was deposited on a Fuji Dri-Chem slide LDH-PIII. (c) For other experiment, washed platelets were preincubated with the solvent control (0.1% DMSO) or Ir-6 (100 μM) for 10 min and subsequently washed two times with Tyrode's solution. Collagen (1 μg/mL) was then added to trigger platelet aggregation. Data are presented as the mean ± standard error of the mean (n = 4). Profiles in (a) and (c) represent four independent experiments.

aggregation are reversible and noncytotoxic. Furthermore, the LDH release results revealed that incubating platelets for 20 min with Ir-6 (20, 50, and 100 μM) did not significantly increase LDH activity or exhibit cytotoxic effects on the platelets (Figure 3(b)), demonstrating that Ir-6 does not affect platelet permeability or induce platelet cytolysis.

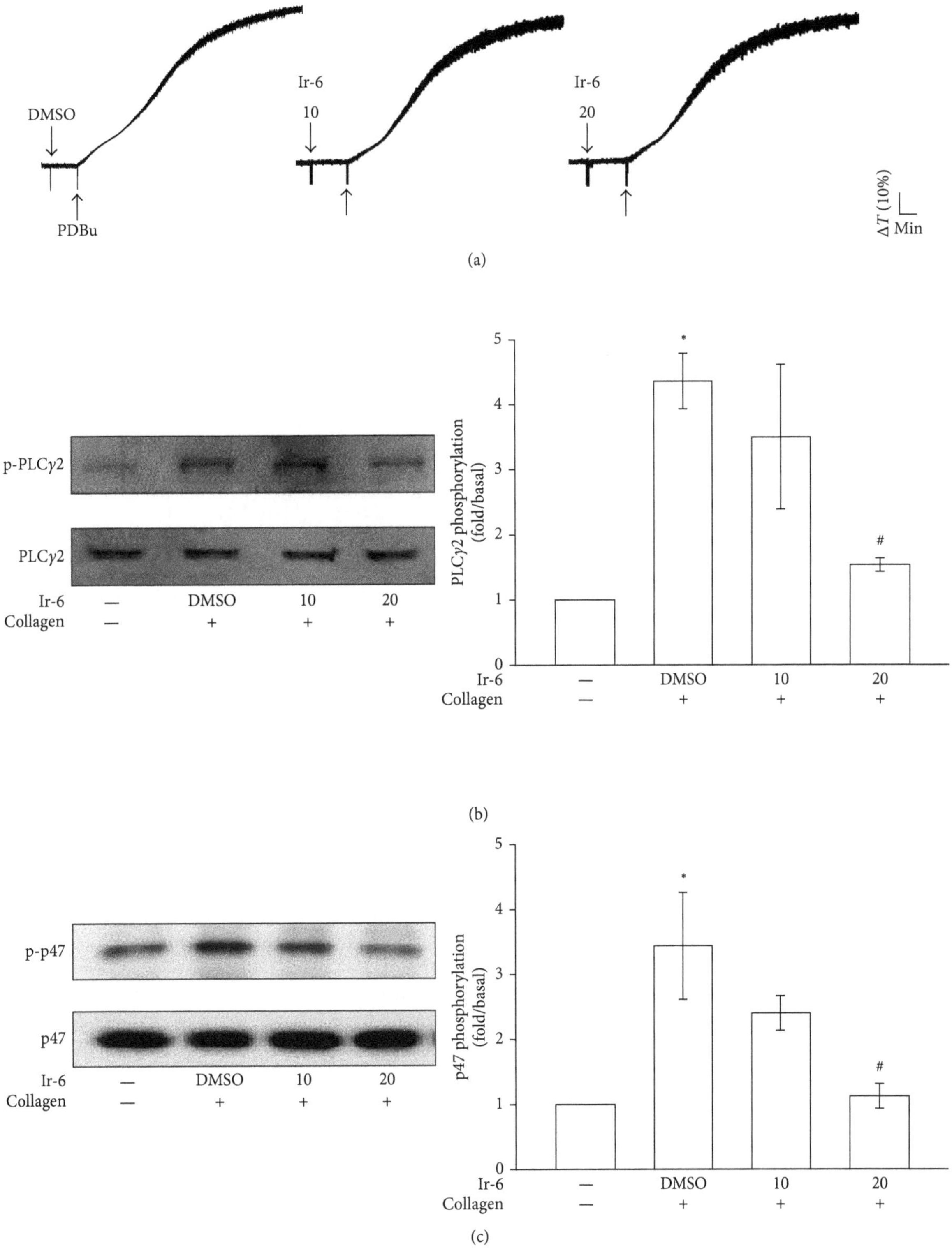

Figure 4: Continued.

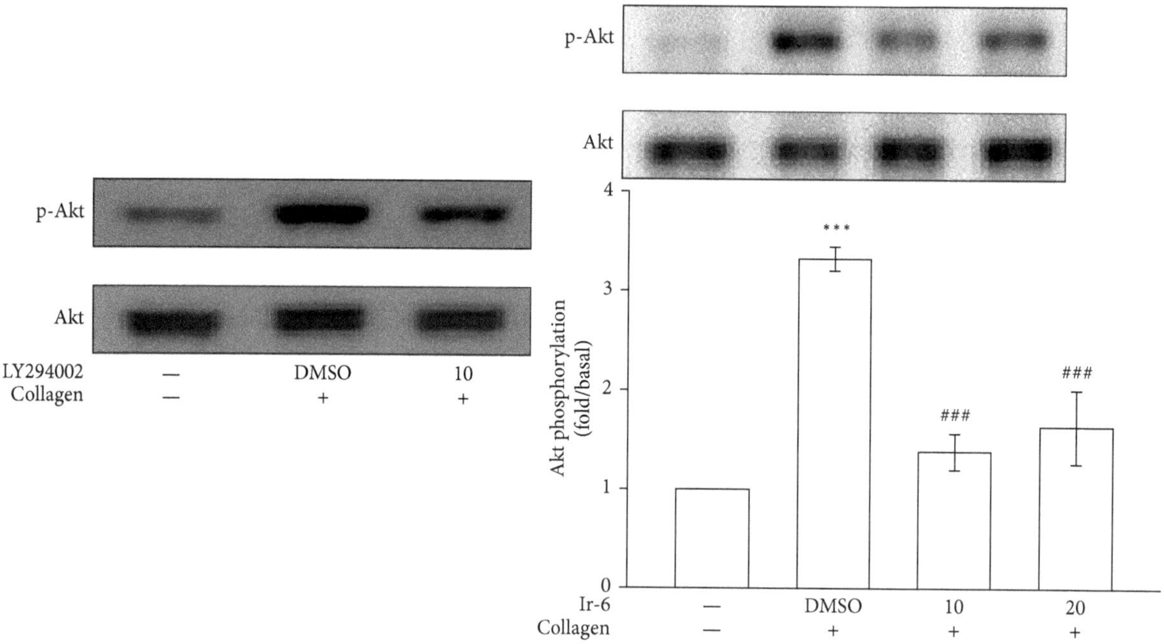

(d)

FIGURE 4: Effects of Ir-6 on PLCγ2, PKC, and Akt activation in human platelets. Washed platelets were preincubated with the solvent control (0.1% DMSO), LY294002 (10 μM), or Ir-6 (10 or 20 μM). Subsequently, either collagen (1 μg/mL) or PDBu (150 nM) was added to trigger (a) platelet aggregation, (b) PLCγ2 phosphorylation, and (c) PKC activation (pleckstrin phosphorylation), or (d) Akt activation. Platelets were collected, and their subcellular extracts were analyzed to determine the levels of phosphorylation of (b) PLCγ2, (c) PKC, and (d) Akt. Data are presented as the mean ± standard error of the mean ($n = 4$). $^{*}P < 0.05$ and $^{***}P < 0.001$, compared with the resting control; $^{\#}P < 0.05$ and $^{\#\#\#}P < 0.001$, compared with the DMSO-treated group. The profiles in (a) and (d, left panel) are representative of four independent experiments.

3.4. Activity of Ir-6 on Regulating PLCγ2–PKC Signaling and Akt Activation. PLCs hydrolyze phosphatidylinositol 4,5-bisphosphate to generate the secondary messengers inositol 1,4,5-trisphosphate (IP_3) and diacylglycerol (DAG). IP_3 triggers relative $[Ca^{2+}]_i$ mobilization, while DAG activates PKC. PKC activation yields a protein, approximately 47 kDa in size, that is predominantly phosphorylated (p47 protein; pleckstrin) and causes the ATP-release reaction [13]. As shown in Figure 4(a), treatment of platelets with 10 or 20 μM Ir-6 did not significantly inhibit aggregation induced by 150 nM PDBu, a PKC activator (Figure 4(a)), indicating that Ir-6 does not directly disrupt PKC activation. Figures 3(a) and 3(b) illustrates the inhibitory effects of Ir-6 on ATP release and relative $[Ca^{2+}]_i$ mobilization induced by collagen. Therefore, we further investigated the effect of Ir-6 on the phosphorylation of the PLCγ2–PKC signaling cascade. At 10 and 20 μM, Ir-6 markedly reduced PLCγ2 phosphorylation as well as PKC activation (pleckstrin phosphorylation) in collagen-stimulated platelets (Figures 4(b) and 4(c)). Akt is a serine/threonine-specific protein kinase that plays a key role in various cellular processes, such as platelet activation, cell proliferation, apoptosis, and cell migration [14]. Both LY294002 (an inhibitor of Akt; 10 μM) and Ir-6 (10 and 20 μM) markedly inhibited collagen-induced Akt phosphorylation (Figure 4(d)), demonstrating the crucial role of inhibition of the Akt signaling pathway in the Ir-6-mediated inhibition of platelet activation.

3.5. Effects of Ir-6 on Inhibiting p38 MAPK, ERK2, and JNK1 Phosphorylation. Several signaling molecules of the MAPK phosphorylation pathway were evaluated to investigate the inhibitory mechanisms of Ir-6 in platelet activation. In eukaryotic organisms, MAPKs (p38 MAPK, ERKs, and JNKs) control major cellular reactions and contribute to various events of cell proliferation, migration, differentiation, and apoptosis. ERKs, JNK1, and p38 MAPK have been identified in platelets [15]. SB203580 (an inhibitor of p38 MAPK; 10 μM), PD98059 (an inhibitor of ERK2; 20 μM), and SP600125 (an inhibitor of JNK1; 10 μM) markedly inhibited p38 MAPK (Figure 5(a)), ERK2 (Figure 5(b)), and JNK1 (Figure 5(c)) phosphorylation in collagen-activated platelets, respectively. Ir-6 reduced the phosphorylation of these three proteins in a concentration-dependent manner. Nevertheless, at 10 μM, Ir-6 nonsignificantly inhibited p38 MAPK phosphorylation (Figure 5).

3.6. Role of OH· Radical on Ir-6-Mediated Inhibition of Platelet Aggregation. An ESR signal *e* of OH· radical formation was observed in both collagen-stimulated platelet suspensions and Fenton reaction solution (cell-free system; Figures 6(a) and 6(b)). A typical OH· signal ($a^N = a^H = 14.8$ G) and a long-lived ($g = 2.005$) radical detectable by using DMPO, a spin trap, were observed in collagen-stimulated platelets but not detected in resting platelets (Figure 6(a)), curve (A). Treatment with 10 or 20 μM Ir-6 did not considerably

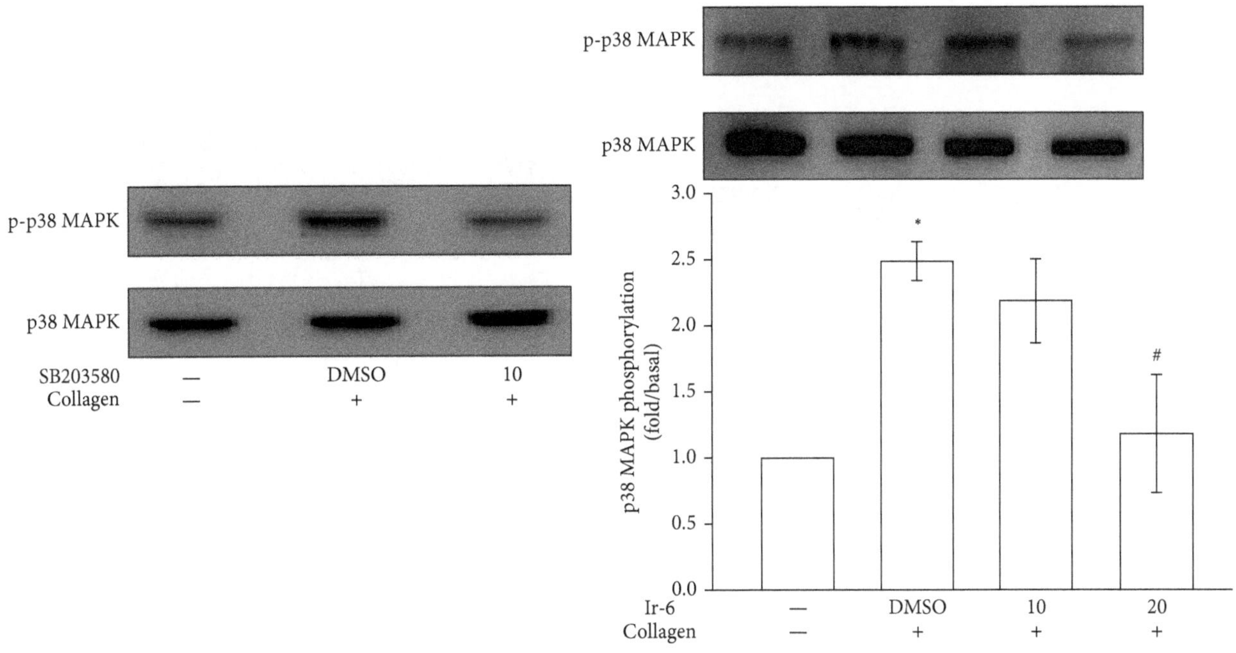

(a)

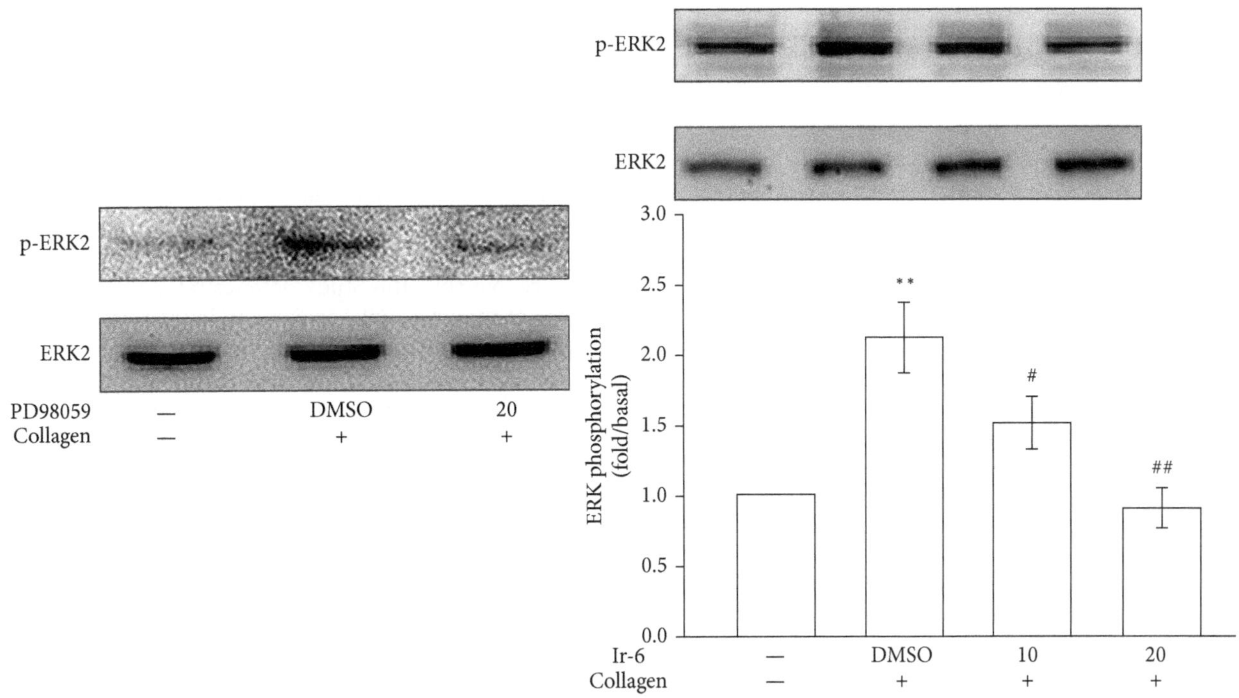

(b)

Figure 5: Continued.

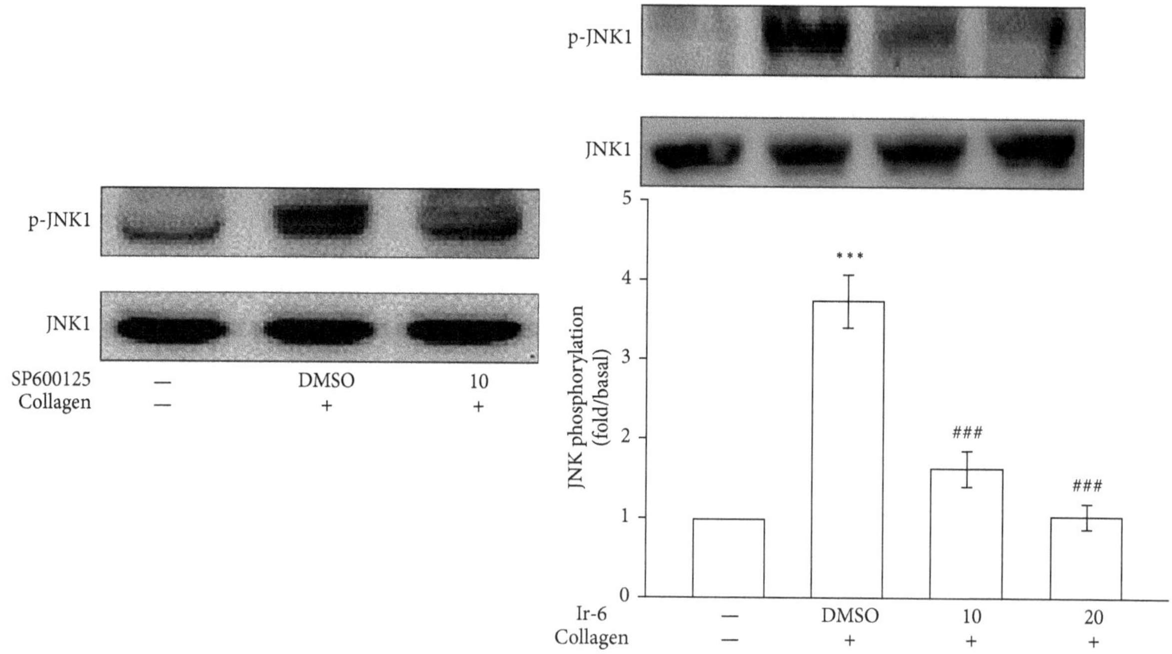

(c)

Figure 5: Inhibitory effects of Ir-6 on the phosphorylation of p38 MAPK, ERK2, and JNK1 in human platelets. The washed platelets (1.2×10^9 cells/mL) were preincubated with the solvent control (0.1% DMSO), SB203580 (10 μM), PD98059 (20 μM), SP600125 (10 μM), or Ir-6 (10 or 20 μM); subsequently, collagen (1 μg/mL) was added to trigger platelet activation. The platelets were collected, and their subcellular extracts were then analyzed to determine the levels of phosphorylation of (a) p38 MAPK, (b) ERK2, and (c) JNK1. Data are presented as the mean ± standard error of the mean ($n = 4$). $^{*}P < 0.05$, $^{**}P < 0.01$, and $^{***}P < 0.001$, compared with the resting control; $^{\#}P < 0.05$, $^{\#\#}P < 0.01$, and $^{\#\#\#}P < 0.001$, compared with the DMSO-treated group. The profiles of left panels are representative of four independent experiments.

diminish the OH signals in both the platelet suspensions activated by collagen and Fenton reaction solution (Figures 6(a) and 6(b)), suggesting that the Ir-6-mediated inhibition of platelet activation may not be regulated by free radical formation.

3.7. Effect of Ir-6 on Bleeding Time. In the tail transection model of mice, after 30 min, the bleeding times were markedly prolonged in mice treated with intraperitoneal administration of 2.0 mg/kg Ir-6 (323.3 ± 55.2 s; $n = 8$), but not in those treated with 1.0 mg/kg (184.3 ± 39.7 s; $n = 8$) in comparison with mice treated with the solvent control (0.1% DMSO-treated group, 150.5 ± 11.9 s; $n = 8$) (Figure 6(c)). Each mouse was monitored whether there was any rebleeding 10 min after the original bleeding stopped.

4. Discussion

Platelets activation contributes a major role on thrombotic events among patients with cancer [16]. Chemotherapeutics approach may increase this effect and stimulate vascular thromboembolic events (VTEs) by inducing platelet aggregation, aggravating endothelial damage, and causing vascular toxicity [17]. Of platinum- (Pt-) based chemotherapeutic agents, cisplatin is widely used to a high incidence of treatment-related VTEs [18]. A combination of gemcitabine and Pt-based therapy increased thrombotic and vascular side effects [19, 20]. Thus, research is currently focusing on the development of new metal-based drugs for the inhibition of platelet activation for treating vascular disease, reducing toxic side effects, and overcoming Pt resistance. Notably, this study demonstrated that in addition to its antitumor activity, Ir-6, an Ir(III) derivative, exhibits potent antiplatelet activity.

Platelets adhere to subendothelial matrix proteins (e.g., collagen) which alter platelet shape and cause the release of their granular contents. Collagen mobilizes $[Ca^{2+}]_i$ to phosphorylate the Ca^{2+}/calmodulin-dependent myosin light chain (20 kDa), which is involved in the secretion of granular contents, such as serotonin and ATP [21], and activate platelet aggregation. Therefore, the degree of inhibition of either $[Ca^{2+}]_i$ mobilization or ATP production is crucial for evaluating the potency of the antiplatelet activity of a compound. In the present study, Ir-6 inhibited platelet aggregation to different degrees, depending on the agonist used to induce aggregation (collagen, U46619, AA, and thrombin), indicating that Ir-6 did not act on the specific individual receptors of these agonists. Therefore, Ir-6 may exert its activity on activated platelets through one or more common signaling pathways.

Platelet activation by collagen substantially alters PLC activation. PLC stimulation results in the production of IP_3 and DAG. Subsequently, DAG activates PKC and consequently induces p47 phosphorylation [13]. PKC activation triggers specific responses simplifying the transmission of

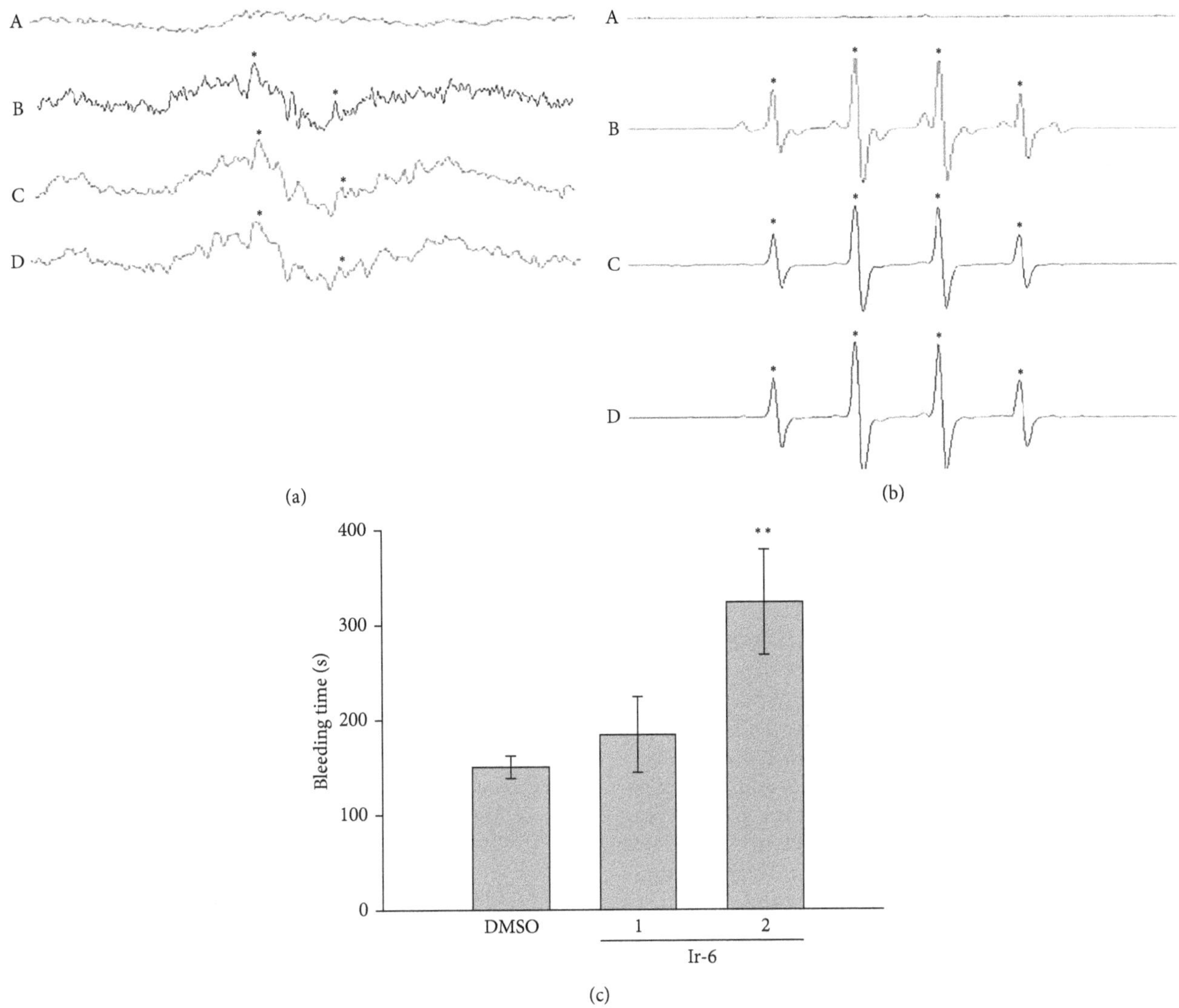

Figure 6: Regulatory activities of Ir-6 on OH· formation in either platelet suspensions or Fenton reaction solution and determination of bleeding time in experimental mice. (a) Washed platelets or (b) the Fenton reaction solution was preincubated with (A) Tyrode's solution (resting control), (B) 0.1% DMSO, (C) 10 μM Ir-6, or (D) 20 μM Ir-6. Collagen (1 μg/mL) was then added into platelet suspensions for the ESR experiments as described in Materials and Methods. Profiles in (a) and (b) are representative of four independent experiments, and an asterisk (*) indicates OH· formation. (c) Mouse tails were transected to measure the bleeding time after 30 min of intraperitoneal administration of either 0.1% DMSO, 1.0 mg/kg Ir-6, or 2.0 mg/kg Ir-6 (all in 50 μL). Data are presented as the mean ± standard error of the mean. $n = 8$. $^{**}P < 0.01$, compared with the DMSO (0.1%)-treated group.

particular upstream signals in individual cellular compartments. The PLCγ family comprises the isozymes PLCγ1 and PLCγ2; PLCγ2 is involved in collagen-dependent signaling in platelets [22]. Ir-6 evidently diminished collagen-induced PLCγ2–PKC activation; however, Ir-6 did not exert direct effects on PKC activation because it did not interfere with PDBu-induced platelet aggregation. This finding suggests that the Ir-6-mediated inhibition of platelet activation involves PLCγ2 downstream signaling. This result may also explain why Ir-6 exhibited higher efficacy in inhibiting platelet activation induced by collagen than in inhibiting that induced by other agonists.

Human platelet activation inhibits through intracellular pathways mediated by cyclic AMP (cAMP) and cyclic GMP (cGMP), and hence these nucleotides are considered to be crucial modulators of platelet activation [23]. Cyclic nucleotides inhibit most of the platelet responses and reduce $[Ca^{2+}]_i$ levels by enhancing Ca^{2+} uptake, thereby suppressing PLC and PKC activation [23]. Therefore, cAMP and cGMP synergistically inhibit platelet activation. Moreover, neither SQ22536 nor ODQ significantly reversed the Ir-6-mediated inhibition of collagen-induced platelet aggregation. Therefore, Ir-6-mediated mechanisms do not involve the enhancement of cyclic nucleotide synthesis in platelets.

Akt, a downstream effector of phosphoinositide 3-kinase (PI3K) has been reported to show defects in agonist-induced platelet activation when it is deleted in mice, which advocates that Akt normalizes platelet activation; such regulation potentially has consequences in thrombosis [14, 24]. Therefore, specific inhibitors of the Akt isoforms, such as individual PI3K isoforms, may be attractive antithrombotic therapy targets [14]. Cytosolic phospholipase A_2 ($cPLA_2$) is a substrate of p38 MAPK activity induced by agonists such as von Willebrand factor (vWF) and thrombin [25]. Consequently, p38 MAPK is critical for $cPLA_2$ stimulation as well as AA release [26]. This observation may explain why Ir-6 had weaker activity in inhibiting p38 MAPK activation as well as thrombin- or AA-stimulated platelet aggregation. Activation of ERK is also an important event involved in platelet aggregation necessitating prior ATP release, which activates P_2X_1-mediated Ca^{2+} influx, thereby enhancing the phosphorylation of myosin light-chain kinase [25]. JNK1, another most recently identified MAPK in platelets; therefore, its activation and role are poorly understood. Several agonists such as thrombin, vWF, collagen, and ADP [23] activate JNK1. In addition, a previous study confirmed an increased bleeding time, decreased integrin $\alpha_{IIb}\beta_3$ activation, and severe granule secretion impairment in $JNK^{-/-}$ platelets [27]. Therefore, the inhibition of JNK phosphorylation may play a crucial role in platelet activation. Consistent with these outcomes, the present results demonstrated that Ir-6 markedly inhibits collagen-induced JNK1 phosphorylation.

Reactive oxygen species (hydrogen peroxide) and free radical species (i.e., OH·) act as secondary signals that are involved in platelet activation [28]. In the current study, our ESR spectrometry investigations provided direct evidence demonstrating that Ir-6 does not significantly affect OH· formation in both activated platelets and Fenton reaction solution. Prolongation of hemostatic platelet plug formation (bleeding time) was observed in Ir-6-treated experimental mice. The observed results of bleeding times suggested that the prolongation of bleeding time in humans does not predict the risk of hemorrhage or surgical bleeding. These results question the rationale behind the use of bleeding time for the clinical evaluation of antiplatelet compounds [29].

5. Conclusion

In conclusion, the present findings reveal that Ir-6, a novel iridium (III) compound, inhibits platelet activation by preventing signaling molecules, such as the PLCγ2–PKC, and subsequently suppresses Akt and JNK1 activation. These alterations inhibit the processes associated with ATP release, $[Ca^{2+}]_i$ mobilization, and surface P-selectin expression and ultimately inhibit platelet aggregation. However, additional studies are required to investigate other unidentified mechanisms involved in the Ir-6-mediated inhibition of platelet activation. Nevertheless, Ir-6 can be used as a chemotherapeutic agent in cancer treatment. In addition, it can be used as an antiplatelet agent for treating thromboembolic disorders or disrupting the interplay between platelets and tumor cells that contributes to tumor cell growth and progression.

Authors' Contributions

Ren-Shi Shyu, Themmila Khamrang, and Joen-Rong Sheu contributed equally to this paper.

Acknowledgments

This work was supported by grants from the Ministry of Science and Technology of Taiwan (MOST 104-2622-B-038-003, MOST 104-2320-B-038-045-MY2, MOST 106-2320-B-038-012), Cathay General Hospital (CGH-MR-A106020), Cathay General Hospital–Taipei Medical University (107CGH-TMU-07), and the University Grants Commission, India (MRP-MAJOR-CHEM-2013-5144; 69/2014 F. no. 10-11/12UGC).

References

[1] T. Jayakumar, C. H. Yang, P. Geraldine, T. L. Yen, and J. R. Sheu, "The pharmacodynamics of antiplatelet compounds in thrombosis treatment," *Expert Opinion on Drug Metabolism and Toxicology*, vol. 12, no. 6, pp. 615–632, 2016.
[2] C. Belloc, H. Lu, C. Soria, R. Fridman, Y. Legrand, and S. Menashi, "The effect of platelets on invasiveness and protease production of human mammary tumor cells," *International Journal of Cancer*, vol. 60, no. 3, pp. 413–417, 1995.
[3] B. Felding-Habermann, T. E. Ooole, J. W. Smith et al., "Integrin activation controls metastasis in human breast cancer," *Proceedings of the National Academy of Sciences of the United States of America*, vol. 98, no. 4, pp. 1853–1858, 2001.
[4] A. Boucharaba, C. M. Serre, S. Gres et al., "Platelet-derived lysophosphatidic acid supports the progression of osteolytic bone metastases in breast cancer," *Journal of Clinical Investigation*, vol. 114, no. 12, pp. 1714–1725, 2004.
[5] J. Yellol, S. A. Pérez, A. Buceta et al., "Novel C,N-cyclometalated benzimidazole ruthenium(II) and iridium(III) complexes as antitumor and antiangiogenic agents: a structure–activity relationship study," *Journal of Medicinal Chemistry*, vol. 58, no. 18, pp. 7310–7327, 2015.
[6] F. Schmitt, K. Donnelly, J. K. Muenzner et al., "Effects of histidin-2-ylidene vs. imidazol-2-ylidene ligands on the anticancer and antivascular activity of complexes of ruthenium, iridium, platinum, and gold," *Journal of Inorganic Biochemistry*, vol. 163, pp. 221–228, 2016.
[7] C. W. Hsia, M. Velusamy, J. T. Tsao et al., "New therapeutic agent against arterial thrombosis: an iridium(III)-derived organometallic compound," *International Journal of Molecular Sciences*, vol. 18, p. 2616, 2017.
[8] K. G. Shyu, M. Velusamy, C. W. Hsia et al., "Novel iridium (III)-derived organometallic compound for the inhibition of human platelet activation," *International Journal of Molecular Medicine*, vol. 41, no. 5, pp. 2589–2600, 2018.
[9] G. Volpi, C. Garino, L. Salassa et al., "Cationic heteroleptic cyclometalated iridium complexes with 1-pyridylimidazo[1,5-alpha]pyridine ligands: exploitation of an efficient intersystem crossing," *Chemistry*, vol. 15, no. 26, pp. 6415–27, 2009.
[10] J. R. Sheu, C. R. Lee, C. H. Lin et al., "Mechanisms involved in the antiplatelet activity of *Staphylococcus aureus* lipoteichoic acid in human platelets," *Thrombosis and Haemostasis*, vol. 83, no. 5, pp. 777–784, 2000.
[11] D. S. Chou, G. Hsiao, M. Y. Shen, Y. J. Tsai, T. F. Chen, and J. R. Sheu, "ESR spin trapping of a carbon-centered free

radical from agonist-stimulated human platelets," *Free Radical Biology and Medicine*, vol. 39, no. 2, pp. 237–248, 2005.
[12] P. Harrison and E. M. Cramer, "Platelet α-granules," *Blood Reviews*, vol. 7, no. 1, pp. 52–62, 1993.
[13] W. D. Singer, H. A. Brown, and P. C. Sternweis, "Regulation of eukaryotic phosphatidylinositol-specific phospholipase C and phospholipase D," *Annual Review of Biochemistry*, vol. 66, no. 1, pp. 475–509, 1997.
[14] D. S. Woulfe, "Akt signaling in platelet and thrombosis," *Expert Review of Hematology*, vol. 3, no. 1, pp. 81–91, 2010.
[15] F. Bugaud, F. Nadal-Wollbold, S. Levy-Toledano, J. P. Rosa, and M. Bryckaert, "Regulation of c-jun-NH2 terminal kinase and extracellular-signal regulated kinase in human platelets," *Blood*, vol. 94, no. 11, pp. 3800–3805, 1999.
[16] G. Y. Lip, B. S. Chin, and A. Blann, "Cancer and the prothrombotic state," *Lancet Oncology*, vol. 3, no. 1, pp. 27–34, 2002.
[17] P. Ferroni, D. Della-Morte, R. Palmirotta et al., "Platinum-based compounds and risk for cardiovascular toxicity in the elderly role of the antioxidants in chemoprevention," *Rejuvenation Research*, vol. 14, no. 3, pp. 293–308, 2011.
[18] M. Jafri and A. Protheroe, "Cisplatin-associated thrombosis," *Anticancer Drugs*, vol. 19, no. 9, pp. 927–929, 2008.
[19] S. Barni, R. Labianca, G. Agnelli et al., "Chemotherapy-associated thromboembolic risk in cancer outpatients and effect of nadroparin thromboprophylaxis: results of a retrospective analysis of the PROTECHT study," *Journal of Translational Medicine*, vol. 9, no. 1, p. 179, 2011.
[20] C. A. Dasanu, "Gemcitabine: vascular toxicity and prothrombotic potential," *Expert Opinion on Drug Safety*, vol. 7, no. 6, pp. 703–716, 2008.
[21] K. Kaibuchi, K. Sano, M. Hoshijima, Y. Takai, and Y. Nishizuka, "Phosphatidylinositol turnover in platelet activation; calcium mobilization and protein phosphorylation," *Cell Calcium*, vol. 3, no. 4-5, pp. 323–335, 1982.
[22] A. Ragab, S. Séverin, M. P. Gratacap et al., "Roles of the C-terminal tyrosine residues of LAT in GP VI-induced platelet activation: insights into the mechanism of PLC gamma 2 activation," *Blood*, vol. 110, no. 7, pp. 2466–2474, 2007.
[23] U. Walter, M. Eigenthaler, J. Geiger, and M. Reinhard, "Role of cyclic nucleotide-dependent protein kinases and their common substrate VASP in the regulation of human platelets," *Advances in Experimental Medicine and Biology*, vol. 344, pp. 237–249, 1993.
[24] J. Chen, S. De, D. S. Damron, W. S. Chen, N. Hay, and T. V. Byzova, "Impaired platelet responses to thrombin and collagen in AKT-1-deficient mice," *Blood*, vol. 104, no. 6, pp. 1703–1710, 2004.
[25] F. Adam, A. Kauskot, J. P. Rosa, and M. Bryckaert, "Mitogen-activated protein kinases in hemostasis and thrombosis," *Journal of Thrombosis and Haemostasis*, vol. 6, no. 12, pp. 2007–2016, 2008.
[26] I. Canobbio, S. Reineri, F. Sinigaglia, C. Balduini, and M. Torti, "A role for p38 MAP kinase in platelet activation by von Willebrand factor," *Thrombosis and Haemostasis*, vol. 91, no. 1, pp. 102–110, 2004.
[27] F. Adam, A. Kauskot, P. Nurden et al., "Platelet JNK1 is involved in secretion and thrombus formation," *Blood*, vol. 115, no. 20, pp. 4083–4092, 2010.
[28] B. Wachowicz, B. Olas, H. M. Zbikowska, and A. Buczyński, "Generation of reactive oxygen species in blood platelets," *Platelets*, vol. 13, no. 3, pp. 175–182, 2002.
[29] S. E. Lind, "The bleeding time does not predict surgical bleeding," *Blood*, vol. 77, pp. 2547–2552, 1991.

18

Synthesis, Structure, DNA Interaction, and SOD Activity of Three Nickel(II) Complexes Containing L-Phenylalanine Schiff Base and 1,10-Phenanthroline

Peiran Zhao,[1] Shanshan Zhai,[1] Jianfang Dong,[1] Lei Gao,[2] Xinru Liu,[1] Lei Wang,[1] Jinming Kong,[3] and Lianzhi Li[1]

[1]*School of Chemistry and Chemical Engineering, Liaocheng University, Liaocheng 252059, China*
[2]*Zhong Yuan Academy of Biological Medicine, Liaocheng People's Hospital, Liaocheng 252000, China*
[3]*School of Environmental and Biological Engineering, Nanjing University of Science and Technology, 200 Xiaolingwei, Nanjing 210094, China*

Correspondence should be addressed to Lianzhi Li; lilianzhi1963@163.com

Academic Editor: Anastasios Keramidas

Three hexacoordinated octahedral nickel(II) complexes, [Ni(sal-L-phe)(phen)(CH_3OH)]·CH_3OH (**1**), [Ni(naph-L-phe)(phen)(CH_3OH)] (**2**), and [Ni(*o*-van-L-phe)(phen)(CH_3OH)]·5CH_3OH (**3**) (sal-L-phe = a Schiff base derived from salicylaldehyde and L-phenylalanine, naph-L-phe = a Schiff base derived from 2-hydroxy-1-naphthaldehyde and L-phenylalanine, *o*-van-L-phe = a Schiff base derived from *o*-vanillin and L-phenylalanine, and phen = 1,10-phenanthroline), have been synthesized and characterized by elemental analysis, IR spectra, and single-crystal X-ray diffraction. The interactions of these complexes with CT-DNA were studied by UV-Vis absorption spectroscopy, fluorescence spectroscopy, circular dichroism spectroscopy, and viscosity measurements. The binding constant (K_b) values of $1.82 \times 10^4\,M^{-1}$ for **1**, $1.96 \times 10^4\,M^{-1}$ for **2**, and $2.02 \times 10^4\,M^{-1}$ for **3** suggest that each of these complexes could bind with DNA in a moderate intercalative mode. Complex **3** exhibited a stronger interaction with CT-DNA than complexes **1** and **2**. In addition, the superoxide scavenging activity of these complexes was investigated by the nitrotetrazolium blue chloride (NBT) light reduction method, and the results showed that they exhibited a significant superoxide scavenging activity with the IC_{50} values of $4.4 \times 10^{-5}\,M$ for complex **1**, $5.6 \times 10^{-5}\,M$ for complex **2**, and $3.1 \times 10^{-5}\,M$ for complex **3**, respectively.

1. Introduction

Bioinorganic chemistry usually studies the interaction of inorganic elements with the organism at the molecular level [1]. The interaction between small molecules and biological macromolecules has become an important research topic in bioinorganic chemistry; especially the interaction between transition metal complexes and DNA has aroused the widespread interest [2, 3]. This helps us not only to understand the life processes at the molecular level but also to promote the development of chemistry discipline itself.

Amino acid Schiff bases are usually composed of amino acids with different aldehyde or ketone carbonyl groups. Schiff base is a multidentate ligand, which plays an important role in medicinal and pharmaceutical areas [4, 5]. In recent years, studies on the amino acid Schiff bases and their metal complexes are very active. Synthesis, characterization, structure, and thermodynamic and kinetic properties of this kind of compounds have been reported, and their antibacterial, anti-inflammatory, and anticancer activities have been widely studied [6]. Therefore, it is an important research topic to study the relationship among structure, property, and biological activity of these amino acid Schiff base complexes [7].

Phenylalanine is an essential amino acid for human and other organisms. Amino acid with a side chain aromatic ring such as phenylalanine contributes mainly to the stabilization of proteins through hydrophobic interactions and the formation of hydrophilic environments [8]. In addition, it is a precursor of many aromatic compounds and of alkaloids such as curare and morphine.

Nickel is an essential trace element for the human body, and the metalloenzymes containing nickel(II) play important physiological functions in the organisms [9]. Nickel(II) complexes have been reported to act as anticonvulsant and antiepileptic agents or vitamins; they have also presented antibacterial, antifungal, antimicrobial, antioxidant, and antiproliferative/anticancer activities [10–20]. Therefore, the research of nickel(II) complexes has attracted more and more attention and become more and more important in the field of bioinorganic and coordination chemistry [21]. Recently, studies on amino acid Schiff base ligands and nickel (II) complexes have been included [22, 23].

Deoxyribonucleic acid (DNA) is an important biomolecule. The biological information stored in the DNA is expressed by means of replication, transcription, and translation. This information has the ability to guide cell growth, metabolism, and mutation. Many small molecules exert their anticancer activities by binding with DNA, thereby altering DNA replication, blocking the division of cancer cells, and resulting in the cell death [24]. The interactions of nickel(II) complexes with DNA/BSA have been reported previously [25–28]. For example, nickel(II) complexes of benzoic acid (2-hydroxy-benzylidine)-hydrazide ligands bind to DNA base pairs via intercalation and π-π stacking interactions [29]. Nickel(II) complexes containing *N*-substituted heterocyclic thiosemicarbazones have been found to exhibit remarkable DNA/protein binding and antioxidant activities [30, 31]. In addition, nickel(II) complexes have attracted the attention of researchers because of their superoxide scavenging activities and broad biological activities [32, 33].

In view of the above considerations, we have synthesized three new nickel(II) complexes of Schiff base derived from the reactions of phenylalanine and aromatic aldehydes. These complexes have been characterized by elemental analysis, infrared (IR) spectra, and single-crystal X-ray diffraction. Furthermore, the interactions of these complexes with calf thymus DNA (CT-DNA) and their superoxide scavenging activities have been investigated by using spectroscopies.

2. Experimental

2.1. Materials and Physical Measurements. L-phenylalanine was obtained from Beijing Jingke Hongda Biotechnology Co., Ltd. Salicylaldehyde, *o*-vanillin, and 2-hydroxy-1-naphthaldehyde were purchased from Alfa Aesar. CT-DNA was obtained from Beijing Biodee Biotechnology Co., Ltd. Ethidium bromide (EB) was gained from Sigma. Tris(hydroxymethyl)aminomethane (Tris), nickel(II) acetate, 1,10-phenanthroline, potassium hydroxide, and anhydrous methanol were commercially available analytical reagents.

Tris-HCl (10 mM) (10 mM NaCl, pH = 7.1) buffer solution was prepared with doubly distilled water. A solution of CT-DNA in the buffer gave a ratio of absorbance at 260 and 280 nm (A_{260}/A_{280}) of 1.8-1.9, reflecting that the DNA was sufficiently free of protein [34, 35].

IR spectra were recorded as KBr pellets on a Nicolet 5700 FT-IR instrument in the frequency range of 400–4000 cm^{-1}. Both electronic and fluorescence spectra of the complexes were recorded using a Shimadzu UV-2550 spectrophotometer and a PerkinElmer LS55 spectrofluorometer, respectively. Circular dichroism (CD) spectra were obtained on a Jasco J-810 spectropolarimeter.

2.2. Synthesis of the Complexes

*2.2.1. Synthesis of the Complex [Ni(sal-L-phe)(phen)(CH_3OH)]·CH_3OH (**1**).* L-phenylalanine (0.1652 g, 1 mmol) and potassium hydroxide (0.056 g, 1 mmol) were dissolved in methanol (20 mL) at 323 K and added to a methanol solution (5.0 mL) of salicylaldehyde (0.11 ml, 1 mmol) and stirred for 1 h. Then, the solution (2.0 mL) of nickel(II) acetate tetrahydrate (0.25 g, 1 mmol) was added dropwise and stirred sequentially for 2 h. In the end, a methanol solution (4.0 mL) of 1,10-phenanthroline (0.180 g, 1 mmol) was added and continuously stirred for 3 h. The resulting solution was filtered, and then, the filtrate was placed at room temperature for two weeks; the green block single crystals suitable for X-ray diffraction analysis were gained. Anal. calc. (%) for $C_{30}H_{29}N_3O_5Ni$ (Mr = 570.27): C, 63.19; H, 5.13; and N, 7.37%. Found: C, 63.31; H, 5.20; and N, 7.26%. IR (KBr, 4000–400 cm^{-1}): 3384.9 (s, ν_{O-H}), 1636.0 (s, $\nu_{C=N}$), 1584.7 (m, ν_{COO^-}), 1384.7(m, ν_{COO}), 847.1 (w, ν_{Ni-O}), and 739.0 (s, ν_{Ni-N}). The mass spectrum and ^{1}H NMR of the Schiff base **1**, H_2(sal-L-phe), are shown in Figures S1 and S4, respectively.

*2.2.2. Synthesis of the Complex [Ni(naph-L-phe)(phen)(CH_3OH)](**2**).* The solution of L-phenylalanine (0.1652 g, 1 mmol), potassium hydroxide (0.056 g, 1 mmol), and 2-hydroxy-1-naphthaldehyde (0.1722 g, 1 mmol) in anhydrous methanol (20 mL) was heated at 323 K for 1 h. Then, a solution (2.0 mL) of nickel(II) acetate tetrahydrate (0.25 g, 1 mmol) was added dropwise and stirred continuously for 2 h. Finally, a methanol solution (5.0 mL) of 1,10-phenanthroline (0.20 g, 1.0 mmol) was added and continuously stirred for 3 h. The resulting solution was filtered, and the brown crystals suitable for X-ray diffraction analysis were obtained from the filtrate kept at normal atmospheric temperature for two weeks. Anal. calc. (%) for $C_{33}H_{27}N_3O_4Ni$ (Mr = 588.28): C, 67.38; H, 4.63; and N, 7.14%. Found: C, 67.29; H, 4.71; and N, 7.08%. IR (KBr, 4000–400 cm^{-1}): 3423.4 (s, ν_{O-H}), 1618.8 (s, $\nu_{C=N}$), 1537.1 (m, $\nu_{COO}-$), 1313.0 (m, $\nu_{COO}-$), 852.9 (w, ν_{Ni-O}), and 728.2 (s, ν_{Ni-N}). The mass spectrum and ^{1}H NMR of the Schiff base **2**, H_2(naph-L-phe), are shown in Figures S2 and S5, respectively.

*2.2.3. Synthesis of the Complex [Ni(o-van-L-phe)(phen)(CH_3OH)]·5CH_3OH (**3**).* Synthetic routes for the preparation of complexes **1**, **2** and **3** are shown in Scheme 1. Complex **3** was prepared following the same procedure as for complex **2** except that an anhydrous methanol solution (2 mL) of *o*-vanillin (0.1522 g, 1 mmol) was used instead of 2-hydroxy-1-naphthaldehyde. The reaction solution was filtered, and the filtrate was kept at normal atmospheric temperature for two weeks; then, the light green crystals suitable for X-ray

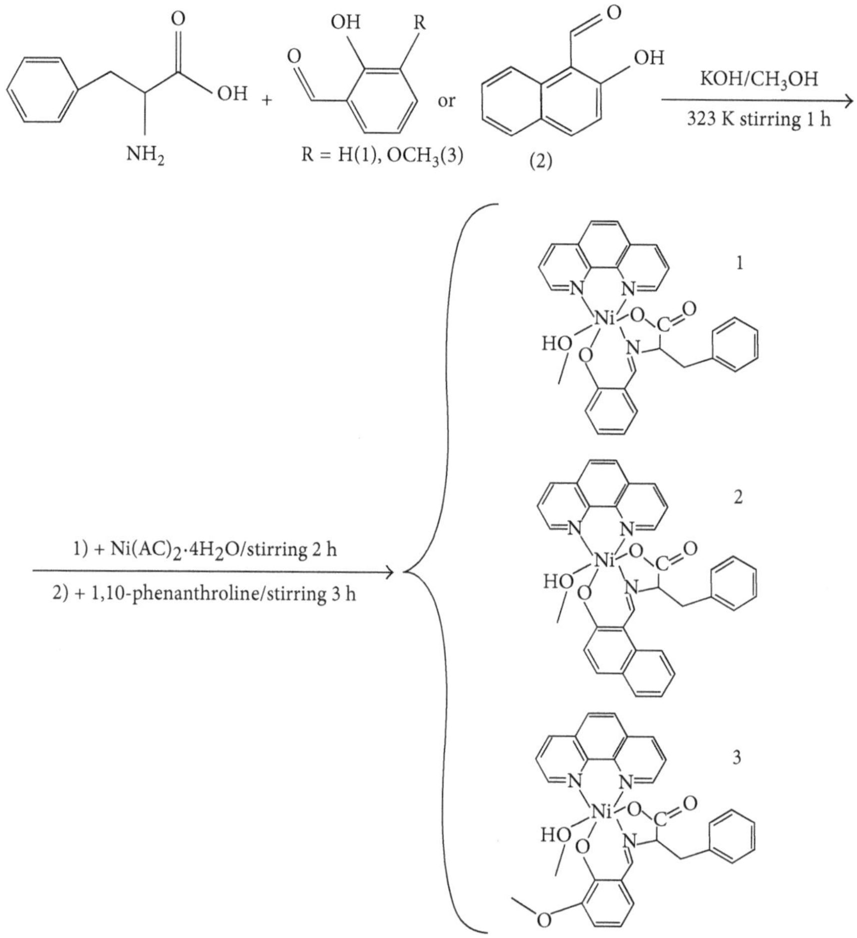

SCHEME 1: Synthetic routes for the preparation of complexes **1**, **2**, and **3**.

diffraction analysis were obtained. Anal. calc. (%) for $C_{95}H_{101}N_9O_{20}Ni_3$ (Mr = 1864.98): C, 61.18; H, 5.46; and N, 6.76%. Found: C, 61.02; H, 5.50; and N, 6.64%. IR (KBr, 4000–400 cm^{-1}): 3378.5 (s, ν_{O-H}), 1632.3 (s, $\nu_{C=N}$), 1440.8 (m, ν_{COO-}), 1082.9 (m, ν_{COO-}), 855.1 (w, ν_{Ni-O}), and 545.5 (s, ν_{Ni-N}). The mass spectrum and ^1H NMR of the Schiff base **3**, H_2(*o*-van-L-phe), are shown in Figures S3 and S6, respectively.

2.3. X-Ray Crystallographic Analysis. X-ray single-crystal diffraction measurements for these nickel(II) complexes were recorded on a Bruker Smart 1000 CCD area-detecting diffractometer. Diffraction intensities for the nickel(II) complexes were collected using graphite-monochromated Mo Kα radiation (λ = 0.071073 nm) at 298(2) K with the ω-2θ scan technique. The correction of semiempirical absorption data was performed using SADABS. The structures of these complexes were solved by direct methods using SHELXS-97 and subsequent Fourier difference techniques and refined anisotropically by full-matrix least-squares on F^2 using SHELXL-97 [36]. The H atoms were assigned with common isotropic displacement factors. All nonhydrogen atoms were refined with anisotropic thermal parameters. Further crystallographic data, experimental details, and refinement results for structural analyses of complexes are listed in Table 1.

Crystallographic data for the structural analysis of these nickel(II) complexes have been deposited in the Cambridge Crystallographic Data Center (CCDC no. 929664 (**1**), no. 1826530 (**2**), and no. 933859 (**3**)). Copies of the information may be obtained free of charge from http://www.ccdc.cam.ac.uk/conts/retrieving.html or from the CCDC, 12 Union Road, Cambridge, CB2 1EZ, UK (email: deposit@ccdc.cam.ac.uk).

2.4. DNA Binding Experiment. The binding experiments of these complexes with CT-DNA were conducted in 10 mM Tris-HCl/10 mM NaCl (pH = 7.1) buffer solution. The determination of UV-Vis absorption spectra was carried out by adding the increasing amounts of DNA (from 0 to 9.0×10^{-5} M) to each of these complexes with a fixed concentration of 1.5×10^{-5} M. The spectra were measured in the wavelength range of 200–500 nm.

For the fluorescence spectra, we have prepared the different concentrations of the samples with Tris-HCl buffer solution via keeping constant concentration of the EB-DNA system ($c_{EB} = 10\,\mu M$ and $c_{DNA} = 10\,\mu M$) and increasing the concentration of complexes **1**, **2**, and **3**, respectively. The mixed solutions were allowed to stand for 30 min to

TABLE 1: Crystallographic and structure refinement data for complexes **1**, **2**, and **3**.

Complex	1	2	3
Empirical formula	$C_{30}H_{29}N_3O_5Ni$	$C_{33}H_{27}N_3O_4Ni$	$C_{95}H_{101}N_9O_{20}Ni_3$
Formula weight	570.27	588.28	1864.98
Wavelength (Å)	0.71073	0.71073	0.71073
Crystal system	Monoclinic	Triclinic	Triclinic
Space group	C2/c	P-1	P-1
a (Å)	26.741(3)	10.8240(8)	12.7068(10)
b (Å)	18.1651(17)	11.9041(9)	13.5015(14)
c (Å)	11.8350(9)	13.0879(11)	29.9214(17)
α (°)	90	67.226(10)	83.667(2)
β (°)	108.528(2)	73.566(2)	82.7470(10)
γ (°)	90	64.094(10)	63.9790(10)
v (Å^3)	5450.9(8)	1385.43(19)	4567.4(6)
Z	8	2	2
D_{calc} (Mg·m^{-3})	1.390	1.410	1.356
F(000)	2384	612	1956
Absorption coefficient (mm^{-1})	0.756	0.744	0.687
Crystal size (mm)	0.43 × 0.38 × 0.35	0.23 × 0.20 × 0.12	0.25 × 0.21 × 0.18
θ range	2.66 to 25.02	2.19 to 25.01	2.57 to 25.02
Index ranges	$-25 \le h \le 31$	$-12 \le h \le 12$	$-15 \le h \le 10$
	$-21 \le k \le 21$	$-14 \le k \le 11$	$-16 \le k \le 16$
	$-13 \le l \le 14$	$-15 \le l \le 13$	$-35 \le l \le 35$
Reflections collected	13562	7109	29803
Unique reflections	4816	4791	16112
R_{int}	0.0432	0.0335	0.1085
Max/min transmission	0.7777, 0.7368	0.9160, 0.8475	0.8863, 0.8470
Data, restraint, parameters	4816, 3, 366	4791, 0, 363	16112, 2613, 1155
Goodness-of-fit on F^2	1.094	0.952	1.010
Final R indices ($I > 2\sigma$ (I))	$R_1 = 0.0432$, $wR_2 = 0.0946$	$R_1 = 0.0566$, $wR_2 = 0.1350$	$R_1 = 0.0867$, $wR_2 = 0.1790$
R indices (all data)	$R_1 = 0.0930$, $wR_2 = 0.1248$	$R_1 = 0.0820$, $wR_2 = 0.1456$	$R_1 = 0.1954$, $wR_2 = 0.2540$
Largest diff. peak and hole (e·Å^{-3})	0.392, −0.287	1.099, −0.500	0.788, −0.500

equilibrium, and the fluorescence emission spectra were recorded from 530 to 700 nm at an excitation wavelength of 510 nm. Each spectrum was measured at a scan speed of 300 nm·min^{-1}, and the widths of both the excitation and emission slits were set to 10.0 nm.

For CD spectra, a series of samples were prepared by adding the increasing concentrations of these complexes to constant concentration of DNA (10.0 mM) in Tris-HCl buffer solution. All the sample solutions were scanned in the range of 220–320 nm with a scan speed of 200 nm·min^{-1}. Each determination of the CD spectrum was the average of three scans with 1.0 nm path resolution and 1 s response time. The optical chamber of the CD spectrometer was deoxygenated with dry nitrogen before use and kept in a nitrogen atmosphere during experiments. The final spectra were background-corrected by subtracting the corresponding buffer spectra.

Viscosity experiments were carried out on an Ubbelodhe viscometer at a constant temperature of 30°C in a thermostated bath. A series of samples were prepared using Tris-HCl buffer solution. The concentration of CT-DNA (10.0 mM) was kept constant, while the ratios r of [c(complex)/c(DNA)] were 0, 0.02, 0.04, 0.06, 0.08, and 0.10, respectively. Each sample was incubated in water bath for 30 min, and the flow times of the solutions through the capillary were repeatedly measured three times using a digital stopwatch. The DNA viscosity was calculated according to $\eta = (t - t_0)/t_0$, where t_0 is the time required for the buffer solution to flow through the capillary. The data were calculated as $(\eta/\eta_0)^{1/3}$ versus r ($r = c_{(complex)}/c_{(DNA)}$), where η_0 is the relative viscosity of the DNA solution without complex and η is the viscosity of DNA in the presence of complexes.

2.5. SOD Activity Assay. The superoxide dismutase (SOD) activities of these complexes were determined by measuring their scavenging degree for the superoxide anion radical. In the absence and presence of different concentrations of these complexes, the standard test solutions containing 6.2 μM riboflavin, 83.0 mM *N,N,N′,N′*-tetramethylethylenediamine (TMEDA), and 10.0 mM NBT in 10.0 mM phosphate buffer (pH = 7.8) were irradiated with constant-intensity cold light at room temperature. The absorbances at 560 nm were recorded by using a UV-Vis absorption spectrophotometer in an interval of 1 min for 10 min.

3. Results and Discussion

3.1. IR Spectra of the Complexes. In IR spectra of complexes **1**, **2**, and **3** (Figure 1), the broad and strong absorption at 3384.9 cm^{-1} for **1**, 3423.4 cm^{-1} for **2**, and 3378.5 cm^{-1} for **3** is due to the O-H stretching vibration. The very sharp single absorption at 1636.0 cm^{-1} for **1**, 1618.8 cm^{-1} for **2**, and

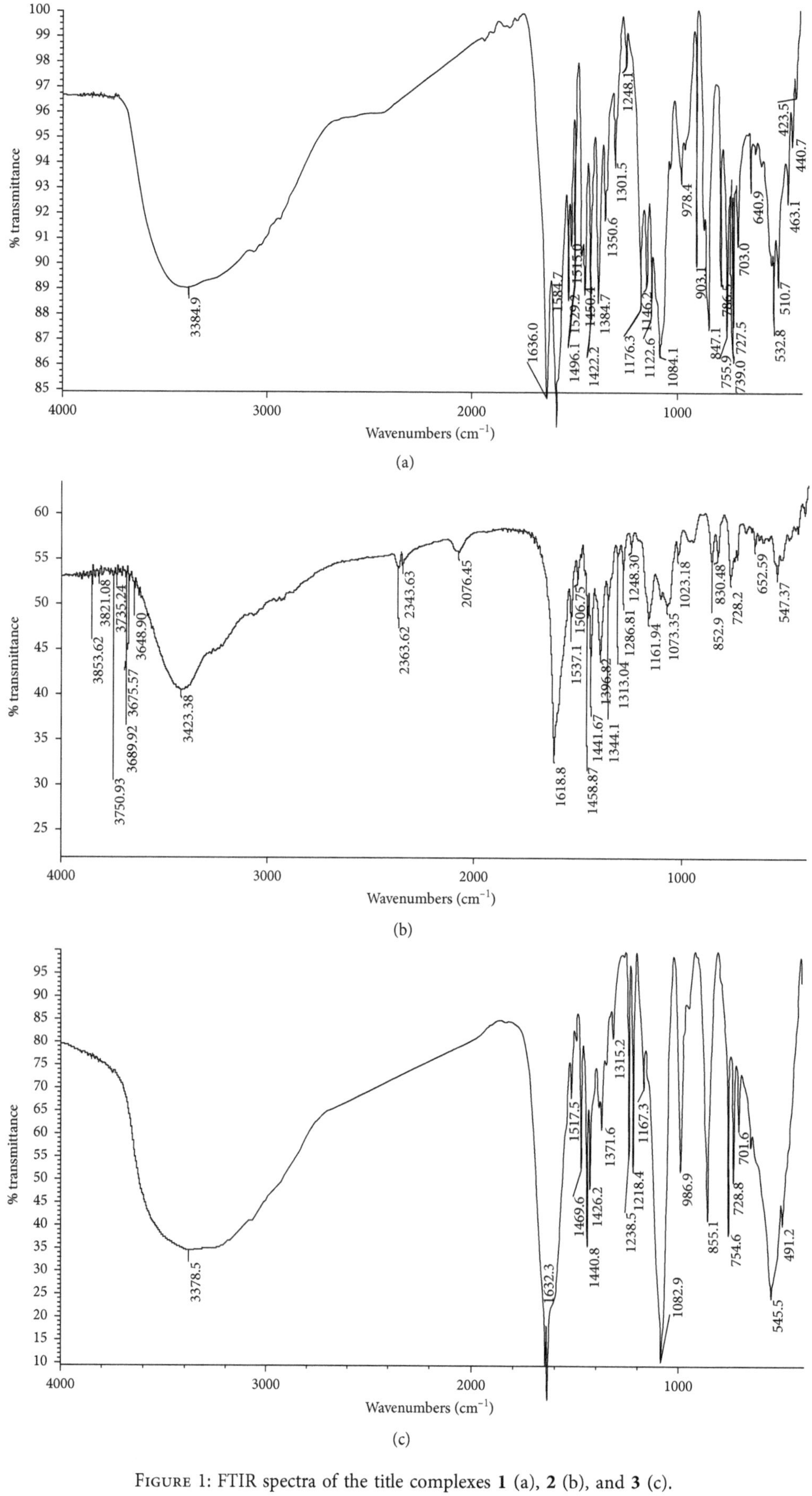

FIGURE 1: FTIR spectra of the title complexes **1** (a), **2** (b), and **3** (c).

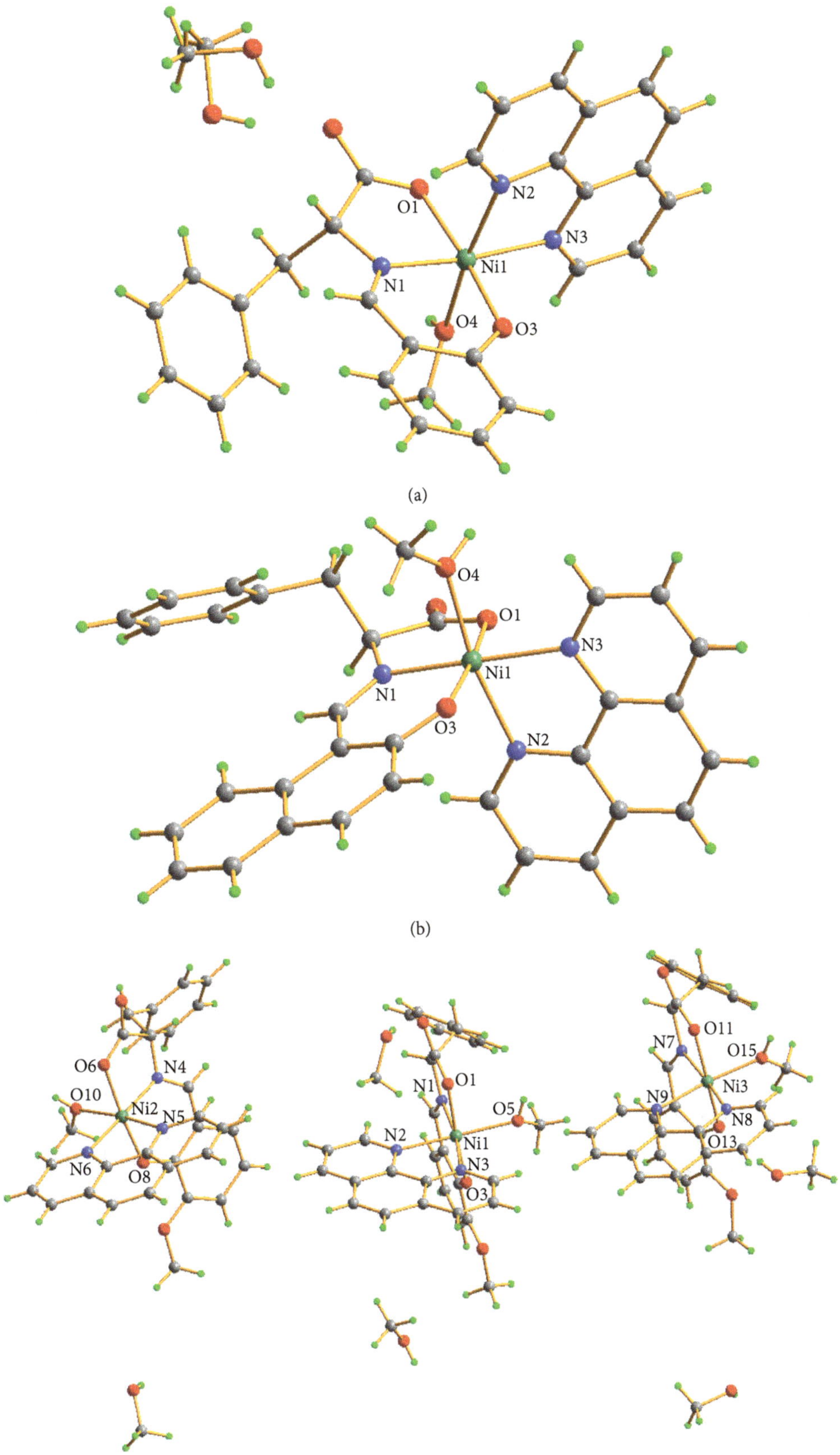

Figure 2: Molecular structures of complexes **1** (a), **2** (b), and **3** (c).

Table 2: Selected bond lengths (Å) and bond angles (°) for complexes **1**, **2**, and **3**.

		Complex 1			
Ni(1)–O(3)	1.990(2)	Ni(1)–N(1)	1.991(3)	Ni(1)–O(1)	2.068(3)
Ni(1)–N(3)	2.075(3)	Ni(1)–N(2)	2.116(3)	Ni(1)–O(4)	2.117(3)
O(3)–Ni(1)–N(1)	92.43(11)	O(3)–Ni(1)–O(1)	173.63(10)	N(1)–Ni(1)–O(1)	81.29(11)
O(3)–Ni(1)–N(3)	93.55(11)	N(1)–Ni(1)–N(3)	172.77(11)	O(1)–Ni(1)–N(3)	92.64(11)
O(3)–Ni(1)–N(2)	91.84(11)	N(1)–Ni(1)–N(2)	96.26(12)	O(1)–Ni(1)–N(2)	87.84(10)
N(3)–Ni(1)–N(2)	79.50(12)	O(3)–Ni(1)–O(4)	89.99(11)	N(1)–Ni(1)–O(4)	92.72(11)
O(1)–Ni(1)–O(4)	91.34(11)	N(3)–Ni(1)–O(4)	91.33(11)	N(2)–Ni(1)–O(4)	170.75(11)
		Complex 2			
Ni(1)–N(1)	1.987(3)	Ni(1)–O(3)	1.997(3)	Ni(1)–O(1)	2.068(3)
Ni(1)–N(3)	2.083(3)	Ni(1)–O(4)	2.100(3)	Ni(1)–N(2)	2.129(3)
N(1)–Ni(1)–O(3)	89.99(12)	N(1)–Ni(1)–O(1)	81.80(12)	O(3)–Ni(1)–O(1)	171.26(10)
N(1)–Ni(1)–N(3)	175.09(12)	O(3)–Ni(1)–N(3)	94.34(11)	O(1)–Ni(1)–N(3)	93.75(11)
N(1)–Ni(1)–O(4)	90.64(13)	O(3)–Ni(1)–O(4)	92.52(12)	O(1)–Ni(1)–O(4)	90.56(12)
N(3)–Ni(1)–O(4)	91.50(13)	N(1)–Ni(1)–N(2)	98.70(12)	O(3)–Ni(1)–N(2)	90.17(11)
O(1)–Ni(1)–N(2)	88.13(11)	N(3)–Ni(1)–N(2)	78.99(12)	O(4)–Ni(1)–N(2)	170.29(12)
		Complex 3			
Ni(1)–N(1)	2.000(6)	Ni(1)–O(3)	2.003(5)	Ni(1)–O(1)	2.053(5)
Ni(1)–N(3)	2.076(6)	Ni(1)–N(2)	2.131(6)	Ni(1)–O(5)	2.146(5)
Ni(2)–N(4)	1.982(6)	Ni(2)–O(8)	1.992(5)	Ni(2)–O(6)	2.059(5)
Ni(2)–N(6)	2.075(7)	Ni(2)–O(10)	2.113(5)	Ni(2)–N(5)	2.116(6)
Ni(3)–N(7)	1.995(7)	Ni(3)–O(13)	2.008(5)	Ni(3)–O(11)	2.056(6)
Ni(3)–N(8)	2.084(7)	Ni(3)–O(15)	2.099(6)	Ni(3)–N(9)	2.114(7)
N(1)–Ni(1)–O(3)	92.3(2)	N(1)–Ni(1)–O(1)	82.0(2)	O(3)–Ni(1)–O(1)	173.3(2)
N(1)–Ni(1)–N(3)	169.9(2)	O(3)–Ni(1)–N(3)	93.2(2)	O(1)–Ni(1)–N(3)	92.9(2)
N(1)–Ni(1)–N(2)	92.6(2)	O(3)–Ni(1)–N(2)	96.1(2)	O(1)–Ni(1)–N(2)	87.8(2)
N(3)–Ni(1)–N(2)	78.4(2)	N(1)–Ni(1)–O(5)	95.3(2)	O(3)–Ni(1)–O(5)	90.31(18)
O(1)–Ni(1)–O(5)	86.70(19)	N(3)–Ni(1)–O(5)	93.2(2)	N(2)–Ni(1)–O(5)	169.7(2)
N(4)–Ni(2)–O(8)	92.7(2)	N(4)–Ni(2)–O(6)	81.3(2)	O(8)–Ni(2)–O(6)	174.0(2)
N(4)–Ni(2)–N(6)	171.7(2)	O(8)–Ni(2)–N(6)	93.3(2)	O(6)–Ni(2)–N(6)	92.7(2)
N(4)–Ni(2)–O(10)	93.9(2)	O(8)–Ni(2)–O(10)	90.41(19)	O(6)–Ni(2)–O(10)	89.3(2)
N(6)–Ni(2)–O(10)	91.6(2)	N(4)–Ni(2)–N(5)	95.0(2)	O(8)–Ni(2)–N(5)	90.8(2)
O(6)–Ni(2)–N(5)	90.5(2)	N(6)–Ni(2)–N(5)	79.3(3)	O(10)–Ni(2)–N(5)	170.9(3)
N(7)–Ni(3)–O(13)	91.9(2)	N(7)–Ni(3)–O(11)	80.3(3)	O(13)–Ni(3)–O(11)	171.9(2)
N(7)–Ni(3)–N(8)	169.6(3)	O(13)–Ni(3)–N(8)	96.3(2)	O(11)–Ni(3)–N(8)	91.6(2)
N(7)–Ni(3)–O(15)	94.9(3)	O(13)–Ni(3)–O(15)	89.3(2)	O(11)–Ni(3)–O(15)	89.1(2)
N(8)–Ni(3)–O(15)	91.6(3)	N(7)–Ni(3)–N(9)	95.4(3)	O(13)–Ni(3)–N(9)	91.4(2)
O(11)–Ni(3)–N(9)	91.6(2)	N(8)–Ni(3)–N(9)	78.1(3)	O(15)–Ni(3)–N(9)	169.6(3)

1632.3 cm^{-1} for **3** can be assigned to the imine group (C = N) stretching frequency of the coordinated Schiff base ligand. Two moderate absorptions at 1584.7 cm^{-1} and 1384.7 cm^{-1} for complex **1**, 1537.1 cm^{-1} and 1313.0 cm^{-1} for complex **2**, and 1440.8 cm^{-1} and 1082.9 cm^{-1} for complex **3** are attributed to asymmetric and symmetric stretching vibrations of the COO^- group, respectively [37]. The frequency separation ($\Delta\nu = \nu(COO^-)_{asy} - \nu(COO^-)_{sym}$) is greater than 200 cm^{-1}, which suggests that the carboxylate is unidentate [38]. Furthermore, the absorption bands at 847.1 cm^{-1} and 739.0 cm^{-1} for **1**, 852.9 cm^{-1} and 728.2 cm^{-1} for **2**, and 855.1 cm^{-1} and 545.5 cm^{-1} for **3** may reasonably be assigned to Ni–O and Ni–N vibrations, respectively.

3.2. The Structure of the Nickel(II) Complexes ***1**, **2**, and **3**.* Figure 2 is the molecular structure of complexes **1**, **2**, and **3**. X-ray single-crystal diffraction analysis showed that complex **1** belongs to the monoclinic system, C2/c space group, and both complexes **2** and **3** belong to the triclinic system, P-1 space group. In each asymmetric structure unit of complex **1**, the equatorial coordination to Ni(II) atom is provided by N3 atom from the phenanthroline ligand and O1, N1, and O3 atoms from the Schiff base ligand. The two axial sites of the octahedron are occupied by N2 atom from the phenanthroline ligand and O4 atom from the methanol ligand, with the trans angle of N2–Ni–O4 = 170.75(11)°. The distance from Ni1 atom to the equatorial plane is 0.0753(14) Å, and the distances from O4 and N2 to the equatorial plane are 2.117(3) Å and 2.116(3) Å, respectively. The plane of the 1,10-phenanthroline ligand is almost perpendicular to the octahedral equatorial plane where Ni(1) lies, and the angle is 87.234(54)°. The Schiff base ligand forms a five-membered ring of Ni1–O1–C1–C2–N1 and a six-membered ring of Ni1–N1–C10–C11–C12–O3 surrounding the Ni1 atom, in which the dihedral angle formed by the plane of the two rings is 3.821(80)°. It is shown that these two rings are not really coplanar, but the stability of the complex is increased.

In complex **2**, O4 atom from the methanol ligand and N2 atom from the phenanthroline ligand lie in the axial positions of the octahedral geometry with the trans angle of 170.29(12)°. And N3 atom from the phenanthroline ligand

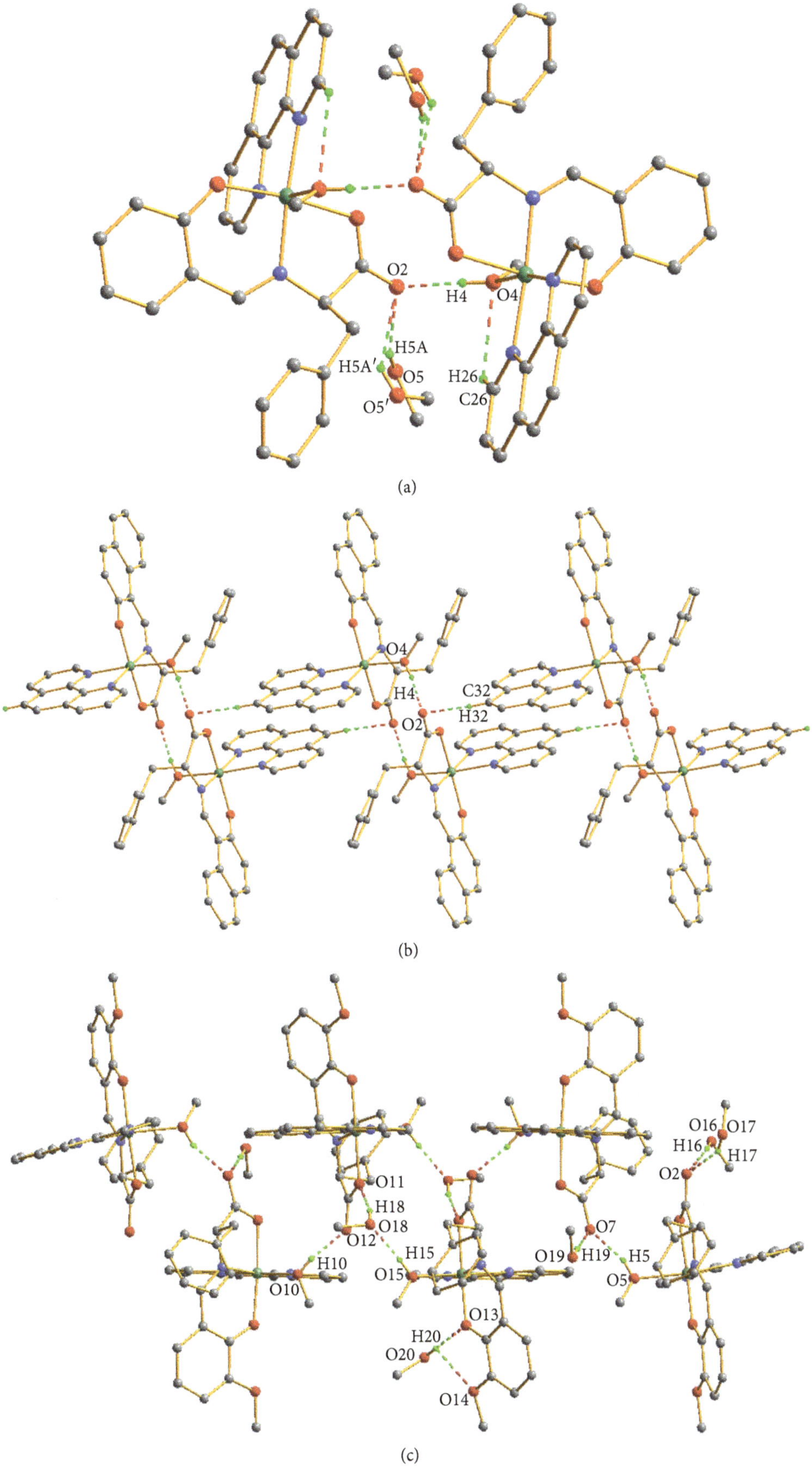

FIGURE 3: 2D supramolecular network in complexes **1** (a), **2** (b), and **3** (c) formed by intermolecular hydrogen-bonding interactions. Some H atoms and C atoms were omitted for clarity.

TABLE 3: Hydrogen bond lengths (Å) and bond angles (°) for complexes **1**, **2**, and **3**.

D–H···A	d(D–H)	d(H–A)	d(D–A)	∠(DHA)
Complex **1**				
O4–H4···O2[#1]	0.82	1.79	2.609(4)	173.4
C26–H26···O4	0.93	2.60	3.141(5)	117.9
O(5′)–H(5A′)···O(2)	0.82	2.15	3.175(5)	136.8
O(5)–H(5A)···O(2)	0.82	2.02	2.765(8)	149.9
Complex **2**				
O4–H4···O2[#1]	0.82	1.76	2.576(4)	170.4
C32–H32···O2[#3]	0.93	2.36	3.274(5)	166.8
Complex **3**				
O(16)–H(16)···O(2)	0.82	2.03	2.828(10)	163.2
O(15)–H(15)···O(18)[#4]	0.82	1.83	2.640(10)	171.2
O(5)–H(5)···O(7)[#5]	0.82	1.93	2.697(8)	154.2
O(10)–H(10)···O(12)[#5]	0.82	1.78	2.588(8)	169.9
O(17)–H(17)···O(2)[#6]	0.82	2.00	2.733(9)	148.3
O(19)–H(19)···O(7)[#6]	0.82	2.03	2.853(9)	177.6
O(18)–H(18)···O(11)[#6]	0.82	1.86	2.659(9)	166.1
O(20)–H(20)···O(13)[#7]	0.82	2.06	2.884(12)	179.0
O(20)–H(20)···O(14)[#7]	0.82	2.55	2.966(12)	112.5

Symmetry codes for the complex: #1: $0.5-x$, $0.5-y$, and $1-z$; #2: $2-x$, $-y$, and $1-z$; #3: $1-x$, $-y$, and $1-z$; #4: $-x+1$, $-y+1$, and $-z$; #5: $-x+2$, $-y+1$, and $-z+1$; #6: $x-1$, y, and z; #7: x, y, and $z+1$.

and O1, N1, and O3 atoms from the Schiff base ligand are located in the equatorial plane of the octahedron. The Schiff base ligand coordinates to Ni1 atom to form one five-membered ring and one six-membered ring, thus increasing the stability of the complex.

There are three structurally identical complex molecules in the structural unit of complex **3**, in which each individual molecule has a Ni atom coordinated with a tridentate Schiff base ligand, a bidentate phenanthroline ligand, and a methanol molecule. In each individual complex molecule, the axial position of the octahedron is occupied by the O atom on the methanol ligand and the N atom on the phenanthroline ligand. The three atoms in the Schiff base and a N atom of phenanthroline occupy its octahedral equatorial plane position, respectively. For example, the equatorial coordination to Ni1 is provided by N3 atom from the phenanthroline ligand and O1, N1, and O3 atoms from the Schiff base ligand, while N2 atom from the phenanthroline ligand and O5 atom from the methanol ligand are located in two axial sites of the octahedron with the trans angle of N2–Ni1–O5 = 169.7(2)°. The Schiff base ligand forms a five-membered ring and a six-membered ring surrounding the Ni(II) atom, and the dihedral angles formed by the planes of the two rings are 3.970(317)°, 9.895(283)°, and 12.639(434)°, respectively, in each individual molecule, which increases the stability of the complex.

In view of all these observations, each asymmetric structural unit in complexes **1**, **2**, and **3** has formed a six-coordination environment with a distorted octahedral structure. The selected bond lengths and angles around the metal center are given in Table 2. There are some intermolecular C–H···O and O–H···O hydrogen bonds linking the molecules to form a dimer or 2D structures as shown in Figure 3. In the crystal of complex **1**, the unordered methanol molecule is linked to the main molecule by O5′–H5A′···O2 and O5–H5A···O2 intrahydrogen bonds. And two main molecules in the crystal form a dimer by an O4–H4···O2 intermolecular hydrogen bond. In the crystal of complex **2**, a one-dimensional structure is formed through the intermolecular hydrogen bonds O4–H4···O2 and C32–H32···O2. In the crystal of complex **3**, five methanol molecules connect with three main molecules through O16–H16···O2, O17–H17···O2, O19–H19···O7, O18–H18···O11, O20–H20···O13, and O20–H20···O14 intermolecular hydrogen bonds, together with O15–H15···O18, O5–H5···O7, and O10–H10···O12 to form a two-dimensional structure. Hydrogen bond lengths and bond angles for the complexes are shown in table 3.

3.3. DNA Binding Studies

3.3.1. UV-Vis Absorption Spectra. Electronic absorption spectroscopy is one of the commonly used methods for studying the interaction of small molecules with DNA. In the presence of DNA, the microenvironment of the ligand of the complex is affected, thus resulting in changes in the absorption intensity and wavelength of the complex. In general, UV-Vis absorption spectrum does not change appreciably when the metal complex interacts with DNA by electrostatic or trench interaction. But the absorption peak of the complex occurs red-shifted and the absorption intensity decreases when the metal complex interacts with DNA by intercalation [39]. This phenomenon can be explained by the generation of π-electron stacking action due to the ligand inserted into the DNA base, in which the π^* orbital within the ligand is coupled to the π orbital of the base, resulting in a reduction of the transition energy of $\pi \rightarrow \pi^*$ and generation of the red shift phenomenon. UV absorption spectra of these complexes with different concentrations of CT-DNA are shown in Figure 4. The absorbances at 268 nm of these complexes had an obvious hypochromic effect and a slightly

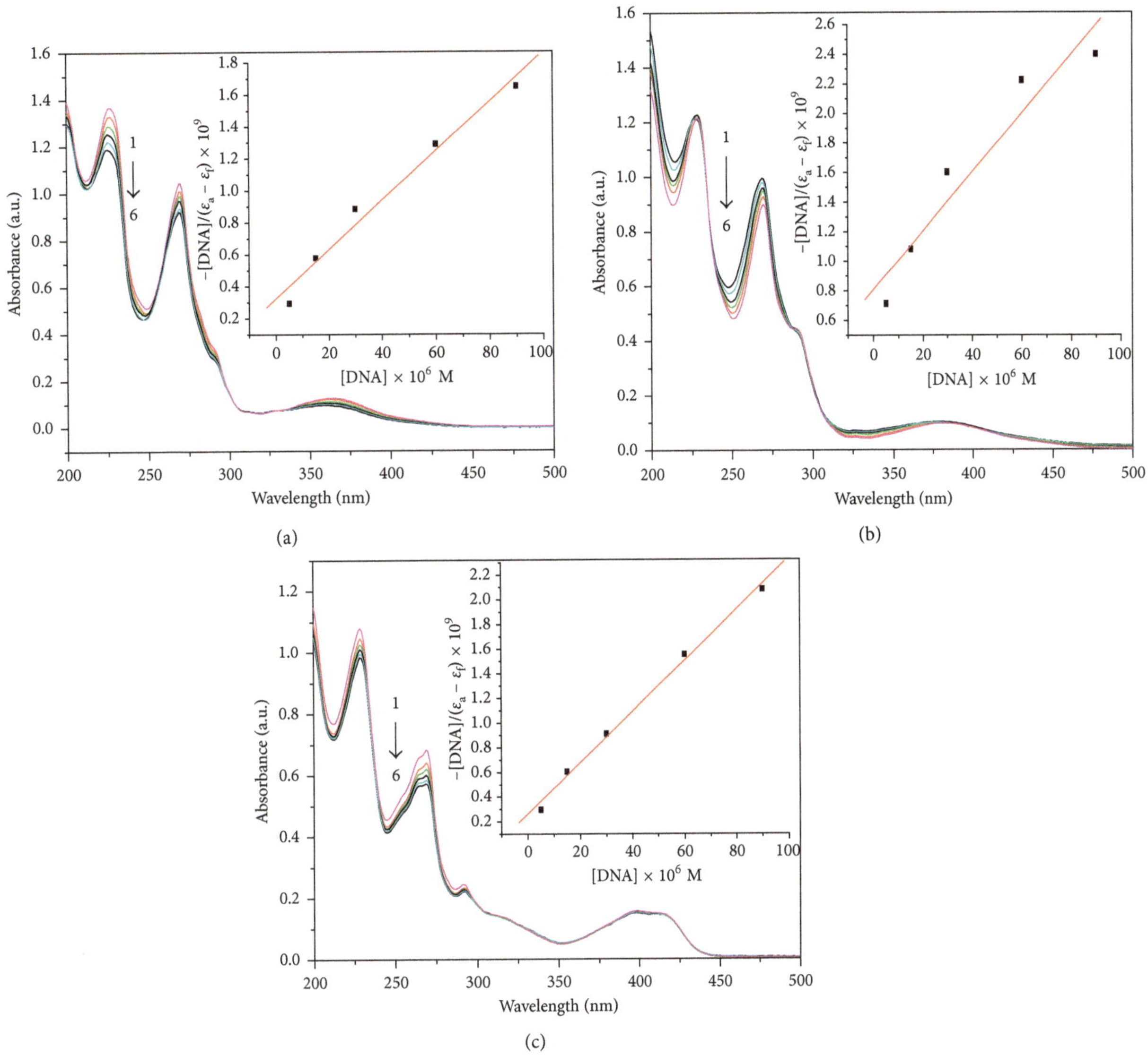

FIGURE 4: UV-Vis spectra of complexes **1** (a), **2** (b), and **3** (c) in the absence and presence of CT-DNA; the complex concentration was 1.5×10^{-5} M, and DNA concentrations were 0, 0.5×10^{-5}, 1.5×10^{-5}, 3.0×10^{-5}, 6.0×10^{-5}, and 9.0×10^{-5} M corresponding to the curves from 1 to 6, respectively. The arrow shows the intensity change on increasing the DNA concentration.

red shift, showing a binding property of the three complexes with DNA, which also indicates that these complexes interact with CT-DNA mainly through the intercalative mode [40, 41].

In order to measure the strength of DNA binding with the complex quantitatively, the intrinsic binding constant (K_b) of the complex with CT-DNA can be calculated by using the following equation [42]:

$$\frac{[\mathrm{DNA}]}{\varepsilon_a - \varepsilon_f} = \frac{[\mathrm{DNA}]}{\varepsilon_b - \varepsilon_f} + \frac{1}{K_b(\varepsilon_b - \varepsilon_f)}, \quad (1)$$

where [DNA] is the concentration of DNA in base pairs, ε_a is the molar extinction coefficient of the complex in the presence of DNA, and ε_f and ε_b are the molar extinction coefficients for the free complex and the fully DNA-bound complex, respectively. K_b can be obtained from the ratio of the slope to the intercept in the plot of $[\mathrm{DNA}]/\varepsilon_a - \varepsilon_f$ versus [DNA]. In this experiment, the calculated intrinsic binding constant K_b values were 1.82×10^4, 1.96×10^4, and $2.02 \times 10^4\ \mathrm{M}^{-1}$ for complexes **1**, **2**, and **3**, respectively. These values suggest that the binding strength of all the three nickel(II) complexes to DNA is weaker than that of ethidium bromide (EB) to DNA ($K_b = 3.3 \times 10^5\ \mathrm{L{\cdot}mol^{-1}}$) [43], indicating that these complexes have a weaker binding ability to DNA, which may be due to the deformed octahedral structure of these complexes.

3.3.2. Fluorescence Spectroscopy. Ethidium bromide (EB) is a classical embedded fluorescent probe that studies the interaction of drug molecules with DNA. It is weakly fluorescent by itself, but the chromophore is embedded in base pairs of DNA molecules to make fluorescence intensity enhancement. When a complex is added to the EB-DNA system, it can interact with DNA and compete with EB for DNA

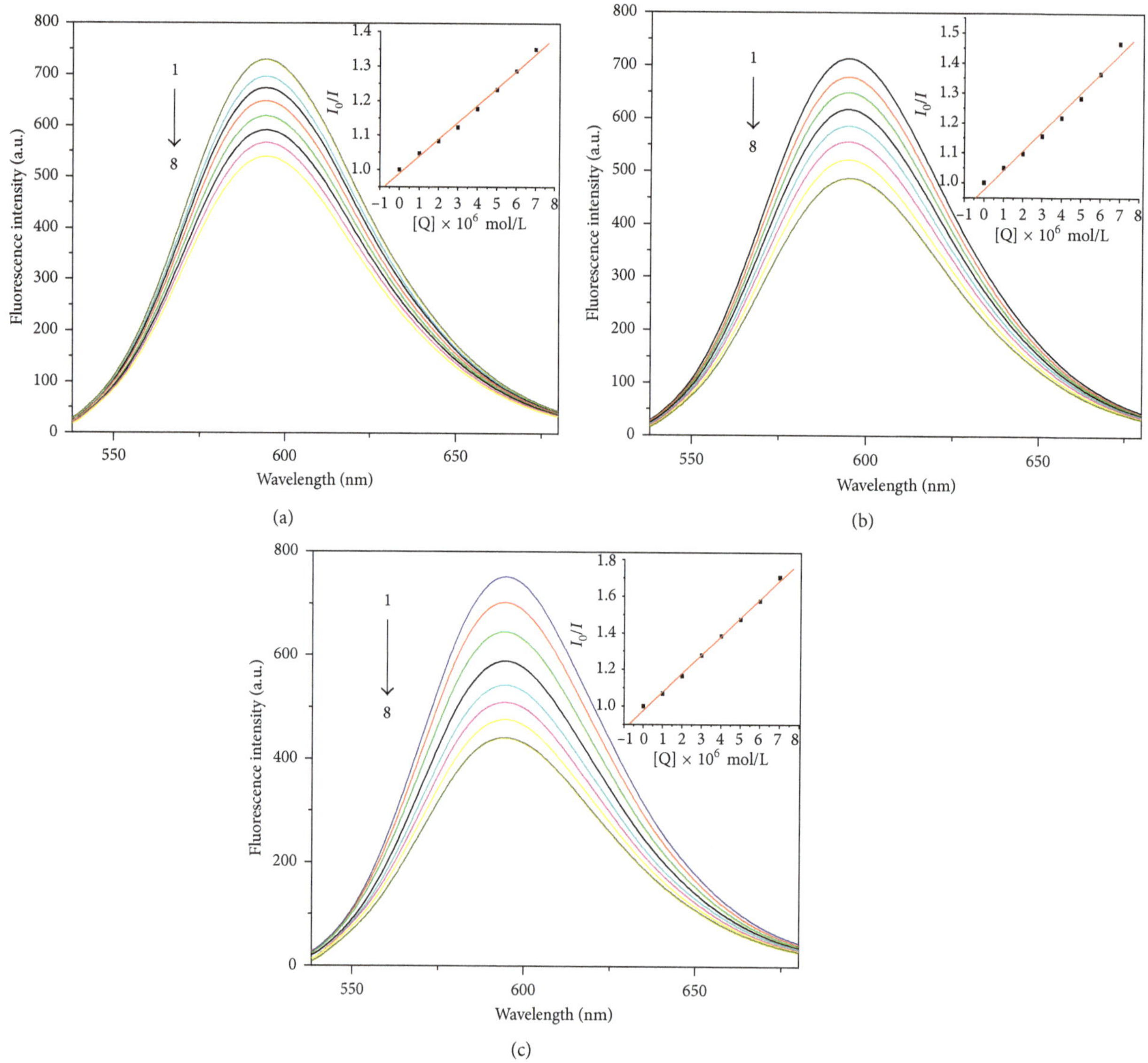

Figure 5: Fluorescence quenching spectra of EB bound to CT-DNA by complexes **1** (a), **2** (b), and **3** (c). $\lambda_{ex} = 510$ nm; DNA and EB concentrations were 1.0×10^{-5} and 1.0×10^{-5} M, and the complex concentrations were 0, 1.0, 2.0, 3.0, 4.0, 5.0, 6.0, and 7.0×10^{-5} M corresponding to the curves from 1 to 8, respectively. The arrow shows the intensity change on increasing the complex concentration.

binding, resulting in a decrease in the fluorescence intensity of the EB-DNA system. Therefore, we can deduce the interaction mode of the complex with DNA by observing the degree of change of the fluorescence system. As shown in Figure 5, the fluorescence intensities of the EB-DNA system at 595 nm decreased significantly with the increase of the concentration of the three complexes, indicating that the complexes can displace EB from the CT-DNA by competitive binding. In order to quantitatively study the fluorescence quenching degree of the complexes, the fluorescence quenching constant K_{sq} can be calculated by the Stern–Volmer equation [44]:

$$\frac{I_0}{I} = 1 + K_{sq} \cdot r, \tag{2}$$

where I_0 is the fluorescence intensity of the EB-DNA system when the complex is not added, I is the fluorescence intensity of the EB-DNA system after adding different concentration of the complexes, K_{sq} is the linear Stern–Volmer constant, and r ([complex]/[CT-DNA]) is the ratio of concentration of the complex to that of CT-DNA. The Stern–Volmer fluorescence quenching curves are shown in Figure 6. The K_{sq} values for these complexes were determined: $K_{sq}(\mathbf{1}) = 0.493$, $K_{sq}(\mathbf{2}) = 0.648$, and $K_{sq}(\mathbf{3}) = 1.011$, respectively.

3.3.3. CD Spectroscopy. CD spectroscopy is one of the most important means to study the interaction of small molecules with DNA and explore the change of DNA conformation by measuring the difference in the absorption of left and right circularly polarized light [45]. As shown in Figure 7, there were two distinct peaks at 273 nm and 245 nm in the CD spectrum of CT-DNA. The positive peak at 273 nm was caused by the π-π stacking of the CT-DNA base pair and the negative peak at

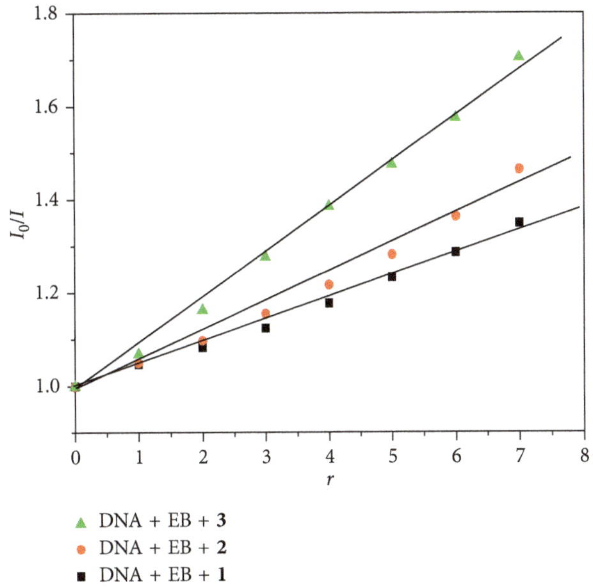

FIGURE 6: The Stern–Volmer fluorescence quenching curves.

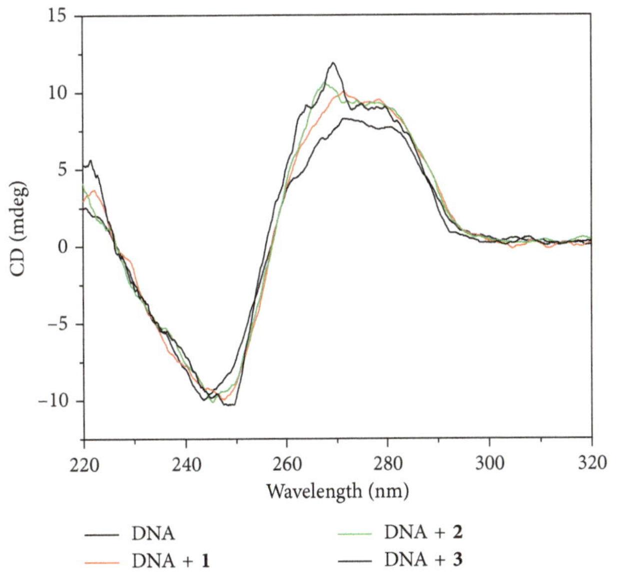

FIGURE 7: Effect of complexes **1**, **2**, and **3** on CD spectra of CT-DNA. DNA concentration was 1.0×10^{-4} M, and the complex concentration was 4.0×10^{-5} M.

245 nm was due to the helicity, which are typical CD spectra of B-DNA conformation [46]. When the three complexes were added to DNA solution, respectively, each CD spectrum of DNA changed obviously. The positive peak at 273 nm increased, and the negative peak at 245 nm remained unchanged. The degree of positive peak variation by complex **3** is stronger than that of complex **2** and complex **1**. The results indicated that these complexes were inserted into the base pair of DNA. This change affects the π-π stacking between the DNA base pairs and makes the conformation of DNA changed. This result is consistent with that of the UV-Vis absorption spectra and the fluorescence spectra.

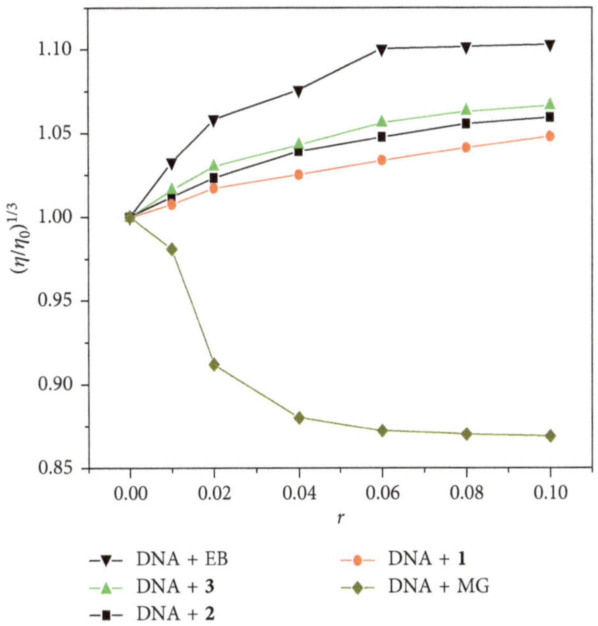

FIGURE 8: Effect of increasing amounts of complexes **1**, **2**, and **3**, EB, and methyl green on the relative viscosity of CT-DNA (1.0×10^{-4} M at pH = 7.4).

3.3.4. Viscosity Measurement. Viscosity determination is a unique method for determining the binding mode of the complex to DNA. The distance between adjacent base pairs increases when the complex interacts with DNA, and the insertion process of the small molecules of the complex is achieved. The double helix of the DNA is elongated and the viscosity increases. In contrast, the double-stranded structure of DNA is kinked and the viscosity is reduced when the complex is inserted into the base pair of DNA by partial insertion or groove bonding, while the effect of the groove surface mode or electrostatic action has a smaller effect on viscosity. The changes of the viscosity of DNA solution with increasing concentrations of the complexes are shown in Figure 8. With the increase of the amounts of these complexes, the relative viscosities of DNA solutions increased, which were consistent with the change of viscosity of DNA solution caused by intercalator EB and contrary to the trend of methylene green (MG). The results also indicated that the interaction between complex **3** and DNA is slightly stronger than that of complexes **2** and **1**. This is consistent with the results of the abovementioned spectral studies.

From the abovementioned results, we can conclude that the DNA binding ability of these three complexes is in the order of complex **3** > complex **2** > complex **1**. The difference among the three complexes is using different aldehydes for synthesizing Schiff bases, salicylaldehyde for complex **1**, 2-hydroxy-1-naphthaldehyde for complex **2**, and *o*-vanillin for complex **3**. In 2-hydroxy-1-naphthaldehyde, the benzene group replaces the hydrogen on the benzene ring of salicylaldehyde, and the methoxy group replaces the hydrogen on the benzene ring of salicylaldehyde in *o*-vanillin. As we know, the methoxy group and benzene group are electron-donating groups, and the electron-donating ability of the

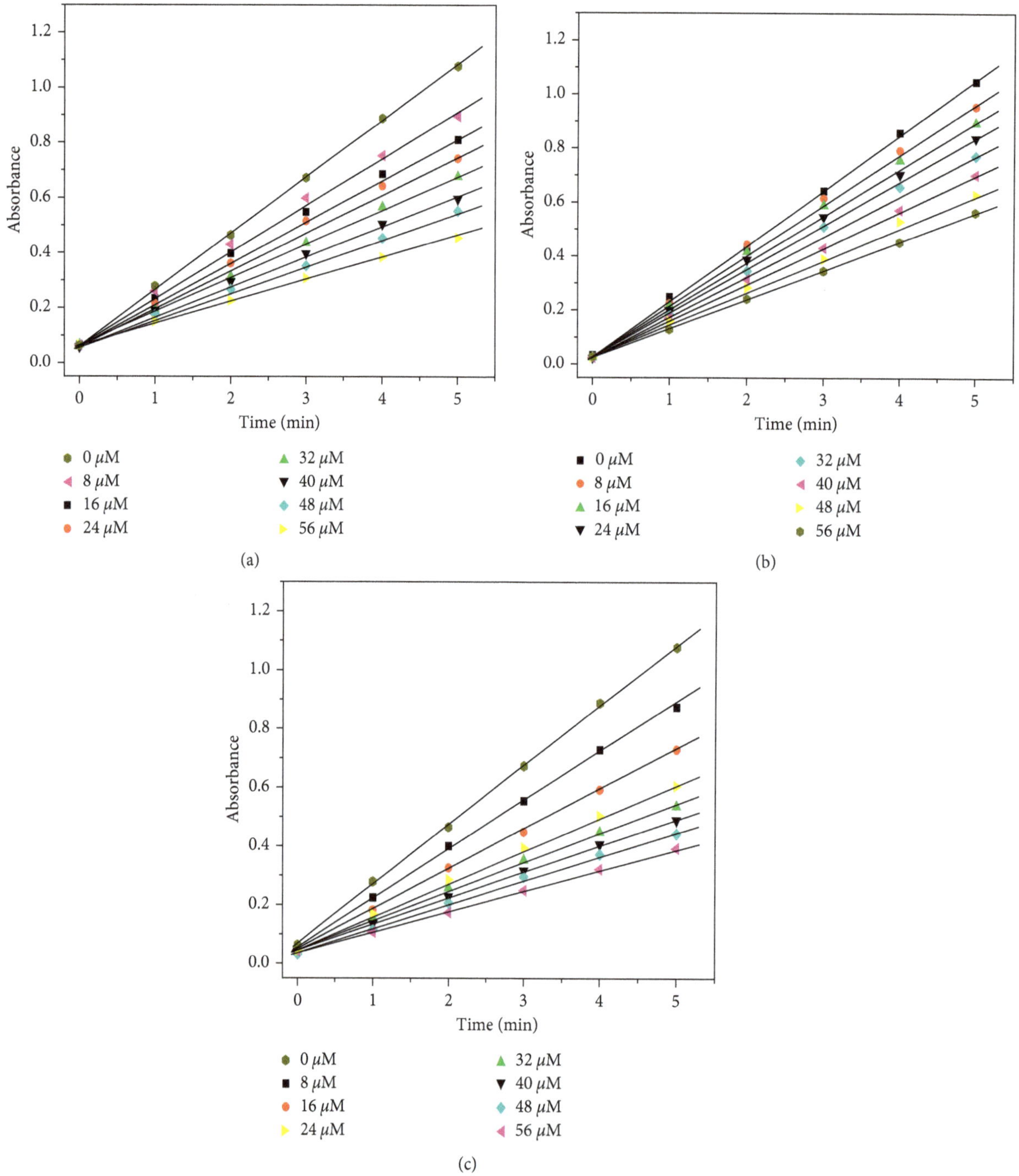

Figure 9: Effect of the concentrations of complexes **1** (a), **2** (b), and **3** (c) on the absorbance (A_{560}) of the reaction system with time. [NBT] = 1.0×10^{-4} mol·L^{-1}, [VB_2] = 6.2×10^{-6} mol·L^{-1}, [TMEDA] = 8.3×10^{-4} mol·L^{-1}, [complex] = 0, 0.8×10^{-5}, 1.6×10^{-5}, 2.4×10^{-5}, 3.2×10^{-5}, 4.0×10^{-5}, 4.8×10^{-5}, and 5.6×10^{-5} mol·L^{-1}.

methoxy group is bigger than that of the benzene group. Therefore, the Schiff base ligand provides electrons more easily to coordinate to Ni(II) in complex **3** than that in complex **2** and complex **1**; thus, the second ligand (1,10-phenanthroline) has higher electron density in complex **3** than that in complex **2** and complex **1**. The interaction of the complexes with DNA is mainly by intercalation of the planar 1,10-phenanthroline ligand into the DNA base pairs, producing the strong π-π stacking interaction. Complex **3** with higher electron density of the 1,10-phenanthroline ligand can interact with DNA stronger than complex **1** and complex **2**.

3.4. SOD Activity. The nitrotetrazolium blue chloride (NBT) light reduction method is a commonly used method to determine the activity of SOD. Riboflavin (VB_2) reacted with tetramethylethylenediamine to produce the superoxide

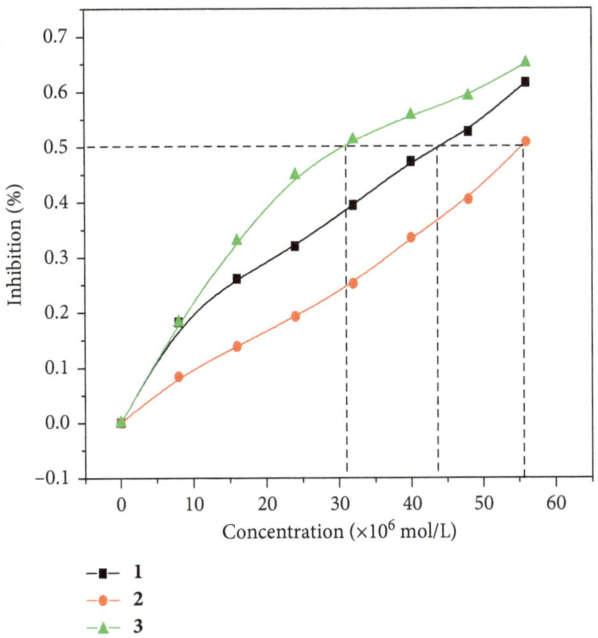

FIGURE 10: The inhibitory rates of complexes **1**, **2**, and **3** varying with concentration.

anion radical under light conditions: $VB_2 + (CH_3)_2NCH_2CH_2N(CH_3)_2 \rightarrow O_2^{\cdot-}$. The superoxide anion radical can reduce NBT to a blue-violet compound named formazan. The absorbance at 560 nm was proportional to the concentration of formazan. Therefore, a straight line can be obtained by measuring the absorbance A_{560} at different times. The slope of the line (k value) can reflect the speed of generation of formazan. Figure 9 shows that, with the increase of the concentration of the complex, the slope of the line gradually decreased, indicating that the greater the concentration, the greater the inhibition rate. The inhibitory rate (η) of the complex to $O_2^{\cdot-}$ can be determined by the following formula [47]:

$$\eta = \left(1 - \frac{k'}{k}\right) \times 100\%, \quad (3)$$

where k is the slope of the straight line without complex and k' is the slope of the straight line after adding the complex. The 50% of activity (IC_{50}) indicates the molar concentration of the tested complex which caused a 50% scavenging effect on superoxide radicals. Figure 10 shows the inhibition rate of the complex to $O_2^{\cdot-}$ at various concentrations of complexes **1**, **2**, and **3**. The obtained IC_{50} values from Figure 10 were 4.4×10^{-5} M for complex **1**, 5.6×10^{-5} M for complex **2**, and 3.1×10^{-5} M for complex **3**, respectively. The results showed that these complexes have a certain SOD activity, and the superoxide anion radical scavenging effect of complex **3** is higher than that of complex **1** and complex **2**.

4. Conclusion

Three new nickel(II) complexes, $[Ni(sal\text{-}L\text{-}phe)(phen)(CH_3OH)]\cdot CH_3OH$ (**1**), $[Ni(naph\text{-}L\text{-}phe)(phen)(CH_3OH)]$ (**2**), and $[Ni(o\text{-}van\text{-}L\text{-}phe)(phen)(CH_3OH)]\cdot 5CH_3OH$ (**3**), have been synthesized. These crystal structures were determined by single-crystal X-ray diffraction and characterized by elemental analysis and IR spectra. The interactions of these complexes with CT-DNA were studied by spectroscopies. The results showed that these complexes could interact with CT-DNA by the way of insertion into the CT-DNA, which was due to the good planarity of the ligand 1,10-phenanthroline in these complexes. Furthermore, the SOD activities of these complexes were studied by the NBT photoreduction method, implying that the three complexes had some SOD activities. The investigation can help us understand their mechanism of action and provide basis for designing new transition metal complexes with better activities.

Acknowledgments

This project was supported by the Natural Science Foundation of Shandong Province of China (nos. Y2004B02 and ZR2016HB73) and the Undergraduate Scientific-Technological Innovation and Culture Innovation Foundation of Liaocheng University (no. 26312160514).

References

[1] Y. J. Liu and C. H. Zeng, "Synthesis and DNA interaction studies of ruthenium(II) complexes with isatino[1,2-b]-1,4,8,9-tetraazatriphenylene as an intercalative ligand," *Transition Metal Chemistry*, vol. 34, no. 4, pp. 455–462, 2009.

[2] H. Liu, L. Li, Q. Guo, J. Dong, and J. Li, "Synthesis, crystal structure, DNA and BSA binding properties of a chromium (III) complex with L-glycine Schiff base and 1,10-phenanthroline," *Transition Metal Chemistry*, vol. 38, no. 4, pp. 441–448, 2013.

[3] L. Li, Q. Guo, J. Dong, T. Xu, and J. Li, "DNA binding, DNA cleavage and BSA interaction of a mixed-ligand copper(II) complex with taurine Schiff base and 1,10-phenanthroline," *Journal of Photochemistry and Photobiology B: Biology*, vol. 125, pp. 56–62, 2013.

[4] M. J. Genin, C. Biles, B. J. Keiser et al., "Novel 1,5-diphenylpyrazole nonnucleoside HIV-1 reverse transcriptase inhibitors with enhanced activity versus the delavirdine-resistant P236L mutant: lead identification and SAR of 3- and 4-substituted derivatives," *Journal of Medicinal Chemistry*, vol. 43, no. 5, pp. 1034–1040, 2011.

[5] M. Verma, S. N. Pandeya, K. N. Singh, and J. P. Stables, "Anticonvulsant activity of Schiff bases of isatin derivatives," *Acta Pharmaceutica*, vol. 54, no. 1, pp. 49–56, 2004.

[6] K. Sztanke, A. Maziarka, A. Osinka, and M. Sztanke, "An insight into synthetic Schiff bases revealing antiproliferative activities in vitro," *Bioorganic & Medicinal Chemistry*, vol. 21, no. 13, pp. 3648–3666, 2013.

[7] R. Ziessel, "Schiff-based bipyridine ligands. Unusual coordination features and mesomorphic behaviour," *Coordination Chemistry Reviews*, vol. 216-217, no. 36, pp. 195–223, 2001.

[8] R. S. Kumar and S. Arunachalam, "DNA binding and antimicrobial studies of polymer–copper(II) complexes, containing 1,10-phenanthroline and L-phenylalanine ligands," *European Journal of Medicinal Chemistry*, vol. 44, no. 5, pp. 1878–1883, 2009.

[9] M. Can, F. A. Armstrong, and S. W. Ragsdale, "Structure, function, and mechanism of the nickel metalloenzymes, CO dehydrogenase, and acetyl-CoA synthase," *Chemical Reviews*, vol. 114, no. 8, pp. 4149–4174, 2014.

[10] G. Morgant, N. Bouhmaida, L. Balde, N. E. Ghermani, and J. D'Angelo, "Low-temperature (100 K) crystal structures of pentaaqua(5-nitrosalicylato) complexes of magnesium(II), zinc(II), cobalt(II) and nickel(II): a π–π stacked and hydrogen bonded 3D supramolecular architecture," *Polyhedron*, vol. 25, no. 11, pp. 2229–2235, 2006.

[11] P. Bombicz, E. Forizs, J. Madarasz, A. Deak, and A. Kalman, "Inclusion compounds containing a drug: structure and thermal stability of the first clathrates of nitrazepam and isothiocyanato ethanol complexes of Co(II) and Ni(II)," *Inorganica Chimica Acta*, vol. 315, no. 2, pp. 229–235, 2001.

[12] O. Z. Yesilel, M. S. Soylu, H. Olmez, and O. Buyukgungor, "Synthesis and spectrothermal studies of vitamin B13 complexes of cobalt(II) and nickel(II) with 4-methylimidazole: crystal structure of $[Ni(HOr)(H_2O)(4\text{-}Meim)_3]_2{\cdot}5H_2O$," *Polyhedron*, vol. 25, no. 15, pp. 2985–2992, 2006.

[13] M. Alexiou, I. Tsivikas, C. Dendrinou-Samara et al., "High nuclearity nickel compounds with three, four or five metal atoms showing antibacterial activity," *Journal of Inorganic Biochemistry*, vol. 93, no. 4, pp. 256–264, 2003.

[14] R. Kurtaran, L. T. Yildirim, A. D. Azaz, H. Namli, and O. Atakol, "Synthesis, characterization, crystal structure and biological activity of a novel heterotetranuclear complex: [NiLPb(SCN)2(DMF)(H2O)]2, bis-{[μ-*N,N*′-bis(salicylidene)-1,3-propanediaminato-aqua-nickel(II)](thiocyanato)(μ-thiocyanato)(μ-*N,N*′-dimethylformamide)lead(II)}," *Journal of Inorganic Biochemistry*, vol. 99, no. 10, pp. 1937–1944, 2005.

[15] R. del Campo, J. J. Criado, E. Garcia et al., "Thiourea derivatives and their nickel(II) and platinum(II) complexes: antifungal activity," *Journal of Inorganic Biochemistry*, vol. 89, no. 1-2, pp. 74–82, 2002.

[16] W. Luo, X. Meng, X. Sun et al., "Synthesis, crystal structure and bioactivity of a novel linear trinuclear nickel(II) complex," *Inorganic Chemistry Communications*, vol. 10, no. 11, pp. 1351–1354, 2007.

[17] K. C. Skyrianou, E. K. Efthimiadou, V. Psycharis, A. Terzis, D. P. Kessissoglou, and G. Psomas, "Nickel-quinolones interaction. Part 1-nickel(II) complexes with the antibacterial drug sparfloxacin: structure and biological properties," *Journal of Inorganic Biochemistry*, vol. 103, no. 12, pp. 1617–1625, 2009.

[18] K. C. Skyrianou, F. Perdih, A. N. Papadopoulos, I. Turel, D. P. Kessissoglou, and G. Psomas, "Nickel–quinolones interaction. Part 5-biological evaluation of nickel(II) complexes with first-, second- and third-generation quinolones," *Journal of Inorganic Biochemistry*, vol. 105, no. 10, pp. 1273–1285, 2011.

[19] Z. Afrasiabi, E. Sinn, W. Lin, Y. Ma, C. Campana, and S. Padhye, "Nickel (II) complexes of naphthaquinone thiosemicarbazone and semicarbazone: synthesis, structure, spectroscopy, and biological activity," *Journal of Inorganic Biochemistry*, vol. 99, no. 7, pp. 1526–1531, 2005.

[20] A. Buschini, S. Pinelli, C. Pellakani et al., "Synthesis, characterization and deepening in the comprehension of the biological action mechanisms of a new nickel complex with antiproliferative activity," *Journal of Inorganic Biochemistry*, vol. 103, no. 5, pp. 666–677, 2009.

[21] E. Jabri, M. B. Carr, R. P. Hausinger, and P. Karplus, "The crystal structure of urease from *Klebsiella aerogenes*," *Science*, vol. 268, no. 5213, pp. 998–1004, 1995.

[22] M. A. Neelakantan, K. Balamurugan, C. Balakrishnan, and L. Subha, "Interaction of amino acid Schiff base metal complexes with DNA/BSA protein and antibacterial activity: spectral studies, DFT calculations and molecular docking simulations," *Applied Organometallic Chemistry*, vol. 32, no. 4, p. e4259, 2018.

[23] F. Sevgi, U. Bagkesici, A. N. Kursunlu, and E. Guler, "Fe(III), Co(II), Ni(II), Cu(II) and Zn(II) complexes of Schiff bases based-on glycine and phenylalanine: synthesis, magnetic/thermal properties and antimicrobial activity," *Journal of Molecular Structure*, vol. 1154, pp. 256–260, 2018.

[24] S. Tabassum, M. Ahmad, M. Afzal, M. Zaki, and P. K. Bharadwaj, "Synthesis and structure elucidation of a copper(II) Schiff-base complex: in vitro, DNA binding, pBR322 plasmid cleavage and HSA binding studies," *Journal of Photochemistry and Photobiology B: Biology*, vol. 140, pp. 321–331, 2014.

[25] Q. Wei, J. Dong, P. Zhao et al., "DNA binding, BSA interaction and SOD activity of two new nickel(II) complexes with glutamine Schiff base ligands," *Journal of Photochemistry and Photobiology B: Biology*, vol. 161, pp. 355–367, 2016.

[26] P. Zhao, Q. Wei, J. Dong, F. Ding, J. Li, and L. Li, "Synthesis, structure and spectroscopic studies on DNA binding, BSA interaction of a nickel(II) complex containing L–methionine Schiff base and 1,10-phenanthroline," *Journal of Coordination Chemistry*, vol. 69, no. 16, pp. 2437–2453, 2016.

[27] Z. Li, H. Yan, G. L. Chang, M. Hong, J. M. Dou, and M. J. Niu, "Cu(II), Ni(II) complexes derived from chiral Schiff-base ligands: synthesis, characterization, cytotoxicity, protein and DNA binding properties," *Journal of Photochemistry and Photobiology B: Biology*, vol. 163, pp. 403–412, 2016.

[28] M. J. Niu, M. Hong, G. L. Chang, X. Li, and Z. Li, "A comparative study of cytotoxicity and interaction with DNA/protein of five transition metal complexes with Schiff base ligands," *Journal of Photochemistry and Photobiology B: Biology*, vol. 148, pp. 232–241, 2015.

[29] P. Krishnamoorthy, P. Sathyadevi, P. T. Muthiah, and N. Dharmaraj, "Nickel and cobalt complexes of benzoic acid (2-hydroxy-benzylidene)-hydrazide ligand: synthesis, structure and comparative in vitro evaluations of biological perspectives," *RSC Advances*, vol. 2, no. 32, pp. 12190–12203, 2012.

[30] E. Ramachandran, D. S. Raja, J. L. Mike, T. R. Wagner, M. Zeller, and K. Natarajan, "Evaluation on the role of terminal *N*-substitution in 6-methoxy-2-oxo-1,2-dihydroquinoline-3-carbaldehyde thiosemicarbazones on the biological properties of new water-soluble nickel(II) complexes," *RSC Advances*, vol. 2, no. 22, pp. 8515–8525, 2012.

[31] N. Selvakumaran, N. S. P. Bhuvanesh, A. Endo, and R. Karvembu, "Synthesis, structure, DNA and protein binding studies, and cytotoxic activity of nickel(II) complexes containing 3,3-dialkyl/aryl-1-(2,4-dichlorobenzoyl)thiourea ligands," *Polyhedron*, vol. 75, no. 12, pp. 95–109, 2014.

[32] J. S. Pap, B. Kripli, T. Váradi, M. Giorgi, J. Kaizer, and G. Speier, "Comparison of the SOD-like activity of hexacoordinate Mn(II), Fe(II) and Ni(II) complexes having isoindoline-based ligands," *Journal of Inorganic Biochemistry*, vol. 105, no. 6, pp. 911–918, 2011.

[33] A. E. M. Ramadan, "Macrocyclic nickel(II) complexes: syn-

thesis, characterization, superoxide scavenging activity and DNA-binding," *Journal of Molecular Structure*, vol. 1015, no. 19, pp. 56–66, 2012.

[34] J. Marmur, "A procedure for the isolation of deoxyribonucleic acid from microorganisms," *Journal of Molecular Biology*, vol. 3, no. 2, pp. 208–218, 1961.

[35] M. E. Reichmann, S. A. Rice, C. A. Thomas, and P. Doty, "A further examination of the molecular weight and size of desoxypentose nucleic acid," *Journal of the American Chemical Society*, vol. 76, no. 11, pp. 3047–3053, 1954.

[36] G. M. Sheldrick, *SHELXS-97 and SHELXL-97, Program for Crystal Structural Solution and Refinement*, Brucker Analytical Instruments Inc., Madison, WI, USA, 1997.

[37] J. Vanco, J. Marek, Z. Trávnícek, E. Racanská, J. Muselík, and O. Svajlenová, "Synthesis, structural characterization, antiradical and antidiabetic activities of copper(II) and zinc(II) Schiff base complexes derived from salicylaldehyde and beta-alanine," *Journal of Inorganic Biochemistry*, vol. 102, no. 4, pp. 595–605, 2008.

[38] I. Turel, I. Leban, and N. Bukovec, "Crystal structure and characterization of the bismuth(III) compound with quinolone family member (ciprofloxacin). Antibacterial study," *Journal of Inorganic Biochemistry*, vol. 66, no. 4, pp. 241–245, 1997.

[39] S. A. Tysoe, R. J. Morgan, A. D. Baker, and T. C. Strekas, "Spectroscopic investigation of differential binding modes of .DELTA.- and .LAMBDA.-Ru(bpy)$_2$(ppz)$^{2+}$ with calf thymus DNA," *Journal of Physical Chemistry*, vol. 97, no. 8, pp. 1707–1711, 1993.

[40] J. M. Kelly, A. B. Tossi, D. J. McConnell, and C. OhUigin, "A study of the interactions of some polypyridylruthenium(II) complexes with DNA using fluorescence spectroscopy, topoisomerisation and thermal denaturation," *Nucleic Acids Research*, vol. 13, no. 17, pp. 6017–6034, 1985.

[41] F. Xue, C. Z. Xie, Y. W. Zhang et al., "Two new dicopper(II) complexes with oxamido-bridged ligand: synthesis, crystal structures, DNA binding/cleavage and BSA binding activity," *Journal of Inorganic Biochemistry*, vol. 115, pp. 78–86, 2012.

[42] A. Wolfe, G. H. Shimer Jr., and T. Meehan, "Polycyclic aromatic hydrocarbons physically intercalate into duplex regions of denatured DNA," *Biochemistry*, vol. 26, no. 20, pp. 6392–6396, 1987.

[43] R. E. Stothkamp, "Fluorescence measurements of ethidium binding to DNA," *Journal of Chemical Education*, vol. 71, no. 1, pp. 77–79, 1994.

[44] J. R. Lakowicz and G. Weber, "Quenching of protein fluorescence by oxygen. Detection of structural fluctuations in proteins on the nanosecond time scale," *Biochemistry*, vol. 12, no. 21, pp. 4171–4179, 1973.

[45] R. W. Woody, "Circular dichroism," *Methods in Enzymology*, vol. 246, pp. 34–71, 1995.

[46] T. Antony, T. Thomas, A. Shirahata, and T. J. Thomas, "Selectivity of polyamines on the stability of RNA-DNA hybrids containing phosphodiester and phosphorothioate oligodeoxyribonucleotides," *Biochemistry*, vol. 38, no. 33, pp. 10775–10784, 1999.

[47] X. Y. Le, S. R. Liao, X. P. Liu, and X. L. Feng, "Synthesis, structure and SOD-like activity of a ternary Cu(II) complex with 1,10-phenanthroline and L-valinate," *Journal of Coordination Chemistry*, vol. 59, no. 9, pp. 985–995, 2006.

Permissions

All chapters in this book were first published in BCA, by Hindawi Publishing Corporation; hereby published with permission under the Creative Commons Attribution License or equivalent. Every chapter published in this book has been scrutinized by our experts. Their significance has been extensively debated. The topics covered herein carry significant findings which will fuel the growth of the discipline. They may even be implemented as practical applications or may be referred to as a beginning point for another development.

The contributors of this book come from diverse backgrounds, making this book a truly international effort. This book will bring forth new frontiers with its revolutionizing research information and detailed analysis of the nascent developments around the world.

We would like to thank all the contributing authors for lending their expertise to make the book truly unique. They have played a crucial role in the development of this book. Without their invaluable contributions this book wouldn't have been possible. They have made vital efforts to compile up to date information on the varied aspects of this subject to make this book a valuable addition to the collection of many professionals and students.

This book was conceptualized with the vision of imparting up-to-date information and advanced data in this field. To ensure the same, a matchless editorial board was set up. Every individual on the board went through rigorous rounds of assessment to prove their worth. After which they invested a large part of their time researching and compiling the most relevant data for our readers.

The editorial board has been involved in producing this book since its inception. They have spent rigorous hours researching and exploring the diverse topics which have resulted in the successful publishing of this book. They have passed on their knowledge of decades through this book. To expedite this challenging task, the publisher supported the team at every step. A small team of assistant editors was also appointed to further simplify the editing procedure and attain best results for the readers.

Apart from the editorial board, the designing team has also invested a significant amount of their time in understanding the subject and creating the most relevant covers. They scrutinized every image to scout for the most suitable representation of the subject and create an appropriate cover for the book.

The publishing team has been an ardent support to the editorial, designing and production team. Their endless efforts to recruit the best for this project, has resulted in the accomplishment of this book. They are a veteran in the field of academics and their pool of knowledge is as vast as their experience in printing. Their expertise and guidance has proved useful at every step. Their uncompromising quality standards have made this book an exceptional effort. Their encouragement from time to time has been an inspiration for everyone.

The publisher and the editorial board hope that this book will prove to be a valuable piece of knowledge for researchers, students, practitioners and scholars across the globe.

List of Contributors

Jinhuan Jiang and Jiang Pi
State Key Laboratory of Quality Research in Chinese Medicines, Macau University of Science and Technology, Macau, China

Jiye Cai
Department of Chemistry, Jinan University, Guangzhou, China

Fritzgerald Kogge Bine, Nyiang Kennet Nkungli, Tasheh Stanley Numbonui and Julius Numbonui Ghogomu
Department of Chemistry, Faculty of Science, Research Unit of Noxious Chemistry and Environmental Engineering, University of Dschang, Dschang, Cameroon

Samuel Tetteh
Department of Chemistry, School of Physical Sciences, College of Agriculture and Natural Sciences, University of Cape Coast, Cape Coast, Ghana

Ismael Acosta-Rodríguez and Juana Tovar Oviedo
Universidad Autónoma de San Luis Potosí, Facultad de Ciencias Químicas, Centrode Investigaci ón y de Estudios de Posgrado, Laboratorio de Micología Experimental, Av. Dr. Manuel Nava No. 6, Zona Universitaria, C.P. 78320, San Luis Potosí, Mexico

Juan F. Cárdenas-González and Adriana S. Rodríguez Pérez
Unidad Académica Multidisciplinaria Zona Media, Universidad Autónoma de San Luis Potosí, Carretera San Ciro de Acosta Km. 4, Ejido Puente del Carmen. C.P. 79617, Río Verde, San Luis Potosí, Mexico

Víctor M. Martínez-Juárez
Área Académica de Medicina Veterinaria y Zootecnia, Instituto de Ciencias Agropecuarias, Universidad Autónoma Del Estado de Hidalgo, Zona Universitaria, Rancho Universitario Km 1. C.P. 43600, Tulancingo de Bravo Hidalgo, Mexico

Yanmin Huang, Erbin Kong, Junyan Zhan, Shuang Chen, Chunfang Gan, Zhiping Liu, Liping Pang and Jianguo Cui
Key Laboratory of Beibu Gulf Environment Change and Resources Utilization, College of Chemistry and Material Science, Guangxi Teachers Education University, Nanning 530001, China

Jianguo Cui
Guangxi Colleges and University Key Laboratory of Beibu Gulf Oil and Natural Gas Resource Effective Utilization, Qizhou University, Qizhou, China

K. Savithri and H. D. Revanasiddappa
Department of Studies in Chemistry, University of Mysore, Manasagangotri, Mysuru, Karnataka 570 006, India

Li Wen, Wei Li and Qiao Cai
School of Public Health, Guangdong Pharmaceutical University, Guangzhou Higher Education Mega Centre, Guangzhou 510006, China

Qing Li
Guangdong Provincial Center for Disease Control and Prevention, Guangzhou 511430, China

Yong-Ming Cai
School of Medical Information Engineering, Guangdong Pharmaceutical University, Guangzhou Higher Education Mega Centre, Guangzhou 510006, China

Ghadah M. Al-Senani and Foziah F. Al-Fawzan
Department of Chemistry, College of Science, Princess Nourah Bint Abdul rahman University, Riyadh, Saudi Arabia

Diogo José Horst
Federal University of Technology-Paraná, 84016-210 Ponta Grossa, PR, Brazil

Sergio Mazurek Tebcherani
Department of Production Engineering, Federal University of Technology-Paraná, 84016-210 Ponta Grossa, PR, Brazil

Evaldo Toniolo Kubaski
Department of Materials Science, State University of Ponta Grossa, 84030-900 Ponta Grossa, PR, Brazil

Rogério de Almeida Vieira
Department of Earth and Exact Sciences, Federal University of São Paulo, 89972-270 São Paulo, SP, Brazil

Kee-Yeon Kum
Department of ConservativeDentistry, Seoul National University Dental Hospital, Seoul Dental Hospital for Disabled, Dental Research Institute, Seoul National University School of Dentistry, Seoul, Republic of Korea

Seok Woo Chang
Department of Conservative Dentistry, School of Dentistry, Kyung Hee University, Seoul, Republic of Korea

Ilaria Gamba
Universidad de La Laguna, San Cristóbal de La Laguna, Santa Cruz de Tenerife, Spain

Anju Arya, Khushbu Gupta and Tejpal Singh Chundawat
Department of Applied Sciences, e North Cap University, Sector 23-A, Gurugram 122017, Haryana, India

Dipti Vaya
Department of Chemistry, Amity University, Gurugram, Haryana, India

Snežana Rajković and Biljana Đ. Glišić
Department of Chemistry, Faculty of Science, University of Kragujevac, R. Domanovića 12, 34000 Kragujevac, Serbia

Beata Warżajtis and Urszula Rychlewska
Faculty of Chemistry, Adam Mickiewicz University, Umultowska 89B, 61-614 Poznán, Poland
Marija D. Živković
Department of Pharmacy, Faculty of Medical Sciences, University of Kragujevac, Svetozara Markovića 69, 34000 Kragujevac, Serbia

Miloš I. Djuran
Serbian Academy of Sciences and Arts, Knez Mihailova 35, 11000 Belgrade, Serbia

Ismail Althagafi
Chemistry Department, College of Applied Sciences, Umm Al-Qura University, Mecca, Saudi Arabia

Nashwa M. El-Metwaly and Marwa G. Elghalban
Chemistry Department, Faculty of Science, Mansoura University, Mansoura, Egypt

Thoraya A. Farghaly
Department of Chemistry, Faculty of Science, University of Cairo, Giza 12613, Egypt

Abdalla M. Khedr
Chemistry Department, Faculty of Science, Tanta University, Tanta, Egypt

Omer Y. Bakather and Ahmad Kayvani Fard
Department of Chemical Engineering, Jazan University, Jazan 45142, Saudi Arabia

Omer Y. Bakather
Department of Chemical Engineering, Hadhramout University, Hadhramout, Yemen

Ahmad Kayvani Fard and Muataz Ali Atieh
Qatar Environment and Energy Research Institute, Hamad Bin Khalifa University, Qatar Foundation, Doha, Qatar
College of Science and Engineering, Hamad Bin Khalifa University, Qatar Foundation, Doha, Qatar

Ihsanullah
Department of Chemical Engineering, King Fahd University of Petroleum & Minerals, Dhahran 31261, Saudi Arabia

Majeda Khraisheh
Chemical Engineering Department, Qatar University, Doha, Qatar

Mustafa S. Nasser
Gas Processing Centre, College of Engineering, Qatar University, Doha, Qatar

Seok Woo Chang
Department of Conservative Dentistry, School of Dentistry, Kyung Hee University, Seoul, Republic of Korea

Ren-Shi Shyu
Division of Nephrology, Department of Internal Medicine, Min-Sheng General Hospital, Taoyuan 330, Taiwan

Ren-Shi Shyu, Themmila Khamrang, Joen-Rong Sheu and Chih-Wei Hsia
Graduate Institute of Medical Sciences, College of Medicine, Taipei Medical University, Taipei 110, Taiwan

Themmila Khamrang
Department of Chemistry, North Eastern Hill University, Shillong 793022, India

Duen-Suey Chou
Department of Pharmacology, School of Medicine, College of Medicine, Taipei Medical University, Taipei 110, Taiwan

Chao-Chien Chang
Department of Cardiology, Cathay General Hospital, Taipei 106, Taiwan

Peiran Zhao, Shanshan Zhai, Jianfang Dong, Xinru Liu and Lei Wang and Lianzhi Li
School of Chemistry and Chemical Engineering, Liaocheng University, Liaocheng 252059, China

Lei Gao
Zhong Yuan Academy of Biological Medicine, Liaocheng People's Hospital, Liaocheng 252000, China

Jinming Kong
School of Environmental and Biological Engineering, Nanjing University of Science and Technology, 200 Xiaolingwei, Nanjing 210094, China

Index

CPSIA information can be obtained
at www.ICGtesting.com
Printed in the USA
LVHW060850030122
707669LV00002B/54

9 781641 165402